Chemistry in the World

First Edition

by Kirstin Hendrickson

Arizona State University

cognella
San Diego, CA

First published in the United States of America in 2011 by Cognella, a division of University Readers, Inc.

15 14 13 12 11 1 2 3 4 5

Printed in the United States of America

ISBN: 978-1-935551-35-5

cognella™

www.cognella.com 800.200.3908

For my students, from whom I have learned more than I could ever teach.

About the Cover

The pictures on the cover of this text are some examples of the chemistry that surrounds us in the real world. At top left, poppies bloom in the Sonoran desert of Arizona in the spring as plants increase their growth rate. Not only does this depend on the chemical process of photosynthesis, it also increases the rate at which the greenhouse gas carbon dioxide is removed from the atmosphere. At top right and bottom left, we see two examples of water in nature—in liquid form as it cascades over Upper Mesa Falls in Idaho, and in solid form as snow in Bryce Canyon in Utah. There's also water vapor in the atmosphere, where it serves the important role of helping to keep the Earth warm and habitable. Water has many interesting and unique chemical properties, and fresh surface water, as in this waterfall, is rare and valuable. At bottom right, aspen leaves in the San Francisco Peaks near Flagstaff, Arizona, turn vibrant shades of yellow in the fall. As seasons change, changing intensity of light prompts trees to adjust concentrations of light-harvesting chemicals in the leaves. This leads to the yellows, oranges, and reds that decorate the trees as the air turns crisp.

Photo credits: Sonoran desert poppies, Arizona © Scott Lefler, 2008; Upper Mesa Falls, Idaho © Scott Lefler, 2008; Bryce Canyon, Utah © Scott Lefler, 2009; aspens in the San Francisco Peaks, Arizona © Scott Lefler, 2006.

Contents

Acknowledgments xiii

To the Student and the Instructor: How to Use this Book 1

Unit 1: All Around Us and Inside Us 3

Chapter 1: Chemistry Is Everywhere 5

1.1 Classifications of Matter 7
1.2 The Periodic Table 14
1.3 Thinking About Chemicals, Chemical Reactions, and Chemical Equations 16
1.4 Balancing Chemical Equations 18
1.5 Nomenclature 20
1.6 The Last Word—Chemical Free? 23

Chapter 2: Chemistry Around Us—The Atmosphere 29

2.1 Composition of the Atmosphere 30
2.2 Chemistry of High Altitude 31
2.3 Atomic and Molecular Mass 32
2.4 Counting Atoms and Molecules—The Mole 33
2.5 Molar Mass 35
2.6 Interconversions: Grams to Moles to Molecules and Atoms 37
2.7 Stoichiometric Equivalents 38
2.8 Subatomic Particles 40
2.9 The Last Word—Land Animals Breathe Air, Aquatic Animals Breathe Water? 45

Chapter 3: Chemistry Inside Us—The Elements of Life 53

3.1 Carbon-Based Life Forms 53
3.2 Electron Shells 55
3.3 Lewis Dot Structures 59

3.4 Covalent Bonding 60
3.5 Multiple Covalent Bonds 62
3.6 So Why Are We Carbon Based? 65
3.7 Molecular Shapes 67
3.8 The Paradox of Bond Formation 70
3.9 The Last Word—Love and Other Chemical Reactions 72

Unit 2: Community Chemistry 79

Chapter 4: Water—The Solvent of Life 81

4.1 Solutions and Solubility 81
4.2 Dipole Moments 82
4.3 Hydrogen Bonds 88
4.4 Unique Properties of Water 90
4.5 Ionic Compounds 94
4.6 Electrolytes 98
4.7 Water Distribution and Safety 100
4.8 Hard Water 102
4.9 Bottled Versus Tap Water—Social Implications 103
4.10 The Last Word—Superwaters 105

Chapter 5: The Chemistry of Acids and Bases 111

5.1 Acids 111
5.2 Polyatomic Ions 114
5.3 Bases 115
5.4 Acids, Bases, and Taste 117
5.5 Molarity and Concentration 118
5.6 The pH of a Solution 120
5.7 Neutralization 121
5.8 Chemistry of Baking 123
5.9 Acid and Microbes 124
5.10 Acid Spills and Industrial Acid Disasters 126
5.11 The Last Word—Fermentation 127

Chapter 6: The Chemistry of Pollution 135

6.1 Common Pollutants—Carbon Monoxide 135
6.2 Common Pollutants—NOx 137
6.3 Common Pollutants—Volatile Organic Compounds and Polycyclic Aromatic Hydrocarbons 138
6.4 Common Pollutants—Ozone 139
6.5 Common Pollutants—Particulate Matter (PM_{10} and $PM_{2.5}$) 140
6.6 Common Pollutants—Sulfur Dioxide 140
6.7 Common Pollutants—Mercury and Lead 141
6.8 Pollutant Ramifications—Acid Rain 143

6.9 The Industrial Revolution 146
6.10 Pollution Regulation 147
6.11 Air Quality Index 149
6.12 Changing Perceptions of Risk 151
6.13 The Last Word—Smoking 153

Unit 3: Personal Chemistry 161

Chapter 7: Pharmaceuticals and Poisons 163
7.1 From Tea to Tablets 163
7.2 Pharmaceutical Design 168
7.3 Wonder Drugs 171
7.4 Drug Solubility 172
7.5 Receptors and Pharmacophores 174
7.6 Early Pharmaceutical Regulation 176
7.7 Through the Looking Glass—Healer or Killer? 178
7.8 What Does "Natural" Mean, Anyway? 184
7.9 The Last Word—The Paracelsus Principle 187

Chapter 8: The Chemistry of What We Eat 197
8.1 Energy and Chemical Bonds 197
8.2 Entropy 199
8.3 Energy of Chemical Reactions 201
8.4 How Molecules Provide Energy 204
8.5 Protein 205
8.6 Protein in Food 208
8.7 Carbohydrates 209
8.8 Carbohydrate Digestion 211
8.9 Carbohydrate Absorption 212
8.10 Fats 215
8.11 Types of Fatty Acids 218
8.12 Cholesterol and Steroid Hormones 222
8.13 Soap 224
8.14 Vitamins 225
8.15 Minerals 228
8.16 The Last Word—Vitamin Enriched! 229

Chapter 9: The Body as a Beaker—Nutrition, Health, and Wellness 237
9.1 Body Fat—Too Much, Too Little, and Just Right 237
9.2 Body Composition Metrics 240
9.3 Nutritional Recommendations 241
9.4 What We're Eating—The Typical American Diet 244
9.5 Trying to Change What We're Eating—Weight-Loss Diets 245
9.6 The Low-Fat Diet 246

9.7 Stimulant Diet Pills 247
9.8 Low-Carbohydrate Diet 249
9.9 Sugar Substitutes 250
9.10 Sustainability 253
9.11 The Other Half of the Equation—Caloric Expenditure 256
9.12 Exercise 257
9.13 Muscle Types 261
9.14 The Last Word—Vitamin and Mineral Supplements 264

Unit 4: Global Chemistry 275

Chapter 10: Changing the World with Chemistry 277

10.1 Refrigeration 277
10.2 UV Light and EMR 280
10.3 The Ozone Layer 285
10.4 CFCs and UV Light 287
10.5 The Montreal Protocol 290
10.6 Refrigerants—Moving Forward 292
10.7 The Nitrogen Cycle 293
10.8 The Haber-Bosch Process 294
10.9 Risks and Benefits of the Haber-Bosch Process 296
10.10 The Precautionary Principle Revisited 297
10.11 The Last Word—Sunlight in the Troposphere 299

Chapter 11: "Current" Events—Electrochemistry 307

11.1 Redox 308
11.2 Galvanic Cells 312
11.3 Electrolytic Cells 316
11.4 Photovoltaic Cells 319
11.5 Plants—Natural Solar Power Generators 322
11.6 Human Batteries—Respiration 324
11.7 Anaerobic Metabolism—Fermentation 326
11.8 The Last Word—Lightning 328

Chapter 12: From Alchemists to Oppenheimer and Beyond—Nuclear Chemistry 337

12.1 Rise of Nuclear Chemistry 337
12.2 Radioactive Decay 339
12.3 Radioactive Decay Series 344
12.4 Nuclear Fission 345
12.5 The Atom Bomb 347
12.6 Nuclear Fusion 350
12.7 Nuclear Energy 351

12.8 Half-Life 353
12.9 Storing Nuclear Waste 356
12.10 Nuclear Disasters 358
12.11 The Future of Nuclear Power 359
12.12 Health Effects of Irradiation 360
12.13 Nuclear Medicine 362
12.14 Carbon Dating 364
12.15 Irradiation of Food 364
12.16 The Last Word—Microwaves 365

Chapter 13—Energy, Chemistry, and Society **373**
13.1 An Important Introductory Note from the Author 373
13.2 Our Global Energy Dependence and Fossil Fuels 374
13.3 The Greenhouse Effect 377
13.4 The Carbon Cycle 381
13.5 Measuring Atmospheric Carbon 383
13.6 Evidence of Climate Change 387
13.7 Climate Modeling 391
13.8 Causality 393
13.9 Why Focus on CO_2? 395
13.10 Making Sense of Rhetoric 395
13.11 The Kyoto Protocol 397
13.12 Global and National Strategies 398
13.13 Community and Individual Strategies 401
13.14 The Last Word—Stomping or Tiptoeing? 402

Glossary **409**

Credits **427**

Index **435**

Acknowledgments

Authoring this book has been a work in progress for the last ten or more years, though the actual writing took (thankfully!) somewhat less than that. There are many without whom this text would never have transitioned from dream to reality—you know who you are, and I thank you. Mostly, to my husband, Scott, for being the catalyst, and for providing the gorgeous photographs that bring the ideas to life.

To the Student and the Instructor: How to Use this Book

This text was written to provide a perspective on socially important topics in the field of chemistry, which as the student will soon ascertain, is a broad field indeed! Of course, even an applications-based approach (as this book is) must by necessity introduce sufficient principle and theory to allow for meaningful discussion. It was a primary goal in writing this text to include only those principles absolutely necessary to understanding observable chemical phenomena and relevant chemical issues. At every step of the way, I strove to make readily apparent the answer to the question, *When am I ever going to use this again?* For this reason, I recommend reading this book more like a literature than a science text—it tells a story. Granted, the story is occasionally punctuated by equations and symbols, but it's a story nevertheless. Rather than trying to internalize the applications in terms of the chemical principles, use the applications to make sense of the principles—the big picture will help you understand the details.

Each chapter introduces a topic or group of related topics in which we see chemistry on a day-to-day basis, in our lives or on the news. Principles are addressed where needed in the course of discussing the topic. Where relevant, practice problems are provided in boxes labeled *Try This*. It's worth taking a moment to try a problem on your own before moving on—there are many additional practice problems at the end of the chapter. Conceptual questions are occasionally presented in boxes labeled *Concept Check*. It should be fairly easy to answer the question posed in the box after reading a particular section—if not, that's an indication that you may have missed a main point from the previous paragraphs. Occasionally, *Concept Check* boxes will refer to older concepts that are being brought in to the current discussion, and in such cases, the answers will refer you to the relevant chapter.

Problems at the end of the chapter are divided into two sections. *Questions and Topics for Discussion* are intended to stimulate thought—in many cases, there are no right or wrong answers. These questions are meant to be suitable for essay or paragraph assignments, stimulation of thinking, or in-class discussions. *Problems* are mathematical or factual in nature—in particular, math problems occur in sets of two (grouped together) for ease of studying and assigning work.

The last section in each chapter is called *The Last Word*, and is an offshoot of something I've done in my classes that has been very popular with my students. In class, I call it "Five Minutes of Really Cool Chemistry," and I'll talk about something nearly relevant and always neat that is based upon the principles we've been discussing. *The Last Word* is my adaptation of that into written form—these sections are meant to be interesting vignettes that relate to the concepts addressed in the chapter. They're not chapter summaries (which do, incidentally, appear separately at the end of each chapter), but are hopefully a few moments' worth of relaxation and wonder—a few minutes of really cool chemistry.

Enjoy the journey!

Unit 1: All Around Us and Inside Us

Chemistry Is Everywhere 1

GETTING STARTED—INTRODUCTION AND ORGANIZATION OF THE TEXT

The word *chemistry*, especially in the title of a class or book, conjures up images of goggled scientists at their lab benches, surrounded by odd bubbling solutions in beakers and flasks. Depending upon the power of your imagination, this scene might even be punctuated in your mind by the occasional explosion. In any case, as you imagine the white-coated chemist going about his business, mixing and stirring and writing down equations rife with symbols and letters of the Greek alphabet, I practically guarantee the two things you are thinking to yourself are *What in the world does this have to do with real life?* and *When am I ever going to use this information!?* The answers to those (very real, very important) questions are EVERYTHING, and EVERY DAY!

In order to really see the applicability of chemistry to real life, we have to stop imagining it as something that takes place in a laboratory and start looking for it in the real world. Chemists work in laboratories because they are attempting to study what goes on everywhere ELSE, not because the lab is the only place chemistry happens! In this book, we'll look for chemistry everywhere EXCEPT the lab. We'll learn that there's chemistry all around us—in leaves changing colors in the fall, in the atmosphere (which keeps our planet warm and habitable the way a greenhouse protects roses from winter frost), in nuclear power plants, and in the food we eat every day.

Image 1.1. (From top to bottom) 1) Leaves change color in the fall as the light-absorbing chemical chlorophyll (which gives them their green color) is no longer produced, leaving behind other light-absorbing chemicals such as carotenes (yellow) and anthocyanin (red). 2) Just as a greenhouse made of glass traps heat to keep plants warm during cooler months, so the gases of the atmosphere trap heat from the Sun, keeping the planet warm and habitable. 3) Nuclear reactions have the potential to release massive amounts of energy that can be used to generate power. 4) The food we eat is made up of chemicals that provide us with energy and raw material for building muscle and other tissue.

Image 1.2. The complex interplay between chemistry and society is apparent in this picture. We affect the chemistry of the oceans by adding environmental pollutants and toxins that disrupt the ecosystem. As a result, the altered chemistry of the ocean affects us. For example, mercury builds up in high-level predator fish that we use as food (tuna and shark are among these), rendering them toxic to us if consumed more than occasionally.

This isn't just a chemistry textbook. It also addresses ways in which chemistry, chemicals, and chemists affect society. However, this is more complex than it sounds. It's very important to think about the relationship between chemistry—any science, really—and society as a reciprocal one. To be sure, chemistry affects society: pollution, for instance, has a clear and unequivocal negative impact upon those who have to breathe it. However, society also affects chemistry, in that our actions produce and release chemicals into the environment, and we formulate legislation that affects what can and can't be done chemically. In this text we will be exploring the nature of this relationship and some of the ways in which chemistry and society affect each other.

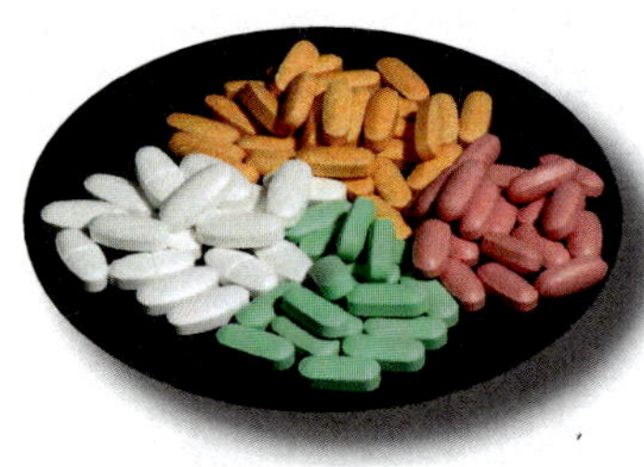

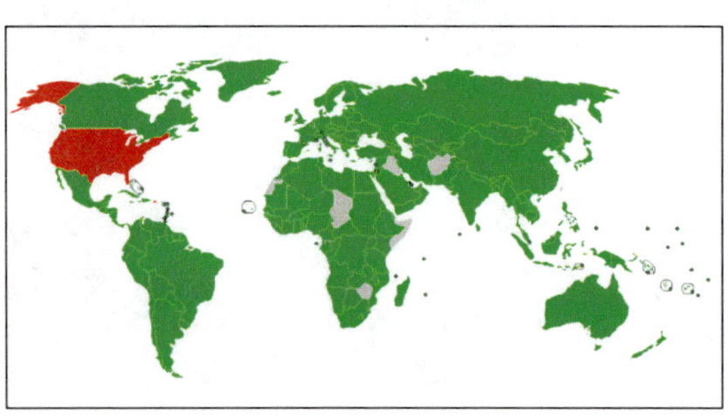

Image 1.3. Chemistry on the individual level includes such topics as pharmaceuticals (above) and food (bottom right). Cities need to consider the chemistry of local air pollution (top right) and water safety (purification plant, top left). As a global community, we are all concerned with and affected by climate change. The countries shown in green (bottom left) have signed the Kyoto Protocol for reducing carbon emissions. (Why hasn't the United States signed? See Chapter 13.)

It's important to note that society is a multifaceted concept. Individuals, local communities, and the entire global population are different levels of society. Any level can affect any other (and be affected by any other) through chemistry. For instance, making appropriate choices regarding nutrition and exercise positively impacts personal health, but also reduces the need for medical care, which decreases insurance rates for all policy holders. A community that installs a light-rail system benefits from a reduction in local pollution, but there are also global benefits through reduced carbon emissions.

In order to address the reciprocal relationship between chemistry and each level of society, this text is divided into four units. In the first, we'll be discussing some introductory chemical principles in the context of the chemistry we encounter every day. The second unit will address chemical concerns within local communities—things like air pollution and water safety. In Unit 3, we'll shift our focus to the chemistry that takes place within our own bodies. Finally, in Unit 4, we'll expand our thinking outward to encompass issues of global concern. At every step along the way, however, we will be focusing not simply on chemical concepts, but on how we affect the chemistry and how the chemistry affects us.

1.1 CLASSIFICATIONS OF MATTER

Stated very simply, chemistry is the study of **matter**. Matter is defined as anything thatoccupies space and has mass, so it is what makes up the physical universe. Everything we can touch is made of matter, but because there are several different types of matter, we need to define some general categories.

Let's start with mixtures, because they're actually the simplest kind of matter to understand. **Mixtures** are physical combinations of two or more pure substances in variable proportions. There are two key components to this definition. First, that a mixture is a **physical combination** means that each substance within the mixture retains its own chemical identity and properties, but is interspersed with one or more other substances in space. For instance, making a cake involves mixing sugar, baking soda, flour, spices, eggs, and so forth. Sugar is a chemical with distinct properties (it tastes sweet, it dissolves in water, and so on). Baking soda is another chemical with a separate set of distinct properties. As an ingredient like sugaris added to the batter, it mixes in and spreads out through the batter until all parts of the mixture are equally sweet. The sugar, though dispersed through the batter, is still sugar—as a chemical, it remains unchanged.That's the nature of a mixture.

Image 1.4. Beating together the ingredients for pancakes makes a mixture with tasty potential (above) (© Scott Lefler, 2010). When table salt (NaCl) is stirred into water (H_2O), neither the salt nor the water changes chemical properties—instead, a mixture forms (right).

Saltwater is another example of a mixture. Salt (a common name for the chemical sodium chloride, or NaCl) has distinct properties. Water (a common name for the chemical H_2O) has a different set of properties. They can be stirred together to make saltwater, but the salt is still NaCl, and the water is still H_2O. In other words, physically mixing salt and water means that bits of NaCl are surrounded by bits of H_2O and vice versa, but the individual

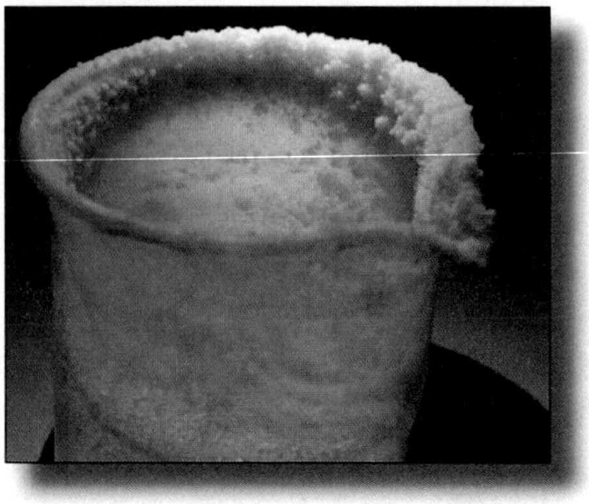

Image 1.5. A beaker full of a saltwater mixture, if put on a burner until all the water evaporates, is left full of salt crust. The evaporating water vapor could be collected and cooled, in which case it would condense back into ordinary liquid water. The salt crust, if crushed, would be indistinguishable from table salt straight from the shaker.

chemicals retain their identities and properties. If we wanted to, we could "un-mix" salt water by boiling it in a pan on the stove, in which case the water would depart as steam (which is still H_2O), and the NaCl would be left behind as a crust on the bottom of the pan. At no point in the mixing or separating process does either chemical change identities or properties.

The second key part of the definition of a mixture is that the substances are combined in variable proportions. This means that it doesn't matter how much salt we mix with a given amount of water—whether we use lots of salt or only a bit, we'll still have a mixture of saltwater.

Chemicals that can be physically combined to produce mixtures are called **pure substances**, andcan be divided into two sub-categories: elements and compounds**. Elements** are the simplest types of matter and are found listed on the **periodic table**(Image 1.6). They cannot be broken into simpler substances by any chemical means. When you look at the periodic table, you'll notice that it's made up of many elements, each with a characteristic **atomic symbol**, which is a letter or combination of letters that represents the element. C, for instance, is the atomic symbol for the element carbon. Elements have different physical and chemical properties. Some, like neon (Ne), are gases at room temperature, while others are liquids or solids. Many, like gold (Au), uranium (U), and mercury (Hg) are metals, while others, like sulfur (S), are not. We'll look more closely at properties of specific elements and at the organization of the periodic table later in this chapter.

Image 1.6. The periodic table of the elements.

1 H 1.01 Hydrogen																1 H 1.01 Hydrogen	2 He 4.00 Helium
3 Li 6.94 Lithium	4 Be 9.01 Beryllium											5 B 10.81 Boron	6 C 12.01 Carbon	7 N 14.01 Nitrogen	8 O 16.00 Oxygen	9 F 19.00 Fluorine	10 Ne 20.18 Neon
11 Na 23.0 Sodium	12 Mg 24.3 Magnesium											13 Al 26.98 Aluminum	14 Si 28.09 silicon	15 P 30.97 Phosphorus	16 S 32.07 Sulfur	17 Cl 35.45 Chlorine	18 Ar 39.95 Argon
19 K 39.10 Potassium	20 Ca 40.08 Calcium	21 Sc 44.96 Scandium	22 Ti 47.88 Titanium	23 V 50.94 Vanadium	24 Cr 52.00 Chromium	25 Mn 54.94 Manganese	26 Fe 55.58 Iron	27 Co 58.93 Cobalt	28 Ni 58.69 Nickel	29 Cu 63.55 Copper	30 Zn 65.39 Zinc	31 Ga 69.72 Gallium	32 Ge 72.59 Germanium	33 As 74.92 Arsenic	34 Se 78.96 Selenium	35 Br 79.90 Bromine	36 Kr 83.80 Krypton
37 Rb 85.47 Rubidium	38 Sr 87.62 Strontium	39 Y 88.91 Yttrium	40 Zr 91.22 Zirconium	41 Nb 92.91 Niobium	42 Mo 95.94 Molybdenum	43 Tc (98) Technetium	44 Ru 101.1 Ruthenium	45 Rh 102.9 Rhodium	46 Pd 106.4 Palladium	47 Ag 107.9 Silver	48 Cd 112.4 Cadmium	49 In 114.8 Indium	50 Sn 118.7 Tin	51 Sb 121.8 Antimony	52 Te 127.6 Tellurium	53 I 126.9 Iodine	54 Xe 131.3 Xenon
55 Cs 132.9 Cesium	56 Ba 137.9 Barium	57 La *138.9 Lanthanum	72 Hf 178.5 Hafnium	73 Ta 180.9 Tantalum	74 W 183.9 Tungsten	75 Re 186.2 Rhenium	76 Os 190.2 Osmium	77 Ir 192.2 Iridium	78 Pt 195.1 Platinum	79 Au 197.0 Gold	80 Hg 200.5 Mercury	81 Tl 204.4 Thalium	82 Pb 207.2 Lead	83 Bi 209.0 Bismuth	84 Po (210) Polonium	85 At (210) Astatine	86 Rn (222) Radon
87 Fr (223) Francium	88 Ra (226) Radium	89 Ac *(227) Actinium	104 Rf (257) Rutheforđium	105 Db (260) Dubnium	106 Sg (263) Seaborgium	107 Bh (262) Bohrium	108 Hs (265) Hassium	109 Mt (266) Meitnerium	110 Ds (271) Darmstadtium								

58 Ce 140.1 Cerium	59 Pr 140.9 Praseodymium	60 Nd 144.2 Neodymium	61 Pm (147) Promethium	62 Sm 150.4 Samarium	63 Eu 152.0 Europium	64 Gd 157.3 Gadolinium	65 Tb 158.9 Terbium	66 Dy 162.5 Dyspromium	67 Ho 164.9 Holmium	68 Er 167.3 Erbium	69 Tm 168.9 Thulium	70 Yb 173.0 Ytterbium	71 Lu 175.0 Lutetium
90 Th 232.0 Thorium	91 Pa (231) Protactinium	92 U (238) Uranium	93 Np (237) Neptunium	94 Pu (242) Plutonium	95 Am (243) Americium	96 Cm (247) Curium	97 Bk (247) Berkelium	98 Cf (249) Californium	99 Es (254) Einsteinium	100 Fm (253) Fermium	101 Md (256) Mendelevium	102 No (254) Nobelium	103 Lr (257) Lawrencium

Image 1.7. Some elements as they occur in nature (clockwise from top left): Neon (Ne) can be used to make glowing signs; mercury (Hg) is a unique liquid metal that is used in thermometers; gold (Au) is a beautiful, butter-yellow metal highly prized for its aesthetic appeal; uranium (U) is used in nuclear reactors to generate energy.

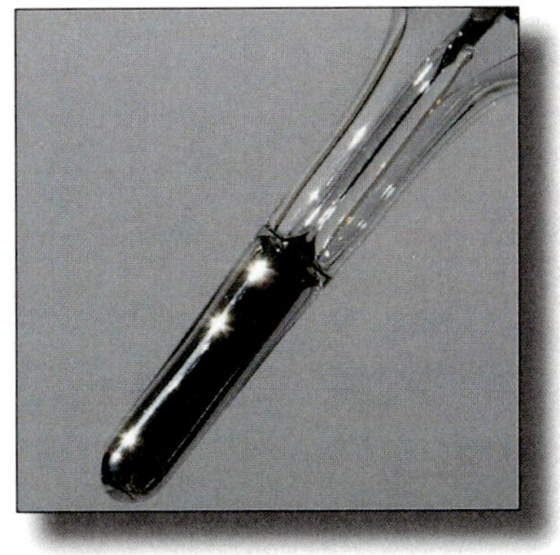

Compounds are chemical combinations of two or more elements in fixed, characteristic proportions. Notice the differences between the definition of a compound and the definition of a mixture. First, where mixtures are physical combinations in which each pure substance retains its identity and properties, compounds are **chemical combinations**, meaning that a**chemical bond** forms between particles of two or more different elements. A chemical bond is a little like glue—it sticks two particles together. The compound has its own identity and properties that are different from the identities and properties of the elements that make it up. For instance, carbon dioxide (CO_2) is a compound made from a chemical combination of the elements carbon and oxygen, but CO_2 is nothing like either carbon OR oxygen—it is its own separate, distinct substance. Another major difference between compounds and mixtures is that compounds must consist of **fixed proportions** of the constituent elements. While we can use any quantity of salt and any quantity of water to make salt water, CO_2 is only CO_2 if it's made up of one particle of carbon and two particles of oxygen. The subscripts in the compound's formula tell us how many particles of each element are chemically combined to form the compound, where the lack of a subscript is always taken to mean *one*. If one particle of carbon and one particle of oxygen combined chemically, the resulting compound would NOT be CO_2.Instead, it would be CO—carbon monoxide—which has very different properties than CO_2.

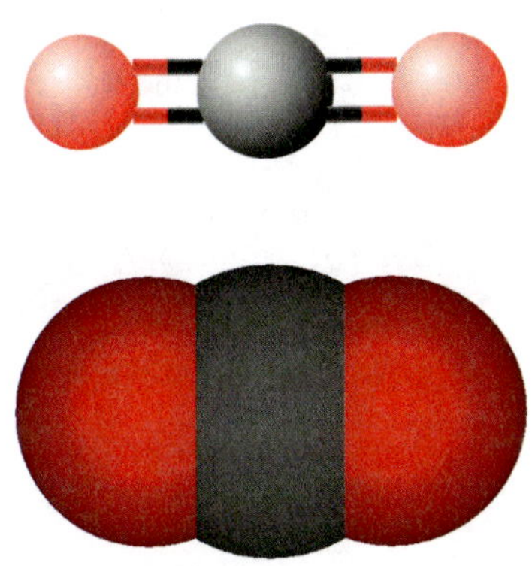

Image 1.8. Two representations of the compound CO_2. The figure on the top shows that an atom of carbon is chemically combined with two atoms of oxygen. This representation is often called a "ball and stick" model, because it represents the atoms as "balls" on either end of chemical bonds, which are represented as "sticks." In reality, bonds form when atoms sort of squish together in space, as represented in the image on the bottom. This is called a space-filling model, and while a bit more accurate, it's harder to draw.

Image 1.9. Some commonly encountered compounds include (clockwise from top left): Water (H_2O), sucrose (table sugar, $C_{12}H_{22}O_{11}$); calcite (the major ingredient in marble, $CaCO_3$); and ethanol (the alcohol in beer, wine, and liquor, C_2H_6O). (Crater Lake picture © Scott Lefler, 2006)

CONCEPT CHECK:

What's the difference between a mixture and a compound?

Answer: A mixture consists of two or more pure substances, physically combined in variable proportions. A compound consists of two or more elements, chemically combined in variable proportions.

CO_2 can be classified further—in addition to being a pure substance, it's also a **molecule**. A molecule is a pure substance made from the chemical combination of two or more **atoms**. An atom is defined as the smallest particle of an element that maintains the identity of that element. In other words, if I had a chunk of the element gold (Au on the periodic table), and I cut it in half, I'd have a smaller chunk of gold. If I cut that smaller chunk in half, and then cut the half in half, and so forth, eventually I'd get to a very, very tiny particle of gold that couldn't be cut in half (or at least, if it were, it wouldn't be gold anymore*). This tiny, indivisible particle of gold is an atom (the word atom comes from the Greek word *atomos*, meaning indivisible). All matter is made up of atoms, some of them uncombined and some of them in chemical combination with other atoms. A molecule of CO_2 is made up of one atom of carbon and two atoms of oxygen.

It's important to note that the words *compound* and *molecule* are not synonymous—it's absolutely possible to have a molecule that is not a compound, as long as it's made of two or more atoms OF THE SAME ELEMENT. For example, most respiring organisms require oxygen, which is found in our atmosphere as the molecule O_2. Of course, atoms of the element oxygen can also be found incorporated into compounds (like CO_2). Uncombined atoms of oxygen aren't found in nature, for reasons that will be discussed in later chapters. O_2 is made of two atoms of oxygen, soeven though it's a molecule, O_2 is an element rather than a compound. In nature, some elements are

***NOTE**

This statement seems paradoxical at first—how can we say that elements are the simplest type of matter, and then say that if we were to divide an atom of an element (gold, for instance), we'd get something that was NOT gold? We'll discuss this more in Chapter 2, but as it turns out, we can't really "cut" an atom and get pieces that hang around at all. If an atom were (theoretically) divided, the resulting pieces would be what are called sub-atomic particles, and would not be stable on their own. So, stated simply, atoms are the smallest particles of matter that are independently persistent.

foundas uncombinedatoms: examples include helium (He) and neon (Ne). Other elements occur most frequently as **diatomic molecules** (molecules consisting of only two atoms) such as O_2 and N_2. Still others occur naturally as**polyatomic molecules**, or molecules composed of many atoms. Sulfur, for instance, is most commonly found in nature as S_8. Atoms, molecules, elements, compounds, and mixtures are visually summarized in Image 1.11.

Image 1.10. The polyatomic molecule S_8 is a crystalline, yellow solid that smells faintly of rotten eggs.

CONCEPT CHECK:

What's the difference between a molecule and a compound?

Answer: A molecule consists of two or more atoms, chemically combined. A compound consists of two or more elements, chemically combined.

TRY THIS:

Which of the following are pure substances, and which are mixtures? Of the pure substances, which are elements and which are compounds? Which are atoms and which are molecules?

Fe, $FeCl_2$, sugar water, silver (Ag on the periodic table), maple syrup, Cl_2.

Answers: Fe is a pure substance, an element, and an atom. $FeCl_2$ is a pure substance, a compound, and a molecule. Sugar water is a mixture. Ag is a pure substance, an element, and an atom. Maple syrup is a mixture. Cl_2 is a pure substance, an element, and a molecule.

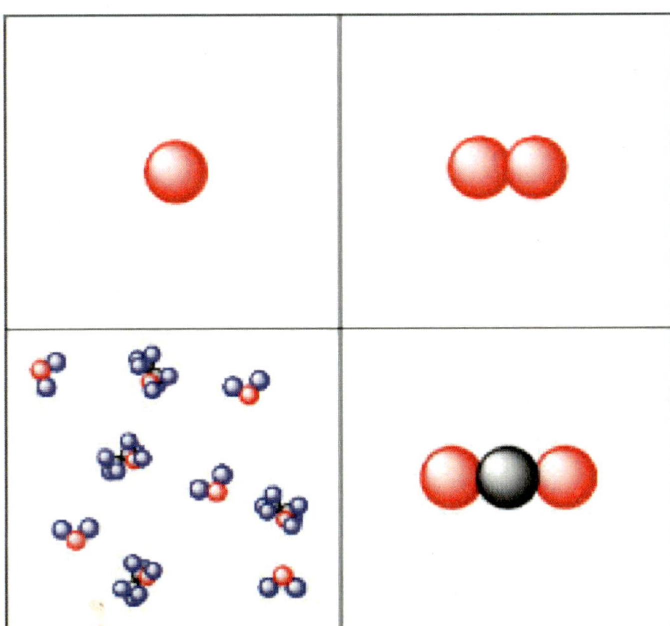

Image 1.11. An atom of oxygen (top left) is a pure substance, an element, and, of course, an atom. As a diatom, oxygen (O_2, top right) is a pure substance, an element, and a molecule. A molecule of carbon dioxide (CO_2, bottom right) is a pure substance, a compound, and a molecule. A mixture of ethanol (C2H_6O) and water contains molecules of both compounds, physically stirred together, but chemically unchanged.

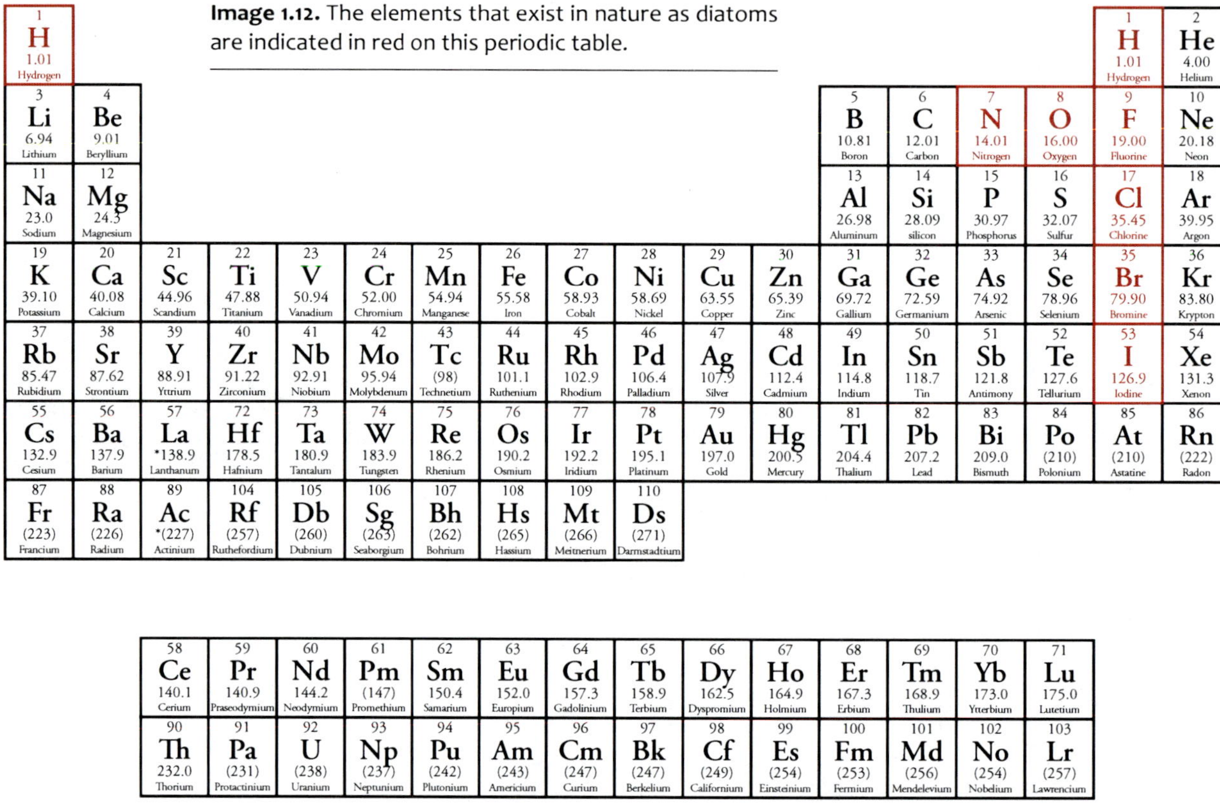

Image 1.12. The elements that exist in nature as diatoms are indicated in red on this periodic table.

One question students often have regarding elements is how to know whether they are found in nature as single atoms, diatomic molecules, or polyatomic molecules. The answer is that frankly, it's a bit complicated! There are a few rules of thumb, though, that are helpful to know. The elements that naturally form diatomic molecules are H, N, O, F, Cl, Br, and I, and it's quite useful to know this about them, because several of these are elements we'll be talking about frequently in this text.

The elements in the far-right column of the periodic table tend to be found as free atoms, and we'll learn why in Chapter 3. One other element whose form in nature is worth mentioning is carbon. Most of the carbon atoms we'll see in this class will be chemically combined with other elements, because carbon is a very common component of compounds in nature. However, elemental carbon is also common—probably more so than you think—and it can be found in several different forms, with VERY different properties. Two of the most frequently encountered forms of elemental carbon are graphite and diamond. These differ only in the way in which carbon atoms are arranged and chemically bonded to one another.* Looking at the arrangement of atoms in these two forms (Image 1.13) helps us to understand some of the differences in their physical properties and appearances. Graphite's structure is comprised of carbon atoms bonded into flat plates. These plates lie on top of one another, but are independent and can slide past each other easily. This sliding capability within the structure of graphite makes it soft and easily spread onto a surface (hence its utility as pencil "lead"). Also, it makes graphite a good lubricant in mechanical applications. Diamond, on the other hand, has carbon atoms

> ***NOTE**
>
> If graphite and diamond are both made of carbon, why is diamond so much more valuable than graphite? The quick answer is that it's much harder and more durable because of the arrangement of atoms. It's also considered aesthetically pleasing. These two factors combine to make it a good candidate for jewelry (among other things), and because it's rare, it comes at a high price. From a chemical perspective, diamond is actually LESS STABLE than graphite, meaning that very slowly, diamond turns into graphite! Don't worry—it doesn't happen in a human lifetime (or even multiple human lifetimes), so your diamonds will still be around for your descendants. Still, it's interesting to think about!

Image 1.13. Note the difference in appearance between graphite (top) and diamond (bottom), both of which are forms of the element carbon.

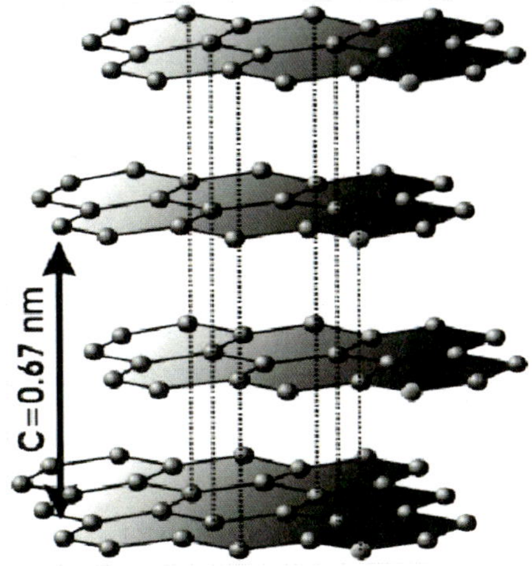

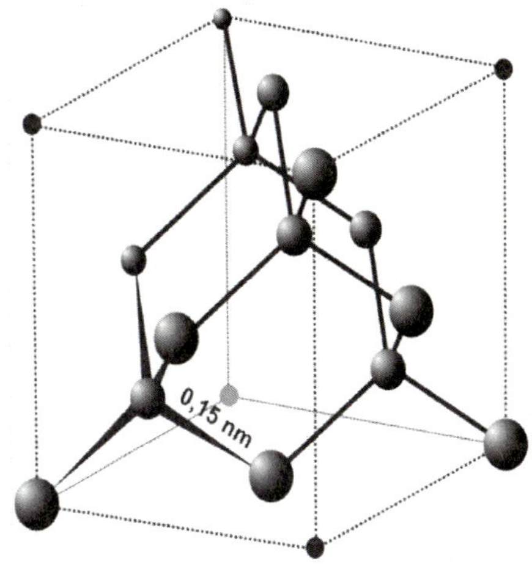

arranged in a crystal lattice in which all atoms are bonded to all nearby atoms. This structure is incredibly rigid, giving diamond its unique hardness.

One of the themes that we will see repeatedly in this class is the pervasiveness of chemistry. Even when we don't think we're talking about chemistry, much of the time we are. For instance, let's ponder jewelry for a moment. This is a topic that seems very far removed from anything we might traditionally associate with chemistry. However, because it occupies space and has mass, jewelry must be made of matter, and chemistry is the study of matter. It becomes a little more complicated when we try to ask ourselves of what KIND of matter jewelry is made. Take, for example, a gold ring. If you look inside a gold ring, you'll see a stamp that says 14K, 18K, or some other number followed by a "K" to indicate the material of the ring.*

Pure gold is an element (Au on the periodic table), and while aesthetically pleasing and rare (which combine to make it valuable), it's far too soft to be used in jewelry designed for daily wear. As such, elemental gold is mixed with other elements like nickel (Ni) and palladium (Pd). A "gold" ring, therefore, is actually made of a mixture rather than a pure substance.

Image 1.14. The stamp inside a gold ring indicates what fraction (parts out of 24) of the metal mixture is the element Au. Photo © Scott Lefler, 2010.

*NOTE

If you're curious, the "K" is for karat, which is just a way of indicating what fraction of the metal mixture is elemental gold. By definition, 24K is elemental gold, or 24/24 parts of the metal are gold. 18K gold has 18/24 parts elemental gold, and 6/24 parts other metals. Metallurgists blend in various metals to affect the color and properties of the resulting alloy (mixture) in addition to strengthening the gold.

Image 1.15. This periodic table has metals, nonmetals, and metalloids indicated.

Metals | Metalloids | Non-metals

1 H 1.01 Hydrogen																1 H 1.01 Hydrogen	2 He 4.00 Helium
3 Li 6.94 Lithium	4 Be 9.01 Beryllium											5 B 10.81 Boron	6 C 12.01 Carbon	7 N 14.01 Nitrogen	8 O 16.00 Oxygen	9 F 19.00 Fluorine	10 Ne 20.18 Neon
11 Na 23.0 Sodium	12 Mg 24.3 Magnesium											13 Al 26.98 Aluminum	14 Si 28.09 silicon	15 P 30.97 Phosphorus	16 S 32.07 Sulfur	17 Cl 35.45 Chlorine	18 Ar 39.95 Argon
19 K 39.10 Potassium	20 Ca 40.08 Calcium	21 Sc 44.96 Scandium	22 Ti 47.88 Titanium	23 V 50.94 Vanadium	24 Cr 52.00 Chromium	25 Mn 54.94 Manganese	26 Fe 55.58 Iron	27 Co 58.93 Cobalt	28 Ni 58.69 Nickel	29 Cu 63.55 Copper	30 Zn 65.39 Zinc	31 Ga 69.72 Gallium	32 Ge 72.59 Germanium	33 As 74.92 Arsenic	34 Se 78.96 Selenium	35 Br 79.90 Bromine	36 Kr 83.80 Krypton
37 Rb 85.47 Rubidium	38 Sr 87.62 Strontium	39 Y 88.91 Yttrium	40 Zr 91.22 Zirconium	41 Nb 92.91 Niobium	42 Mo 95.94 Molybdenum	43 Tc (98) Technetium	44 Ru 101.1 Ruthenium	45 Rh 102.9 Rhodium	46 Pd 106.4 Palladium	47 Ag 107.9 Silver	48 Cd 112.4 Cadmium	49 In 114.8 Indium	50 Sn 118.7 Tin	51 Sb 121.8 Antimony	52 Te 127.6 Tellurium	53 I 126.9 Iodine	54 Xe 131.3 Xenon
55 Cs 132.9 Cesium	56 Ba 137.9 Barium	57 La •138.9 Lanthanum	72 Hf 178.5 Hafnium	73 Ta 180.9 Tantalum	74 W 183.9 Tungsten	75 Re 186.2 Rhenium	76 Os 190.2 Osmium	77 Ir 192.2 Iridium	78 Pt 195.1 Platinum	79 Au 197.0 Gold	80 Hg 200.5 Mercury	81 Tl 204.4 Thalium	82 Pb 207.2 Lead	83 Bi 209.0 Bismuth	84 Po (210) Polonium	85 At (210) Astatine	86 Rn (222) Radon
87 Fr (223) Francium	88 Ra (226) Radium	89 Ac •(227) Actinium	104 Rf (257) Rutherfordium	105 Db (260) Dubnium	106 Sg (263) Seaborgium	107 Bh (262) Bohrium	108 Hs (265) Hassium	109 Mt (266) Meitnerium	110 Ds (271) Darmstadtium								

58 Ce 140.1 Cerium	59 Pr 140.9 Praseodymium	60 Nd 144.2 Neodymium	61 Pm (147) Promethium	62 Sm 150.4 Samarium	63 Eu 152.0 Europium	64 Gd 157.3 Gadolinium	65 Tb 158.9 Terbium	66 Dy 162.5 Dyspromium	67 Ho 164.9 Holmium	68 Er 167.3 Erbium	69 Tm 168.9 Thulium	70 Yb 173.0 Ytterbium	71 Lu 175.0 Lutetium
90 Th 232.0 Thorium	91 Pa (231) Protactinium	92 U (238) Uranium	93 Np (237) Neptunium	94 Pu (242) Plutonium	95 Am (243) Americium	96 Cm (247) Curium	97 Bk (247) Berkelium	98 Cf (249) Californium	99 Es (254) Einsteinium	100 Fm (253) Fermium	101 Md (256) Mendelevium	102 No (254) Nobelium	103 Lr (257) Lawrencium

1.2 THE PERIODIC TABLE

The periodic table of the elements contains a lot of information—we won't go over it all just yet. However, one interesting organizational feature of the table is that it divides elements into three distinct categories based upon their position on the table.

Metals, which are shiny, **ductile** (can be drawn into wires), **conductive** (can conduct electricity), and **malleable** (can be hammered into sheets), are located in the two columns on the left-hand side of the table and in the central, recessed portion (these recessed elements are called*transition metals*). Also, the two rows of elements that appear below the table (as though they were kicked out) are metals. The rows classically displayed under the periodic table actually fit into the center of the table as shown in Image 1.16. There are a few reasons commonly given for the fact that these elements, sometimes called the *rare earth metals*, are conventionally placed under the periodic table. One is that they were not yet discovered when the classical periodic table was developed, so room was not made for them. Another reason is that they make the table absurdly long and difficult to display on wall posters and in text figures!

Nonmetals, which are either gases or solids* at room temperature, have physical properties opposite those of metals. They are dull rather than shiny, brittle rather than ductile or malleable, and non-conductive. Nonmetals are located on the right-hand side of the table. Along a diagonal line between the metals and nonmetals are the **metalloids**, whose physical properties are

***NOTE**

There's one liquid nonmetal—bromine (Br_2).

Image 1.16. An "extended" periodic table of the elements.

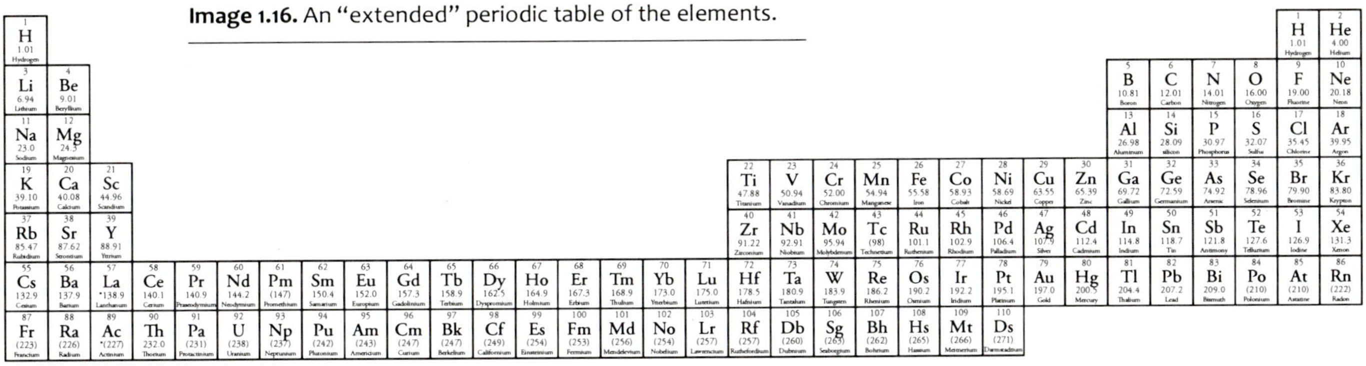

halfway between those of metals and nonmetals, and which are often referred to as *semiconductors*. They are useful in making photovoltaic cells for solar power panels (as we'll see in Chapter 11), as well as in computer applications. The element hydrogen (H) shows up in two places in many periodic tables. This is because hydrogen has some properties in common with metals, and some properties in common with nonmetals. Its technical classification, however, is as a nonmetal.

TRY THIS:

Classify each element as a metal, metalloid, or nonmetal.

Fe, B, I, Kr, Ca, U, Si.

Answers: The metals are Fe (iron), Ca (calcium), and U (uranium). The metalloids are B (boron) and Si (silicon). The nonmetals are I (iodine) and Kr (krypton).

Image 1.17. Copper (top left) has all the properties of metal: it's shiny, ductile, conductive, and malleable. Sodium (top right) is a metal (shown here in liquid paraffin to keep it from reacting with air), but we normally don't think of it as such, because we find it combined with other elements in compounds like table salt. Silicon (bottom right) is a metalloid (semiconductor) commonly used in computer chips. The nonmetals include gases such as chlorine (a toxic yellowish gas, bottom center), and dull, brittle, non-conductive solids such as iodine (bottom left).

As you look at the different elements of the periodic table, you'll notice a number of names among the metals that belong to substances you may never have thought of as metallic, like sodium (Na) and calcium (Ca). Both sodium and calcium are, in their elemental form, shiny metals. They have all the properties of common metals like aluminum and silver, but interestingly, they are soft. Sodium can actually be cut with a butter knife! The reason we don't find these elements in their metallic form in nature is that they are very reactive and readily combine with other elements to form compounds. For instance, you probably encounter the element sodium most commonly as part of the compound sodium chloride (NaCl), which is table salt.

1.3 THINKING ABOUT CHEMICALS, CHEMICAL REACTIONS, AND CHEMICAL EQUATIONS

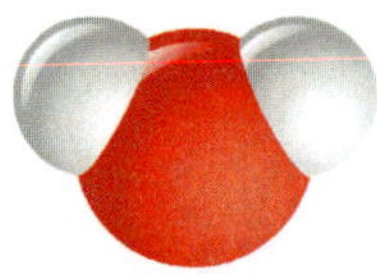

H_2O

Image 1.18. Macroscopic (left), microscopic or molecular (top right), and symbolic (bottom right) representations of the compound water. (Havasu Falls picture © Scott Lefler, 2007.)

It's fundamentally a bit difficult to be a student of chemistry, because unlike some of the other sciences, chemistry doesn't really afford us the opportunity to see and manipulate the object of our examination. In biology, we can touch the fetal pig we're dissecting, or turn over the leaf we're sketching. Even bacteria, which are invisible to the naked eye, can be viewed with the help of a microscope. Chemistry is a little different. Even though we can go to the lab,mix a few chemicals and note a color change or bubbling or some other visible evidence of a reaction, we have to take our instructor or lab manual at face value when they tell us what the chemicals are and what reaction is occurring. We can't see the molecules, and we can't watch bonds break and reform. As a result, the study of chemistry involves employing a variety of visualization aids. Let's take the example of water. We can think about water on a macroscopic scale, in terms of what we actually see—it's a clear liquid. We can also think about water on a microscopic scale, in terms of how the atoms are arranged—an atom of oxygen is bonded to two separate atoms of hydrogen.

Finally, we can symbolically represent water in a concise way that provides information about its chemical composition—water has the formula H_2O. Depending upon what it is we're trying to communicate or what we're trying to imagine, we might pick any one or a combination of these strategies to visualize the compound water. For instance, if I wanted to communicate to you that the chemical methane (CH_4) burns in oxygen (O_2), making carbon dioxide (CO_2) and water (H_2O), it would be sufficient to symbolically represent chemicals with their formulas—no further information is needed.In Chapter 13, however, we'll need to think about the shape of a water molecule in order to understand the greenhouse effect. Because of that, we'll spend a lot of time thinking about its microscopic representation, which helps us to imagine the physical arrangement of the atoms in the molecule and its shape in space.

Let's take a closer look at the chemical reaction mentioned in the previous paragraph. Methane (CH_4) is a member of the class of chemicals collectively called **hydrocarbons**, because they're composed of carbon and hydrogen. We'll be talking about hydrocarbons frequently in this textbook, in part because they all burn in oxygen (in chemistry, burning is called **combustion**) to produce carbon dioxide and water. Have you ever barbecued on a gas grill? If so, you've combusted the hydrocarbon propane (C_3H_8). If you're a camper, you've probably used a camp stove that runs on the hydrocarbon butane (C_4H_{10}). Have you fueled your car recently? That was octane (C_8H_{18}) you put in the tank, unless you drive an alternativefuel vehicle that runs on ethanol (C_2H_5OH).* When we describe the behavior of methane in oxygen, we are describing a **chemical reaction**, which is a rearrangement of atoms and molecules to form new molecules. The original chemical species are called the **reactants**, and the resulting chemical species are called the **products**. In the example above, methane and oxygen are reactants, while carbon dioxide and water are products.

In addition to describing a reaction in words, we can represent it symbolically. Let's take the example of a very simple chemical reaction with only two reactants and one product. Sulfur (S) combines with oxygen (O_2) to produce sulfur dioxide (SO_2). Represented microscopically, the reaction looks like this:

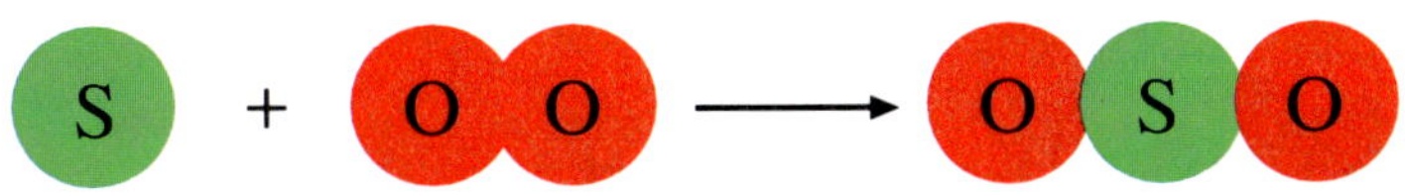

While this is a convenient way to visualize the rearrangement of atoms and molecules taking place in the reaction, it quickly becomes prohibitive to draw microscopic representations of reactants and products, especially in more complicated reactions. **Chemical equations** are symbolic representations of chemical reactions. The chemical equation for the reaction shown above would look like this:

$$S + O_2 \rightarrow SO_2$$

The arrow indicates that the reactants sulfur and oxygen* (S and O_2) are rearranging to form the product sulfur dioxide (SO_2). Aloud, this would read: *sulfur and oxygen react to form sulfur dioxide.*

You may be asking yourself right about now: *Why would I want to know how to write out a chemical equation, since I'm not a chemist?* That's a valid

> Try This:
>
> In the reaction of carbon and oxygen to form carbon dioxide, what are the reactants and the products?
>
> Answer: The reactants are carbon and oxygen, and the product is carbon dioxide.

> ***NOTE**
>
> Compounds like ethanol are called hydrocarbon derivatives—these compounds are mostly carbon and hydrogen, but contain small amounts of other elements, such as N, O, or S. Many of them, like ethanol, also combust in oxygen.

> ***NOTE**
>
> Why do we call O_2 *oxygen* when an atom of O is also called *oxygen*? This is a slightly confusing issue, but it's worth addressing. Even though an atom of O is called *oxygen*, atomic oxygen is rarely found on its own in nature—it's almost always combined into molecules. Elemental oxygen in nature is found as the molecule O_2. It's so common, in fact, that rather than saying *molecular oxygen* or *diatomic oxygen*, we simply call it oxygen. Generally, we can figure out what is meant by the word *oxygen* from context. If I say *CO_2 consists of one carbon and two oxygens*, you pretty well know I mean atoms of oxygen. If I say *the atmosphere contains 21 percent oxygen*, you should know that elemental oxygen found in nature is always O_2. If the context leaves doubt as to what is meant, we generally reserve oxygen for O_2, and say *atomic oxygen* or *an atom of oxygen* for O.

question. The answer is that in order to talk meaningfully about chemistry in everyday life, we'll need to speak a little bit of the language of chemistry, which is written in terms of reactants, products, and reactions.

1.4 BALANCING CHEMICAL EQUATIONS

Some chemical reactions end up causing us a bit of trouble when we try to write them out as chemical equations. Take, for instance, the true statement that elemental hydrogen (found in nature as a diatomic molecule, remember?) burns in oxygen to form water. When we write out an equation, we get something that looks like this:

$$H_2 + O_2 \rightarrow H_2O$$

Looking closely, it seems that we have a problem. There are two atoms of hydrogen (the subscript tells us that one molecule of hydrogen consists of two atoms of hydrogen chemically bonded together) and two atoms of oxygen on the reactant side of the equation. However, there are two atoms of hydrogen and only ONE atom of oxygen on the products side. A microscopic representation of the equation makes this very clear:

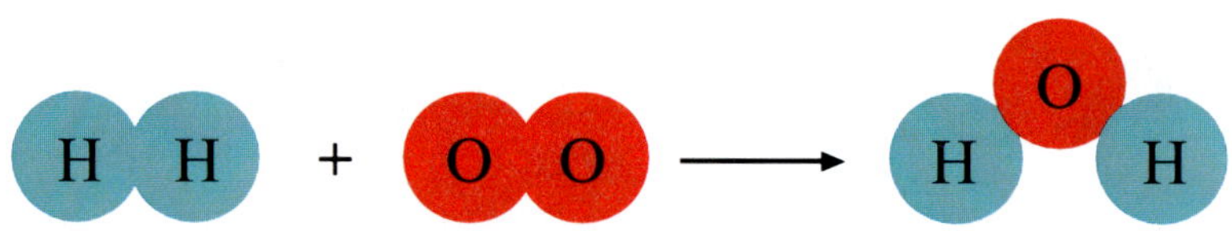

Where did the second atom of oxygen go? The answer is *nowhere*—atoms can't disappear (nor can they appear!)—that's a fundamental law of nature. The above is therefore not an accurate representation of the chemistry, because it's not a **balanced equation**. Balanced equations obey the **Law of Conservation of Matter**, which means that atoms cannot appear, disappear, or change into atoms of other elements through any chemical reaction. In order for our equation to be chemically correct, we must ensure that there is the same number of each type of atom on each side of the equation. In order to do this, we can specify that different AMOUNTS of each reactant or product are involved in the reaction, but under no circumstances can we change the SUBSCRIPTS associated with any element in any molecule.* To specify that there is more than one of any given chemical involved in a reaction, we use a coefficient in front of the species (the lack of a coefficient in front of a species is always taken to mean *one*). In this case, we need two water molecules to balance the oxygen atoms, and two hydrogen molecules to account for the hydrogen needed to form the second water molecule. Our equation, correctly balanced, becomes:

$$2\ H_2 + O_2 \rightarrow 2\ H_2O$$

Doing a quick atomic head count, this gives us four hydrogen atoms on each side of the equation, and two oxygen atoms on each side:

***NOTE**

Tempting though it might be to balance the oxygen atoms by changing the product to H_2O_2, the chemical truth is that hydrogen and oxygen combine to form water, which is H_2O. H_2O_2 (while appearing to balance the equation nicely) is NOT water. It's hydrogen peroxide, which is a very different chemical indeed!

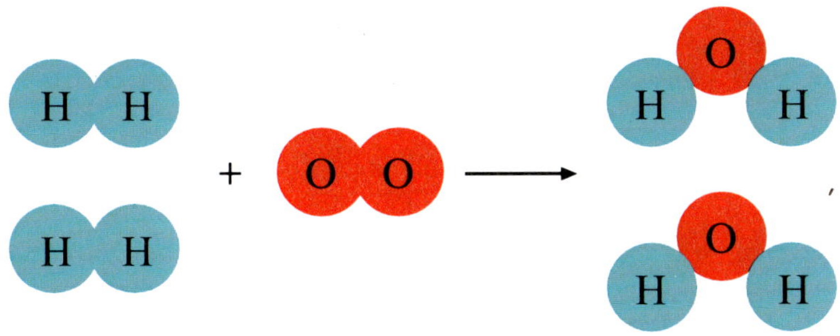

Even though we started with three molecules and ended up with two, we've conserved the numbers and identities of the atoms. In chemical reactions, reactants and products can contain different numbers of molecules, but they must contain exactly the same number and type of atoms, which is why we count atoms rather than molecules when balancing equations.

> **CONCEPT CHECK:**
>
> Why did we have to change the number of hydrogen molecules and water molecules in the equation, but not the number of oxygen molecules?
>
> Answer: By changing the number of water molecules (to get equal numbers of hydrogen atoms on each side of the equation), we ended up with equal numbers of oxygen atoms on each side.

Some equations are a bit more difficult to balance, because they involve more species. Let's return, for instance, to the reaction between methane and oxygen that produces carbon dioxide and water:

$$CH_4 + O_2 \rightarrow CO_2 + H_2O$$

A cursory glance at this equation reveals that it's not balanced. Unlike the hydrogen and oxygen reaction, however, this one is difficult to balance simply by looking at it. A good technique for balancing reactions like this follows a series of steps:

1) Identify an element that appears in ONE compound (do not pick elements that appear alone—save these for last!) on either side of the equation. Balance that element by adjusting the coefficients for molecules containing the element.

Oxygen appears in our reaction as a free element (O_2), so we'll save it for last. Carbon and hydrogen are both good Step 1 candidates, as they appear in only ONE compound on either side. We can therefore start with either carbon or hydrogen.

$$CH_4 + O_2 \rightarrow CO_2 + H_2O$$

Image 1.19. A Bunsen burner combusts methane (CH_4) in oxygen to produce carbon dioxide and water.

Carbon is already in balance; there is one atom of carbon on each side of the equation.

2) Repeat Step 1 for remaining elements OTHER than free elements.

Hydrogen needs to be balanced; we can accomplish this by adding the coefficient 2 in front of the product H_2O.

$$CH_4 + O_2 \rightarrow CO_2 + 2\ H_2O$$

There are now four atoms of hydrogen on each side of the equation.

3) Balance any remaining elements.

Oxygen can now be balanced. There are four total oxygen atoms on the right—two from CO_2, and one from each of the two waters— so we need four total oxygen atoms on the left. We therefore add the coefficient 2 in front of the reactant O_2.

$$CH_4 + 2\ O_2 \rightarrow CO_2 + 2\ H_2O$$

This is a balanced equation.

TRY THIS:

Balance each of the following equations:

___S_8 + ___O_2 → ___SO_3

___N_2 + ___O_2 → ___N_2O

___$C_{10}H_{16}$ + ___Cl_2 → ___C + ___HCl

Answers:

$S_8 + 12\ O_2 \rightarrow 8\ SO_3$

$2\ N_2 + O_2 \rightarrow 2\ N_2O$

$C_{10}H_{16} + 8\ Cl_2 \rightarrow 10\ C + 16\ HCl$

1.5 NOMENCLATURE

Shakespeare said in *Romeo and Juliet* that a rose by any other name would smell as sweet, but as it turns out, naming is critically important in chemistry! You've seen a few chemical formulas coupled with names so far in this chapter, so you may be getting familiar with the way the names tend to sound. For instance, it was mentioned earlier that a compound consisting of sodium (Na) and chlorine (Cl) is called *sodium chloride*, and I'll tell you right now that a compound of rubidium (Rb) and fluorine (F) is called *rubidium fluoride*—there's a pattern there. Just to throw a wrench in the works, however, a compound of carbon and oxygen is NEVER called *carbon oxide*. WHY? The short answer is that there are several ways in which carbon and oxygen can combine,* and using the (incorrect) name *carbon oxide* doesn't help us to distinguish them from one another. Clearly, we need to outline some rules for how to name compounds. Knowing a little bit of **nomenclature** (which means a system for naming) helps us to understand one another when we are discussing chemistry. The easiest compounds to name are **binary compounds**, which are made up of two (and only two) different elements. There are lots of compounds made of more than two elements, but their nomenclature is much more complex. Of course, anything made up of only one element is not a compound at all. Binary compounds may consist of a metal and a nonmetal, or a nonmetal and a nonmetal; there are no compounds of metals with metals.

> ***NOTE**
>
> Two examples include CO (carbon monoxide) and CO_2 (carbon dioxide).

Binary Compounds—Metal and Nonmetal

To name a binary compound made of a metal and a nonmetal, we use the name of the metal followed by the name of the nonmetal, adding the suffix *-ide* to the nonmetal. You may need to drop some letters to do this phonetically—sulfur becomes sulfide, and oxygen becomes oxide.

Example: A compound of the metal barium (Ba) and the nonmetal oxygen (O) is called *barium oxide.*

Example: The compound LiF is called *lithium fluoride.*

Note that chemical formulas for binary compounds of metals and nonmetals are written in the same order in which the name is given—metal first.

When there are subscripts in the chemical formula of a compound, we ignore them in naming the compound.

Example: The compound $AlCl_3$ is called *aluminum chloride.*

The reason we don't reference the relative number of atoms of each element when we name the compound is that metals can only combine with a given nonmetal in ONE possible way. In other words, when lithium and fluorine combine, the ONLY possible combination is LiF. When aluminum and chlorine combine, the ONLY possible combination is $AlCl_3$. How

> **TRY THIS:**
>
> Name each of the following: K_2O, MgS, AlF_3.
>
> Answers: Potassium oxide, magnesium sulfide, aluminum fluoride.

Image 1.20. Binary compounds of a metal and a nonmetal are also called salts. They are solid compounds with some interesting chemical properties that will be discussed further in Chapter 4. Among the binary compound salts we commonly encounter (besides table salt, of course!) are iron oxide (Fe_2O_3), which is rust, and sodium fluoride (NaF), used in toothpaste.

would you know this? As it turns out, there are some rules that we'll learn in Chapter 4 that we can use to figure out what combination is possible for a given metal and nonmetal. In the meantime, though, we'll have to satisfy ourselves with predicting names from formulas. We can't predict formulas from names just yet, because the names don't tell us how many atoms of each element are present.

Binary Compounds—Nonmetal and Nonmetal

To name a binary compound made of two nonmetals (or in the much rarer case of a metalloid and a nonmetal), we name the element closer to the left side of the periodic table first, then the element closer to the right side, adding the suffix *–ide* to the second element. Additionally, we need to make reference to the number of atoms of each element present, using the Greek prefixes listed below:

Number of atoms of an element	Prefix
1	mono-
2	di-
3	tri-
4	tetra-
5	penta-
6	hexa-

Table 1.1. Prefixes used in nomenclature.

Example: The compound N_2O is called *dinitrogen monoxide* (nitrogen is closer to the left side of the periodic table, so it is named first and written first in the formula).

If there's only one of the element closer to the left side (the one we'd write or say first), we leave off the "mono" prefix.

Example: The compound CO_2 is called *carbon dioxide.*

We may need to drop vowels from the end of the Greek prefix when element names begin with vowels in order to make the name phonetically sensible.

Example: The compound CO is *carbon monoxide.*

The two examples above demonstrate why we need prefixes to indicate the relative number of each atom; unlike compounds of metals and nonmetals, nonmetals can combine with other nonmetals in many different ways. The prefixes help us know to which of the many possible combinations of two given nonmetals we refer.

TRY THIS:

Name each of the following: CF_4, PCl_5, CS_2.

Answers: Carbon tetrafluoride, phosphorus pentachloride, carbon disulfide.

TRY THIS:

Predict formulas for each of these names: Silicon tetrafluoride, dinitrogen trisulfide, diboron hexahydride.

Answers: SiF_4, N_2S_3, B_2H_6.

Image 1.21. Binary compounds of two nonmetals or of a metalloid and a nonmetal may be gaseous like the volcanic gas sulfur dioxide, liquid like water, or solid like silicon dioxide (sand). (Costa Rica photos © Kirstin Hendrickson, 2002.)

1.6 THE LAST WORD—CHEMICAL FREE?

One thing you'll notice as you start to look at chemistry in the real world is that chemicals are often treated as something to be avoided. People tend to crave the "natural" and fear or dislike the "unnatural," but the lines drawn to separate the two are often randomly placed and inaccurate. In particular, the word *chemical* is frequently taken to be synonymous with *unnatural*, and treated as something to be avoided. Case in point, I was reading the label on a loaf of bread not long ago while making toast. I noticed that, in an attempt to reassure the segment of society interested in "natural" foods, the bread claimed to be "chemical free."

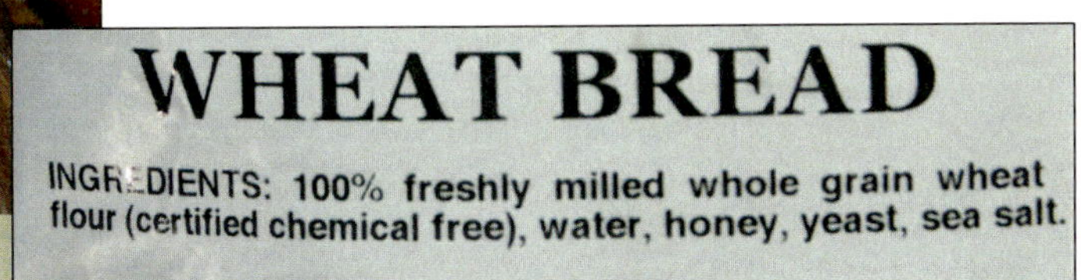

Image 1.22. "Certified chemical free" bread. (Photos © Scott Lefler, 2010)

Of course, *chemical* is simply another way of saying *matter*, so no physical material is chemical free any more than it can be matter free. As we've learned in this chapter, atoms and molecules, elements and compounds are ALL chemicals. Just for emphasis, the bread's ingredients included: whole wheat flour (a mixture of elements and compounds, chief among them amylose [a large molecule consisting of lots of C, H, and O] and cellulose [a different large molecule, also made of C, H, and O]); water (the molecule H_2O); honey (a mixture of elements and compounds, including lots of the sugars fructose and glucose, both of which are compounds consisting of C, H, and O); yeast (an organism composed of a variety of chemicals); and sea salt (again, a mixture of chemicals, but comprised primarily of NaCl). The bread contains nothing BUT chemicals!

As you look around, nothing you can touch is chemical free. The paper in these pages is made from the chemical cellulose (which comes from plants). All the food you eat is composed of various chemicals, just like the bread we've been discussing. The water you drink is a chemical. The air you breathe is a mixture of many chemicals, including the O_2 your cells need.

Of course, a devil's advocate would say that the expression *chemical free* is nothing more than a concise way to say, "free of unnatural chemicals that will cause harm." Touché, but as we'll see later, even the idea that unnatural = BAD and natural = GOOD is fundamentally flawed. Furthermore, it can be quite difficult to draw lines between *natural* and *unnatural.* More than anything, the point is simply that where chemistry and society intersect, society can respond with a fear of the unknown, and that fear may lead to irrational (and scientifically unfounded) behavior and ideas. Discussing these departures from logic makes us aware of them, and we may begin to evaluate the reasoning behind our behavior, the logic that informs our reasoning, and the efficacy of our logic.

SUMMARY OF MAIN POINTS

- Chemistry is all around us. Everything we can see or touch is made of chemicals or is a by-product of a chemical reaction.
- When we think about the interaction between chemistry and society, we have to consider the individual, community, and global levels of society.
- Chemistry is the study of matter, which consists of mixtures and pure substances. Pure substances include elements and compounds.
- A pure substance can be atomic or molecular in nature; molecules can either be elements or compounds.
- The elements on the periodic table can be divided into three broad categories by their position on the table: metals, nonmetals, and metalloids. Each of these categories has characteristic properties.
- When we think about chemicals, we can consider them from macroscopic, microscopic, and/or symbolic perspectives, depending upon what we're trying to communicate.
- Chemical equations are symbolic representations of chemical reactions, consisting of reactants and products separated by an arrow.
- Chemical equations must obey the Law of Conservation of Matter, which states that atoms may not appear, disappear, or change into atoms of other elements during a chemical reaction. In other words, equations must be balanced.
- Balancing equations must be done by adjusting coefficients in front of the reactant or product species; subscripts within chemical formulas may not be changed.
- Binary compounds of a metal and a nonmetal are named without referencing subscripts. The metal is named first, then the nonmetal (with the suffix *-ide*).
- Binary compounds of two nonmetals are named using Greek prefixes to indicate the number of atoms of each element. The element closer to the left side of the periodic table is named first, followed by the one closer to the right side of the periodic table (with the suffix *-ide*).

Questions and Topics for Discussion

1. Provide an example of each of the following relationships with regard to chemistry (do not use examples discussed in the text):
   —Individual behavior affects that individual
   —Individual behavior affects society
   —Behavior of a society affects an individual
   —Behavior of a society affects that society
   —Behavior of a society affects another society.
2. With regard to chemistry, do you think that there are any examples of individual behavior that have NO effect whatsoever upon society? Justify your answer.
3. Using only items you could find around yourself in everyday life (i.e., NOT in a lab), name an object that represents a mixture. Name an object made up of a compound. Name an object made of a pure element.
4. Explain the primary differences between mixtures and pure substances.
5. Provide an example of when the word *nitrogen* would be understood from context to mean an atom of N. Provide an example of when the same word would be understood from context to mean a molecule of N_2. Provide an example of when the meaning might be ambiguous and clarification would be required.
6. What is the primary naming difference between binary compounds of metals with nonmetals and binary compounds of nonmetals with nonmetals? What is the reason for this difference? When might knowing how to name a chemical (or knowing the formula from the name) be really important in real life?
7. Provide a "real-life" example of when, in discussing a substance, we'd need to visualize macroscopically. Microscopically? Symbolically?
8. Why do you think people associate the word *chemical* with something bad? Is there any context in which *chemical* is taken to mean something good? Provide an example.

Problems

1. Define each of the following and provide an example: matter; mixture; pure substance; element; compound; atom; and molecule. Can an element be a molecule?
2. Classify each of the following using as many of the words in Problem 1 as are applicable in each case: CO_2; Ne, N_2; saltwater; glucose ($C_6H_{12}O_6$); and HCl.
3. Classify each of the following using as many of the words in Problem 1 as are applicable in each case: H_2O; S_8; $MgCl_2$; sugar water; C; and Fe.
4. What are the primary physical characteristics shared by metals? By nonmetals? By metalloids?
5. Classify each of the following elements as metal, nonmetal, or metalloid: Rb; fluorine; Xe; Pb; mercury; Si; germanium; K; and Po.
6. Classify each of the following elements as metal, nonmetal, or metalloid: B; Ne; titanium; Cm; tin; At; F; Na; and silver.
7. Identify the products and reactants in the reaction: $2\ C_2H_2 + 5\ O_2 \rightarrow 4\ CO_2 + 2\ H_2O$. Explain how the coefficients ensure a balanced reaction in terms of atoms.
8. Identify the products and reactants in the reaction: $S_8 + 12\ O_2 \rightarrow 8\ SO_3$. Explain how the coefficients ensure a balanced reaction in terms of atoms.
9. What is the Law of Conservation of Matter and how does it affect our writing of chemical equations?
10. Write out a chemical equation for the statement *Elemental nitrogen reacts with elemental hydrogen to produce ammonia (NH_3).* Balance if necessary.

11. Write out a chemical equation for the statement *Elemental nitrogen reacts with oxygen to form nitrogen dioxide.* Balance if necessary.
12. Balance each of the following reactions:

    ___C_3H_8 + ___O_2 → ___CO_2 + ___H_2O (this is the reaction that takes place when you run a gas grill)

    ___HgO → ___Hg + ___O_2 (this is the reaction that helped the early chemist Joseph Priestley discover the element oxygen)

    ___SO_3 + ___H_2O → ___H_2SO_4 (this is how acid rain forms)

    ___Fe + ___O_2 → ___Fe_2O_3 (this is how rust forms).
13. Balance each of the following reactions:

    ___O_2 → ___O_3

    ___$C_6H_{12}O_6$ + ___O_2 → ___H_2O + ___CO_2

    ___Al + ___O_2 → ___Al_2O_3

    ___I_2 + ___$Na_2S_2O_3$ → ___NaI + ___$Na_2S_4O_6$.
14. What of the following MUST be conserved in a chemical equation: number of atoms, number of molecules, identity of atoms, identity of molecules?
15. Name each of the following compounds: CsCl; AgBr; Ag_2O; Al_2O_3; BaF_2; and Na_2S.
16. Name each of the following compounds: $MgCl_2$; NaF; KCl; Na_2O; MgO; and K_2O.
17. Name each of the following compounds: NO; N_2O; NO_2; and H_2O (name systematically!).
18. Name each of the following compounds: SO_2; SO_3; PBr_3; and PCl_3.
19. Name each of the following compounds: $BaBr_2$; CS_2; PCl_5; Al_2S_3; P_2O_3; S_2Cl_2; Na_3N; Mg_3P_2; and CaO.
20. Name each of the following compounds: Rb_2O; LiF; Li_2S; N_2O_4; H_2O_2; K_3N; ClO (this one is a little strange; there's an exception to the naming rule that says we always name O LAST if it appears in a formula); CF_4; and BaS.
21. Provide formulas for each of the following names: dinitrogen pentoxide; iodine monochloride; carbon tetrachloride; and boron trifluoride.
22. Provide formulas for each of the following names: sulfur dioxide; sulfur trioxide; phosphorus pentabromide; and silicon dioxide.

Answers to Problems

1. Matter occupies space and has mass; mixtures are composed of two or more pure substances, physically combined, in variable proportions; pure substances are elements and compounds; elements are the simplest type of matter, listed on the periodic table: they cannot be broken into simpler substances by any chemical means; compounds are a combination of two or more elements, chemically combined, in fixed proportions; atoms are the smallest (indivisible) part of matter of a given element; molecules are two or more atoms, chemically combined, in fixed proportions. Examples will vary. An element can be a molecule (examples include the diatomic molecular elements such as O_2).
3. H_2O is matter, a pure substance, a compound, a molecule; S_8 is matter, a pure substance, and element, a molecule; $MgCl_2$ is matter, a pure substance, a compound, a molecule; sugar water is matter, a mixture; C is matter, a pure substance, an element, an atom; Fe is matter, a pure substance, an element, an atom.
5. Rb is a metal; fluorine is a nonmetal; Xe is a nonmetal; Pb is a metal; mercury is a metal; Si is a metalloid, germanium is a metalloid, K is a metal; Po is a metal.
7. C_2H_2 and O_2 are reactants, CO_2 and H_2O are products. The coefficients are there to balance the equation (4 C atoms on each side, 4 H atoms on each side, 10 O atoms on each side).

9. The Law of Conservation of Matter states that atoms cannot appear, disappear, or change into atoms of any other element by any chemical means. Because of this law, chemical equations must be balanced.
11. $N_2 + 2\ O_2 \rightarrow 2\ NO_2$.
13. $3\ O_2 \rightarrow 2\ O_3$; $C_6H_{12}O_6 + 6\ O_2 \rightarrow 6\ H_2O + 6\ CO_2$; $4\ Al + 3\ O_2 \rightarrow 2\ Al_2O_3$; $I_2 + 2\ Na_2S_2O_3 \rightarrow 2\ NaI + Na_2S_4O_6$.
15. Cesium chloride; silver bromide; silver oxide; aluminum oxide; barium fluoride; and sodium sulfide.
17. Nitrogen monoxide; dinitrogen monoxide; nitrogen dioxide; and dihydrogen monoxide (a funny name for water!).
19. Barium bromide; carbon disulfide; phosphorus pentachloride; aluminum sulfide; diphosphorus trioxide; disulfur dichloride; sodium nitride; magnesium phosphide; and calcium oxide.
21. N_2O_5; ICl; CCl_4; and BF_3

Chemistry Around Us—The Atmosphere 2

Our Earth is surrounded by a shell of gases, which is called the atmosphere. We can't see these gases, so we don't tend to think much about them, but our atmosphere is not only critical to life as we know it, it is also the site of some pretty interesting and important chemistry. We'll discuss some of what's going wrong in the atmosphere in Chapters 10 and 13, but for now, let's focus our attention on the chemistry of a healthy atmosphere.

Image 2.1. The moon through the shell of our atmosphere. Notice that compared to the Earth's radius of curvature, the atmosphere really is quite thin.

The Earth is 12756 km, or 7926 miles, in diameter. Our atmosphere, by comparison, is a mere 62 miles or so in thickness, although it's difficult to pinpoint exactly where the atmosphere stops—since it's gaseous in nature, it just tapers off. An easier way to think about it is that the atmosphere is thickest near Earth's surface and becomes exponentially thinner with increasing altitude. Ninety percent of the atmosphere by mass is below an altitude of 10 miles, and the remaining 10 percent is located between 10 and 62 miles above Earth's surface. For reference, it may be helpful to think about this in terms of the altitude above sea level of some other common things (Image 2.2).

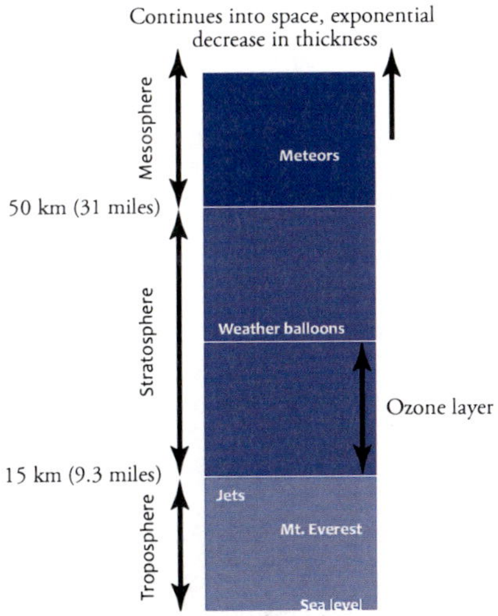

Image 2.2. Regions of the atmosphere and what we find there.

The atmosphere is divided into several major regions by altitude. The two we'll discuss in this text are the troposphere (the region of the atmosphere that we inhabit) and the stratosphere (the location of the ozone layer).

One interesting property of the atmosphere is that it exerts pressure on everything it surrounds. We commonly hear about **barometric pressure**—the ambient atmospheric pressure in a given location at a given time—in reference to the weather. However, we don't often extend that concept to the realization that the atmosphere truly exerts pressure on us—literally pushes against us—all the time. Pressure is defined in physics as a force (which is generated by a mass) exerted over an area, so if the majority of the atmosphere's mass is located in the first 10 or so miles above Earth's surface, then the highest pressures should also be closest to sea level. In fact, air pressure, like the density of the atmosphere, falls off exponentially with increasing altitude.

Even small changes in altitude can have dramatic effects upon air pressure. For instance, at 10,000 feet (the approximate elevation of Quito, Ecuador), the air pressure is about two-thirds of that at sea level. On top of Mt. Everest (just over 29,000 feet), the air pressure is about one-third of that at sea level. One common experience that makes this pressure differential quite apparent is the "popping" sensation in the ears or throat that accompanies a change in altitude and air pressure. For instance, passengers on an airplane are subjected to

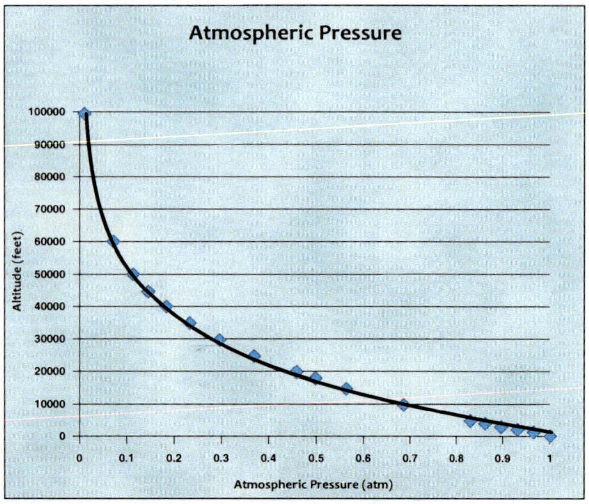

Image 2.3. Atmospheric pressure (shown on the x-axis) decreases exponentially as altitude (y-axis) increases.

a change in air pressure as the plane takes off and begins to climb. The pressure in the cabin is adjusted to the equivalent of about 7000 feet above sea level.* Assuming the plane took off from an elevation lower than 7000 feet, this represents a loss of atmospheric pressure, which creates a pressure imbalance between the middle and outer portions of the ear. The higher pressure inside the ear causes the eardrum to bow outward, which is uncomfortable and affects hearing. To return the eardrum to a relaxed, unbowed state, air must escape the middle ear through a normally collapsed tube (called the Eustachian tube) that opens into the back of the throat. The bubbles of air escaping through this tube produce a popping sensation. In preparation for landing, air pressure in the cabin is increased to the equivalent of that at the plane's destination. This, too, can affect the ears—the higher pressure outside the ear bows the eardrum inward, requiring that air enter the middle ear through the Eustachian tube, once again producing the sensation of a "pop."

***NOTE**

If you're curious, the logic here is that most people feel fine going from sea level to 7000 feet, but many experience symptoms of altitude sickness above that elevation, particularly if they ascend rapidly. Since planes can easily exceed 35,000 feet in elevation, the cabin must be pressurized. Why not just maintain sea-level pressure? Remember, pressure comes from mass, and mass has weight! An airplane pressurized to sea level would weigh more than an airplane pressurized to 7,000 feet, so it improves gas mileage to reduce the pressure inside the plane as much as possible.

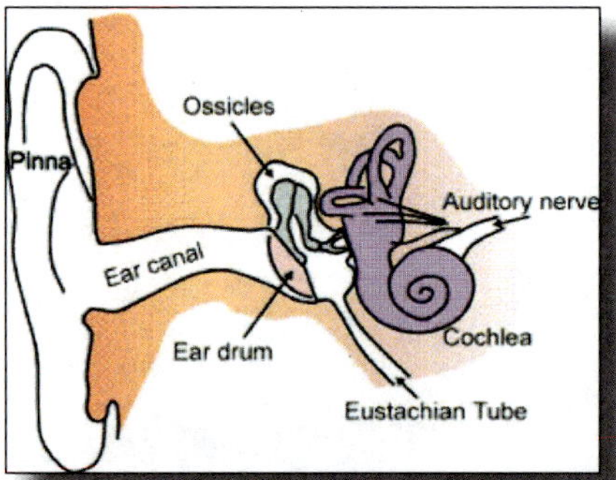

Image 2.4. Air pressure must be the same in the middle ear (the space just to the right of the eardrum in the figure) as in the ear canal to prevent bowing of the eardrum.

2.1 COMPOSITION OF THE ATMOSPHERE

We now know a little about the distribution of the atmosphere, so the next thing we need to think about is its composition. Our atmosphere is composed of a mixture of many different gases, but elemental nitrogen (N_2) is the most prevalent among these. Nitrogen accounts for 78 percent of our atmosphere. Most of the rest of the atmosphere is elemental oxygen (O_2), which accounts for 21 percent of the total. The remaining 1 percent is a mixture of other naturally occurring gases, including some water vapor.

This composition is maintained throughout the troposphere. In the stratosphere other gases are present, and we'll discuss those more in Chapter 10. For now, though, it's sufficient to remember that anywhere we go on Earth (even at high altitude), we're breathing an atmospheric mixture of consistent composition.

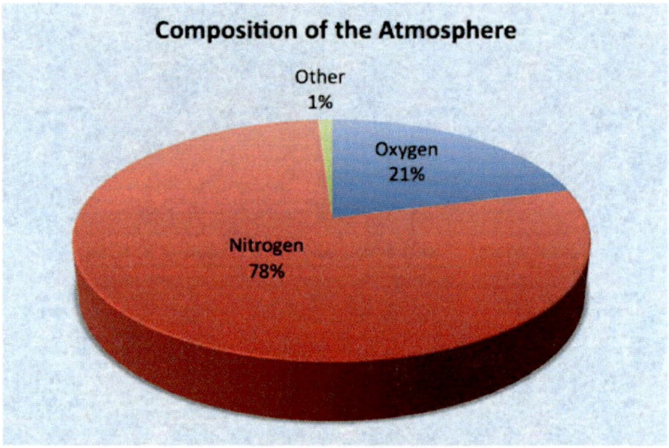

Image 2.5. The composition of Earth's atmosphere. Other gases include Ar, CO_2, H_2, Ne, Kr, and methane (CH_4).

2.2 CHEMISTRY OF HIGH ALTITUDE

If the percentage of O_2 is constant throughout the troposphere, why do people feel short of breath at high altitude and, at sufficient elevation, experience symptoms of altitude sickness?

Where air pressure is high—close to sea level—the atmosphere is more dense, meaning that there are more molecules of gas in a given volume. When we fill our lungs, we fill them with a mix of 78 percent N_2 and 21 percent O_2, and there are lots of molecules of each gas. At high elevation, a lungful of air contains fewer total molecules, so even though the percentages of nitrogen and oxygen are the same, 21 percent of a smaller total means fewer molecules of O_2 per breath. With less oxygen entering the lungs per breath, less oxygen makes it into the bloodstream, and we begin to experience shortness of breath and other symptoms of oxygen deprivation.

The issue of high altitude breathing has some interesting chemistry associated with it. At first glance, the solution to the problem of oxygen deprivation seems simple—just breathe more! Unfortunately, it's not quite that easy. The problem is that every breath we take has TWO functions—to bring in O_2 and to get rid of CO_2. CO_2 is a metabolic waste product that is produced by cells as they burn sugars and other fuels for energy. Even though CO_2 is a waste product that we exhale, that doesn't mean it has no function in the body. It turns out that CO_2 serves the very important purpose of keeping the blood at a constant level of acidity (pH). Acidity will be discussed in more detail in Chapter 5, but right now it's sufficient to know that we have to maintain a constant body acidity in the same way that we have to maintain a constant body temperature (98.6°F) to be comfortable and healthy. By way of example, just think how awful it feels if your body temperature is off by even a few degrees—a fever of 101°F, just slightly higher than normal temperature, will make you feel terribly ill! The role of CO_2 in maintaining acidity comes from its reaction with water, of which there's plenty in the bloodstream. CO_2 released from cells combines with water by the reaction:

$$CO_2 + H_2O \rightarrow H_2CO_3$$

H_2CO_3 is carbonic acid, and we need a constant amount of it in the blood—not too much, not too little—to maintain appropriate acidity. The carbonic acid travels to the lungs, and there it reacts again, regenerating water and CO_2, which is then exhaled:

$$H_2CO_3 \rightarrow CO_2 + H_2O$$

The cells are always making more CO_2, so the level of carbonic acid in the blood stays constant under normal circumstances. However, if we inhale more (to bring in more O_2 at high altitude), we're also EXHALING more, which means we quickly get too low on CO_2 and lose too much acidity. You can try an experiment if you like: If you breathe very rapidly in and out for a while, you'll start to feel a little light-headed. It's not the extra O_2 that's responsible for that

Image 2.6. At 22,485 feet in elevation, Aconcagua (in the Andes mountains of South America) is the highest peak in the western hemisphere. It's popular with climbers, and because of its staggering height, they often experience symptoms of altitude sickness. (Photos © Kirstin Hendrickson, 2002.)

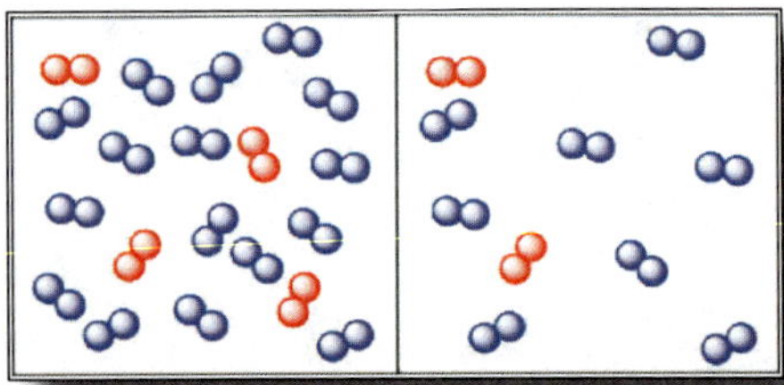

Image 2.7. A representation of what we would take into our lungs at high atmospheric pressure (left) and low atmospheric pressure (right). The percentage of oxygen (red) is the same in each case, but there are fewer total oxygen molecules in the low-pressure lungful of air.

CONCEPT CHECK

Why is the amount of CO_2 in the body important, if it's a waste product?

Answer: CO_2 helps to regulate acidity of the blood.

feeling; it's the lack of CO_2. In fact, you probably know that if someone is hyperventilating (breathing very hard and fast, perhaps because they're panicking or upset), they are at risk of passing out. It sometimes helps to give them a paper bag to breathe in. Why? The rapid breathing is causing the loss of too much CO_2. By trapping that "lost" CO_2 in a bag, where they can re-inhale it with their next breath, they stabilize their CO_2 levels and help to bring their acidity back to normal. Of course, the more a person breathes into a bag, the more oxygen in the bag gets used up, so no one should breathe into a bag for too long.

There's an odd phenomenon at very high altitude that results from the body's competing drives to bring in enough O_2 without losing too much CO_2. When high altitude travelers are sleeping (and breathing becomes reflexive), they often experience what's called *paradoxical respiration* or *Cheyne-Stokes respiration*. This involves a period of hyperventilation followed by a period of apnea, meaning lack of breathing. The hyperventilation brings in plenty of O_2, but also results in the loss of too much CO_2, which lowers the acidity of the blood. The body senses the change in acidity, and responds by shutting off the breathing reflex temporarily, allowing CO_2 levels (and acidity) to increase and normalize. Of course, during this period of apnea, O_2 levels start to fall, and eventually the individual begins to breathe (fast!) to prevent oxygen deprivation. This cycle of hyperventilation and apnea repeats itself all night. It's not dangerous, and it's completely normal in high altitude sleepers, but it is certainly interesting chemistry!

2.3 ATOMIC AND MOLECULAR MASS

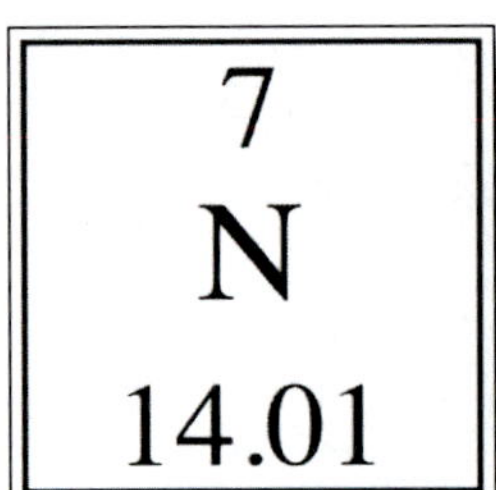

Image 2.8. The element nitrogen has an atomic mass of 14.01.

Thinking about atmospheric pressure raises another interesting issue. Remember that according to physics, pressure is a force exerted over an area, which means that the components of the atmosphere literally push against us and everything else they surround. It stands to reason, then, that the molecules of the atmosphere must have mass, since without mass they wouldn't be able to exert a force. It turns out that all atoms have mass, and we can use the periodic table to determine the mass of an atom of a given element. On the periodic table, there are two numbers associated with any element. For right now, we'll focus on the number located underneath the atomic symbol. This is called the **atomic mass**, and it tells us the mass of an atom of that element. The units of atomic mass are called **atomic mass units (amu)**, where 1 amu = 1.67×10^{-24} g.* Nitrogen, as we see in Image 2.8, has an atomic mass of 14.01 amu.

***NOTE**

We generally don't calculate the mass of atoms or molecules in grams, simply because they are so light—it would be like measuring the length of your pinkie fingernail in miles! Atomic mass units are much more convenient (and appropriate) to the scale of atomic mass.

If atoms have mass and atoms combine chemically to make molecules, then molecules must also have mass. We can calculate the mass of a molecule (the **molecular mass**) by adding up the masses of the atoms comprising that molecule. For example, a molecule of water is made up of one atom of oxygen and two atoms of hydrogen. We'd therefore calculate the mass of water as follows:

Molecular mass of water = mass of oxygen atom + 2(mass of hydrogen atom)
Molecular mass of water = 16.00 amu + 2(1.01 amu) = 18.02 amu

TRY THIS:

What is the atomic mass of H? Of C? Of U?

Answers: H = 1.01 amu, C = 12.01 amu, U = 238 amu.

TRY THIS:

What is the molecular mass of CO_2? Of HCl? Of NaCl?

Answer: CO_2 = 44.01 amu, HCl = 36.46 amu, NaCl = 58.45 amu.

Solution:
Molecular mass of CO_2 = atomic mass of carbon + 2 (atomic mass of oxygen)
Molecular mass of CO_2 = 12.01 amu + 2 (16.00 amu) = 44.01 amu

Molecular mass of HCl = atomic mass of hydrogen + atomic mass of chlorine
Molecular mass of HCl = 1.01 amu + 35.45 amu = 36.46 amu

Molecular mass of NaCl = atomic mass of sodium + atomic mass of chlorine
Molecular mass of NaCl = 23.0 amu + 35.45 amu + 58.45 amu

2.4 COUNTING ATOMS AND MOLECULES—THE MOLE

One more thing worth considering about the atmosphere has to do with its density. If we know that the atmosphere is most dense closest to Earth's surface, then we know that there are more molecules of gas per unit volume at low altitude. This raises an interesting question, however. How, exactly, does one COUNT atoms and molecules? Even a tiny amount of air contains a mind-boggling number of gas molecules—there are nearly a billion molecules of nitrogen and oxygen in 1 mL (1 cm^3) of air! Clearly, counting atoms and molecules in a sample of ANYTHING is going to involve some pretty big numbers.

Sometimes it makes sense to think about certain items in groups rather than as individuals. For instance, eggs tend to come in packages of a dozen (12) at the grocery store. That's because generally, no one needs just ONE egg (so it doesn't make much sense to sell them individually), and most people wouldn't be able to eat 100 eggs before they went bad (so they're not sold in packages of 100). Twelve is a nice size for a package of eggs. If I asked you to count all the eggs at the grocery store, it'd be much easier to count the number of packages (dozens) than to count each individual egg. In much the same way, it would be useful to have a "package size" for atoms and molecules so that instead of counting them individually, we could think of them in groups. A fellow named Lorenzo Romano Amadeo Carlo Avogadro di Quaregna e di Cerreto (he lived from 1776 to 1856, and for simplicity's sake, we'll just call him **Avogadro**) came up with a very reasonable package size that is useful in tabulating the number of atoms or molecules in a sample of air, water, or any substance.

Avogadro decided that atoms and molecules should be counted in groups of 6.02×10^{23}. In the same way that the number 12 is given the nickname *a*

Image 2.9. Avogadro.

dozen, the number 6.02×10^{23} is called a **mole** (it is also called **Avogadro's number**). A mole (mol) of ANYTHING is 6.02×10^{23} of that thing. We could have 6.02×10^{23} atoms of Ne; that would be a mole of Ne. We could have 6.02×10^{23} molecules of water; that would be a mole of water. We could even have 6.02×10^{23} cats (but we're not likely to, thank goodness—that'd be a lot of kitty litter), and that would be a mole of cats. Frankly, because a mole is such a very BIG number, even though we could theoretically have a mole of anything (atoms, molecules, cats, hamburgers, pencils, and so forth), we're not ever likely to find a mole of anything macroscopic, so we generally only use moles to count atoms and molecules.

We can use Avogadro's number to figure out how many atoms or molecules are in a given number of moles of a substance, or vice versa.

Example: How many atoms of Ne are in 2 moles of Ne?

$$2\ mol\ Ne \times \frac{6.02 * 10^{23}\ atoms}{1\ mol} = 1.20 * 10^{24}\ atoms\ of\ Ne$$

Example: If we have 5.34×10^{23} molecules of H_2, how many moles of H_2 is that?

$$5.34 * 10^{23}\ molecules\ of\ H_2 \times \frac{1\ mol}{6.02 * 10^{23}\ molecules} = 0.887\ mol\ H_2$$

Remember that molecules may contain more than one of a given atom, which affects our calculations!

Example: How many molecules of H_2O are in 2.5 mol H_2O? How many ATOMS of O are in that same sample? How many ATOMS of H are in that same sample?

$$2.5\ mol\ H_2O \times \frac{6.02 * 10^{23}\ molecules}{1 mol} = 1.50 * 10^{24}\ molecules\ H_2O$$

$$2.5\ mol\ H_2O \times \frac{6.02 * 10^{23}\ molecules}{1 mol} \times \frac{1\ atom\ O}{1\ molecule\ H_2O} = 1.50 * 10^{24}\ atoms\ O$$

$$2.5\ mol\ H_2O \times \frac{6.02 * 10^{23}\ molecules}{1 mol} \times \frac{2\ atom\ H}{1\ molecule\ H_2O} = 3.01 * 10^{24}\ atoms\ H$$

CONCEPT CHECK:

Why do we tend to count atoms and molecules in moles rather than individually?

Answer: Because there are so many of them in any sample we'd want to evaluate that it becomes much more convenient to count in "packages" of moles.

TRY THIS:

How many moles of C_2H_4 are in a sample of 8.56×10^{24} molecules? How many moles of C atoms are in the sample? How many H atoms are in the sample?

Answers: 14.2 mol C_2H_4, 28.4 mol of C, 3.42×10^{25} atoms of H.

Solution:

$$8.56 * 10^{24}\, molecules\ C_2H_4 \times \frac{1\, mol}{6.02 * 10^{23}\, molecules} = 14.2\, mol\ C_2H_4$$

$$14.2\, mol\ C_2H_4 \times \frac{2C}{1C_2H} = 28.4\, mol\ C$$

$$8.56 * 10^{24}\, molecules\ C_2H_4 \times \frac{4H}{1C_2H_4} = 3.42 * 10^{25}\, atoms\ H$$

2.5 MOLAR MASS

Here's a question: Why on Earth did Avogadro pick such a strange number? Why not a nice round number like 1×10^{24}, or something to that effect? It turns out that the specific number he selected makes determining the mass of a mole of atoms or molecules very easy! We know that atoms have mass, so if we wanted to calculate the mass of a mole of atoms, we'd multiply the mass of one atom by the number of atoms in a mole. This makes for a lot of math every time we want to determine the mass of a mole of something, but Avogadro did a **thought experiment** and came up with a solution. A thought experiment is a way of thinking through a problem or question that is impractical or impossible to test in real life. They're very important to science, because they allow for the development of theories that advance scientific understanding long before technology is available to test those theories.* For his experiment, Avogadro thought about what would happen if he were to pick up, say, one atom of carbon (atomic mass = 12 amu) and put that atom on a very small scale. The scale would read 2×10^{-23} g (12 amu $\times\ 1.67 \times 10^{-23}$ g/amu). If he then added another atom of carbon, the scale would read 4×10^{-23} g, and so on with additional atoms of carbon. He thought about continuing to add carbon atom after carbon atom until the scale read 12 g, and he determined that he'd have to place 6.02×10^{23} carbon atoms on the scale to achieve that mass (Image 2.11).

The beauty of Avogadro's number is that it means *the atomic mass of any element is both an atom of that element's mass in amu AND a mole of atoms of that element's mass in grams*! This holds true not just for carbon, but for all other elements as well. One atom of nitrogen weighs 14.01 amu, and a mole of nitrogen atoms weighs 14.01 grams. The mass of a mole of a chemical is called the chemical's **molar mass**, so for instance, the molar mass of nitrogen is 14.01 g/mol.

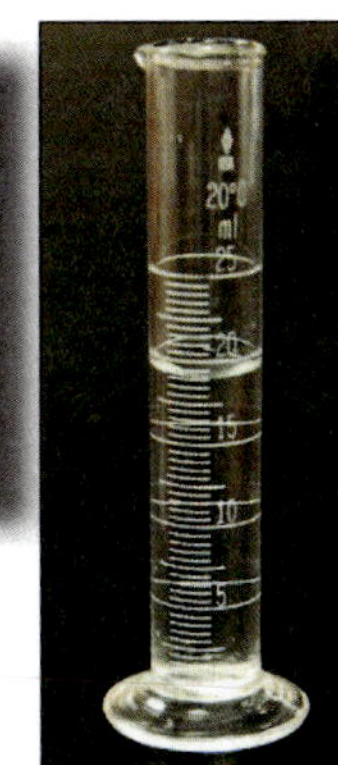

Image 2.10. How much is a mole? A mole of O_2 would fill just about 6 one-gallon water bottles. A mole of water is a mere 18 mL. This is because liquids are much more dense than gas, and the same number of molecules takes up less room. (Photos © Scott Lefler, 2010.)

*NOTE

If a thought experiment isn't tested in "real life," can it really be accurate? The answer, as it turns out, is absolutely! In addition to Avogadro's thought experiment yielding us excellent and useful information, Einstein developed his theory of relativity as a thought experi-ment. There are examples of accurate theories developed in this manner throughout the history of science.

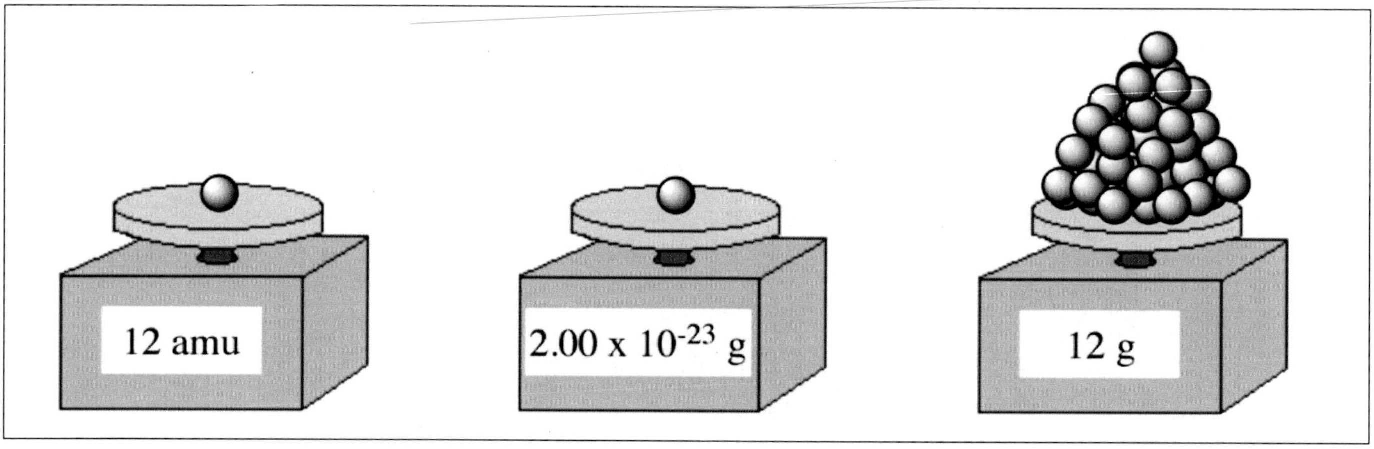

Image 2.11. Avogadro's thought experiment. If one atom of carbon weighs 12 amu, or 2.00×10^{-23} g, it would take many, MANY atoms of carbon (6.02×10^{23} to be exact) to get a scale to read 12 g.

This same principle holds true for molecules. We can add up the masses of the atoms in a molecule to find the mass of the molecule, and we can also add up the molar masses of atoms in a molecule to find the molar mass of the molecule.

Example: What is the molar mass of water?

Molar mass of water = molar mass of oxygen + 2(molar mass of hydrogen)
Molar mass of water = 16.00 g/mol + 2(1.01 g/mol) = 18.02 g/mol

TRY THIS:

What is the molar mass of carbon? Of CO_2?

Answers: 12.01 g/mol, 44.01 g/mol

Solution:
Check periodic table for molar mass of carbon (or any element).

Molar mass of CO_2 = molar mass of carbon + 2(molar mass of oxygen)
Molar mass of CO_2 = 12.01 g/mol + 2(16.00 g/mol) = 44.01 g/mol

We can use a chemical's molar mass to convert between the number of moles of a substance and the mass of that substance.

Example: How many moles of water are in 35.6 g of water?

$$35.6\, g\, H_2O \times \frac{1\, mol\, H_2O}{18.02\, g} = 1.98\, mol\, H_2O$$

Example: What is the mass of 2.5 mol of C_2H_2?

$$2.5\, mol\, C_2H_2 \times \frac{26.04\, g}{1\, mol\, C_2H_2} = 65.1\, g\, C_2H_2$$

TRY THIS:

What is the mass of 1.5 mol of nitrogen atoms?

Answer: 21.02 g

Solution:

$$1.5\ mol\ nitrogen \times \frac{14.01g}{1\ mol\ nitrogen} = 21.02\ g\ nitrogen$$

TRY THIS:

How many moles of N are in 54.5 g N_2O?

Answer: 2.48 mol N

Solution:

$$54.5\ g\ N_2O \times \frac{1\ mol\ N_2O}{44.02g} \times \frac{2N}{1N_2O} = 2.48\ mol\ N$$

2.6 INTERCONVERSIONS: GRAMS TO MOLES TO MOLECULES AND ATOMS

So far, we've used molar mass to convert between grams and moles, and Avogadro's number to convert between moles and molecules or atoms.

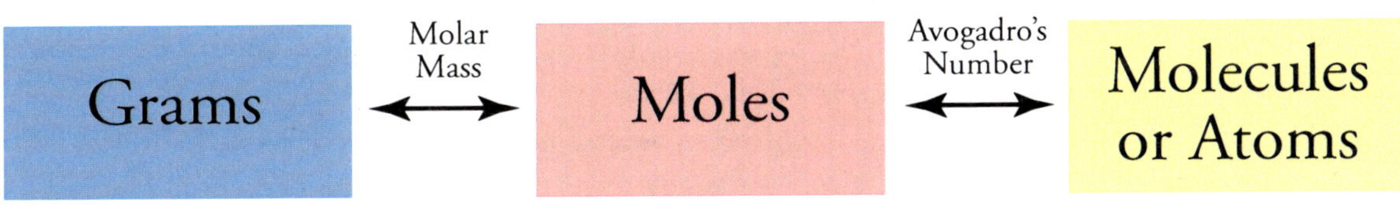

By combining these conversion factors, we can figure out how many molecules or atoms are in a given mass of a substance, or vice versa.

Example: How many atoms of H are in 24.3 g H_2O?

$$24.3\ g\ H_2O \times \frac{1\ mol\ H_2O}{18.02g} \times \frac{6.02 * 10^{23}\ molecules}{1\ mol} \times \frac{2\ atoms\ H}{1\ molecule\ H_2O} = 1.62 * 10^{24}\ atoms\ H$$

Example: What is the mass of 8.5×10^{20} molecules of CH_4?

$$8.5 * 10^{20}\ molecules\ CH_4 \times \frac{1\ mol}{6.02 * 10^{23}\ molecules} \times \frac{16.05g}{1\ mol\ CH_4} = 2.27 * 10^{-2}\ g$$

TRY THIS:

How many atoms of S are in a 64.0 g sample of SO_2? What is the total mass of the S?

Answers: 6.01×10^{23} atoms S, 32.03 g S.

Solution:

$$64.0\ g\,SO_2 \times \frac{1\,mol\,SO_2}{64.07\,g} \times \frac{6.02 * 10^{23}\,molecules}{1\,mol} \times \frac{1S}{1\,SO_2} = 6.01 * 10^{23}\,atoms\,S$$

$$64.0\ g\,SO_2 \times \frac{1\,mol\,SO_2}{64.07\,g} \times \frac{1S}{1SO_2} \times \frac{32.07\,g}{1\,mol\,S} = 32.03\ g\,S$$

2.7 STOICHIOMETRIC EQUIVALENTS

Now that we've discussed the concept of moles, we can look at chemical equations in a slightly different way. Formerly, we'd defined the coefficients in a chemical equation as the number of each species involved in the reaction, but they really just represent a ratio. For example, let's look at the familiar reaction below:

$$2\ H_2 + O_2 \rightarrow 2\ H_2O$$

The coefficients tell us that two molecules of hydrogen react with one molecule of oxygen to form two molecules of water. However, we could just as easily imagine four molecules of hydrogen reacting with two molecules of oxygen (this would be like having the reaction written above happen twice). The products in this case would be four molecules of water. The coefficients in a chemical equation are **stoichiometric equivalents**, which simply means they provide us with the ratios in which reactants and products appear in a chemical reaction. In the case of the reaction above, it is therefore possible to react ANY number of oxygen molecules, as long as there are twice as many hydrogen molecules present. Also from the balanced reaction, the number of water molecules formed would be the same as the number of hydrogen molecules reacted. For instance, for three oxygen molecules to react, the stoichiometric equivalent required would be six molecules of hydrogen, and the reaction would form six molecules of water. Because a mole is just a number, we could react a mole of oxygen molecules (6.02×10^{23} molecules) if we wanted to. We'd need TWO moles of hydrogen molecules, though, to have a stoichiometric equivalent of hydrogen (1.20×10^{24} hydrogen molecules), and we'd make two moles of water.

Amount of H_2	**Amount of O_2**	**Amount of H_2O**
2 molecules	1 molecule	2 molecules
4 molecules	2 molecules	4 molecules
6 molecules	3 molecules	6 molecules
1.20×10^{24} molecules	6.02×10^{23} molecules	1.20×10^{24} molecules
2 moles	1 mole	2 moles
1 mole	0.5 mole	1 mole

Table 2.1. Stoichiometric equivalents in the reaction of H_2 and O_2 to form water.

Looking at the last line of Table 2.1, we notice something odd. We're not used to seeing fractional numbers in chemical reactions; it makes no sense to think about half an atom or half a molecule. However, a mole is a large number that can easily be split in half (just as half of a dozen is 6), so it's not inappropriate at all to talk about portions of moles when we refer to stoichiometric equivalents. If only one mole of hydrogen were to react, it would react with HALF a mole of oxygen (that's still a LOT of oxygen—it's 3.01×10^{23} molecules), to produce a mole of water.

From now on, whenever we see a chemical equation, we should remember that the coefficients merely represent ratios—reactions can involve just a few molecules or many moles of molecules, as long as the ratios are satisfied. Because we can use molar mass to relate moles to grams, we can even think about stoichiometric equivalents in terms of masses. To do this, we use the mass of the first substance to determine the number of moles of that substance, then use stoichiometry to find the equivalent number of moles of the second substance. Finally, we find the mass of the second substance using its molar mass:

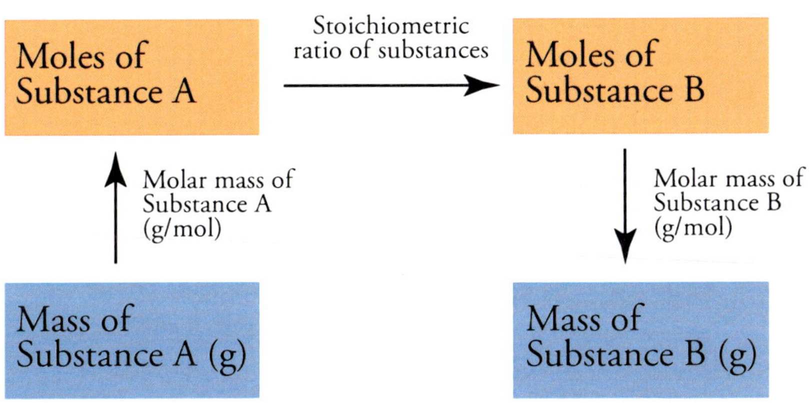

Example: What mass of H_2 is required to react with 65 g O_2 in the reaction $2\,H_2 + O_2 \rightarrow 2\,H_2O$, and what mass of water is formed?

$$65\ g\ O_2 \times \frac{1\ mol\ O_2}{32.0\ g} \times \frac{2\ mol\ H_2}{1\ mol\ O_2} \times \frac{2.02\ g\ H_2}{1\ mol} = 8.2\ g\ H_2$$

$$65\ g\ O_2 \times \frac{1\ mol\ O_2}{32.0\ g} \times \frac{2\ mol\ H_2O}{1\ mol\ O_2} \times \frac{18.02\ g\ H_2O}{1\ mol} = 73.2\ g\ H_2O$$

CONCEPT CHECK

Why can we react half a mole of O_2, even though we can't have the coefficient 0.5 in a chemical equation?

Answer: The coefficients in chemical equations must represent the ratio of each species that reacts. These numbers must be integer values so they are valid for individual atoms and molecules as well as for moles.

Try This:

For the equation $2\,S + 3\,O_2 \rightarrow 2\,SO_3$, fill in the table of stoichiometric equivalencies:

Amount of S	Amount of O_2	Amount of SO_3
4 atoms		
	12 molecules	
		2 moles
	1.5 moles	
		44 g

Answer:

Amount of S	Amount of O_2	Amount of SO_3
4 atoms	6 molecules	4 molecules
8 atoms	12 molecules	8 molecules
2 moles	3 moles	2 moles
1 mole	1.5 moles	1 mole
18 g (0.55 moles)	26 g (0.83 moles)	44 g (0.55 moles)

2.8 SUBATOMIC PARTICLES

The last puzzle piece we need in order to make sense of atomic mass is to know where that mass comes from. Atoms are made of subatomic particles, which were mentioned briefly in Chapter 1 as the bits you'd (theoretically) divide an atom into if you were able to cut it into pieces, which you can't actually do by any chemical means. Subatomic particles are responsible for making an atom what it is (chemically and physically), but they are not really found on their own, separate and apart from atoms, so we are very correct to say atoms are indivisible.

There are three subatomic particles that go into making up atoms: **protons, neutrons,** and **electrons.** Each particle can be defined in terms of its mass and its charge (Table 2.2).

Particle	Mass (g)	Mass (amu)	Charge (C)	Charge (relative)
Proton	1.67×10^{-24} g	1 amu	$+1.6 \times 10^{-19}$ C	+1
Neutron	1.67×10^{-24} g	1 amu	0	0
Electron	9.11×10^{-28} g	~0 amu	-1.6×10^{-19} C	-1

Table 2.2. The subatomic particles. Masses are given in grams and amu, charge is given in Coulombs (C), as well as in relative terms.

There are a few interesting points to note with regard to these particles. First, as per our discussion of the masses of atoms and molecules, it doesn't really make sense to measure the mass of these very light particles in grams. Instead, we use atomic mass units (amu). Note that protons and neutrons have the same mass—they weigh in at 1 amu each. Electrons do technically have mass, but because they are so light relative to protons and neutrons (nearly 2000 times lighter), their mass contributes virtually nothing to the overall mass of an atom. An atom would have to have well over 1000 electrons for them to come anywhere close to contributing significant mass, and no atom has that many! For this reason, we consider the electron essentially massless in our discussion of subatomic particles. A second point worth noting is that the charge (measured in Coulombs) of a proton is equal in magnitude but opposite in sign to that of an electron. In other words, they have the same amount of charge, but protons are positively charged and electrons are negatively charged. Because of this, it's easier to refer to them as having "units" of charge, where a proton has one unit of positive charge and an electron has one unit of negative charge. This simplifies discussions of charge significantly. A brief and relevant bit of information at this juncture is that opposite charges attract, and like charges repel each other. As such, protons are repelled by other protons, and are attracted to (and by) electrons. These attractions and repulsions are called **electrostatic forces.**

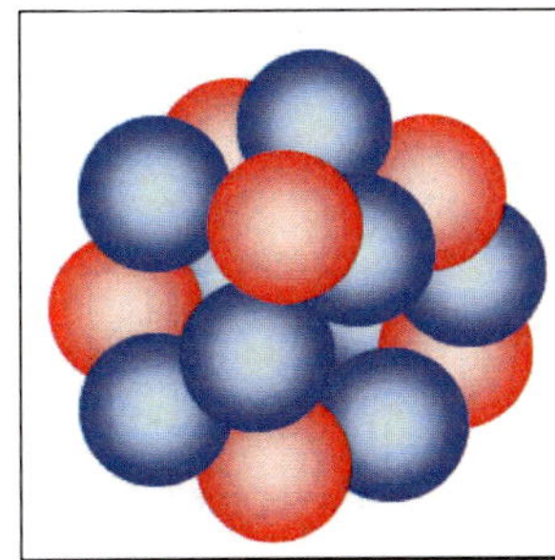

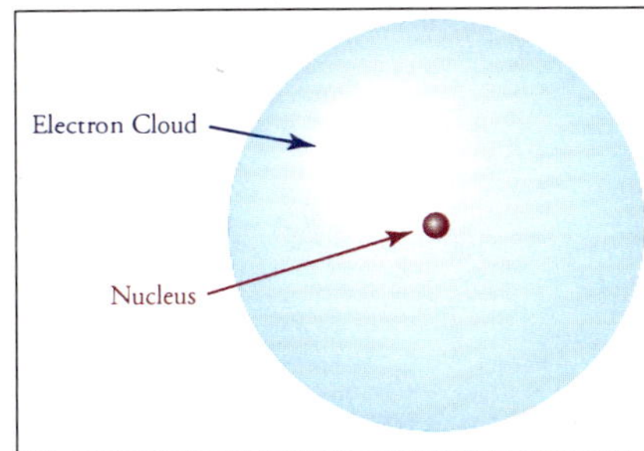

Image 2.12. The nucleus consists of densely packed protons and neutrons (top) and is located at the center of an atom, surrounded by an electron cloud.

The distribution of subatomic particles within an atom is very organized. Atoms have a very tiny, central portion called a **nucleus**, which is made up of protons and neutrons. Because only protons and neutrons contribute significantly to an atom's mass, and because all an atom's protons and neutrons are located in the nucleus, the mass of an atom is localized in its nucleus. The nucleus is surrounded by a large, virtually massless "cloud" of electrons.*

*NOTE

Because electrons move around a lot, we say the cloud is large even if we're talking about an atom with very few electrons.

Concept Check

Where is most of an atom's mass? Where is most of its volume?

Answer: Most of the mass is in the nucleus. Most of the volume is in the electron cloud.

Armed with this information, we can now begin to make sense of the subatomic structure of an atom. Since mass comes from protons and neutrons (each of which weighs 1 amu), we can use the **mass number** of an atom to figure out how many total protons and neutrons are in the atom's nucleus. An atom's mass number is the number of protons plus the number of neutrons in the atom's nucleus. Because there must be whole numbers of protons and neutrons, mass number is always a whole number. We can find the mass number simply by rounding the atomic mass to the nearest whole number.

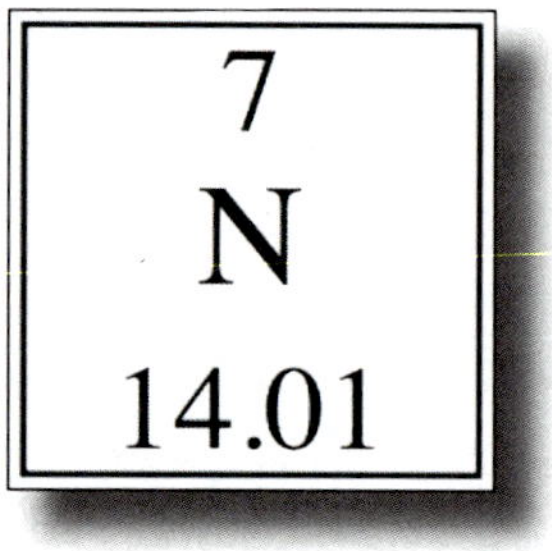

Image 2.13. Nitrogen's atomic number is 7.

Carbon, for instance, has an atomic mass of 12.01, from which we determine that its mass number is 12, and that tells us carbon has twelve mass-contributing particles (protons and/or neutrons) in its nucleus. In order to know more about how many of these particles are protons and how many are neutrons for a particular atom, we have to go back to the periodic table.

Recall that each element has two numbers associated with it: the larger of the two (always underneath the atomic symbol) is the atomic mass. The smaller of the two (positioned above the atomic symbol) is called the **atomic number**. The atomic number indicates the number of protons in the nucleus of an atom of any element. Nitrogen, for instance, with an atomic number of seven, has seven protons. Further, because we know that its atomic mass is 14.01, from which we determine that its mass number is 14, we can calculate the number of neutrons in the nucleus of a nitrogen atom:

Mass number = number of protons + number of neutrons
14 = 7 + number of neutrons
Number of neutrons = 7

To determine how many electrons an atom has, we use the charges of subatomic particles. Since both protons and electrons have charge, we can determine the total charge on an atom by adding up the sum of positive charges (from protons) and the sum of negative charges (from electrons). It is important to know that any atom that has not reacted chemically will be **neutral**, which is to say that the sum of its positive and negative charges will be zero. Remember that protons have a charge of +1 and electrons have a charge of -1. Therefore, the number of protons and the number of electrons in an atom must be the same. For example, nitrogen has 7 protons (sum of positive charge is +7), so there must be 7 electrons in nitrogen (sum of negative charge is -7) for the net charge to be zero.

TRY THIS:

What is the number of protons, neutrons, and electrons in O? In F?

Answer:
O has 8 protons, 8 neutrons, 8 electrons.

F has 9 protons, 10 neutrons, 9 electrons.

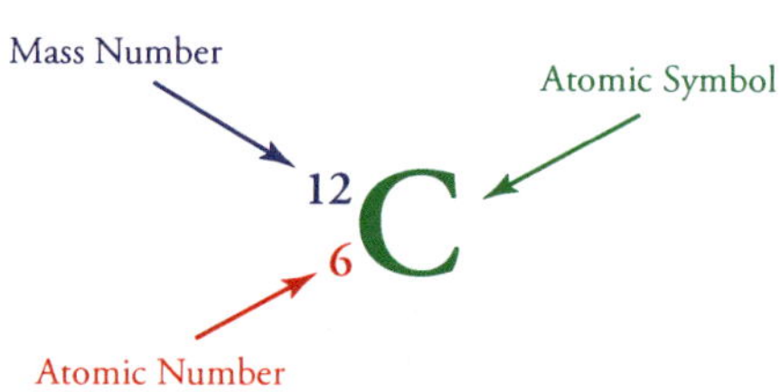

Figure 2.14. A shorthand representation of carbon.

Aside from the periodic table, we can extract information such as atomic number and mass number from shorthand representations of elements (Image 2.14). These representations allow us to determine quickly (without needing to refer to a periodic table) how many of each subatomic particle make up an atom of a given element.

TRY THIS:

Without referring to a periodic table, how many protons, neutrons, and electrons are in $^{31}_{15}P$? In $^{200}_{80}Hg$?

Answer: P has 15 protons, 16 neutrons, and 15 electrons. Hg has 80 protons, 120 neutrons, and 80 electrons.

*NOTE

If you're curious, the reason that some mass numbers on the periodic table are in parentheses is that these elements are radioactive, and have several forms. The mass number in parentheses is the one that corresponds to the most stable form of the element. This gets into the topic of isotopes, which we'll discuss more fully in Chapter 12.

There are a few things worth noting at this point. First, as you look at a periodic table, notice that atomic masses are not unique: Po and At, for instance, both have atomic masses of 210.* Further, you'll notice that while atomic masses *generally* increase left to right and top to bottom on the periodic table, this trend is not absolute. I (number 53), for instance, has a smaller atomic mass than Te (number 52). These inconsistencies with regard to atomic mass relate to the fact that while the number of neutrons generally increases as the number of protons increases, there are exceptions to this trend. Atomic numbers, however, follow much stricter rules. From left to right and top to bottom on the periodic table, atomic numbers increase sequentially. They are always integer values (because they indicate the number of protons)♦ and they are always unique to a given element—there are no two elements with the same atomic number. Because of this, the atomic number is the most important identifying feature of an atom: carbon, and no element other than carbon, is atomic number 6; an atom whose nucleus contains 10 protons must be an atom of Ne.

It's worth noticing the relationship between number of protons and number of neutrons in atoms of various elements. The lighter elements tend to have the same number of protons as neutrons (carbon, for instance, has six of each), while the heavier elements have more (sometimes MANY more) neutrons than protons. Why? To answer this question, we need to know a little about what holds an atom's nucleus together. The nucleus is a tight mass of protons and neutrons, which is a bit counterintuitive given what we know about protons. If protons are all positively charged and like charges repel, how is it possible to pack many positively charged particles into a very tiny space? It turns out that**nucleons** (a collective term for protons and neutrons) are strongly attracted to one another over VERY short distances by a very strong force (which is cleverly named the **strong force**). As a result, even though protons are electrostatically repelled by other protons, if they're

♦NOTE

If mass number is the number of protons plus number of neutrons, and both of those subatomic particles have masses of 1 amu, how is it possible to have fractional mass numbers? Again, this is an isotope issue. It turns out that all elements (not just the radioactive ones) have several forms and the number of neutrons varies from form to form. The mass number is actually the weighted average mass of an element. We round mass numbers to the nearest integer value (for now), because that gives us the mass of the most common form of any element, and allows us to determine how many neutrons the most common form has.

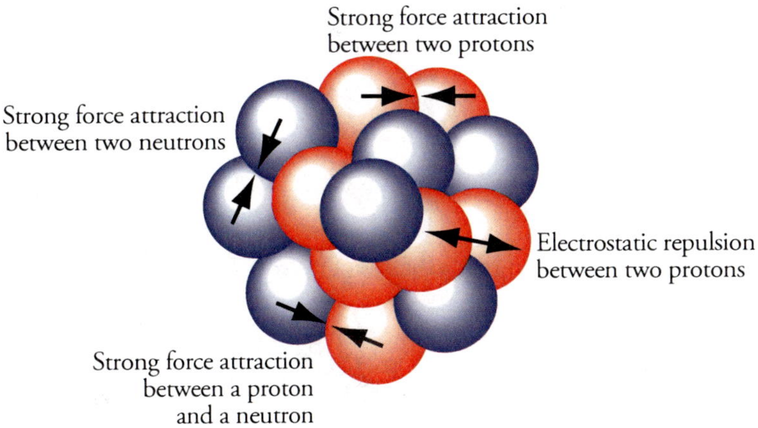

Image 2.15. Neutrons attract each other and protons via the strong nuclear force. Protons attract other protons via the strong force, but also repel them electrostatically. The more neutrons in a nucleus, the more attraction between nucleons without additional repulsion.

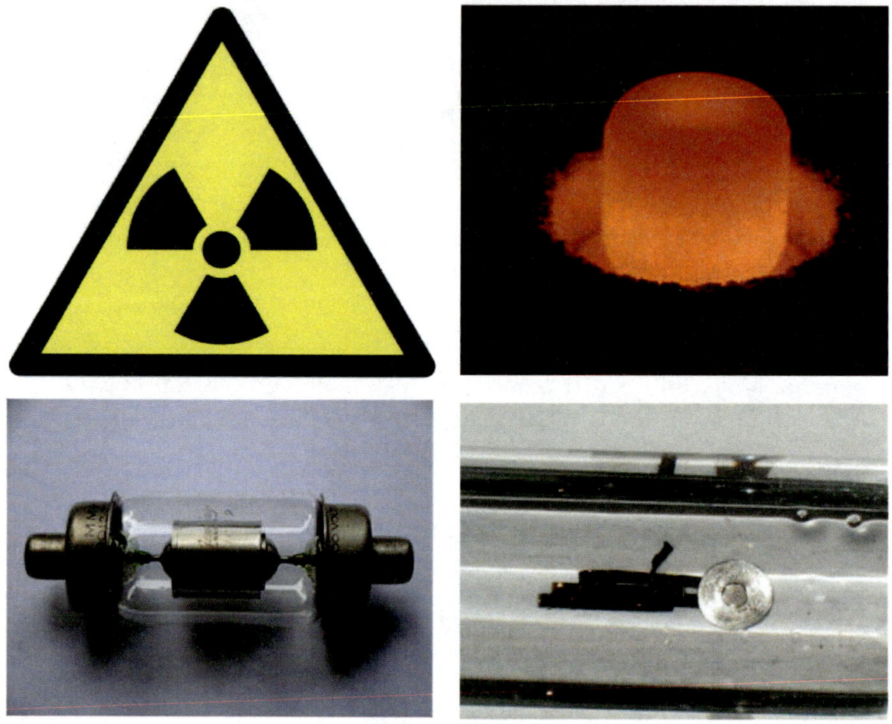

Image 2.16. Radioactive elements include (clockwise from top right): plutonium (Pu), polonium (Po), and uranium (U). We'll discuss radioactivity further in Chapter 12.

packed in VERY tightly, they're also attracted by and to other protons, due to the strong force. Neutrons are not electrostatically attracted to or repelled by protons, but they are held strongly to each other and to protons over very small distances by the strong force.

This explains two nuclear phenomena. First, the nucleus is tiny because it has to be very tightly packed in order for strong forces to be able to hold the protons together. Second, the more protons in a nucleus, the harder it is to hold all those like-charged particles together, since there's more electrostatic repulsion and the radius of the nucleus is greater, which weakens the strong forces. Neutrons act like nuclear "glue"; they contribute additional strong force without increasing electrostatic repulsion, so bigger nuclei have more neutrons relative to the number of protons. If a nucleus is big enough, no number of

Image 2.17. The size difference between an atom in an apple and the apple itself is the same as the size difference between an apple and the planet Earth!

neutrons is really sufficient to hold all those protons together. This makes very large nuclei fundamentally unstable, and they tend to fall apart. This phenomenon is called radioactive decay, and while it can be observed in small elements, it's ubiquitous in the large ones. In fact, every element with an atomic number equal to or greater than that of bismuth (83) is radioactive!

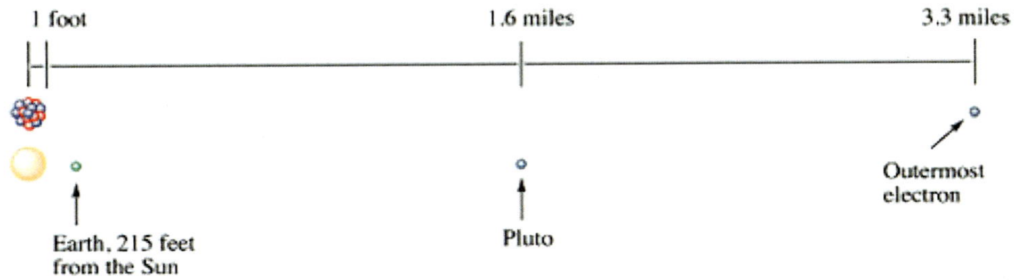

Image 2.18. A comparison of the amount of empty space in the solar system and in an atom.

> **CONCEPT CHECK**
>
> Why do large atoms have more neutrons relative to the number of protons as compared to small atoms?
>
> Answer: Additional neutrons help hold the nucleus together (more strong forces) without adding more electrostatic repulsion.

A few final words about atomic structure: We all know atoms are small (so by extension, subatomic particles must be smaller), but it's really hard to conceive of exactly HOW small. One interesting way to think about the size of atoms is to imagine an apple. If we were to expand an apple to the size of the Earth, each atom in the apple would end up approximately the size of the original apple.

Further, we tend to think about matter (and therefore, atoms, since matter is made up of atoms) as being solid, but it turns out that since almost all the mass in an atom is clustered into a tiny central nucleus, most of the atom is actually empty space! In order to conceive of this, imagine scaling down the solar system until the Sun was a mere 1 foot in diameter. Earth would be 215 feet away from the Sun and the (ex-)planet Pluto would be 1.6 miles from the Sun. That's a lot of emptiness, which fits with our notion of the solar system as mostly empty space.* However, if we looked at an atom of gold on the same scale, expanding the nucleus to a diameter of 1 foot, the outermost electron would be 3.3 miles away—almost double the radius of the solar system! Clearly, atoms consist of a LOT of empty space!

> ***NOTE**
>
> If you're interested in this scaled model of the solar system, the nearest star other than our Sun would be Alpha Centauri, at over 10,000 miles away!

One more interesting bit of information: The nucleus is SO dense that a matchbox full of nuclei (separated from their electron clouds so that they'd take up less space) would weigh 2.5 billion tons (that's 5 trillion pounds)!

2.9 THE LAST WORD—LAND ANIMALS BREATHE AIR, AQUATIC ANIMALS BREATHE WATER?

A whole chapter's worth of discussing the atmosphere leads to an interesting question—what do humans (and respiring land animals) breathe? Most people would answer *air*. Well, yes, we DO breathe air, but as it turns out, we don't

NEED to breathe air. We NEED O_2, and the "air" (the rest of the atmosphere) is nothing more than the substance with which that O_2 happens to be mixed. As it turns out, we're perfectly capable of breathing a substance other than air and staying alive and well, as long as that substance has an appropriate percentage of O_2 mixed in! Let's think about fish for a moment: like land animals, fish also require O_2. They do not, however, breathe air. Fish get their oxygen from water, which has a small amount of atmospheric O_2 dissolved (mixed) into it. Unlike our atmosphere, however, which is 21 percent oxygen, the percentage of O_2 dissolved in water is in the range of 0.5–3 percent. Fish handle this just fine; their gills are adapted to allow them to pull O_2 from the water extremely efficiently. We are not nearly as efficient at removing O_2 from mixtures, and as such would not be able to survive by breathing water. Oddly enough, though, we are able to breathe liquid as long as that liquid dissolves O_2 at a higher percentage than water does! The liquid *perfluorodecalin* ($C_{10}F_{18}$, Image 2.19) is much better at dissolving O_2 than water is—perfluorodecalin can dissolve up to 40 percent O_2, making it a perfectly acceptable medium for humans (or other land animals) to breathe.

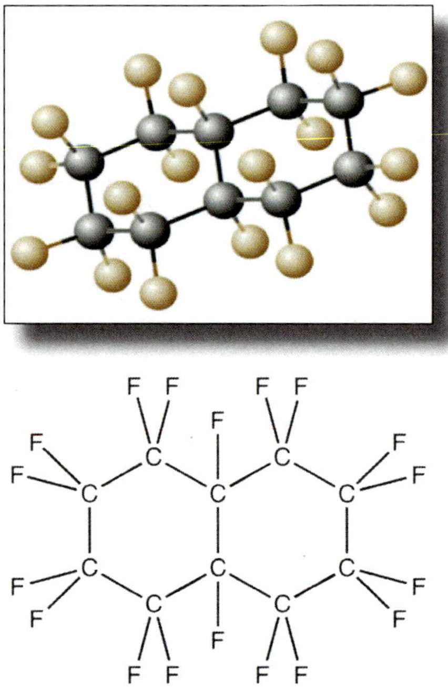

Image 2.19. The compound perfluorodecalin is composed of the elements carbon and fluorine.

Aside from the fact that it's a scientific curiosity, what is the function of breathing a liquid? It turns out that there are several medical applications being explored. First, breathing a liquid medium could reduce the incidence of IRDS (infant respiratory distress syndrome), a condition resulting from the tendency of premature infants' underdeveloped lungs to collapse. Infants suspended in perfluorodecalin would get the oxygen they need without the risk of lung collapse associated with breathing air. Further, liquids do not expand and contract significantly (like gases do) when exposed to changes in pressure. As such, liquid breathing media are being investigated for deep-sea diving applications. Additionally, since perfluorodecalin is a liquid, it has the ability to "wash" the lungs as it delivers oxygen, and might find applications in medicine for removing debris and foreign matter. Finally, because of its ability to dissolve high concentrations of oxygen, perfluorodecalin has the potential to function as an ingredient in artificial blood. (The job of the blood is to carry oxygen from the lungs to the tissues using an O_2-carrying molecule called *hemoglobin*). All around, it's pretty interesting stuff!

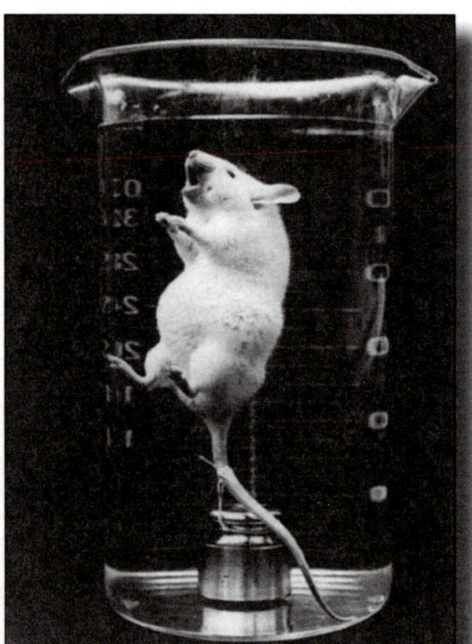

Image 2.20. The mouse is alive and healthy (if not necessarily happy) submerged in perfluorodecalin. The weight tied to its tail keeps it under the surface of the fluid, which is quite dense and in which the mouse would otherwise float (Image reproduced by permission of The Royal Society of Chemistry).

SUMMARY OF MAIN POINTS

- Earth is surrounded by a thin shell of gases. This is the atmosphere.
- The atmosphere exerts pressure, and pressure varies inversely with altitude.
- The atmosphere is composed primarily of nitrogen and oxygen and its composition is nearly uniform throughout the troposphere.
- Atoms of elements have mass as indicated by their atomic masses on the periodic table. Atomic masses are measured in units of amu.
- The mass of a molecule is determined by adding the masses of its constituent atoms.
- Atoms and molecules are counted in moles, where a mole is 6.02×10^{23} (Avogadro's number) of ANYTHING.
- The mass of a mole of an element is indicated by its atomic mass on the periodic table, where molar masses are measured in units of grams.
- The molar mass of a molecule is determined by adding the molar masses of its constituent atoms.
- Avogadro's number can be used to convert from atoms or molecules to moles and vice versa.
- Molar mass can be used to convert from mass in grams to moles and vice versa.
- Stoichiometric equivalents are quantities of reactant or product that satisfy the ratios indicated by coefficients in chemical equations in atoms/molecules, moles, or grams.
- The atom is made of protons and neutrons (in a dense, low-volume nucleus) and electrons (in a nearly massless, high-volume cloud around the nucleus).
- The mass number is the number of protons plus neutrons in an atom's nucleus.
- The atomic number is an atom's identifying feature and is equal to the number of protons.
- The number of electrons in an elemental atom is equal to the number of protons in that atom.

Questions and Topics for Discussion

1. Why is the air on mountaintops sometimes described as "thin"? How is the air there similar to the air at sea level? How is it different?
2. What kind of relationship do altitude and atmospheric pressure have? Is it a linear correlation (doubling altitude cuts pressure in half), or is it exponential? At what altitude is pressure approximately 50 percent of its sea level value? At what altitude is it approximately 10 percent of its sea level value?
3. Why does it make sense that we can add the masses of the atoms in a molecule to determine the mass of the molecule? What does this show us about chemical reactions (making molecules) and conservation of mass?
4. Why are atomic and molecular masses measured in amu rather than grams? What is an amu?
5. How is the concept of a mole similar to the concept of a dozen? How are they different? What sorts of things can we measure in moles?
6. What was Avogadro's thought experiment? Do a little Internet research to find another famous thought experiment (there are many). Why do scientists sometimes do thought experiments? Are they dependable?
7. Why is the atomic mass for any element equal to both the mass of an atom of that element in amu and its molar mass in grams? Is that merely coincidence?
8. Why can we add the molar masses of the elements in a molecule to determine the molar mass of the molecule?
9. What does "stoichiometric equivalent" mean? Do these values refer to individual atoms/molecules, moles, or both?

10. Are the stoichiometric coefficients in a chemical equation representative of the ratio of masses in the equation? Why or why not? How can we use these coefficients to determine stoichiometrically equivalent masses?
11. Why is an electron considered massless?
12. Where is most of the mass in an atom? Where is most of its volume? What does this mean with regard to how "full" or "empty" the space in an atom is?
13. Explain how to use the periodic table to find the number of protons, neutrons, and electrons in an atom.

Problems

1. Using Image 2.3 in the chapter, what is the atmospheric pressure at 10,000 feet? At 65,000 feet?
2. Using Image 2.3 in the chapter, at what altitude is the atmospheric pressure at 0.8 atm? At 0.25 atm?
3. What percent of our atmosphere is made of N_2? What percent is O_2?
4. What is the atomic mass of hydrogen? Of helium? What is the mass of an atom of hydrogen in grams?
5. What is the atomic mass of polonium (Po)? Of radium (Ra)? What is the mass of an atom of radium in grams?
6. Calculate the molecular mass of the following: $CaCO_3$ (calcite), Fe_2O_3 (rust), H_2O_2 (hydrogen peroxide, a sterilizer).
7. Calculate the molecular mass of the following: C_2H_6O (alcohol), $C_6H_{12}O6$ (glucose, a sugar); and NO_2 (nitrogen dioxide, an atmospheric pollutant).
8. Calculate how many atoms are in each of the following samples: 1.5 moles C; 10.2 moles He; and 0.75 moles H.
9. Calculate how many atoms are in each of the following samples: 7.2 mol N; 1.3 mol Rn; and 0.24 mol Ca.
10. Calculate how many molecules are in each of the following samples: 0.60 mol H_2O; 1.2 mol NaCl; and 23.5 mol $C_4H_{10}O$ (ether, an anesthetic).
11. Calculate how many molecules are in each of the following samples: 12.6 moles NaF (an ingredient in toothpaste); 1.5 moles N_2O (nitrous oxide or laughing gas); and 0.65 moles CO.
12. Determine the number of moles in each of the following: 5.4×10^{22} atoms Xe; 1.22×10^{23} molecules CO_2; and 3.5×10^{25} eggs (OK, we're not likely to have this many eggs, but the point remains that you COULD count eggs in moles ...).
13. Determine the number of moles in each of the following: 3.2×10^{25} atoms Ti; 7.2×10^{24} molecules H_2O; and 3.5×10^{20} cats.
14. How many atoms of C are in each of the following samples: 1.2 moles CH_2O (formaldehyde, a preservative); 5.7 moles CH_4O (wood alcohol); and 10.4 moles $C_2H_4O_2$ (vinegar)?
15. How many atoms of O are in each of the following samples: 4.3 moles CH_2O_2 (formic acid, one of the irritants in ant bites); 0.50 moles H_2SO_4 (sulfuric acid, a component of acid rain); and 1.3 moles $NaHCO_3$ (baking soda)?
16. What is the molar mass of S? Of SO_3 (volcanic gas)?
17. What is the molar mass of Cl? Of CCl_2F_2 (Freon, responsible for ozone depletion)?
18. Calculate the mass of each of the following: 1.2×10^{23} molecules of CO_2; 5.4×10^{25} atoms of K; 3.5 moles CO_2; and 0.65 moles K.
19. Calculate the mass of each of the following: 6.5×10^{22} molecules of HNO_3 (nitric acid, a component of acid rain); 4.4×10^{25} atoms of B; 0.23 moles HNO_3; and 12.5 moles B.
20. Calculate the number of moles in each of the following samples: 118 g H_2O; 75 g $C_6H_{12}O_6$ (glucose, a sugar); and 0.3 g NaCl.

21. Calculate the number of moles in each of the following samples: 212 g NO_2 (nitrogen dioxide, an atmospheric pollutant); 61 g O_3 (ozone); and 6.2 g O_2.
22. Calculate the number of molecules in each of the following samples: 118 g H_2O; 75 g $C_6H_{12}O_6$ (glucose, a sugar); and 0.30 g NaCl.
23. Calculate the number of molecules in each of the following samples: 212 g NO_2 (nitrogen dioxide, an atmospheric pollutant); 61 g O_3 (ozone); and 6.2 g O_2.
24. Calculate the number of atoms (total) in each of the following samples: 118 g H_2O; 75 g $C_6H_{12}O_6$ (glucose, a sugar); and 0.30 g NaCl.
25. Calculate the number of atoms (total) in each of the following samples: 212 g NO_2 (nitrogen dioxide, an atmospheric pollutant); 61 g O_3 (ozone); and 6.2 g O_2.
26. For the reaction 3 H_2 + N_2 → 2 NH_3, fill out the following table with stoichiometrically equivalent quantities:

Amount of H_2	**Amount of N_2**	**Amount of NH_3**
6 molecules		
	5.25×10^{23} molecules	
		2 moles
0.75 mole		

27. For the reaction C_2H_4 + 3 O_2 → 2 CO_2 + 2 H_2O, fill out the following table with stoichiometrically equivalent quantities:

Amount of C_2H_4	**Amount of O_2**	**Amount of CO_2**	**Amount of H_2O**
	9 molecules		
2.9×10^{24} molecules			
			2 moles
		0.33 mole	

28. For the reaction 3 H_2 + N_2 → 2 NH_3, fill out the following table with stoichiometrically equivalent quantities:

Amount of H_2	**Amount of N_2**	**Amount of NH_3**
6.5 g		
		32.4 g
	125 g	

29. For the reaction C_2H_4 + 3 O_2 → 2 CO_2 + 2 H_2O, fill out the following table with stoichiometrically equivalent quantities:

Amount of C_2H_4	**Amount of O_2**	**Amount of CO_2**	**Amount of H_2O**
		35.4 g	
	1.2 g		
			518.2 g

30. How many protons, neutrons, and electrons are in an atom of each of the following: K, Cm, Pt, Kr?
31. How many protons, neutrons, and electrons are in an atom of each of the following: Li, Mn, Br, Pb?

32. Without using the periodic table, how many protons, neutrons, and electrons are in ${}^{65}_{30}Zn$? In ${}^{98}_{43}Tc$? In ${}^{24}_{12}Mg$?
33. Without using the periodic table, how many protons, neutrons, and electrons are in ${}^{210}_{84}Po$? In ${}^{184}_{74}W$? In ${}^{32}_{16}S$?
34. Write out shorthand representations $\left({}^{\text{mass number}}_{\text{atomic number}}symbol\right)$ of each of the following: Nickel (Ni); selenium (Se); and astatine (At).
35. Write out shorthand representations $\left({}^{\text{mass number}}_{\text{atomic number}}symbol\right)$ of each of the following: Barium (Ba); tin (Sn); and zinc (Zn).

Answers to Problems

1. 0.69 atm, ~0.06 atm.
3. 78 percent, 21 percent.
5. 210 amu, 226 amu, 3.77×10^{-22} grams.
7. 46.08 amu, 180.18 amu, 46.01 amu.
9. 4.3×10^{24} atoms N, 7.8×10^{23} atoms Rn, 1.4×10^{23} atoms Ca.
11. 7.59×10^{24} molecules NaF, 9.0×10^{23} molecules N_2O, 3.9×10^{23} molecules CO.
13. 53 moles Ti, 12 moles H_2O, 5.8×10^{-4} moles cats.
15. 5.2×10^{24} atoms O, 1.2×10^{24} atoms O, 2.3×10^{24} atoms O.
17. 35.45 g/mol, 120.91 g/mol.
19. 6.8 g, 790 g, 14.5 g, 135 g.
21. 4.61 moles NO_2, 1.3 moles O_3, 0.19 moles O_2.
23. 2.78×10^{24} molecules NO_2, 7.8×10^{23} molecules O_3, 1.1×10^{23} molecules O_2.
25. 8.34×10^{24} atoms, 2.3×10^{24} atoms, 2.2×10^{23} atoms.
27. For the reaction $C_2H_4 + 3\ O_2 \rightarrow 2\ CO_2 + 2\ H_2O$, fill out the following table with stoichiometrically equivalent quantities:

Amount of C_2H_4	**Amount of O_2**	**Amount of CO_2**	**Amount of H_2O**
3 molecules	9 molecules	6 molecules	6 molecules
2.9×10^{24} molecules	8.7×10^{24} molecules	5.8×10^{24} molecules	5.8×10^{24} molecules
1 mole	3 moles	2 moles	2 moles
0.165 mol	0.495 mol	0.33 mole	0.33 mole

29. For the reaction $C_2H_4 + 3\ O_2 \rightarrow 2\ CO_2 + 2\ H_2O$, fill out the following table with stoichiometrically equivalent quantities:

Amount of C_2H_4	**Amount of O_2**	**Amount of CO_2**	**Amount of H_2O**
11.3 g	38.6 g	35.4 g	14.5 g
0.35 g	1.2 g	1.1 g	0.45 g
403.5 g	1380 g	1257 g	518.2 g

31. Li has 3p, 4n, 3e. Mn has 25p, 30n, 25e. Br has 35p, 45n, 35e. Pb has 82 p, 125n, 82e.
33. Po has 84p, 126n, 84e. W has 74p, 110n, 74e. S has 16p, 16n, 16e.
35. ${}^{137}_{56}Ba$, ${}^{119}_{50}Sn$, ${}^{65}_{30}Zn$.

Sources and Further Reading

Guo, Z. L., Lu, G. P., Ren, T., *et al.* (2009). Partial liquid ventilation confers protection against acute lung injury induced by endotoxin in juvenile piglets. *Respiratory Physiology and Neurobiology,* 167(3), 221–26.

Shaffer, T. H., Wolfwon, M. R., Greenspan, J. S., *et al.* (1996). Liquid ventilation in premature lambs; uptake, biodistribution and elimination of perfluorodecalin liquid. *Reproduction, Fertility and Development*, 8(3), 409–16.

Spence, R. K., Norcross, E. D., Costabile, J., *et al.* (1994). Perfluorocarbons as blood substitutes—Early years—Experience with fluosol DA-20-percent in the 1980s. *Artificial Cells, Blood Substitutes, and Immobilization Biotechnology*, 22(4), 955–63.

Suy, G. Y., Chung, M. P., Parks, S. J., *et al.* (1999). Partial liquid ventilation with perfluorocarbon improves gas exchange and decreases inflammatory response in oleic-acid-induced lung injury in beagles. *Journal of Korean Medical Science*, 14(6), 613–22.

Chemistry Inside Us—The Elements of Life

3

Of all the elements on the periodic table, only a handful of them are represented to any significant extent in living organisms. Whether an organism is composed of one cell (such as a bacterium) or many (like us), cells are made mostly of just four elements: C, H, O, and N. A few additional elements show up in trace amounts, collectively comprising a mere 4 percent of the human body by mass.

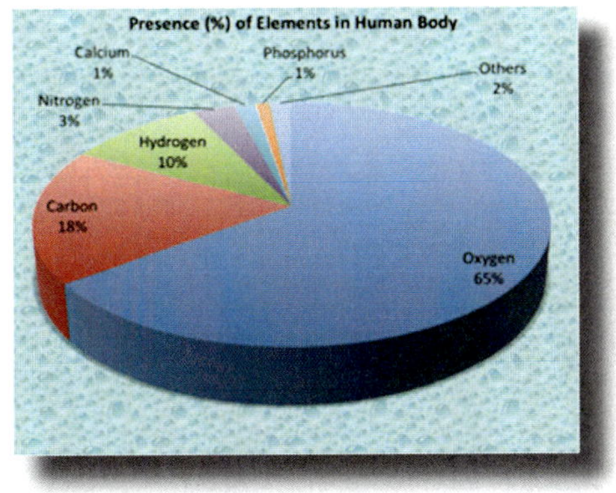

Image 3.1. The elements of life, percent by mass.

3.1 CARBON-BASED LIFE FORMS

You've probably heard before that we are "carbon based life forms." If this is true, then why does oxygen account for 65 percent of our elemental breakdown by mass? As it turns out, we (like all living organisms) are composed largely of water, and water contains oxygen. We can look at the elemental makeup of the

Image 3.2. The major and trace elements of life.

Elements of Life - **Major**

Elements of Life - **Trace**

1 H 1.01 Hydrogen																1 H 1.01 Hydrogen	2 He 4.00 Helium
3 Li 6.94 Lithium	4 Be 9.01 Beryllium											5 B 10.81 Boron	6 C 12.01 Carbon	7 N 14.01 Nitrogen	8 O 16.00 Oxygen	9 F 19.00 Fluorine	10 Ne 20.18 Neon
11 Na 23.0 Sodium	12 Mg 24.3 Magnesium											13 Al 26.98 Aluminum	14 Si 28.09 silicon	15 P 30.97 Phosphorus	16 S 32.07 Sulfur	17 Cl 35.45 Chlorine	18 Ar 39.95 Argon
19 K 39.10 Potassium	20 Ca 40.08 Calcium	21 Sc 44.96 Scandium	22 Ti 47.88 Titanium	23 V 50.94 Vanadium	24 Cr 52.00 Chromium	25 Mn 54.94 Manganese	26 Fe 55.58 Iron	27 Co 58.93 Cobalt	28 Ni 58.69 Nickel	29 Cu 63.55 Copper	30 Zn 65.39 Zinc	31 Ga 69.72 Gallium	32 Ge 72.59 Germanium	33 As 74.92 Arsenic	34 Se 78.96 Selenium	35 Br 79.90 Bromine	36 Kr 83.80 Krypton
37 Rb 85.47 Rubidium	38 Sr 87.62 Strontium	39 Y 88.91 Yttrium	40 Zr 91.22 Zirconium	41 Nb 92.91 Niobium	42 Mo 95.94 Molybdenum	43 Tc (98) Technetium	44 Ru 101.1 Ruthenium	45 Rh 102.9 Rhodium	46 Pd 106.4 Palladium	47 Ag 107.9 Silver	48 Cd 112.4 Cadmium	49 In 114.8 Indium	50 Sn 118.7 Tin	51 Sb 121.8 Antimony	52 Te 127.6 Tellurium	53 I 126.9 Iodine	54 Xe 131.3 Xenon
55 Cs 132.9 Cesium	56 Ba 137.9 Barium	57 La *138.9 Lanthanum	72 Hf 178.5 Hafnium	73 Ta 180.9 Tantalum	74 W 183.9 Tungsten	75 Re 186.2 Rhenium	76 Os 190.2 Osmium	77 Ir 192.2 Iridium	78 Pt 195.1 Platinum	79 Au 197.0 Gold	80 Hg 200.5 Mercury	81 Tl 204.4 Thalium	82 Pb 207.2 Lead	83 Bi 209.0 Bismuth	84 Po (210) Polonium	85 At (210) Astatine	86 Rn (222) Radon
87 Fr (223) Francium	88 Ra (226) Radium	89 Ac *(227) Actinium	104 Rf (257) Rutherfordium	105 Db (260) Dubnium	106 Sg (263) Seaborgium	107 Bh (262) Bohrium	108 Hs (265) Hassium	109 Mt (266) Meitnerium	110 Ds (271) Darmstadtium								

58 Ce 140.1 Cerium	59 Pr 140.9 Praseodymium	60 Nd 144.2 Neodymium	61 Pm (147) Promethium	62 Sm 150.4 Samarium	63 Eu 152.0 Europium	64 Gd 157.3 Gadolinium	65 Tb 158.9 Terbium	66 Dy 162.5 Dyspromium	67 Ho 164.9 Holmium	68 Er 167.3 Erbium	69 Tm 168.9 Thulium	70 Yb 173.0 Ytterbium	71 Lu 175.0 Lutetium
90 Th 232.0 Thorium	91 Pa (231) Protactinium	92 U (238) Uranium	93 Np (237) Neptunium	94 Pu (242) Plutonium	95 Am (243) Americium	96 Cm (247) Curium	97 Bk (247) Berkelium	98 Cf (249) Californium	99 Es (254) Einsteinium	100 Fm (253) Fermium	101 Md (256) Mendelevium	102 No (254) Nobelium	103 Lr (257) Lawrencium

body in three different ways: by mass, by number of atoms, and by number of atoms with water removed (Table 3.1).

Element	Percent by Mass	Percent by Number of Atoms	Percent by Number of Atoms (no H_2O)
Carbon	18%	9%	39%
Hydrogen	10%	63%	52%
Oxygen	65%	26%	1.6%
Nitrogen	3%	1.25%	5.6%

Table 3.1. Elemental composition of the human body.

Glucose—a sugar	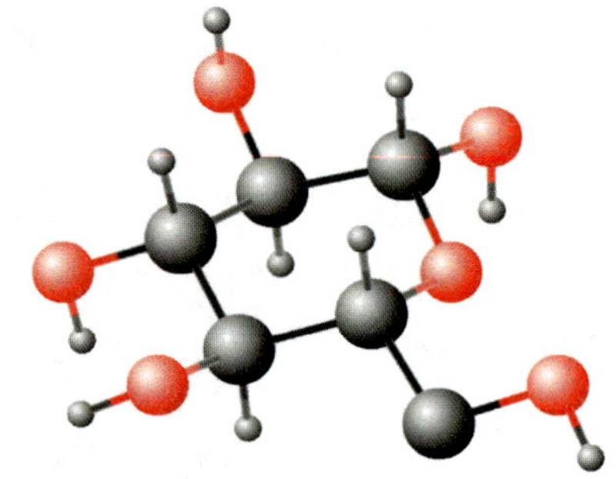
Alanine—a component of protein	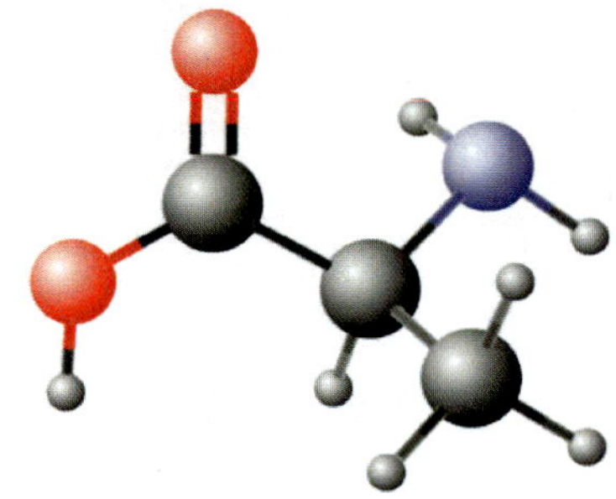
Adenine—a component of the genetic material DNA	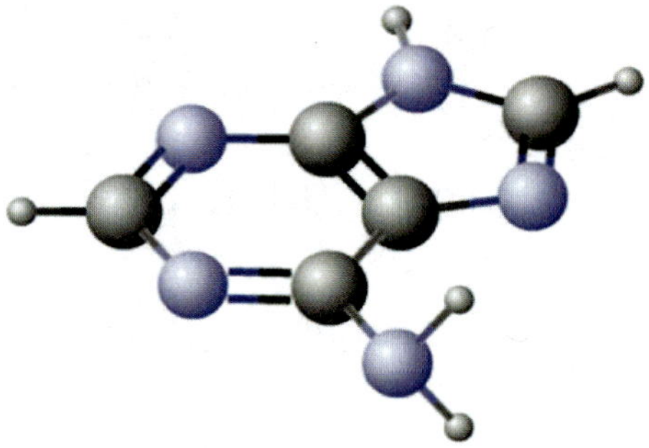

Image 3.3. Some of the molecules of life. In each of these molecules, C (gray), H (white), O (red), and N (blue) figure prominently.

Discounting water, we are primarily carbon (by mass) and hydrogen (by sheer quantity). The large molecules that make up cells, which in turn make up organisms, have carbon "backbones."

So why carbon? What's so special about this element, that it is the foundation upon which we are all built? In order to answer this question, we need to begin to look at the nature of chemical bonding. And in order to do that, we need to first go deeper into the structure of the atom itself.

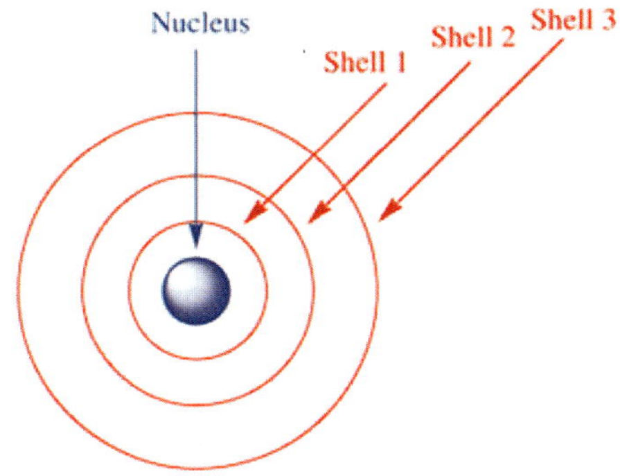

Image 3.4. The first three electron shells (not shown relative to size of nucleus). Radius increases with increasing shell number.

3.2 ELECTRON SHELLS

In the last chapter, we discussed the "electron cloud" that surrounds the nucleus of an atom, accounting for most of the volume, but only the tiniest fraction (so tiny, in fact, that we ignore it!) of the mass. In reality, describing electrons as forming a nebulous "cloud" around the nucleus does nothing to help us understand what is, in actuality, a very organized arrangement of electrons within an atom.* Electrons are distributed in **shells** around the nucleus, where shells are numbered according to their radius (shell 1 is closest to the nucleus).

*NOTE

Actually, there's no one best way to describe the arrangement of electrons in an atom—subatomic particles are strange beasts, and it's nearly impossible to use macroscopic, familiar images as metaphors for describing them with any rigor. For this reason, it's pretty common to think about atoms in many ways, picking analogies as relevant to emphasize a particular point. In considering the size of the nucleus relative to the volume occupied by electrons, we do well to imagine the electrons as a cloud. In conceptualizing the arrangement of electrons within shells in an atom, we tend to picture a little "solar system." It's probably best to remember that these are all analogies, and while they each add to our understanding of atomic structure, no one of them is better or more correct than any other.

Because electrons are negatively charged, they're attracted to (and by) protons. For this reason, even though each electron in an atom must be located within a shell, electrons are most stable in shell 1 because of its proximity to the nucleus. You can think about the attractive force a bit like gravity; if you were to hold a ball (electron) away from the Earth (nucleus of an atom), the ball would be attracted to the Earth due to the force of gravity (electrostatic force between protons and electrons). If you let go of the ball, it would fall toward Earth's surface, getting as close as it could … unless something (like a table) got in its way. The shells are a bit like tables; electrons "fall" as closely to the nucleus as they can get, but they can't get closer than shell 1, so they stop there. Departing a bit from our ball, Earth, and table analogy (as with all atomic structure conceptualizations, each analogy only gets us so far…), it turns out that shells can only hold a certain number of electrons. After that, they are full, and no more electrons can "fall" into them. Shell 1, for instance, can only hold 2 electrons. As such, if we were considering an atom with one electron—that would be hydrogen, which has one proton and therefore one electron—we'd expect to find that electron in shell 1. For an atom with two electrons (helium), we'd expect to find them BOTH in shell 1. However, an atom with three electrons would have two of them in shell 1, and the third would be located in shell 2. Electrons ALWAYS "fall" as closely as they can get to the nucleus, but they can't enter full shells!

How can we tell when a shell is full? Conveniently, the periodic table provides us with a guide. Rows on the periodic table correspond to shells, so we can count the number of elements in any row and determine how many electrons it takes to fill the correspondingly numbered shell. Row 1 on the periodic table contains two elements, which reminds us that shell 1 holds two electrons. Row 2 contains eight elements, which reminds us that shell 2 holds eight electrons. Row 3 also contains eight elements, so shell 3, like shell 2, holds eight electrons. Things get a little weird after row 3. *Technically*, shell 3 actually holds 18 electrons, but it acts full with eight, as do shells 4 and up. Very briefly, shell 3 finishes filling partway through the filling of shell 4—the first row of transition metals represent the remainder of shell 3, the second row of transition metals are the remainder of shell 4, and so on.

To avoid the confusion associated with some of these higher-numbered shells, we use three simple rules:

1) Aside from shell 1 (which is full with two electrons), each shell is (or acts as if it is) full when it contains eight electrons.
2) Because rows on the periodic table correspond to shells, elements at the right end of each row have a filled outer electron shell.
3) For elements in row 3 such as Ar, shell 3 holds up to eight electrons. For elements near the end of row 4 (after the transition metals) such as Kr, shell 3 holds 18 electrons.

Let's look at the second rule a bit closer. Helium (He) is at the end of row 1. We know that shell 1 holds two electrons, and remembering that He has two electrons, He must have a full shell 1. Neon (Ne) is at the end of row 2, and has a total of ten electrons. The first two have to be in shell 1, and the remaining eight that can't fit in shell 1 are left to fill shell 2, which is consequently full. We see the same thing with argon (Ar): It has 18 electrons, so the first two are found in shell 1; there are eight in shell 2; and the remaining eight go into shell 3, filling it. Even though things get a bit confusing in terms of our electron count after row 3 on the periodic table, the pattern regarding the elements at the end of each row holds true. These elements ALWAYS have full outermost shells, which contain eight electrons (helium being the one exception). Krypton, for instance, has 36 total electrons. The first two are in shell 1. The next eight are in shell 2. Krypton is toward the end of row 4, after the transition elements, so shell 3 is filled with a total of 18 electrons. We've accounted for 28 of krypton's electrons so far, so the last eight must be in shell 4, filling it.

We can use the information about rows and their corresponding shells to figure out where the electrons are in any element, not just those at the ends of rows. Elements in row 2 will have their first two electrons in shell 1 and any

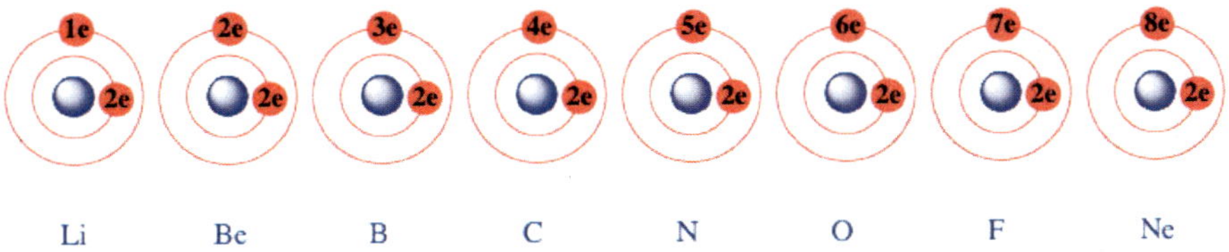

Image 3.5. Electron configurations of the row 2 elements.

> ## TRY THIS:
>
> Draw electron configurations for H, He, Na, Ca.
>
> Answers:
>
> 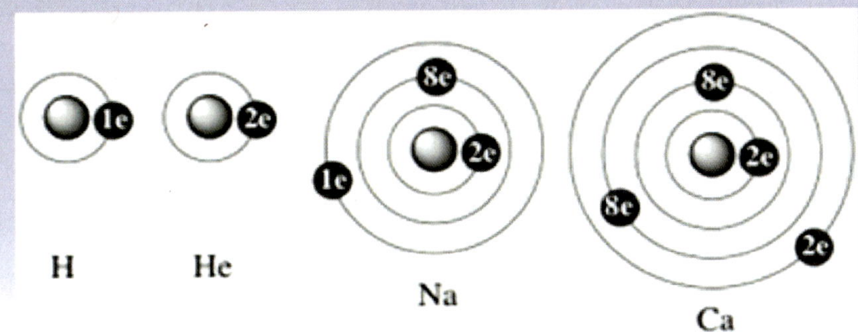
>

remaining electrons in shell 2. Notice that of the elements in row 2, however, only Ne has a filled shell 2.

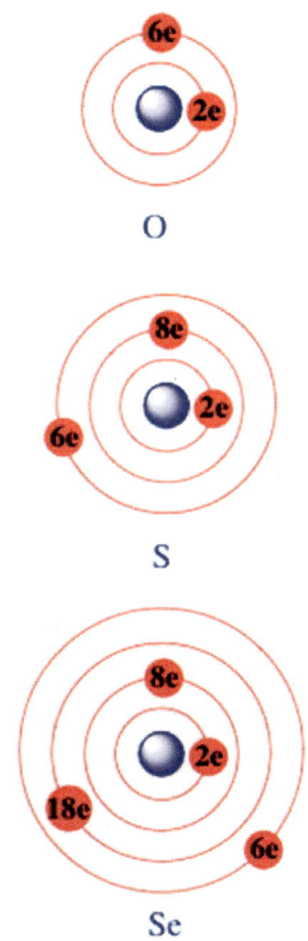

Image 3.6. Regardless of the total electron count, all elements in any column on the periodic table have the same number of valence electrons.

Valence

Aside from helping us to know where specifically the electrons are within an atom, what can we do with our understanding of configuration? It turns out that only the electrons in an atom's outermost shell participate in bonding and chemical reactions. In terms of understanding the chemistry of carbon, for instance, it's more important to know that it has four electrons in its outer shell (these are called **valence electrons**) than that it has six electrons total. Furthermore, a closer look at the periodic table provides another revelation: All elements within a given column on the periodic table have the same number of valence electrons, even though their total electron counts differ! We already saw this trend in the column at the far right of the periodic table: neon, argon, and the rest of the elements in this column all have full outer shells of eight electrons. In the same manner, oxygen, sulfur, selenium, and the rest of that column's elements have six valence electrons.

Many periodic tables number the columns; this provides a convenient way of de-

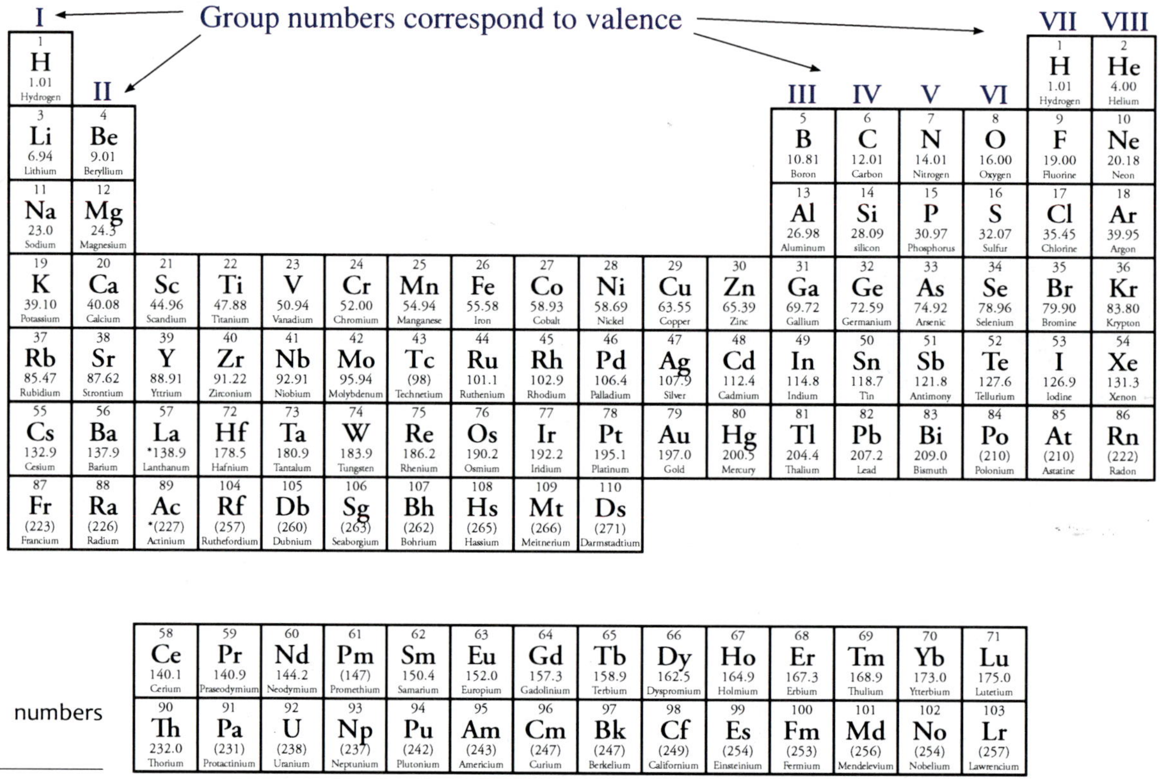

Image 3.7. Group numbers valence.

termining an atom's valence without having to sketch a full electron configuration. Elements in the first column (called group 1) have one valence electron, elements in the second column (group 2) have two valence electrons, and so forth. Even if a periodic table's columns aren't numbered, it's still possible to count the columns from left to right (skipping over the transition metals) to determine an atom's valence.

If valence electrons are the only ones that participate in bonding and reactions, and the number of valence electrons is, as stated earlier, more important

Image 3.8. From left to right, lithium, sodium and potassium are all soft, reactive metals. They're generally stored in ether or paraffin to keep them away from air and water, with which they react.

than the total number of electrons in predicting how an atom acts chemically, then it should come as no surprise that elements in the same group have similar chemical properties. Lithium, sodium, potassium, and the other members of group 1 show striking chemical similarities (they are all, for instance, very reactive soft metals that catch fire or explode when exposed to water).

***NOTE**

Other groups have names as well. For instance, group 1 elements are called *alkali metals*, group 2 elements are called *alkaline earth metals*, and group 6 elements are called *chalcogens*. Aside from group 8's name, though, the only one you'll probably encounter commonly is the name for the group 7 elements, which are called the *halogens*.

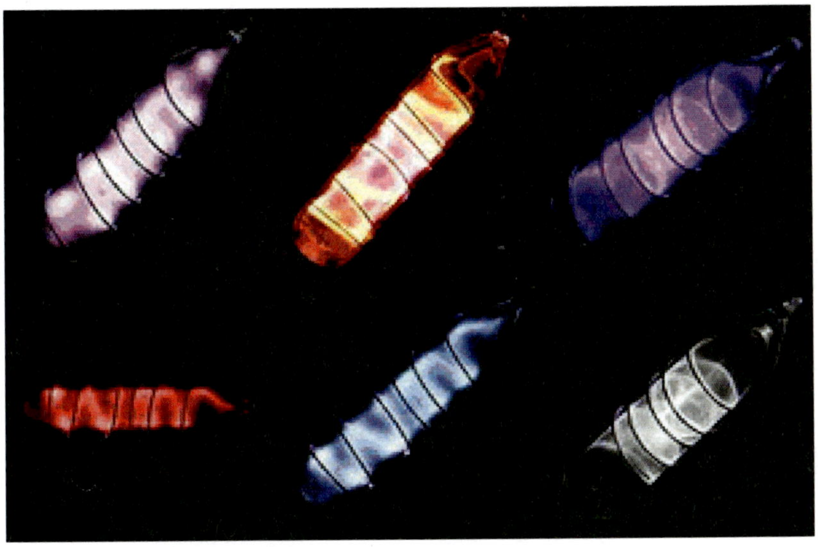

Image. 3.9 Glowing tubes of the noble gases (counterclockwise from top left): helium, neon, argon, krypton, xenon, radon. Each of these gases glows when subjected to electricity, which is why neon is used to make signs.

The same is true for groups 2–8. Group 8, in particular, is interesting; like any other group, these elements behave similarly to one another in that they do NOT tend to react or form bonds. In fact, their tendency not to form chemical bonds with other elements has resulted in the group's name—they're called the *noble gases* (for their "refusal" to associate with "lesser" elements)!*

The noble gases were discovered much later than some of the other elements, simply because they don't tend to react or form compounds. This inertia on their part is even reflected in their elemental names. Aside from helium (so named because it's made in the Sun, which is *helios* in Greek), the noble gases are neon ("new one"), argon ("lazy one"), krypton ("hidden one"), xenon ("strange one"), and radon ("radioactive one"). It's no coincidence that the noble gases are both unreactive AND have full electron shells—as it happens, chemical reactivity is a product of valence. Full outer shells are very stabilizing for atoms; those that have them (the noble gases) have no reason to react, and so they don't. *Atoms that do not have full valence shells react chemically in order to achieve them.* The chemical reactions we'll see in this class, including formation of bonds and combining of atoms into molecules, are driven by the atoms' energetic need to obtain full valence shells. Because all shells (except shell 1) are full when they contain eight electrons, the tendency of an atom to react so as to fill its valence is called the **octet rule**.

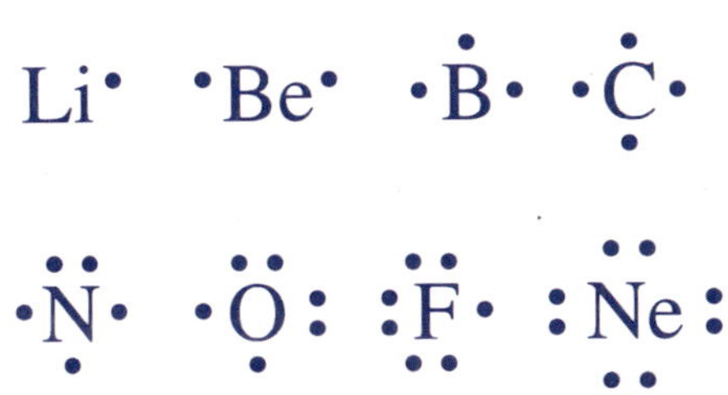

Image 3.10. Lewis dot structures of several elements.

Image 3.11. Correctly and incorrectly drawn Lewis dot structures for fluorine. The "imaginary box" around an atomic symbol helps us to place electrons symmetrically and quickly count valence electrons (and holes in the octet!).

3.3 LEWIS DOT STRUCTURES

At this point, it's clear that in order to have a meaningful discussion about chemical reactivity, the electron count we're most interested in is not an atom's total number of electrons, but its valence. A chemist named Gilbert Lewis developed (as a grad student!) a very elegant way to keep track of and portray an atom's valence; the element's atomic symbol is written circumscribed by dots representing the valence electrons. These shorthand representations are referred to as **Lewis dot structures** (Image 3.10).

While there are not strict rules regarding the placement of electrons in Lewis dot structures (lithium's single electron, for instance, could be placed to the left of the symbol or to the right, above or below), it is nevertheless common practice to imagine an invisible square surrounding the atomic symbol, with electrons distributed symmetrically around the square (Image 3.11). Electrons typically aren't paired on any one side of the square until all sides have at least one electron (note the Lewis dot structures for carbon and nitrogen in Image 3.10). This convention has *nothing whatsoever* to do with the electrons' actual

TRY THIS:

Draw Lewis dot structures for: H, Si, S, Cl.

Answers:

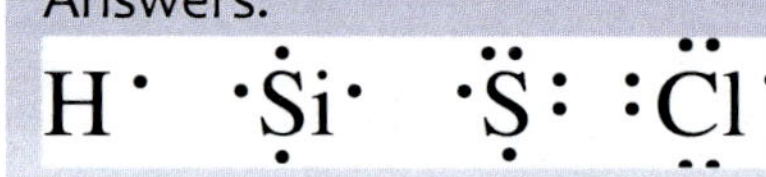

location within the valence shell—it's merely shorthand. However, it is useful because rapidly determining an atom's valence from a Lewis dot structure is much easier if electrons are placed sensibly rather than scattered. Compare the two structures in Image 3.11—it's much quicker to glance at the correctly drawn Lewis dot structure on the left and note that fluorine has seven valence electrons than it is to try to count the dots in the improperly drawn structure on the right.

3.4 COVALENT BONDING

One really convenient characteristic of Lewis dot structures is that aside from

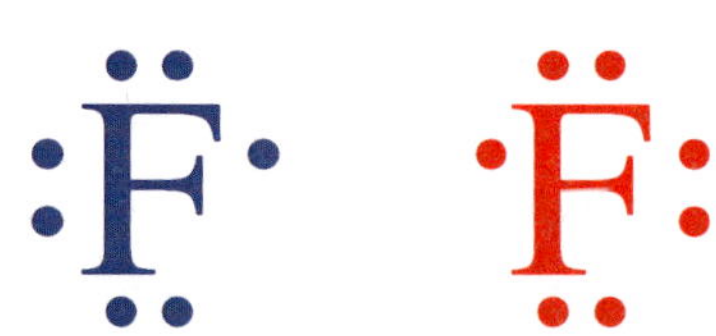

allowing us to deduce the number of valence electrons rapidly, they also make "holes" in the valence readily apparent. In the Lewis dot structure of fluorine (Image 3.11), there is one electron "hole," precluding a full octet. Fluorine would be more stable if it had a full outer shell like its group 8 neighbor, neon. Unfortunately for fluorine, electrons can't be conjured out of empty space.

However, by reacting chemically, fluorine can gain an electron and complete its octet. Let's imagine that two atoms of fluorine were to come upon each other; they each have a "hole" in their valence, and they each need one more electron.

As it happens, electrons don't have to belong to a SINGLE atom—they can be shared by a pair of atoms as long as the two are in close enough

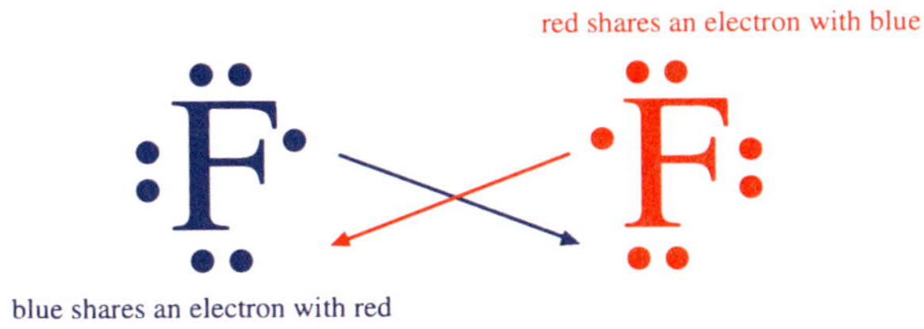

proximity that their valence shells overlap. As a result, the fluorine on the left (we'll call it "blue fluorine" for the sake of this example) can let the fluorine on the right ("red fluorine") share one of blue fluorine's electrons. If that were

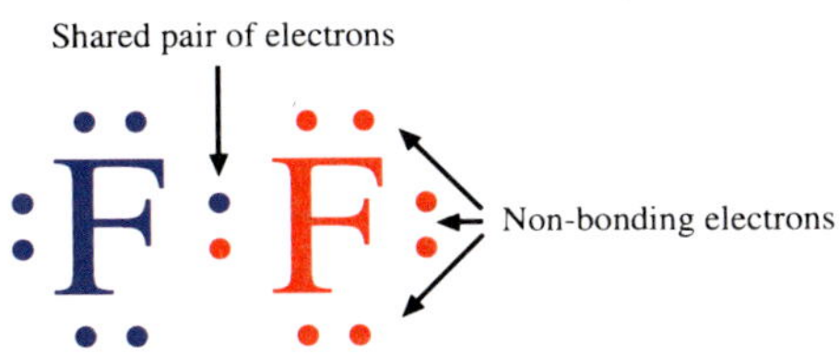

to happen, red fluorine would have a full octet—seven electrons of its own, and one that technically belongs to blue fluorine, but blue fluorine is sharing. This is great for red fluorine, but what about blue fluorine? In a lovely *quid pro quo* (a Latin expression meaning a fair trade) exchange, red fluorine shares one of its electrons with blue fluorine.

As long as the two fluorine atoms stay close to one another in space, they each get the benefit of an additional electron in their valence, and they each have complete octets.

The shared pair of electrons is a physical connection between the two atoms—almost like they're holding hands—and it keeps them stuck together. This is a **chemical bond**, which, because it's made up of electrons contributing to BOTH fluorines' valences, is also called a **covalent bond**. It's very important to point out that neither fluorine atom has truly *gained* an electron (which would unbalance the protons and electrons in the fluorine atom and result in it no longer being neutral in charge). Instead, each atom maintains ownership of its individual electrons—the shared electrons are more *lent* than

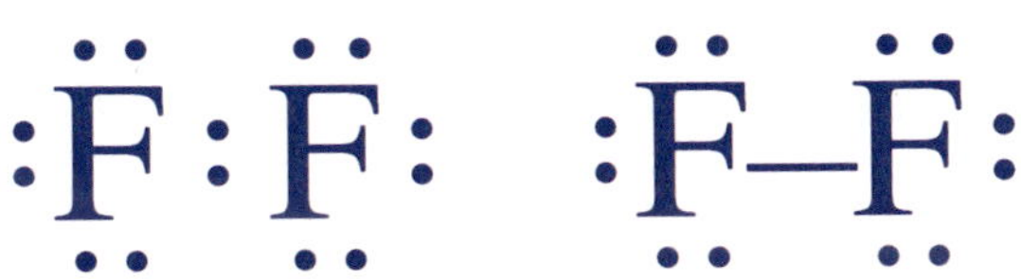

given. If we wanted to do a quick electron head count at this point, we'd say that each fluorine atom still has seven "personal" electrons, but that as a result of the shared electron pair, each fluorine also has a total of eight valence electrons—a full octet. The diagram of the bonded fluorine atoms

***NOTE**

Now you know why F is one of the elements found as a diatomic molecule, or F_2. As a lone atom, fluorine has an unfilled octet, but in a molecule, its octet is filled!

above is a Lewis dot structure of a molecule; the molecule in this case is F_2.* When drawing the Lewis dot structure for a molecule, we have the option of showing a pair of dots between the atoms to represent the shared electrons; or alternately, we can use a line to indicate the bond.

A line (bond) is ALWAYS interpreted to mean a shared PAIR of electrons.

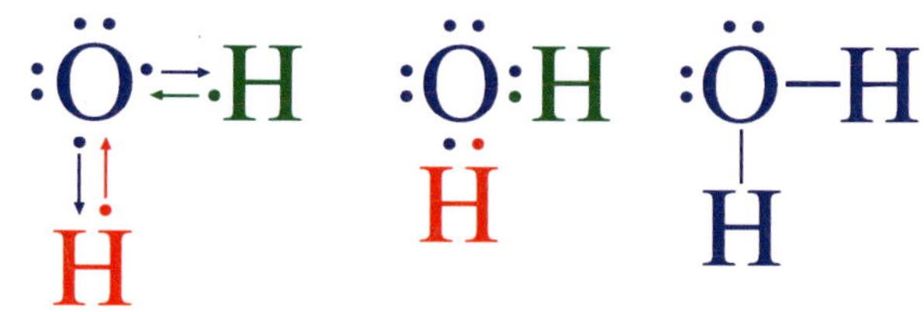

Also, even if we're showing the bond as a line rather than a pair of dots, we always show the remainder of each atom's electrons (the nonbonding electrons, also called **lone pairs**). This helps us keep track of our electrons and each atom's valence.

3.5 MULTIPLE COVALENT BONDS

Sometimes atoms need to fill more than one hole in their valence. Oxygen, for instance, has a valence of six electrons and therefore needs to fill two holes. As you might guess, oxygen needs to form TWO bonds. One possibility is for oxygen to bond to two different atoms, as in water:

CONCEPT CHECK:

How do we know how many bonds an atom will need to make?

Answer: An atom will make enough bonds to fill the hole(s) in its valence—one hole means one bond, two holes means two bonds, and so forth.

Another possibility is for oxygen to bond twice with one other atom, as it might with another atom of oxygen to form the molecule O_2. Because two pairs of electrons are shared by the atoms in the O_2 molecule, the bond is referred to as a double bond, and can be portrayed either as four dots between the atoms, or two parallel lines.

It's also possible for atoms to form triple bonds, sharing three pairs of electrons. Take the example of nitrogen: because each atom of N has three holes in its valence, it needs to form three bonds. This can be achieved by forming a single bond to each of three different atoms (as in NH_3), by forming a double bond to one atom and a single bond to another, or by forming a triple bond to one other atom (as in N_2).

In predicting Lewis structures of molecules, there are a few things worth remembering. First, if a molecule consists of only two atoms, we make as many bonds as necessary (one, two, or three—there's actually no such thing as a quadruple bond) between the atoms to fill their valences. If a molecule consists of more than two atoms, it will generally fall into one of two categories. One possibility is that the molecule will consist of two elements, with one atom of one element, and more than one atom of the second element (as in CO_2, for instance, or CH_4). In this case, the correct Lewis structure will have the element of which there is only one atom at the center of the structure (C is at the center of CH_4, and is bonded to four hydrogen atoms). If, alternately, a molecule consists of three different elements, the atoms will be bonded together in the order in which the formula is written (HCN, for instance, is an H bonded to a C bonded to an N). This second case is much less common. There are molecules, of course, much more complicated and extensive than the examples shown here, but drawing them is beyond the scope of this text.

With a molecule like NH_3 or N_2, it's pretty easy to put together a Lewis structure intuitively. However, as molecules get bigger, the sharing can be pretty complicated and counterintuitive, and it makes sense to have a systematic approach. Let's look at the steps necessary to drawing a correct Lewis structure for a more complicated molecule, using the example of SO_2:*

***NOTE**

How are you to know when a molecule is complicated enough to warrant following the steps? It's good practice to use them every time, at least until you get used to drawing molecules.

1) Count up the total number of valence electrons contributed by each atom (the total in the molecule MUST be the sum of the individual valences).

Sulfur has a valence of 6 electrons; oxygen has a valence of 6 electrons.
Total electrons in $SO_2 = 6 + 6 + 6 = 18$ *electrons.*

2) Put together a skeleton structure of the molecule, following the hints for connectivity given above. Don't worry about double and triple bonds or nonbonding pairs for right now; just connect everything with single bonds.

SO_2 *consists of one sulfur atom and two oxygen atoms, so sulfur should go in the middle.*

O–S–O

3) Subtract the number of electrons used to make bonds in Step 2 from the total to determine the number of electrons that still need to be placed.

Two single bonds so far $= 2 \times 2 = 4$ *electrons*
$18 - 4 = 14$ *electrons remaining*

4) Place remaining electrons in pairs around OUTER atoms; do not overfill octets. If outer octets fill before electrons run out, place left over electrons on central atom. Be as symmetrical as possible.

Twelve of the remaining electrons can be placed around the oxygen atoms. Two electrons are left over, so they are placed on the central sulfur.

5) Count electrons around each atom to determine whether octets are full, remembering that each bond consists of a pair of electrons that counts toward the valence of both atoms. Where octets are left unfilled, share pairs from outer atoms with inner atom to fill all octets. Be as symmetrical as possible.

Each oxygen has a full octet, but sulfur has only six electrons. One of the pairs currently on an oxygen atom will need to be shared with sulfur. There's no way to do this symmetrically, so we simply pick an oxygen atom; it doesn't matter which one.

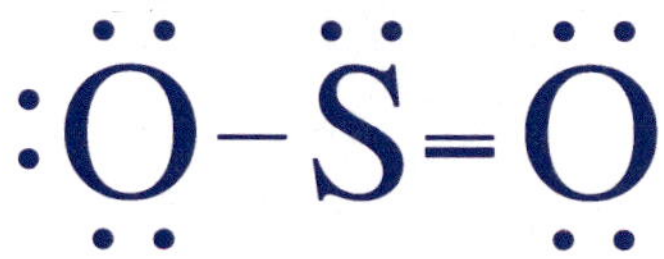

6) Verify that each atom has a filled octet.

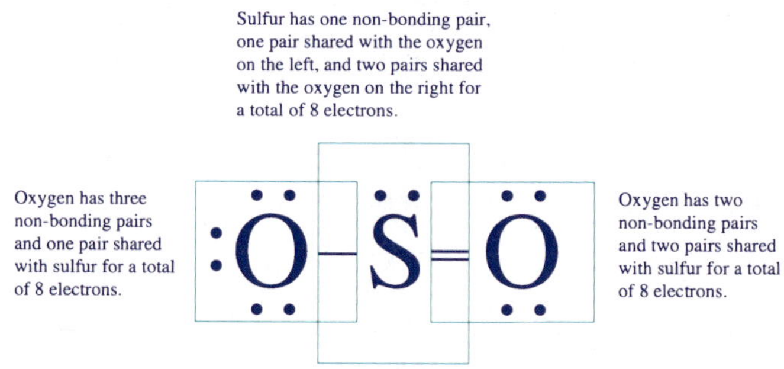

Our SO_2 molecule is now complete.

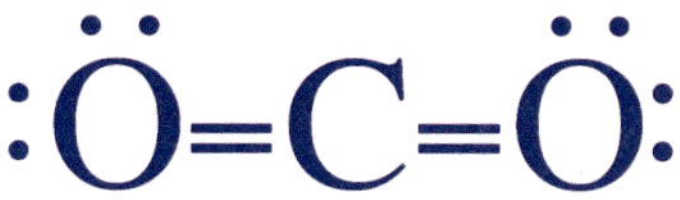

Correctly drawn structure

Incorrectly drawn structure

Notice that the correct Lewis structure of SO_2 is not what we might have anticipated; based upon their valence holes, we anticipate atoms of sulfur and oxygen to make two bonds. The sharing is quite a bit more complex here. This sort of thing happens sometimes, and rather than try to figure out when it's going to happen and when it's not, we're better off simply using the steps described above to generate the correct Lewis structure every time. The most important thing to realize about this example is that even though the bonding didn't come out exactly the way we might have expected, ALL the octets are filled. Furthermore, the case above is definitely less common

TRY THIS:

Draw Lewis dot structures for: HCl, H2, HCN, CH4.

Answers:

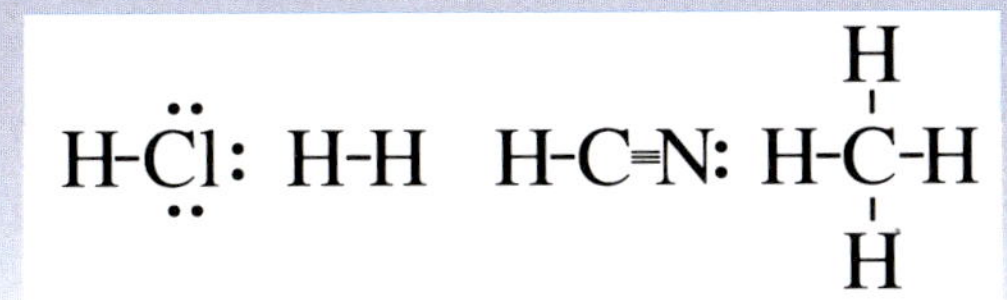

than cases in which the number of bonds is exactly what we expect, so it's fine to anticipate that NORMALLY we'll see oxygen forming two bonds, and so forth.

One more helpful hint: correct Lewis dot structures tend to be as symmetrical as possible. In determining the structure of CO_2, for instance, we can fill valences of all atoms by drawing double bonds between the carbon and each oxygen atom; OR we can draw a single bond between carbon and one oxygen, and a triple bond between carbon and the other oxygen. The correct Lewis structure is the symmetrical one, in which CO_2 has two double bonds.

3.6 SO WHY ARE WE CARBON BASED?

We should return to the questions posed at the beginning of this chapter. First off, given the incredible variety of elements in the universe, why are so few of them major components of living systems?

The answer to this question requires us to look a bit more closely at the nature of covalent bonding. When fluorine, for instance, needs to fill the hole in its valence, it enters into a relationship with another atom (of fluorine or some other element). The second atom shares an electron with fluorine, and in return, is able to share one of fluorine's electrons. The nature of a covalent bond is that it requires an *even exchange*. Oxygen, in order to fill its valence by sharing two electrons belonging to other atoms, must allow those other atoms to share two of oxygen's own electrons. Nitrogen shares three electrons belonging to other atoms, and in return allows them to share in three of its own. In the case of carbon, all electrons end up being shared (carbon gets four shared with it, and in exchange must share its own four electrons).

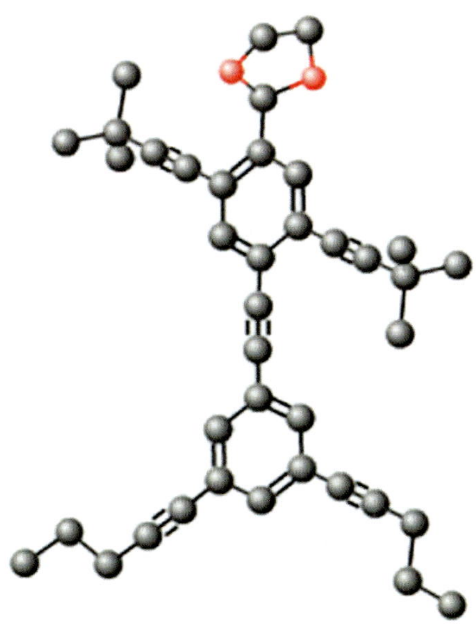

Image 3.12. A LITERAL "carbon backbone"—a molecule specially designed in a lab (see Sources and Further Reading) to look like a running person with oxygen (red) for eyes! Hydrogen atoms have been left out of this picture of the structure for ease of viewing. Researchers sometimes build "specialty" molecules like this to engender interest in chemistry while simultaneously refining synthetic (molecule-building) techniques that can be used in drug design and other applications.

> ***NOTE**
>
> Remember, hydrogen has a sort of "split personality"—it acts a bit like a group 1 element and a bit like a group 7 element. Remember too, though, that while it technically belongs to group 1, it is formally classified as a nonmetal. In its ability to form covalent bonds, we see hydrogen displaying its nonmetallic character.

If we were to continue along this line of reasoning, backing up through row 2 of the periodic table, we'd run into a problem when we got to boron. Boron has five holes in its valence, so it would need to have five electrons shared with it (forming five bonds). However, boron CAN'T form five bonds, because it only has three electrons of its own to offer; the most bonds it could form would be three. Beryllium would need to form six bonds to fill its valence but can't, because it has only two electrons of its own to offer up for bonding, and lithium is in a real pickle—it needs to form seven bonds, but has only one electron of its own to share. Groups 1, 2, and 3 on the periodic table clearly have problems when it comes to filling their octets by forming covalent bonds; they simply don't have enough electrons of their own to share. As a result, the

> **CONCEPT CHECK:**
>
> Why can't group I, II, and III elements fill their valences by forming lots of covalent bonds?
>
> Answer: Covalent bonds require an atom to share one of its electrons for each bond made—an atom can't make more bonds than it has valence electrons.

metals do not use covalent bonds to fill their octets. They still react chemically—all atoms need to obtain full valence shells—they just do so slightly differently. We'll discuss the chemical reactivity of the metals in Chapter 4. Suffice it to say, we'll only see covalent bonds forming between nonmetals (or metalloids) and nonmetals.*

Even though we won't talk about the chemical reactivity of metals just yet, it's worth mentioning that the covalent bond is the most stable, strongest kind of bond there is. As a result, compounds made up of nonmetallic elements are

Bonding Patterns Available to Group IV, V, VI, and VII Elements (based upon the number of covalent bonds they can each form)				
F	H–F:			
O	H–O–H	O=O		
N	H–N–H (with H on N)	H–N=CH_2	:N≡N:	
C	H–C–H (with H above and below C)	O=C with two H on C	N≡C–H	O=C=O

Table 3.2. The elements of life have a number of possible bonding patterns in which they can engage.

much more long-lived. A biological system depends upon stability—cells can't have their structural components falling apart! Thus, you'll notice that the major elements of life are all nonmetals, and the reason is that these elements form stable, long-lasting bonds with each other.

The second question posed at the beginning of the chapter was, of all the elements that are major components of living systems, why does carbon form the structural "backbone" of the molecules of life? Given that elements form bonds based upon the number of valence holes they have to fill (group 7 elements form one bond, group 6 elements form 2 bonds, and so forth), the group 4 elements are of particular interest. Carbon (and the other group 4 elements) form more bonds than any other nonmetals. As a result, they display the greatest versatility and variety of bonding patterns (Table 3.2). Furthermore, their prodigious bonding ability allows for the development of extensive networks of bonds and the formation of very large molecules. In light of this reasoning, we can make sense of the fact that the biomolecules depend upon a backbone made up of a group 4 element—versatility of bonding patterns and strength of bonds are both characteristics very desirable in the molecules of life!

There's just one question left: why carbon rather than another group 4 element? The answer to this is simply that carbon is the only group 4 nonmetal. Silicon is metalloid, as is germanium, and the others in the column are metallic. Even though metalloids can form covalent bonds, the very strongest bonds are formed between nonmetals and nonmetals. Of the group 4 elements, carbon forms the best life-promoting bonds.

3.7 MOLECULAR SHAPES

Actually, the story of carbon's special stability and bond-forming ability goes even deeper than simply that it forms four bonds. We've been representing molecules so far in this chapter as nothing more than a collection of atoms, bonded to one another in a certain order. Actually, though, there are two very important things we need to remember about chemical bonds:

1) Bonds are made up of pairs of electrons.
2) Electrons repel other electrons.

This takes us pretty naturally to the realization that bonds will repel each other, and should separate from one another as much as possible. At first glance, it seems as though that's what we've been portraying. After all, we haven't been showing methane (CH_4) with the bonds all smashed together like the picture below on the left; we've been showing the bonds spaced out as shown on the right:

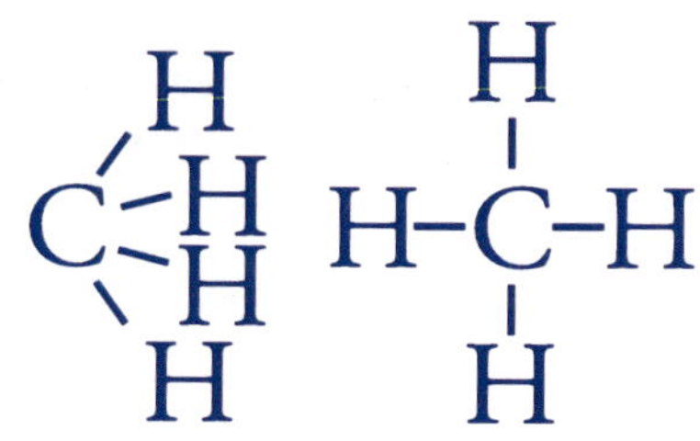

However, even a reasonable Lewis dot structure of methane (like the one above) doesn't take advantage of an important fact about molecules: they exist in three-dimensional space. Our Lewis dot structure portrays the bonds as separated as possible on a two-dimensional page, but in three dimensions, bonds can get much further apart than 90°. In fact, four objects are maximally spaced when placed at the corners of a four-sided pyramid. When its bonds are separated as much as possible, methane actually looks like this:

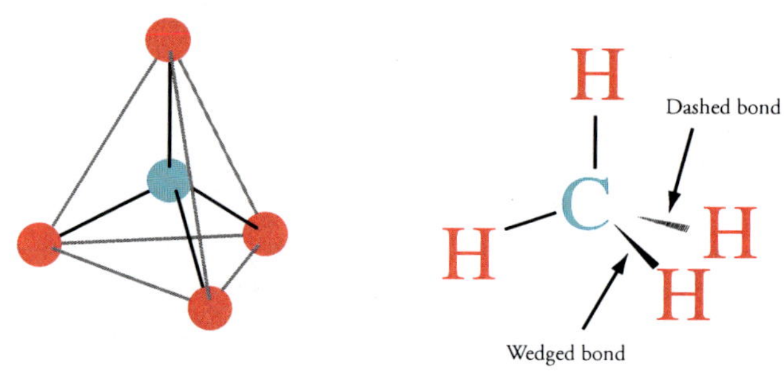

The "wedged" and "dashed" bonds are there to remind us that not all of the hydrogen atoms attached to the central carbon are in the plane of the paper; any two of them can be, but of the other two, one must be in front of and the other behind the plane. In this perspective representation, the wedged bond is taken to mean in front of the plane of the page, and the dashed bond is taken to mean behind. If you were to calculate the angle from one corner of the pyramid to the center (where the carbon is) back to another corner, you'd find that the bond angle is 109.5°; this arrangement clearly affords the electrons more space than they'd have in only two dimensions. Using perspective to portray shape is just one more way of communicating information about molecules. Our old Lewis dot structure of methane told us everything we needed to know about connectivity, but didn't accurately represent shape, since methane isn't flat. This is actually very important—the pyramid formed by a carbon bonded to four other atoms (this geometry is called **tetrahedral**) is very versatile and stable. If you need reminding of just HOW stable this shape can be, go back and look at Image 1.13 in Chapter 1: The structure of diamond is nothing more than a tetrahedral arrangement of carbon atoms, each of which is bonded to four other carbons.

Let's revisit some of the molecules for which we drew Lewis dot structures earlier in the chapter. The structure of ammonia (NH_3) has a central nitrogen with three bonds to hydrogen and a lone pair of electrons. Just like the carbon in methane, the nitrogen in ammonia has four groups of electrons around the central atom (a lone pair is an electron group, because in the same way that a pair of electrons sticks together in a bond, a nonbonding pair sticks together). Thus, ammonia will have the same tetrahedral arrangement of electron groups that we saw with methane. The only difference will be that instead of an atom at one of the corners of the pyramid, there will be a lone pair of nonbonding electrons there:

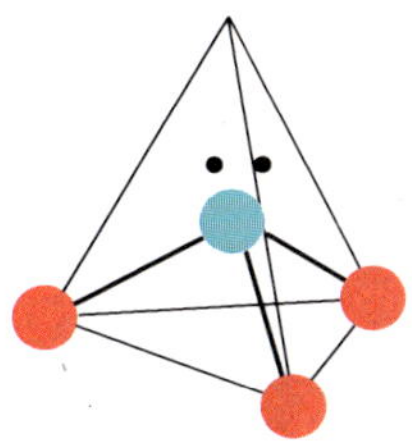

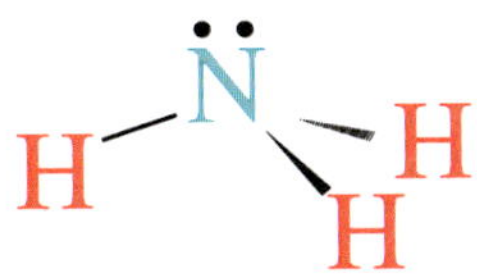

We can portray ammonia using wedge-dash notation just as we did with methane. Remember, molecules are three-dimensional objects that can be turned over and upside down; there's no reason the lone pair of electrons has to be shown on TOP of the nitrogen in ammonia; the molecule has the freedom to turn any which way. It's often easiest to get the perspective drawing to "look right" if we draw the molecule this way, but it's completely legitimate to turn it around and draw it from any perspective you like!

One more thing about ammonia: Notice that the electrons in each bond are pulled slightly out and away from the central atom, but nitrogen's lone pair is held closer, simply because nitrogen isn't sharing it. This means that the lone pair is exerting a little extra repulsive force (electrostatic forces are strongest at close range), so the electrons in the N–H bonds are repelled more by the lone pair than they would be by a bond. The result is that the bond angle from one of the H's to the central N to a second H is slightly reduced as compared to the perfect tetrahedral shape formed by methane; ammonia has bond angles of about 107°. We still consider it to have tetrahedral geometry, but it's worth remembering that the bond angles are a bit constricted.

Water also has four electron groups around its central atom, so like methane and ammonia, the groups will arrange themselves naturally into a tetrahedron, giving the atoms in water a "bent" shape:

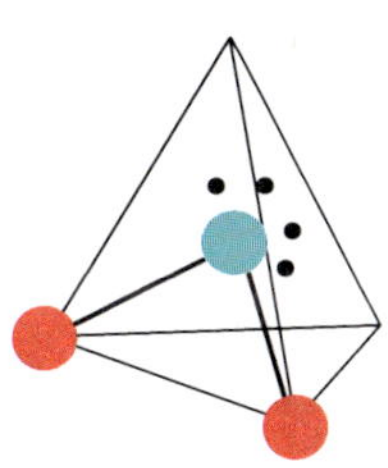

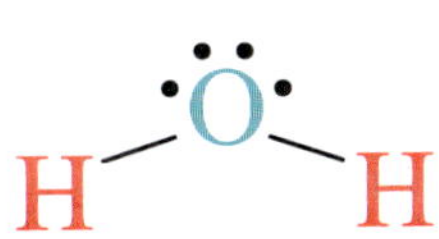

For ease of drawing, we generally don't use wedged and dashed lines when portraying water; it's simpler to show the two O–H bonds in the plane of

CONCEPT CHECK:

Why don't we use wedged and dashed lines to draw water?

Answer: In any tetrahedral arrangement, two of the bonds are in the plane of the paper. Since water only has two bonds, we can show them both in the plane.

the paper, remembering that the lone pairs (which we never put on the ends of wedged or dashed lines, because there are no "bonds" to lone pairs) are in front of and behind the plane. Incidentally, the lone pairs in water affect the H–O–H bond angle the same way they do in ammonia—they condense it. With two lone pairs, water's bond angle is even more affected than that of ammonia, at around 104.5°.

It's possible for molecules to have fewer than four electron groups around a central atom. Consider the molecule COH_2 (this is formaldehyde, if you're curious), which has a central C attached to an O and two H's. To fill octets, the C and O must share a double bond:

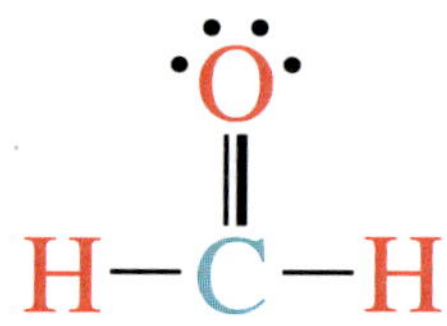

In this case, there are three electron groups around the central C; the two electron pairs that make up the C–O double bond must stay together between the C and O. They are treated as one group, since they can't separate from each other. The furthest from one another that three electron groups can get in three-dimensional space is 120°, forming an equilateral triangle:

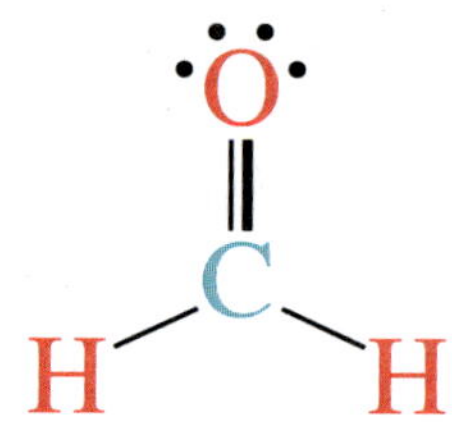

Image 3.13. Carbon can produce molecular backbones with an incredible variety of shapes. This is norethisterone, a synthetic progestin (pregnancy hormone) used in oral contraceptives. Most hydrogen atoms have been left off this image of the molecule for ease of viewing the carbon backbone. Note areas of tetrahedral shape, trigonal planar shape (along double bonds) and linear shape (near the C–C triple bond).

We don't need wedges and dashes to portray a molecule with geometry like this; the **trigonal planar** shape created by three electron groups around a central atom is in a single plane. Incidentally, the same bond condensation effect that we saw in ammonia and water takes place in trigonal planar molecules. SO_2, with two bonds and a lone pair, has a bond angle slightly less than 120°.

There's another important shape family that we see in the molecules of life, which is observed in molecules with two electron groups around a central atom. These molecules, including carbon dioxide, have a linear shape. Portraying a linear molecule does not require wedge-dash perspective, as the molecule exists in a single plane:

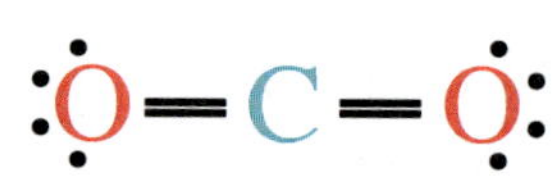

Considering molecular shape is just one more way to think about molecules and helps to explain some of the chemistry we observe in the world, such as the ubiquity of carbon.

3.8 THE PARADOX OF BOND FORMATION

Our discussion of molecular shape arose from the principle that like charges repel, so electrons in bonds or lone pairs repel other electrons in bonds or lone pairs. One might even wonder why bonds can form at ALL, given that they (by necessity) bring two electrons (for a single bond) or four (for a double bond) or six (for a triple bond) into close proximity. Don't the electrons in a bond repel each other, and if so, how and why does the bond form? The answer has to do with stability, but can be explained using a simple cost-benefit analysis. Let's imagine that you're trying to make a decision—for example, perhaps you're considering buying a new car. This action has costs associated with it: a higher payment, higher insurance premiums, and maybe a greater risk of getting the car stolen if it's something really nice. There are also definite benefits associated with buying a new car. Just to name a few, you'd have a nicer vehicle to drive around in, your gas mileage would probably improve, and maybe you'd impress someone special. In any case, while considering your purchase, you'd probably spend some time thinking about whether the benefits were worth the costs, or whether in the end, it simply wasn't a good exchange. In the same way, most of what happens in chemistry has stability costs (the system becomes less stable) and stability benefits (the system becomes more stable). If the sum of the stability lost is greater than the sum of stability gained in a process, then the system becomes less stable overall, which is not good. Conversely, if the stability benefits outweigh the stability costs, the system will become MORE stable, which is favorable. Think about it this way—in terms of electron configurations in an atom, electrons always fill the lowest numbered shell they can. That gets them as close as possible to the nucleus, where they're most stable. In the same way, a dropped ball falls toward the Earth because it is most stable on the ground. Chemical reactions aren't magic. They happen for ONE reason, and ONE reason only—because the result is more stable than the starting point. We'll discuss this concept a little more when we talk about energy in Chapter 8.

What does this have to do with bond formation? When a bond forms, the atoms involved gain something—they get their valence shells filled. That's a definite stability benefit to each atom. The electrons involved in forming the bond ALSO get something. Even though they're brought close together (a stability cost), they get sandwiched into the valence of TWO nuclei, to which they're attracted. Since proximity to a nucleus stabilizes electrons, being close to two nuclei is more stabilizing than being close to one! Imagine two hydrogen atoms getting ready to bond. Before bonding, each atom would have an unfilled valence shell, and each electron would be stabilized by one nucleus:

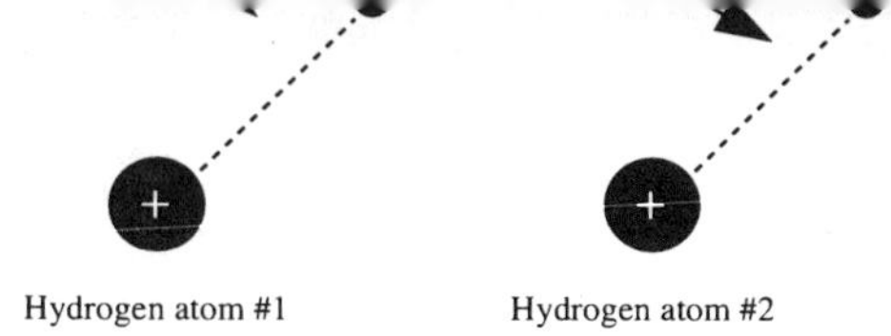

After bonding, each atom would have a full valence shell, and each electron would be stabilized by two nuclei:

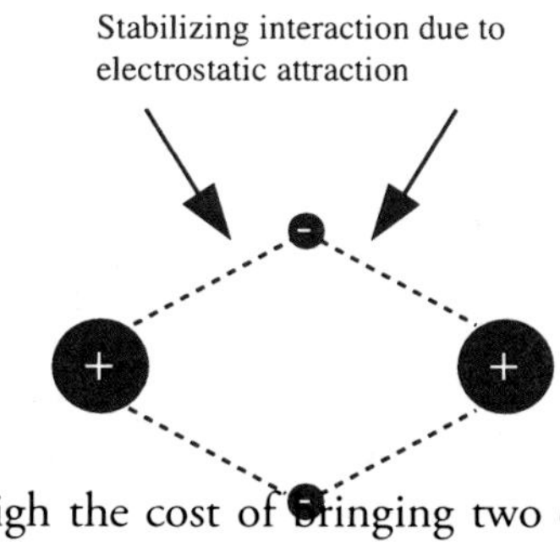

These benefits outweigh the cost of bringing two electrons close together. The only other possible problem with this arrangement is that in bringing the atoms close enough to bond, the positively charged nuclei are also brought close together. Is this a stability cost? Do they repel each other? As it turns out, the answer is no. The electrons sandwiched between them "shield" the nuclei from each other, preventing them from feeling the repulsive force of a like charge. All in all, forming a bond is stabilizing for atoms. That's why it happens!

3.9 THE LAST WORD— LOVE AND OTHER CHEMICAL REACTIONS

The entire Unit 1 of this text has been focused on one main point: chemistry is everywhere. It's a much bigger part of our lives than we sometimes think it is. Still, though, as higher animals that think, feel, reason, and dream, surely not EVERYTHING that goes on inside us can be reduced to chemistry, right? Well, let's think about that.

It turns out that the science of attraction—why we are drawn to or develop crushes on certain individuals—has a lot of chemistry to it! Researchers have been trying for some time to determine whether humans have *pheromones*, as many other animals do. Pheromones are chemical signals that help animals to communicate with each other. They can indicate pregnancy, availability to mate, health status, and so forth. These chemicals are not detected with the nose in most animals, but are instead sensed by a specialized organ called the *vomeronasal organ*. While the presence of a vomeronasal organ in humans has been debated, there is now some evidence for its existence, and considerable evidence that we use our regular sense of smell to make decisions chemically. For instance, there are several studies that show we use sense of smell to help us select the partner with whom we are most likely to have healthy children! How does this work?

It's been known for quite some time that one function of the immune system (responsible for protecting us from infection) is to produce proteins called *major histocompatibility complexes (MHCs)*. These proteins are located on the surface of every body cell and help that cell recognize threats from bacterial and viral invaders. The ability to recognize a wide variety of potential invaders leads to the ability to fight off infection, and is therefore directly linked to survival. Because of this, individuals are healthiest and most likely to survive if they have the greatest possible variety of MHC proteins. Since the genes for these proteins are passed from parent to child, the healthiest child will result from parents who have very different MHC genes. Because MHCs are microscopic proteins on the surface of cells, though, it's impossible to look at another person and tell what their MHC profile is. How, then, do we make certain we find a partner with whom we will have the healthiest possible children?

Researchers have recently determined two interesting bits of information that help answer this question. First, it appears that our MHC profile is reflected to some extent in our natural body odor—people with similar MHC profiles smell alike. Second, we instinctively respond to another person's scent, and research has shown we are most attracted to individuals who smell differently than we do! To determine this, researchers asked a group of men to wear the same t-shirt to bed for several nights in a row (they were asked to use no scented hygiene products during this time, to ensure natural body odor would be the only thing on the shirt). A group of women was then brought in. Each woman was asked to smell several shirts: three from men who had been genetically tested and determined to have similar MHC profiles to hers, and three from men whose MHC profiles were different. Women strongly preferred the shirts worn by men whose MHC profiles were different from their own. Turns out, attraction is chemical!

OK, so initial attraction involves chemicals, but what about love? Is this complex emotion really just chemistry at work? As a matter of fact, a lot of what we think of as LOVE really is chemistry—everything from the "butterflies in the stomach" feeling of a great first date to the deep sense of contentment and security we feel when cuddling with a loved one. The "butterflies" feeling comes primarily from dopamine, a chemical released in the pleasure centers of the brain as a sort of "reward" molecule. Dopamine reacts with neurons in these pleasure centers to produce feelings of excitement, focused attention,

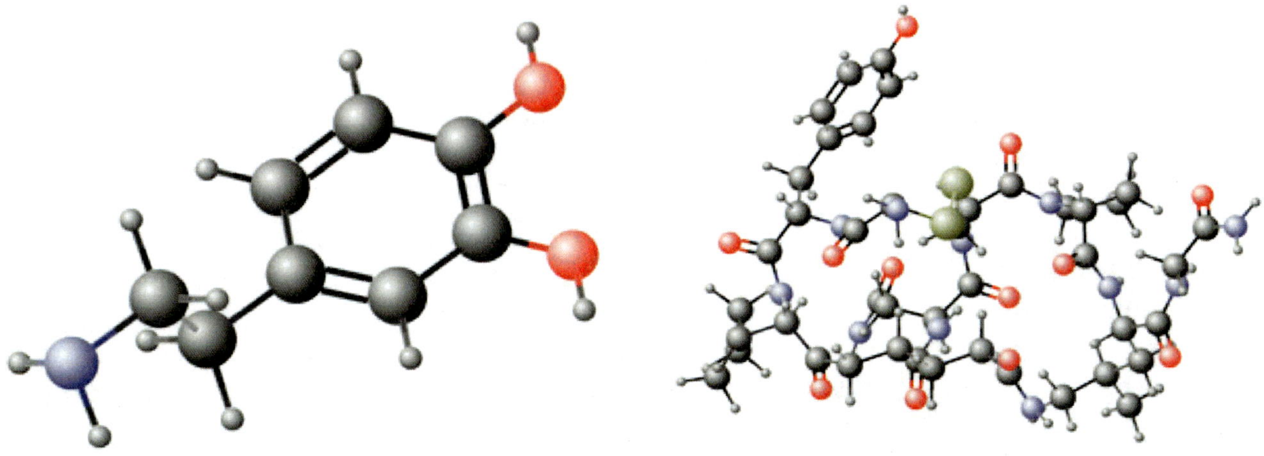

Image 3.14. Dopamine (left) and oxytocin (right). The two greenish atoms in oxytocin are sulfur.

and intense energy. As a result, a person who is falling in love feels great, thinks about their sweetheart constantly, and can easily stay up all night on a date or on the phone.

As we fall more deeply in love and begin the process of bonding emotionally with a partner, a different chemical comes into play. This one, oxytocin, is produced and released in response to cuddling and physical affection, and results in a feeling of closeness and love. It's also the chemical, incidentally, that floods through the brain of a new mother, allowing her to fall instantly and completely in love with her baby! Knowing the chemistry behind these emotions doesn't make them any less magical or real, but it's interesting to know where they come from. Furthermore, next time you refer to a great romantic connection by saying, "we've got good chemistry," you'll know you're absolutely right—literally!

SUMMARY OF MAIN POINTS

- Living organisms are composed primarily of the elements C, H, O, and N.
- Electrons are organized into shells around the nucleus; the shell closest to the nucleus is Shell 1.
- Radii of shells increase with shell number; electrons in Shell 1 are most stable.
- Shells fill with electrons from the inside out; the number of electrons in a shell is the same as the number of atoms in the corresponding row in the periodic table.
- Most shells are (or act) full with eight electrons.
- Only electrons in the outermost shell are available for chemical reactions; these are the valence electrons.
- Elements with the same number of valence electrons (same periodic group) react similarly.
- Atoms react chemically to fill their valence—this is the octet rule.
- Lewis dot structures portray just an atom's valence electrons.
- Atoms form covalent bonds to fill their octets.
- Lewis dot structures of molecules show shared electron pairs.
- The major elements of life are nonmetals because of their ability to form strong, covalent bonds with each other.
- Electrons in bonds and nonbonding pairs spread out as much as possible in three-dimensional space—this results in the characteristic shapes of molecules.
- Shapes of molecules can be portrayed using perspective bonds (wedge-dash).
- The backbone element of life is carbon because it has the greatest variety of bond shapes and structures available to it.
- Bond formation is stabilizing to the atom in may ways—it fills both atoms' valences, it stabilizes electrons with additional nuclei, and nuclei are shielded from one another by electrons.

Questions and Topics for Discussion

1. Given that most of the elements on the periodic table don't show up in living cells to any significant degree, where on Earth are they found? Pick three elements, do a little Internet research and find out where they are commonly found.
2. C, H, O, and N form the majority of the large biomolecules in a living organism. However, there are many trace elements found in any cell. Pick three of these trace elements, do a little Internet research and find out where they're found in the cell or body, and what they're used for.
3. How is it possible that the percent of hydrogen in the human body is much higher than the percent of carbon by number of atoms, but by mass, the percent of carbon is much higher than that of hydrogen? Explain.
4. Explain what the octet rule means.
5. Why do atoms in the same group on the periodic table act similar chemically to one another?
6. Explain what it means to say that atoms react chemically to achieve full valence shells.

7. Why does it make sense to draw Lewis dot structures of atoms with the electrons spaced out around four sides of an imaginary box?
8. When a covalent bond forms, atoms add additional electrons to their valence. However, their charges do not become unbalanced. Why not?
9. What holds the atoms together in a covalently bonded pair?
10. Based upon everything you've learned this chapter, why do you think hydrogen often appears in two different places on the periodic table? What about it is similar to the Group 1 metals? What about it is similar to the Group 7 non-metals? Which is it more like, in your opinion?
11. Explain why the major elements of life are nonmetals, and why of these, carbon is the "backbone" element.
12. Explain why even though there are more hydrogen atoms (by number) in a living cell than carbon atoms, we still say carbon is the "backbone" element of life.
13. Why do electron groups (bonds and nonbonding pairs) spread out as much as possible in space?
14. Imagine making molecular models using balloons. If you were to tie two balloons together, what sort of "bond angle" would they form? Do they bunch up, or spread out?
15. What would happen to the shape of the balloon model in question 13 if you tied three balloons together? Four? What does this tell us about molecular shapes and where they come from?
16. Draw a perspective (wedge-dash) NH_3 so that it looks like the N is pointed up, resting on a tripod of the three H's. Now mentally turn it 180° along a vertical axis, and redraw. Now flip it over (upside down) and redraw. Remember that EVERY molecule is a three-dimensional object that can be turned and flipped! We can draw these from any perspective we like.

Problems

1. List the four major elements of life.
2. List the trace elements of life.
3. Which element is represented in the highest percentage (by mass) in the human body INCLUDING water?
4. Which element is represented in the highest percentage (by number of atoms) in the human body INCLUDING water?
5. Which element is represented in the highest percentage (by number of atoms) in the human body EXCLUDING water?
6. Which element is represented in the highest percentage (by mass) in the human body EXCLUDING water? (Hint: you'll need to use the data for number of atoms, and then take mass into account).
7. What information about an atom is provided by knowing in what row of the periodic table the element is found?
8. How many electrons does shell 1 hold? Shell 2? Shell 3?
9. Using the periodic table as your guide and without drawing electron configuration diagrams, in what shell are the outermost electrons in each of the following: Sr; As; Xe; Na; B; and He.
10. Using the periodic table as your guide and without drawing electron configuration diagrams, in what shell are the outermost electrons in each of the following: Ba; I; Rn; Ar; Al; and Si.
11. Draw electron configurations for the Row 3 elements.
12. Draw electron configurations for the Row 4 elements (do not include transition metals).
13. Draw electron configurations for the first three Group 2 elements. What do they all have in common?
14. Draw electron configurations for the first three Group 7 elements. What do they all have in common?

15. What information about an atom is provided by knowing what in what column of the periodic table the element is found?
16. List elements from the periodic table that Li is most chemically similar to.
17. List elements from the periodic table that N is most chemically similar to.
18. Draw Lewis dot structures for the Row 3 elements.
19. Draw Lewis dot structures for the Row 4 elements (do not include transition metals).
20. Draw Lewis dot structures for the first three Group 2 elements. What do they all have in common?
21. Draw Lewis dot structures for the first three Group 7 elements. What do they all have in common?
22. Draw Lewis structures for each of the following molecules: Br_2; H_2O_2; HF; H_2S; and O_3.
23. Draw Lewis structures for each of the following molecules: I_2; PCl_3; CH_2Cl_2 (put the C in the middle); CO; and PH_3.
24. A single bond is considered an electron "group" for purposes of determining molecular shape. What are the other electron "groups"?
25. What general shape is formed by four electron groups around a central atom?
26. What general shape is formed by three electron groups around a central atom?
27. What general shape is formed by two electron groups around a central atom?
28. What bond angles are found in the shape formed by four electron groups around a central atom?
29. What bond angles are found in the shape formed by four electron groups, one of which is a lone pair of electrons, around a central atom?
30. What bond angles are found in the shape formed by four electron groups, two of which are lone pairs of electrons, around a central atom?
31. What bond angles are found in the shape formed by three electron groups around a central atom?
32. What bond angles are found in the shape formed by three electron groups, one of which is a lone pair of electrons, around a central atom?
33. What bond angles are found in the shape formed by two electron groups around a central atom?
34. Draw each of the following molecules in perspective, showing shape: PH_3; $COCl_2$ (put C in the middle with a double bond to O); H_2S; SO_2; and CO_2.
35. Draw each of the following molecules in perspective, showing shape: CH_2Cl_2; O_3; CCl_4; $COBr_2$ (put C in the middle with a double bond to O); and PCl_3.
36. Provide expected bond angles for each of the following: PH_3; $COCl_2$; H_2S; SO_2; and CO_2.
37. Provide expected bond angles for each of the following: CH_2Cl_2; O_3; CCl_4; $COBr_2$; and PCl_3.
38. Compare the shapes of CO_2 and SO_2. Despite the similarity of the chemical formulas, the shapes are quite different. Why?

Answers to Problems

1. C, H, O, N.
3. Oxygen.
5. Hydrogen.
7. Row tells us in what shell the outermost electrons are found.
9. Sr – shell 5; As – shell 4; Xe – shell 5; Na – shell 3; B – shell 2; He – shell 1.
11.

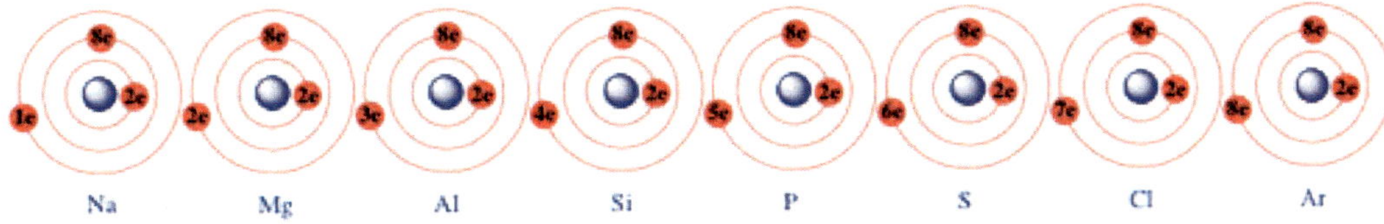

13. Each of the configurations has two electrons in the outer shell.

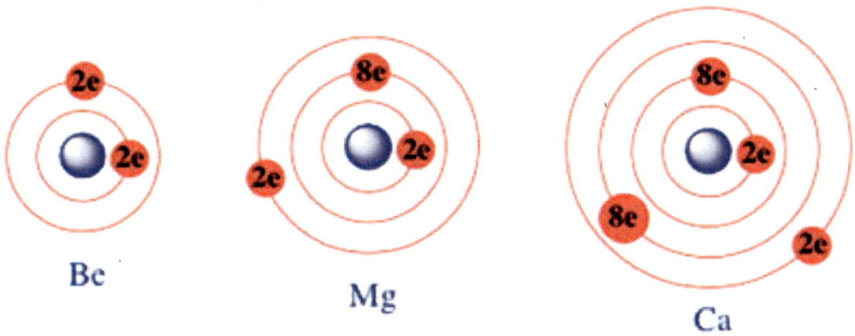

15. Column tells us how many valence electrons the atom has.
17. P, As, Sb, Bi.
19.

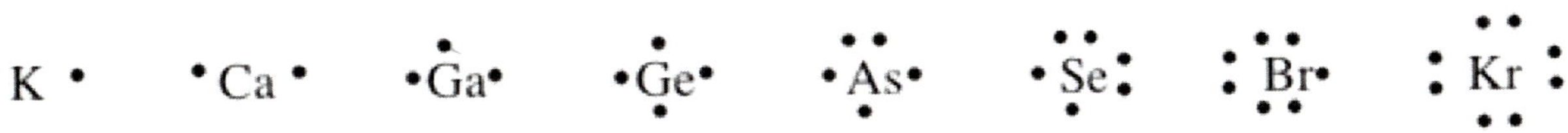

21. Each of the structures has seven valence electrons.

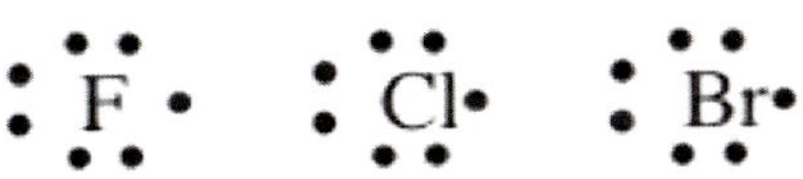

23.

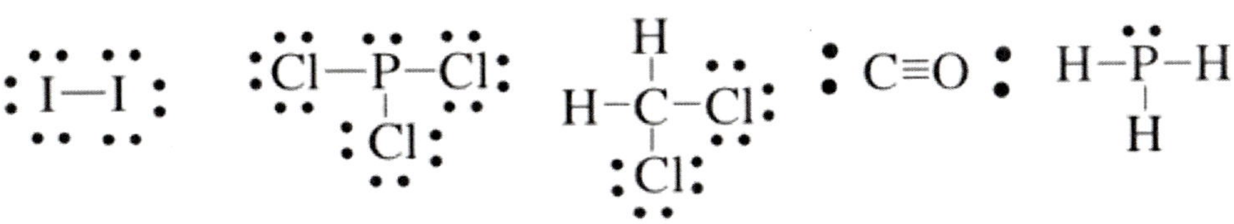

25. Tetrahedral.
27. Linear.
29. ~107°.
31. 120°.
33. 180°.
35.

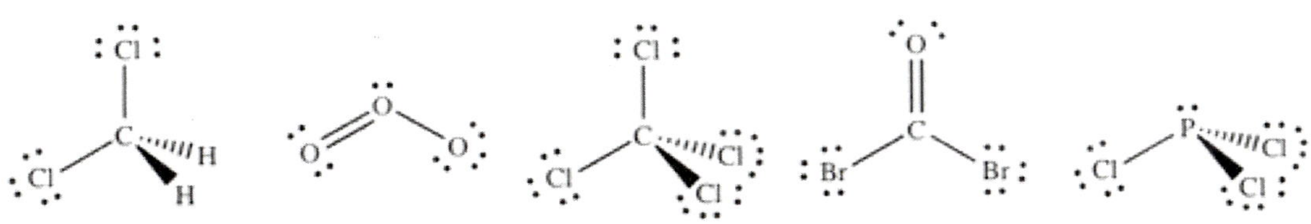

37. 109.5°, <120°, 109.5°, 120°, ~107°.

Sources and Further Reading

Anthropomorphic Molecules

Chanteau, S. H., Tour, J. M. (2003). Synthesis of Anthropomorphic Molecules: The NanoPutians. *Journal of Organic Chemistry,* 68(23), 8750–66.

Pheromones

Lawton, G. (1997). Change of mood about pheromones. *Chemistry and Industry,* 16, 635.

Wright, K. (1994). The sniff of legend; human pheromones? Chemical sex attractants? And a sixth sense organ in the nose? What are we, animals? *Discover,* 15(4), 60–67.

MHCs and Attraction

Herz, R. S. (2002). Sex differences in response to physical and social factors involved in human mate selection – The importance of smell for women. *Evolution and Human Behavior,* 23(5), 359–64.

Jacob, S., McClintock, M. K., Zelano, B., *et al.* (2002). Paternally inherited HLA alleles are associated with women's choice of male odor, *Nature Genetics,* 30, 175–79.

Santos, P. S. C., Schinemann, J. A., *et al.* (2005). New evidence that the MCH influences odor perception in humans: a study with 58 Southern Brazilian students. *Hormones and Behavior,* 47(4), 384–88.

Dopamine, Oxytocin, and Love

Bales, K. L., Mason, W. A., Catana, C., *et al.* (2007). Neural correlates of pair-bonding in a monogamous primate. *Brain Research,* 1184, 245–53.

Stein, D. J., Vythilingum, B. (2009). Love and attachment: the psychobiology of social bonding. *CNS Spectrums,* 14(5), 239–42.

Zeki, S. (2007). The neurobiology of love. *FEBS Letters*, 581(14), 2575–79.

Unit 2: Community Chemistry

$Mg(OH)_2 + 2HCl \rightarrow MgCl_2 + 2H_2O$

4 Water—The Solvent of Life

Image 4.1. A beautiful waterfall like this one in the Eagle Cap Wilderness is just one of the ways in which we experience the interesting and unique chemical that is water (Eagle Cap, Oregon © Scott Lefler, 2009).

If a chemist or biologist were asked what SINGLE chemical is most critical to life, the answer would be *water*. Water forms the internal environment of cells, it is the medium through which cells communicate with one another, it is combined chemically with carbon dioxide to produce new structural material as plants grow, and it serves as the environment for countless aquatic species. There is not a single known living thing that can function without water. In addition to its importance to life, water is also a very strange little molecule in many ways, both chemically and physically. Happily, it's water's very oddities that make it so uniquely suited to supporting life. An exploration of the nature of water will help us learn how and why it serves the function it does and will also help us understand some properties of the chemicals that dissolve in and interact with water.

4.1 SOLUTIONS AND SOLUBILITY

Water is sometimes (via a bit of hyperbole) referred to as the "universal solvent." **Solvents** are chemicals that can dissolve other chemicals, and water is good at dissolving many things (but not everything!).

When a solvent dissolves other chemicals (which are called **solutes**), the resulting mixture is a **solution**. For instance, in a solution of salt water, water is the solvent, and salt is the solute. Like all mixtures, solutions are physical combinations of chemicals, so neither the solute nor the solvent is chemically changed in the process of mixing. Further, like all mixtures, solutions can be made in varying proportions. The proportion of solute to a given volume of solution is called the solution's **concentration**. It's worth knowing that a solution made using water as the solvent is called an **aqueous solution** (aqueous comes from the Latin word for water). Due to water's ubiquity and its prodigious ability to function as a solvent, the most common kind of solution is aqueous.

Image 4.2. (Cartoon © Kirstin Hendrickson, 2010.)

We routinely encounter solutes that dissolve well in water—just to name a few, these include salt, sugar, many coloring and flavoring agents used in cooking, and liquids like vinegar and ethanol. Other things, however, do not dissolve well in water. Oil, for instance, floats on top of the aqueous vinegar solution in a bottle of salad dressing. Shaking the dressing temporarily combines the oil and vinegar-water, but they separate again in fairly short order.

Image 4.3. Dark drops of balsamic vinegar bead up and sink to the bottom of a bowl of olive oil.

Why is it that some molecules dissolve well in water and others do not? The answer to this question has to do with the chemical structures of the solutes and solvents.

4.2 DIPOLE MOMENTS

We learned in the last chapter that atoms form covalent bonds, which are electron-sharing interactions, to fill their octets. It turns out, though, that two bonded atoms don't always share electrons equally. In many cases, one atom "hogs" the electrons, pulling them closer to its nucleus and away from the other atom's nucleus. This electron-pulling behavior comes from an atomic property called **electronegativity**. An atom's electronegativity is a measure of its ability to hold its own electrons tightly; when a very electronegative atom forms a bond, it pulls the shared electrons tightly toward it as well. Electronegativity is a **periodic trend**, meaning that it can be predicted from looking at the periodic table. More electronegative atoms are found at the right end of a row, and at the top of a column. Remembering that hydrogen *technically* belongs

Electronegativity increases toward the right

Electronegativity increases toward the top

1 H 1.01 Hydrogen																1 H 1.01 Hydrogen	2 He 4.00 Helium
3 Li 6.94 Lithium	4 Be 9.01 Beryllium											5 B 10.81 Boron	6 C 12.01 Carbon	7 N 14.01 Nitrogen	8 O 16.00 Oxygen	9 F 19.00 Fluorine	10 Ne 20.18 Neon
11 Na 23.0 Sodium	12 Mg 24.3 Magnesium											13 Al 26.98 Aluminum	14 Si 28.09 silicon	15 P 30.97 Phosphorus	16 S 32.07 Sulfur	17 Cl 35.45 Chlorine	18 Ar 39.95 Argon
19 K 39.10 Potassium	20 Ca 40.08 Calcium	21 Sc 44.96 Scandium	22 Ti 47.88 Titanium	23 V 50.94 Vanadium	24 Cr 52.00 Chromium	25 Mn 54.94 Manganese	26 Fe 55.58 Iron	27 Co 58.93 Cobalt	28 Ni 58.69 Nickel	29 Cu 63.55 Copper	30 Zn 65.39 Zinc	31 Ga 69.72 Gallium	32 Ge 72.59 Germanium	33 As 74.92 Arsenic	34 Se 78.96 Selenium	35 Br 79.90 Bromine	36 Kr 83.80 Krypton
37 Rb 85.47 Rubidium	38 Sr 87.62 Strontium	39 Y 88.91 Yttrium	40 Zr 91.22 Zirconium	41 Nb 92.91 Niobium	42 Mo 95.94 Molybdenum	43 Tc (98) Technetium	44 Ru 101.1 Ruthenium	45 Rh 102.9 Rhodium	46 Pd 106.4 Palladium	47 Ag 107.9 Silver	48 Cd 112.4 Cadmium	49 In 114.8 Indium	50 Sn 118.7 Tin	51 Sb 121.8 Antimony	52 Te 127.6 Tellurium	53 I 126.9 Iodine	54 Xe 131.3 Xenon
55 Cs 132.9 Cesium	56 Ba 137.9 Barium	57 La *138.9 Lanthanum	72 Hf 178.5 Hafnium	73 Ta 180.9 Tantalum	74 W 183.9 Tungsten	75 Re 186.2 Rhenium	76 Os 190.2 Osmium	77 Ir 192.2 Iridium	78 Pt 195.1 Platinum	79 Au 197.0 Gold	80 Hg 200.5 Mercury	81 Tl 204.4 Thalium	82 Pb 207.2 Lead	83 Bi 209.0 Bismuth	84 Po (210) Polonium	85 At (210) Astatine	86 Rn (222) Radon
87 Fr (223) Francium	88 Ra (226) Radium	89 Ac *(227) Actinium	104 Rf (257) Rutherfordium	105 Db (260) Dubnium	106 Sg (263) Seaborgium	107 Bh (262) Bohrium	108 Hs (265) Hassium	109 Mt (266) Meitnerium	110 Ds (271) Darmstadtium								

58 Ce 140.1 Cerium	59 Pr 140.9 Praseodymium	60 Nd 144.2 Neodymium	61 Pm (147) Promethium	62 Sm 150.4 Samarium	63 Eu 152.0 Europium	64 Gd 157.3 Gadolinium	65 Tb 158.9 Terbium	66 Dy 162.5 Dyspromium	67 Ho 164.9 Holmium	68 Er 167.3 Erbium	69 Tm 168.9 Thulium	70 Yb 173.0 Ytterbium	71 Lu 175.0 Lutetium
90 Th 232.0 Thorium	91 Pa (231) Protactinium	92 U (238) Uranium	93 Np (237) Neptunium	94 Pu (242) Plutonium	95 Am (243) Americium	96 Cm (247) Curium	97 Bk (247) Berkelium	98 Cf (249) Californium	99 Es (254) Einsteinium	100 Fm (253) Fermium	101 Md (256) Mendelevium	102 No (254) Nobelium	103 Lr (257) Lawrencium

Image 4.4. The three most electronegative elements are F, O, and N.

on the left side of the table in group 1, and that the group 8 elements are not really relevant to a discussion of electronegativity, since they don't generally form bonds, fluorine emerges as the element furthest to the right and closest to the top of the table—fluorine is the most electronegative element!

While a periodic table can generally be used to determine which of a pair of atoms is more electronegative, it doesn't help us if the pair is situated diagonally with one closer to the top, and one closer to the right. For cases like this, there's a mnemonic device that is useful in remembering how the most common elements rank with regard to electronegativity.* In order of decreasing electronegativity, the common elements fall into the order:

FONClBrISCH—pronounce as a word: "foncliberish"

Even though "foncliberish" isn't a real word, if you can remember it, you can use it to quickly determine which of two elements is more electronegative in difficult cases. Take, for instance, nitrogen and sulfur. Nitrogen is closer to the top of the periodic table, but sulfur is further to the right. The mnemonic helps us remember that nitrogen is more electronegative. It also reminds us of the important fact that fluorine, oxygen, and nitrogen are the three most electronegative elements. Mnemonic aside, it's also worth knowing that carbon and hydrogen are exceedingly similar to one another—almost identical—in electronegativity. There is unequal electron distribution (carbon pulls electrons), but it is extremely minor.

***NOTE**

The word *mnemonic* comes from the Greek word for memory. A mnemonic device is a trick that makes things easy to remember.

TRY THIS:

Pick the element from each pair that is more electronegative: I and Cl; C and H; C and O; H and O; S and C.

Answers: Cl, C, O, O, S.

What does all of this have to do with bonding? When two atoms of the same element (such as two hydrogen atoms) bond to one another, they each pull electrons in the bond toward their own nuclei. However, since two atoms of the same element have the same electronegativity, they'd pull electrons with equal strength. Just as in a game of tug-of-war, the electrons wouldn't go anywhere:

Conversely, when two atoms of different elements with different electronegativities (such as hydrogen and fluorine) bond to one another, the more electronegative atom pulls the electrons in the bond toward its nucleus—AWAY from the nucleus of the less electronegative atom. The electrons are still being shared, but they're not exactly halfway between the nuclei. Instead, they're being "hogged" by one of the atoms:

While shared electron pairs are always called covalent bonds (regardless of whether one atom is pulling the electrons with greater strength), if there is an unequal distribution of electrons, we say the bond is **polarized**. We can also call a bond between atoms of different electronegativities, like the bond between hydrogen and fluorine, a **polar covalent bond**.

Concept Check:

Why would an atom of nitrogen pull electrons toward it when bonded to carbon, but not when bonded to oxygen?

Answer: Nitrogen is more electronegative than carbon, but less electronegative than oxygen. The more electronegative atom always pulls electrons toward itself.

Another ramification is that since fluorine is not only holding on to its own shared electron, but hogging hydrogen's shared electron as well, fluorine's atomic charges become slightly off balance. From the periodic table, we know that fluorine has nine protons, or a charge of +9 in its nucleus. It also has nine electrons, contributing a charge of -9, and making the atom neutral overall. Bonded to hydrogen, though, it's as though fluorine has partially acquired a tenth electron, since it's pulling BOTH shared electrons closer to its own nucleus. This creates a slight charge imbalance on the fluorine atom, giving it a **partial negative charge**. On the other end of the spectrum, the polarized bond affects hydrogen as well. Hydrogen has one proton, for a nuclear charge of +1. It also has one electron, and is normally charge-neutral. However, with fluorine partially taking hydrogen's electron, hydrogen also becomes slightly charge imbalanced, and has a **partial positive charge**. We can indicate these charge imbalances in two ways. A **bond dipole** arrow (a dipole is a separation of charge) can be used to indicate the direction of polarization. The dipole arrow has a plus sign at the less electronegative atom, and points in the direction in which the electrons are being pulled—toward the more electronegative atom:

A nice mnemonic for remembering which way dipole arrows are oriented is the short rhyme *arrows show where electrons go*. Alternately, atoms can be marked with a plus or minus sign preceded by the Greek letter *delta* (δ), which is taken to mean *partial charge*:

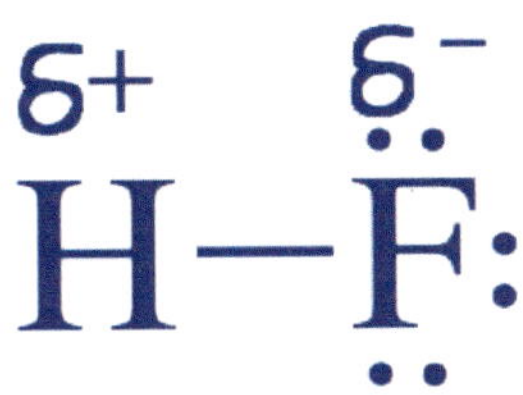

Remember that ANY time two atoms of different elements are bonded together, they'll have different electronegativities, and their bond will be polarized. While the electrons in the bond still contribute to both atoms' valences, the atoms will be partially charged instead of being fully neutral.

In a molecule consisting of a single polar covalent bond like HF, the bond dipole also represents the **net dipole**, or the molecule's overall polarization. For a molecule consisting of more than one polar covalent bond, the bond dipoles must be summed to determine the net dipole. This process requires taking the bond dipoles AND the molecular shape into account. For instance, the molecule CO_2 has two polar covalent bonds, with dipoles that completely oppose each other:

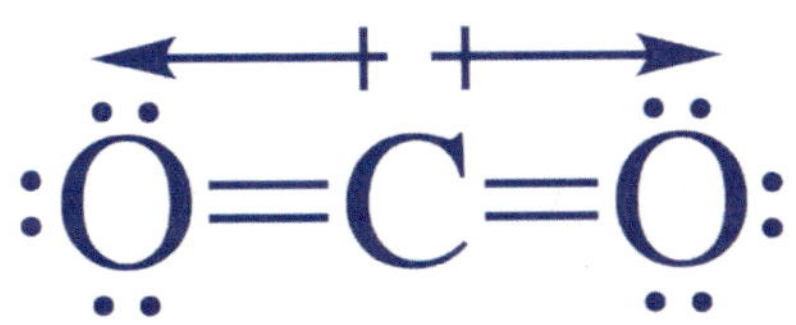

As a result of the molecule's linear shape, the bond dipoles completely cancel one another out on a molecular level, and CO_2 has no net dipole. In other words, CO_2 is completely symmetrical with regard to its charge distribution. Regardless of symmetry, CO_2 still has bond dipoles, so it still has partial charges:

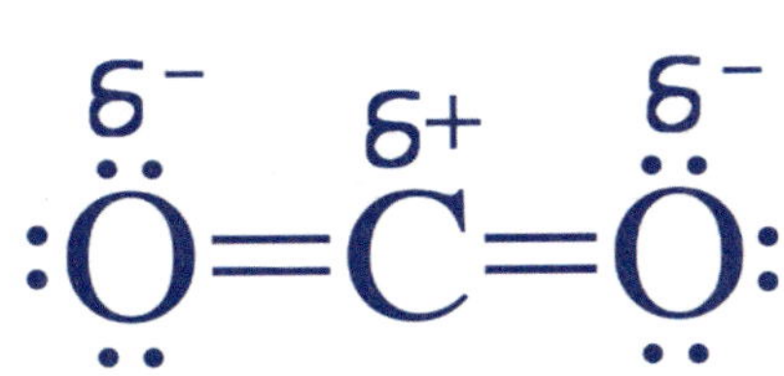

A molecule like CO_2 that has no net dipole is called a **nonpolar** molecule. Some molecules, however, are asymmetrical with regard to their charge distribution. This means the dipoles don't completely cancel each other out, but instead sum to form a net dipole. These are called **polar** molecules. HF, because it has only one bond dipole (which therefore IS its net dipole), is polar. H_2O is also polar, because while it has two bond dipoles, they do not oppose one another completely:

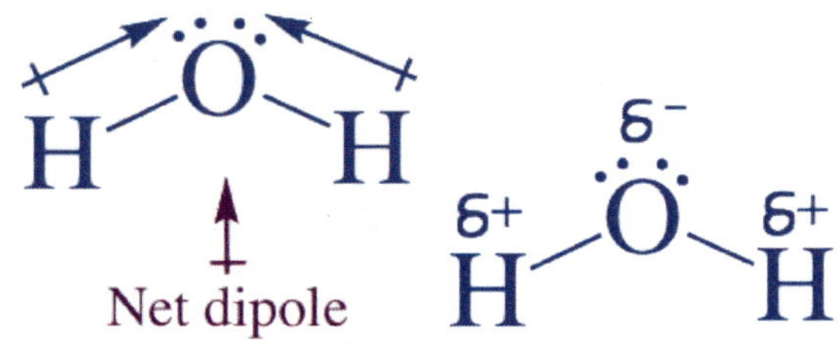

In the image above, the two O–H dipoles oppose each other left and right, but both point diagonally upward. Notice that to sum dipoles correctly, both bonds AND shape must be taken into account—we have to know that water has a bent shape to know that its dipoles don't completely oppose each other like CO_2's do. As a result of the orientation of water's bond dipoles, they sum to produce a net dipole toward the oxygen atom. It's tempting to say there's a net dipole *upward*, but remember that water (like all molecules) is a three-dimensional object that can be turned around and flipped over. If we were to draw water with the oxygen toward the bottom of the page and the hydrogen atoms toward the top, the dipole would point down. Regardless of the orientation, however, the dipole ALWAYS points toward oxygen. A net dipole indicates that with regard to the molecule as a whole, electrons are clustered at one particular end or side—in this case, the side with the oxygen.

TRY THIS:

Draw bond and net dipoles, and indicate partial charge, for: HCl, F_2, SO_2.

Answers:

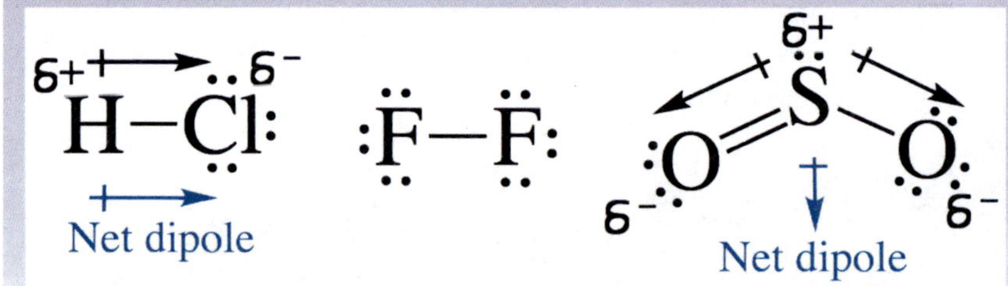

Concept Check:

Why do the dipoles cancel each other out in CO_2, but not in H_2O?

Answer: The shape of CO_2 means the dipoles completely oppose each other, but in H_2O they do not.

When we start talking about molecules with tetrahedral geometry, determining net dipoles requires some three-dimensional visualization. For example, the molecule CCl_4 has four bond dipoles:

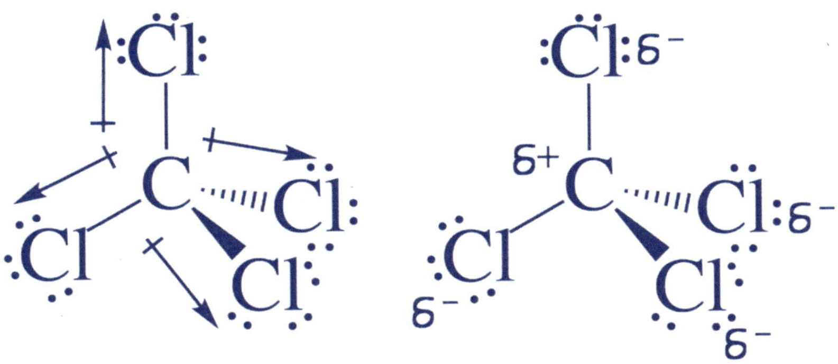

In a two-dimensional image, it's very difficult to picture the sum of the dipoles. However, if we remember that the perspective drawing is nothing more than a way to portray a three-dimensional object, it gets a little easier to imagine what's going on. In three dimensions, our molecule occupies a tetrahedral space, with each chlorine atom at the corner of a four-sided pyramid:

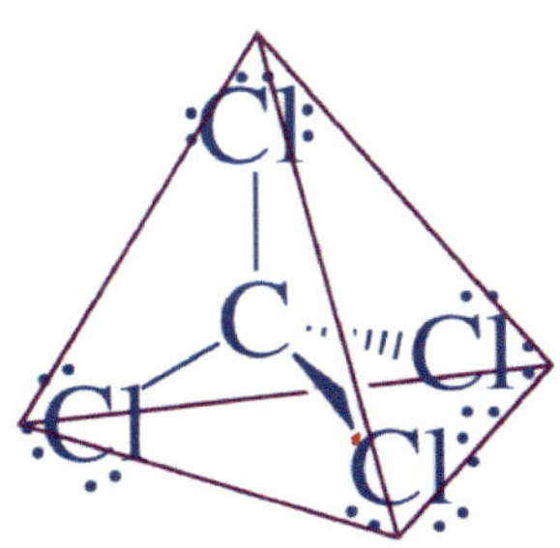

As such, each bond dipole actually points out symmetrically from the center of the molecule. This means the dipoles cancel each other out completely, and CCl_4 is nonpolar.

To make polarity relevant to our discussion of solubility, we have to remember that charge attracts opposite charge. As a result, the partially negative end or side of one polar molecule is attracted to the partially positive end or side of another polar molecule. For this reason, polar compounds tend to be attracted to, and therefore dissolve in, other polar compounds. Nonpolar compounds, lacking regions of partial charge, are not similarly attracted to polar compounds and therefore do not tend to dissolve in them. Fats, for instance, which consist primarily of carbon and hydrogen (the C-H dipole is so small as to be virtually nonexistent on account of the atoms' similar electronegativities), do not dissolve well in water, which is extremely polar. Ammonia (NH_3), however, is very polar, and dissolves well in water. A maxim by which to remember this principle is *like dissolves like*, meaning that polar molecules dissolve other polar molecules. The maxim can also be used in reference to nonpolar molecules, which dissolve other nonpolar molecules. This attraction between nonpolar molecules is not because of electrostatics, since nonpolar molecules don't have areas of partial charge. Instead, it has to do with a number of factors, including their exclusion by water. Because nonpolar molecules don't dissolve well in water (their lack of partial charge means water is not at all attracted to them), the water molecules tend to exclude (push out) nonpolar molecules. The excluded molecules therefore cluster together, dissolving in and mixing with each other. This solution behavior is referred to as a **hydrophobic** effect (hydrophobic means water-fearing). While water doesn't dissolve nonpolar molecules, it's very adept at solvating anything polar, as we'll discuss further.

TRY THIS:

Draw bond and net dipoles, and indicate partial charge, for: NH_3, CH_2Cl_2, CH_4.

Answers:

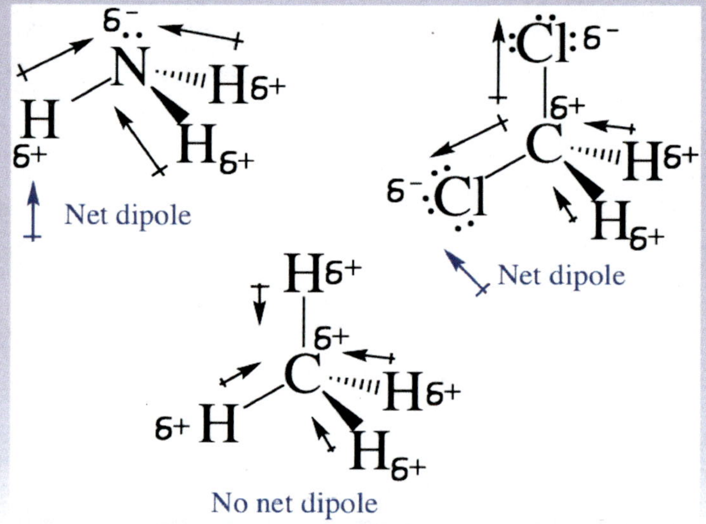

4.3 HYDROGEN BONDS

One of water's molecular oddities is that it has really BIG dipoles compared to most compounds. The most electronegative elements are fluorine, oxygen, and nitrogen. The least electronegative element that still forms covalent bonds is hydrogen.* As a result, the biggest bond dipoles (meaning they have the greatest polarization and largest resulting partial charges) are along F–H, O–H, and N–H bonds. The only compound with an F–H bond is HF, hydrofluoric acid. In practical terms then, the O–H and N–H bonds have the largest dipoles we find routinely in a variety of compounds. The bond dipoles are so significant that the partial charges are VERY large—they're almost complete charges—and the attraction between molecules is very great.♦ In fact, the electrostatic attractions are so strong as to be almost bond-like. Though they are not true bonds, these interactions are called **hydrogen bonds**:

> ***NOTE**
>
> Remember from Chapter 3 that the metals are not able to fill their valences by forming covalent bonds—they bond in a different way that will be discussed later in this chapter.

> **♦NOTE**
>
> How do we indicate larger versus smaller dipoles? The size of a dipole arrow is sometimes used as a visual cue in molecules with more than one type of bond dipole. The symbol δ, however, contains no quantitative information whatsoever. In general, we just keep track mentally—a larger difference in electronegativity means a bond dipole will be large.

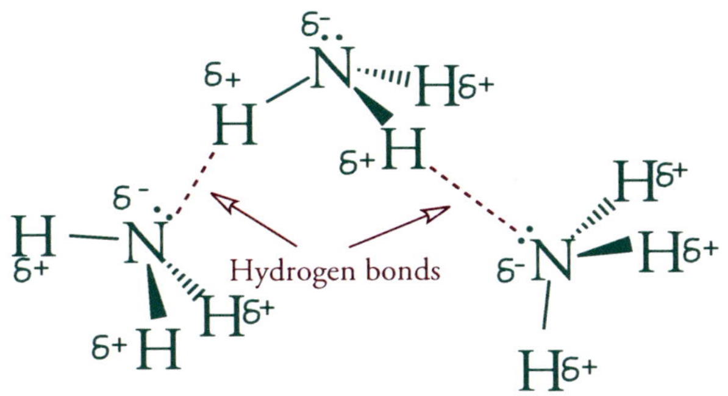

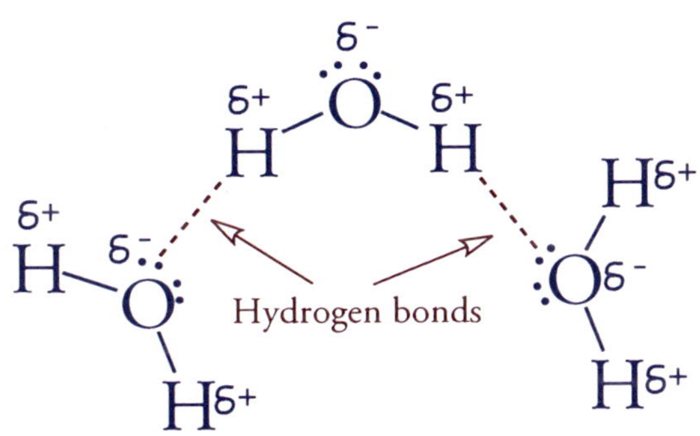

Note that the term "hydrogen bond" doesn't refer to the covalent bond between oxygen or nitrogen and hydrogen in a given molecule; instead, it refers to the attraction between the partially negative oxygen or nitrogen of one molecule and the partially positive hydrogen of another molecule. Additionally, while water and ammonia are two of the simplest molecules that form hydrogen bonds, ANY compound with one or more O–H or N–H bonds is capable of the same interaction.

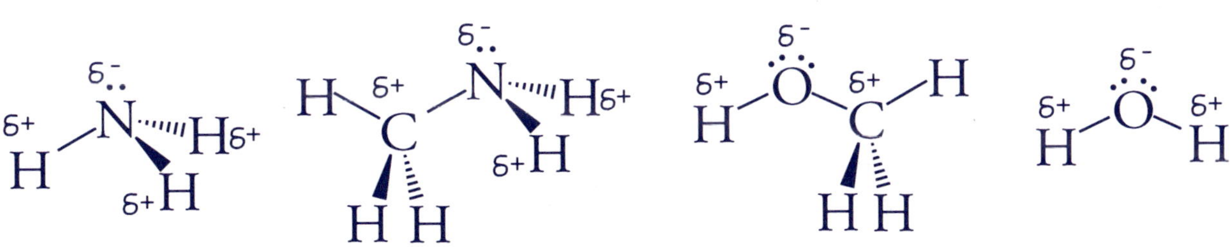

Image 4.5. A selection of compounds capable of hydrogen bonding.

In addition to molecules forming hydrogen bonds to other molecules of the same compound, it's also possible for hydrogen bonds to form between molecules of different compounds, as long as both are capable of hydrogen bonding. Water, for instance, forms hydrogen bonds with ammonia (and any other compound with O–H or N–H bonds):

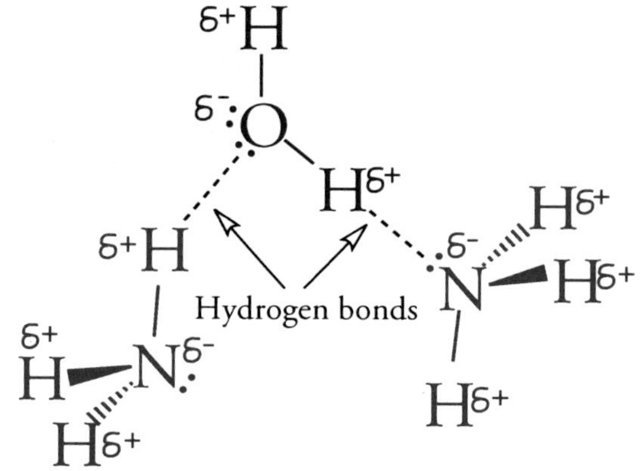

TRY THIS:

Which of the molecules shown is capable of hydrogen bonding?

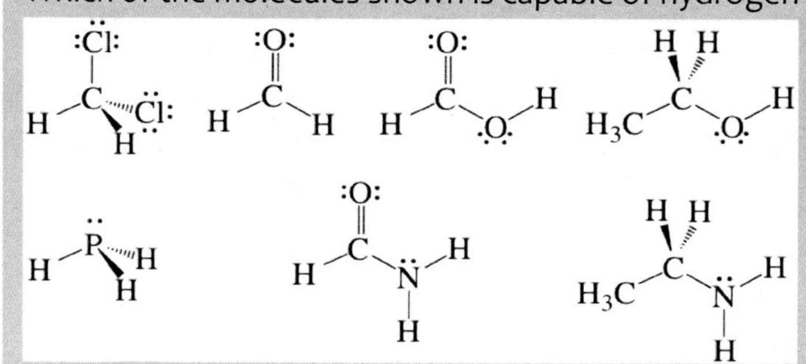

Answers:

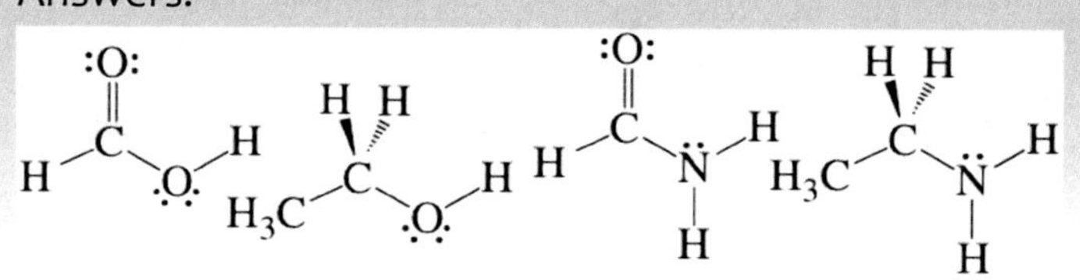

4.4 UNIQUE PROPERTIES OF WATER

Hydrogen bonding has several ramifications on the physical behavior of water. First, as already discussed, it is very good at dissolving polar compounds. We can now add that in particular, water is an excellent solvent for compounds capable of hydrogen bonding. Secondly, though, of all the hydrogen bonding compounds, water is the one with the strongest and most robust hydrogen bonding network: It's composed of nothing but O-H bonds, and O-H bonds produce slightly stronger hydrogen bonds than N-H bonds, since O and H are more disparate in electronegativity. As a result of water's extensive hydrogen bonding network, molecules of water are very strongly attracted to each other. This results in some interesting physical properties. To understand why, we need to think about the three phases of matter—**solid**, **liquid**, and **gas**. Solids form when particles that are very attracted to one another pack tightly with essentially no space between them. The lack of empty space means solids are **incompressible** (they can't be stuffed into a smaller volume), and the strong attractive forces between particles means they don't slide past each other, so they don't take the shape of their container. If particles are slightly less attracted to one another, a liquid forms. In this phase of matter, the particles are still quite close in space, so liquids are also incompressible. However, the weaker attractive forces allow particles to slide past one another, so liquids can take the shape of their container. Gases form when very weakly attracted particles spread out with a lot of empty space between them. They can therefore be compressed, and the particles are free to move around and take the shape of their container. The process of **melting** involves pulling densely packed particles a bit further from each other to make a liquid. This takes energy, which is why we have to heat a solid to melt it. **Boiling** separates those particles from one another even more; again, this represents a pulling apart of attracted particles, which requires energy. Why is energy required to separate particles? Think about water and other polar compounds—the molecules are attracted to one another because of opposite partial charges, so pulling the charges apart requires energy, much as it requires energy to pull apart two magnets. It also takes energy to separate nonpolar compounds, because it turns out they're also attracted to one another, though not electrostatically. The attractions between nonpolar molecules are called **van der Waalsforces**. While a complete discussion of the nature of van der Waals forces is beyond the scope of this text, it's sufficient to know that they are the weak attractions between nonpolar molecules and that they are stronger in larger molecules than they are in smaller ones.*

***NOTE**

The relationship between molecular size and strength of the van der Waals forces explains why methane (CH_4) and ethane (C_2H_6) are gases, while hexane (C_6H_{14}) and octane (C_8H_{18}) are liquids!

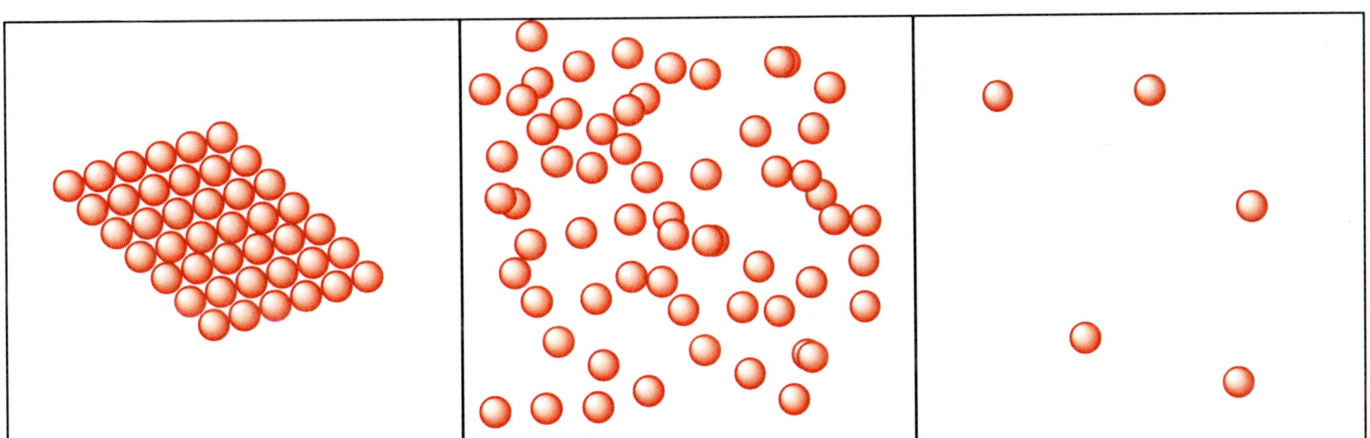

Image 4.6. Solids are densely packed (left), liquids have slightly more space between particles and more freedom of motion (center), and gases have a great deal of space between particles with complete freedom of motion (right).

As a result of attraction between water molecules, it takes a lot of energy to separate the molecules, because separation requires breaking the hydrogen bonds. As a result, water has a very high **melting point** and **boiling point** for its size. Note the vast difference in melting points and boiling points between water and the other similarly sized molecules, NH_3 and CH_2Cl_2:

Compound	Melting Point	Boiling Point
H_2O	0°C	100°C
NH_3	–77.7°C	–33.3°C
CH_2Cl_2	–96.7°C	40°C

Additionally, water is much more **dense** and **cohesive** than many other liquids due to its attraction to itself. Density is a measure of how much matter is packed into a given volume of space. Cohesiveness is a measure of the extent to which a chemical tends to stick to itself rather than spreading out on a surface.

CONCEPT CHECK:

Why does water's hydrogen bonding network contribute to its melting and boiling points, density, and cohesiveness?

Answer: Hydrogen bonding means molecules of water are very attracted to each other and pull each other into a smaller volume of space (density and cohesiveness). They also take lots of energy to break (MP and BP).

Further, water's extensive hydrogen bonding network makes it resistant to temperature change, a property very important to living creatures. Life depends upon cells maintaining a fairly stable temperature, and water's enhanced ability to maintain its temperature despite losing or absorbing small amounts of heat contribute to making it the perfect cellular medium. For a demonstration of water's ability to maintain its temperature, imagine placing a cup of cool water and a similarly sized block of cool metal on the sidewalk in full sunlight on a very hot July day. After an hour in the Sun, you can probably still put a finger in the water without getting burned, but you dare not touch the metal! Despite absorbing the same amount of the Sun's heat, the water's temperature has not changed nearly as much as that of the metal.

Image 4.7. Dewdrops demonstrate water's tendency to bead up due to its high cohesiveness.

Another of water's interesting properties is that, again due to its extensive hydrogen bond network and high cohesiveness, it's very lubricating. This helps reduce friction between cell membranes in multicellular organisms and protects delicate organ tissue from damage due to rubbing against other body organs.

Image 4.8. Solid water floats in a glass of water, soda, or tea. Icebergs float on the ocean.

Perhaps water's most interesting and unique property of all, however, is that unlike almost any other substance, solid water floats on top of liquid water. At first glance, this is unremarkable—we've all enjoyed a tall glass of iced tea on a hot day, and we know the ice floats.

The experience of floating ice is so common to us that we often don't stop to consider just how much it's at odds with what we know about states of matter and density. If solids are closely packed and liquids slightly less so with gases containing a lot of empty space, then a solid should be the most dense physical state of any given substance, and a gas the least dense. As a result, a solid should not float, but SINK in its own liquid! We don't generally observe this, since there are very few substances we have the opportunity to experience in multiple states. However, pure acetic acid (vinegar) solid sinks like a rock in liquid acetic acid.* The same goes for ammonia, ethanol, and essentially any other substance with which we've ever come in contact. So what's different about water that its solid is evidently less dense than its liquid? Again, the answer to this question has to do with hydrogen bonding. The hydrogen-bonding network formed by water when it freezes into ice takes the shape of a series of tetrahedrons, with an atom of oxygen at the center of each. This lattice is very stable—actually, it looks just like the structure of diamond, as we saw in Chapter 1—but it contains a lot of empty space compared to most solids. In order to maximize hydrogen bonding, the water molecules need to space themselves out a bit, which reduces the density of solid water.

> ***NOTE**
>
> Don't try this at home—the vinegar in your kitchen or bathroom is not pure vinegar. In fact, it's mostly water, so the experiment won't work at all. It's quite difficult to obtain pure acetic acid outside the lab.

By comparison, as solid water melts, the crystal lattice begins to break down. There is still a network of hydrogen bonds, but it's not nearly as extensive. As molecules rotate and slide past one another, the network becomes much less orderly than that of ice. As a result, the molecules aren't forced into crystal lattice positions and they actually move CLOSER together than in solid water. This makes water more dense as a liquid than as a solid. It's a good thing, too—in addition to the convenience of being able to float ice in a cold drink, this phenomenon also allows for turnover of lake water. In winter surface water cools, and as it becomes more dense,♦ it descends to the lake bottom. This carries fresh oxygen from the surface into the depths. Incidentally, the crystal lattice of solid water is also directly responsible for the beautiful and symmetrical patterns seen in snowflakes!

> **♦NOTE**
>
> Water is most dense at about 4°C.

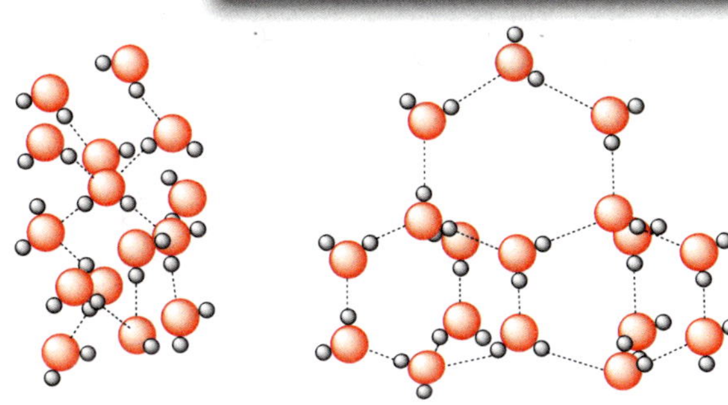

Image 4.9. The structure of liquid water (left) shows considerably closer packing of molecules than that of solid water (right). Notice how similar the crystal lattice is to that of diamond (Chapter 1).

Image 4.10. The amazing complexity and symmetry of snowflakes arises from the hydrogen bonding properties of water.

4.5 IONIC COMPOUNDS

Water's large bond dipoles clearly given it a number of interesting properties and explain its ability to dissolve other polar compounds. However, there's another class of compounds that dissolves readily in water—these are the **salts**. Salts form when there is such a large difference in electronegativity between two atoms that when they try to bond, the more electronegative atom actually completely takes the less electronegative atom's electron. The result isn't a covalent bond, since there are no shared electrons. Instead, the more electronegative atom now has an extra electron, and is negatively charged. A negatively charged atom is called an **anion**. The less electronegative atom is now missing an electron and is positively charged—this is a **cation**. Collectively, cations and anions are called **ions**. Interestingly enough, even though a covalent bond does not form via this electron transfer, the resulting cation and anion are attracted to one another because of their opposite charges. They therefore stick together, held not by a physical bond, but by an electrostatic attraction. We call this interaction between atoms an **ionic bond**. Consider a very electronegative element like fluorine, trying to form a bond with a minimally electronegative element, like sodium. If they tried to share a pair of electrons (one contributed by sodium, one by fluorine), the dipole would be tremendous. It turns out that such a dipole is SO large as to pull the pair of electrons entirely into fluorine's electron cloud, away from sodium. The fluorine atom is now called fluoride—we add the suffix *–ide* to the name of an atom when it becomes an anion. Further, fluoride is electrostatically attracted to the sodium cation (names of atoms don't change when they become cations), and an ionic bond results, producing the compound *sodium fluoride*:

Fluorine literally pulls sodium's electron right out of the electron cloud

The resulting ions are attracted to one another

The ionic compound NaF results—the bond is not physical (covalent), but electrostatic (ionic)

Note that while we can use Lewis dot structures to portray ions just as we can use them for neutral atoms, there's an important difference. When an electron is lost from sodium's outer shell of electrons—shell 3—the new structure for sodium continues to depict the ORIGINAL valence shell, which is now empty. Charges for ions are always included in Lewis dot structures.

When will two atoms form an ionic rather than a covalent bond? As a rule of thumb, ionic bonds will form when metals (which have very low electronegativities) react with non-metals. We saw in Chapter 3 that metals do not form covalent compounds—we now know that they do form ionic compounds. Don't forget that hydrogen, despite its position on the far left of the periodic table, has an electronegativity virtually identical to carbon's and can form covalent bonds.

Let's spend a little more time on the ionization reaction that takes place between sodium and fluorine, because it explains more than simply how metals can be incorporated into compounds. Ionization introduces a new strategy whereby an atom can obtain a complete valence shell of electrons. We discussed in Chapter 3 that we don't expect to see group 1, 2, or 3 elements forming covalent bonds, because they don't have enough electrons to share. Sodium, for instance, would need to form seven bonds to fill its valence. With only a single electron to share, though, forming this many bonds is simply not possible. However, obtaining enough electrons to fill shell 3 is not the only way for sodium to achieve a full octet. The other possibility is to lose one electron, emptying shell 3. This would leave sodium with only its shell 1 and shell 2 electrons, and its new outermost shell (shell 2) would be full. As such, not only does ionization result naturally from the differences in electronegativity between sodium and fluorine, it also leaves them both with complete octets!

What happens if an atom has more than one electron that it needs to lose in order to empty its valence shell? Magnesium, for instance, has two valence electrons. If it were to try to bond to fluorine, the result would be ionization. This would complete fluorine's octet, and form the fluoride anion. However, magnesium would still have an electron left in its valence:

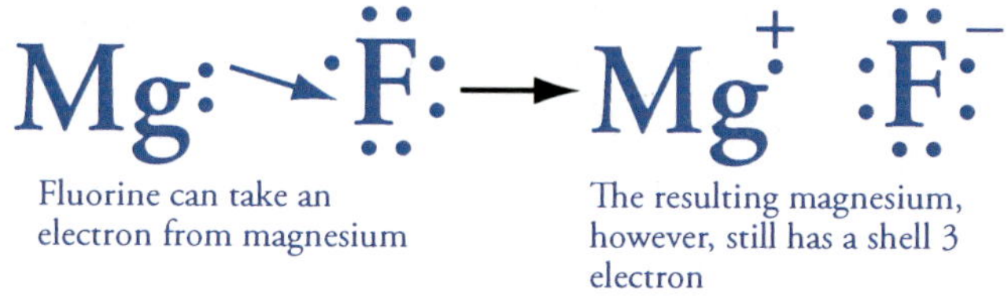

In order to lose that remaining electron, magnesium will ionize a second time. Because we know that atoms are typically found in sizeable samples, we can assume there's plenty of fluorine around. The magnesium will therefore lose a second electron to another, as yet unreacted, fluorine atom:

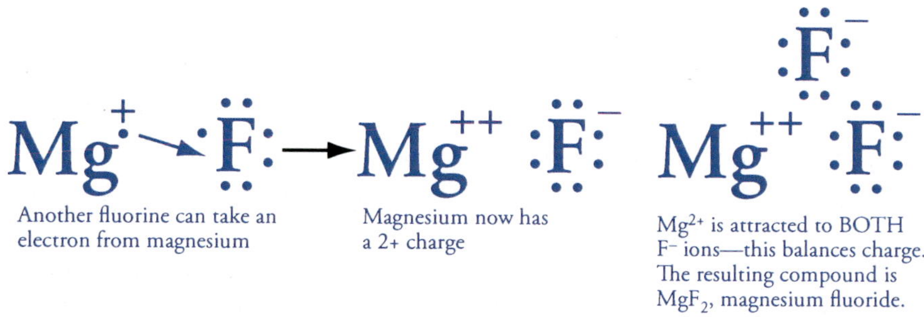

Using the naming conventions we learned in Chapter 1, the product is named without reference to the number of each species involved, so it is called *magnesium fluoride*. In any combination of a metal and a nonmetal, atoms continue to ionize as necessary until ALL atoms involved have gained or lost sufficient electrons to achieve full valence shells. Keeping track of valence mentally allows us to predict reactions without having to rely on Lewis structures. For instance, sodium needs to lose one electron to have a full shell and oxygen needs to gain two. As a result, two sodium atoms would need to react with one oxygen atom:

$$2\ \mathrm{Na} + \mathrm{O} \rightarrow 2\ \mathrm{Na}^{+} + \mathrm{O}^{2-} \rightarrow \mathrm{Na_2O}\ \text{(sodium oxide)}$$

We don't really need to include the intermediate step, showing the charges of the individual ions, but it can be helpful when learning to put together ionic compounds. It is perfectly acceptable to write this reaction in a more simplified form:

$$2\ \mathrm{Na} + \mathrm{O} \rightarrow \mathrm{Na_2O}$$

In another example, since magnesium has two electrons to lose, only one magnesium atom would be necessary to react with one oxygen atom:

$$\mathrm{Mg} + \mathrm{O} \rightarrow \mathrm{Mg}^{2+} + \mathrm{O}^{2-} \rightarrow \mathrm{MgO}\ \text{(magnesium oxide)}$$

Note that each of the compounds above is neutral overall: the charges sum to zero. This is true of all ionic compounds. As such, a quick and easy way of determining how many of the metal or nonmetal atoms are required to fill all octets is simply to balance the charges on the resulting ions. This technique is convenient simply because charges are so predictable: group 1 metals always have one valence electron, so when they ionize, they'll form +1 cations. Similarly, group 2 metals always have two valence electrons, so they'll ionize to form +2 cations. Group 3 elements form +3 cations, but it's difficult to build up that much charge in one place, so the only group 3 element we'll commonly see ionizing is aluminum. Group 4 elements don't ionize—it's simply too difficult to gain OR lose four electrons. Instead, these elements form covalent bonds. From group 5 on, the elements (all nonmetals) can either ionize OR form covalent bonds to fill their octets. Fluorine, for instance, can form one covalent bond (if bonding to another nonmetal), or it can take one electron from a metal and form the fluoride anion. Just as with the metals, though, the charges on these elements (if they ionize) are predictable. Group 5 elements have three valence holes, but don't often ionize for the same reason group 3 elements don't—that's an awful lot of charge. It's generally more favorable for group 5 elements to simply form three covalent bonds. Occasionally, however, we'll see N^{3-}, the nitride anion. Group 6 elements can ionize to form -2 anions, and group 7 elements, with only one valence hole, can ionize to form -1 anions.*

Let's use this information to try to predict the formula of an ionic compound made from aluminum and oxygen. To make our prediction, we need to approach the problem methodically:

1) Determine what the charges on each element will be after ionization.

Al is in group 3—it will form Al^{3+}. O is in group 6—it will form O^{2-}.

2) Determine how many of each ion is required to produce a neutral ionic compound. Mathematically, we're looking for the least common multiple of the charges.

The least common multiple of 2 and 3 is 6. This means we expect the total positive charge in the compound to be +6 (from Al^{3+}), and the total negative charge to be -6 (from O^{2-})—note that these will sum to zero, making our compound neutral. It takes 2 Al^{3+} and 3 O^{2-} to achieve these quantities, so our compound is Al_2O_3.

This shortcut technique gets us to the same answer at which we would have arrived had we used Lewis dot structures.

Another interesting bit of information to come out of our discussion of ionic compounds is that we now know why it's unnecessary to use Greek prefixes in the names of binary compounds of metals and nonmetals. Because the charges on ions determine how many of each will be present in the neutral compound, we don't need prefixes to help us predict the chemical formula.

TRY THIS:

Predict the charges on ions of each of the following (you'll need a periodic table):

Br, S, Rb, Ca.

Answers: Br-, S_2-, Rb+, Ca_2+.

***NOTE**

Unfortunately, it's exceedingly difficult to predict the charges on ionized transition metals. Like all metals, they form cations, but many can form more than one possible cation. Copper, for instance, can ionize to Cu^{+1} or Cu^{+2}. Though we will occasionally discuss these ions, charge will be given. Determining the charges on transition metal cations is far beyond the scope of this text.

TRY THIS:

Predict the ionic compounds formed by combinations of each of the following: K and F, K and S, Ca and F, Ca and S.

Answers: KF, K_2S, CaF_2, CaS.

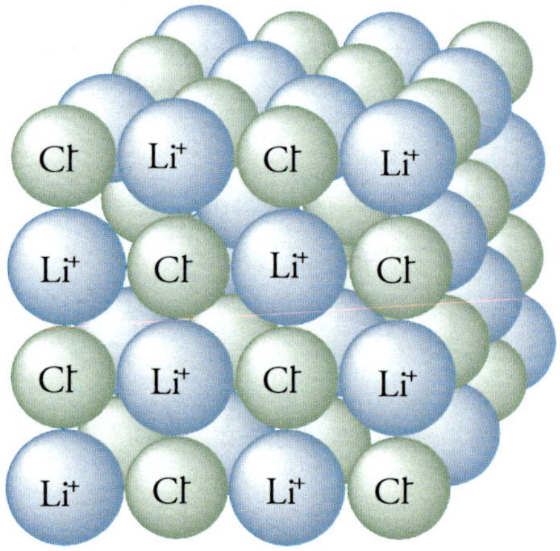

Image 4.11. LiCl is a network of lithium cations and chloride anions.

For example, lithium chloride is composed of lithium (a metal) and chlorine (a nonmetal). Because we're dealing with a metal and a nonmetal, we know this will be an ionic compound. Further, lithium is a group 1 metal, so it will form Li^+. Chlorine is a group 7 nonmetal, so it will form Cl^-. To balance charge, only one of each ion is necessary. The compound, then, is LiCl.

It's worth mentioning that it's not really accurate to envision a single "molecule" of an ionic compound. Rather, if a compound forms (as, say, in the case of LiCl), we can assume that there are MANY Li^+ and MANY Cl^- ions present. As such, each Li^+ cation is actually surrounded on all sides by Cl^- anions, and each Cl^- anion is surrounded on all sides by Li^+ cations. While we refer to "molecules" of a salt, in reality, the formula simply represents the simplest possible "unit" of that salt's composition. LiCl, then, is a shorter (and much more convenient) way of saying *a salt composed of alternating units of lithium cation and chloride anion, on a 1:1 ratio.*

TRY THIS:

Determine the chemical formula of each of the following: beryllium oxide, cesium oxide, lithium iodide, magnesium chloride.

Answers: BeO, Cs_2O, LiI, $MgCl_2$.

4.6 ELECTROLYTES

Given that polar compounds dissolve well in water and hydrogen bonding compounds (VERY polar compounds) dissolve well in water, you might guess that ionic compounds also dissolve well in water. You'd be right. Opposite charges attract, and the more charge there is, the greater the attraction. Thus, water (with its partial charges) is very attracted to ionic compounds, because of their full units of charge. In fact, water is SO attracted to ionic compounds that when stirred into water, the cations and anions in an ionic compound separate from one another and become surrounded by water molecules. This doesn't mean the compound is chemically reacting, nor does it mean the compound doesn't exist anymore. Remember, the nature of an ionic bond is different than that of a covalent bond. There's nothing physically holding the cation and anion together; they simply stick together under normal circumstances because of attraction. Water can also satisfy that attraction of charge. When we make salt water, for instance, we stir NaCl into H_2O. The Na^+ cations separate in space from the Cl^- anions, and both become surrounded by water molecules.

The reason NaCl "disappears" when we dissolve it in water is that the crystal lattice falls apart, and the ions spread out in solution. Why would the

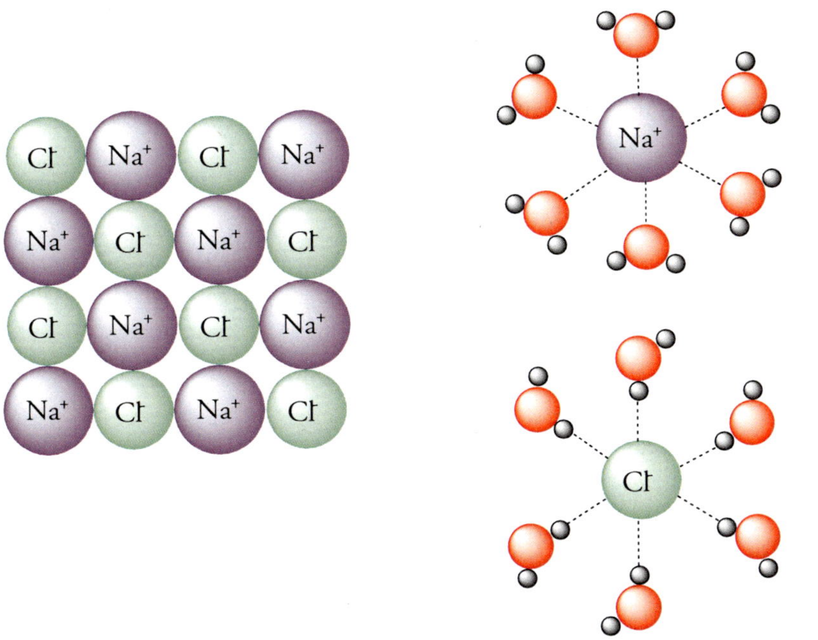

Image 4.12. Solid NaCl (left). In water, the sodium cation separates from the chloride anion. The cation becomes surrounded by the partially negatively charged oxygen atoms of water, and the anion becomes surrounded by the partially positively charged hydrogen atoms of water.

crystal lattice fall apart, allowing the sodium and chloride ions to separate from each other? It's important to understand that from an atomic perspective, there's no such thing as "better" or "worse" charge—Na^+ is electrostatically attracted to Cl^- when Cl^- is the most prevalent negatively charged substance around. In aqueous solution, however, there's far more water (which has partial negative charges on oxygen) than Cl^-, so Na^+ is more likely to become surrounded by water. By the same token, in aqueous solution, it's more likely that chloride will be drawn to the partially positive hydrogen atoms of water than to the sodium cation, by sheer number. However, if we were to boil off the water, the Na^+ and Cl^- would resume their ionic attraction to each other (since there would longer be any water to be attracted to), and solid NaCl would recrystallize.

The separation of cations and anions in aqueous solution is referred to as **dissociation.** Compounds that dissociate in water—ionic compounds—are called **electrolytes**, which are solutes that conduct **electricity** in aqueous solution. Electricity is a current of charged particles moving through a wire or other medium, so only materials with mobile charged particles can conduct electricity. Metals make good conductors because they hold their valence electrons so loosely that the electrons form something of a "communal electron cloud" around the entire sample of metal. For this reason, electrons can easily be pulled to one end or another of a sample of metal, and this movement of electrons is an electrical current. Similarly, ions in aqueous solution are free to move throughout the solvent, and can therefore transmit charge. **Molecular compounds** (those consisting of purely covalent bonds) do not contain ions, and therefore cannot dissociate in solution. This means solutions containing molecular compound solutes are not capable of transmitting electricity. Sucrose (table sugar, $C_{12}H_{22}O_{11}$), for instance, is composed entirely of nonmetals. It is therefore molecular, and while it dissolves well in water (it happens to

Image 4.13. You've probably heard that sports drinks contain electrolytes. This is really just a way of saying they contain salts. A sweating body loses lots of salts (especially NaCl), and these beverages replenish those salts in order to keep the body functioning normally.

have lots of O-H bonds), it does not dissociate. Sugar-water solutions are not electrolytic (we could also say sugar is a **nonelectrolyte**).

As an extension of this discussion, it turns out that pure water is NOT a good conductor of electricity—after all, it's molecular! Still, we all think of water as a good conductor—electrical appliances have labels warning us not to use them in the tub, for instance. What's going on? As it happens, water is SUCH a good solvent that, while it's theoretically possible to purify water completely (generally in the lab), it's simply not practical. Our tap water is far from pure—it contains lots of salts and other dissolved particles. Essentially, any water you run across is going to have many contaminants dissolved in it. For this reason, then, almost all water we encounter is actually an aqueous solution of many things—and is electrolytic.

TRY THIS:

Identify the electrolytes among the following: glucose ($C_6H_{12}O_6$), $CaCl_2$, LiF, NH_3.

Answers: $CaCl_2$, LiF. The others are compounds of nonmetals and are molecular.

4.7 WATER DISTRIBUTION AND SAFETY

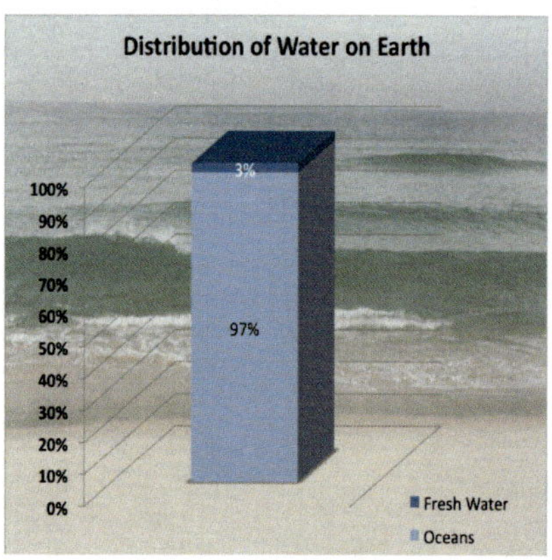

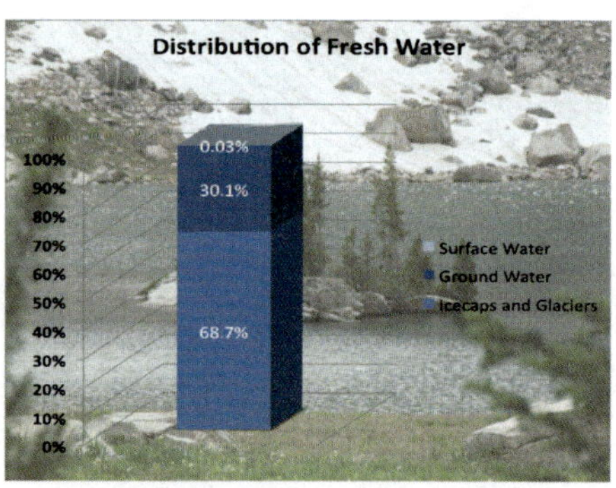

The surface of the planet Earth is approximately 71 percent water, but most of this water is unfit for human consumption. In fact, 97 percent of the planet's water is saline (saltwater), contained in the oceans. Ocean water requires so much processing to be made **potable**—fit for drinking, which in the case of ocean water requires desalination to remove salt as well as purification—as to be very energy inefficient. The remaining 3 percent of the water on Earth is fresh, but even much of this is unusable or inaccessible. Of the freshwater supply, nearly 69 percent is tied up in glaciers and icecaps, which are mainly localized at the poles and unavailable to large portions of the population. Thirty percent of freshwater is **groundwater**—it's found below the surface of the Earth in large underground pools called **aquifers**. This water can be accessed via wells, and is generally potable without extensive treatment. A very small percentage of water (0.03 percent of fresh, or 0.009 percent of the total) is **surface water**, which includes lakes, rivers, swamps, and the like. While surface water can be made drinkable, it requires extensive treatment to remove impurities and contaminants.

Image 4.14. Global water distribution by category. Note that despite its visibility, surface water forms only the tiniest percentage of water on the planet. (Background photo (bottom) © Scott Lefler, 2008)

It is worth mentioning that safe drinking water does not need to be 100 percent pure H_2O. As mentioned above, it is incredibly difficult to purify water to the point of 0 percent contamination. In addition to being impractical to fully purify water for drinking purposes, it's not necessary. Small amounts of dissolved chemicals, including salts of sodium, magnesium, and calcium, do no harm. In any case, the colloquial reference to "pure water" should always be taken to mean potable water—not water that is chemically pure.

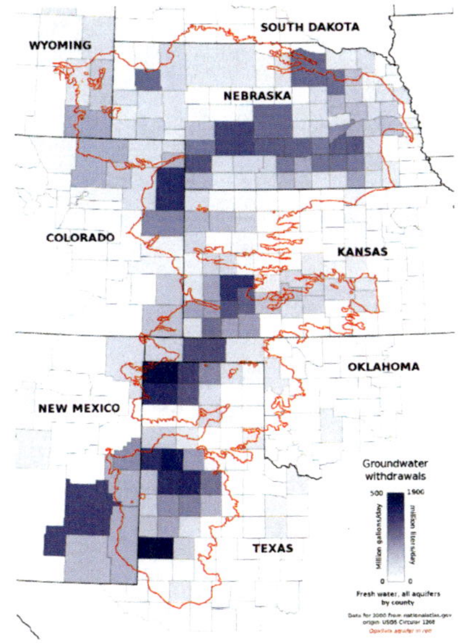

Image 4.15. The Ogallala Aquifer provides for the water needs of much of the Midwest. Shading indicates withdrawals from the aquifer by region. Map from U.S. Geological Survey.

In order to render surface water potable, the water is filtered in a treatment plant to remove dirt, leaves, and other bits of organic matter. Subsequent filtration through a smaller-pore filtering system then removes bacteria, viruses, and other organisms. Disinfectant chemicals such as chlorine are often added as well; depending upon the source of the water, chemical contaminants may be removed using activated carbon, which *adsorbs* (sticks to and removes) the chemicals. Groundwater is generally filtered naturally due to the fact that it has passed through various subterranean layers on its way down to the aquifers, and it typically does not need as extensive a treatment process as surface water. Both surface and groundwater generally have fluoride levels adjusted to optimized concentrations—sufficiently high to produce the benefit to dental health without being concentrated enough to cause staining of the teeth.

Concept Check:

Why does groundwater generally require less processing than surface water?

Answer: Groundwater is filtered as it passes through the layers of the Earth, while surface water is open to contamination.

Residents of the United States are lucky to have convenient, unrestricted access to safe drinking water. Per the United Nations, 18–20 percent of the world's population lack access to a sanitary water supply. This results in millions of deaths each year from diseases associated with unsafe water and unsanitary living conditions, such as cholera, giardia, and malaria. Many global organizations, including the United Nations (UN), UNICEF, and the World Health Organization (WHO), have initiatives focused on increasing global access to safe water, particularly in underdeveloped and developing nations.

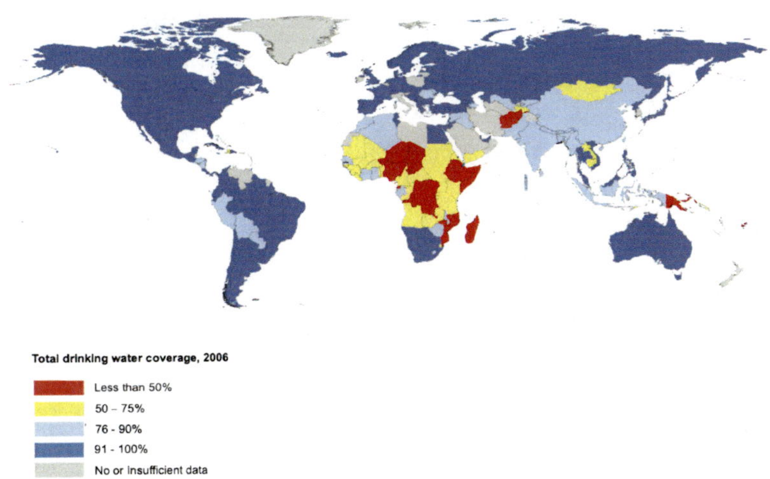

Image 4.16. Percentage of the population with access to safe drinking water (rural and urban areas) by country. Map from United Nations Environmental Programs (UNEP). Copyright © 2008 World Health Organization.

Image 4.17. A calcified faucet. Photo © Scott Lefler, 2010.

4.8 HARD WATER

The Environmental Protection Agency (EPA), a governmental regulatory organization, maintains strict water safety standards, such that tap water in the United States is incredibly safe to drink. Because potable does not mean free from solutes, however, tap water may contain various dissolved compounds. Of particular interest are the cations Ca^{2+} and Mg^{2+}, since these are responsible for the phenomenon of **hard water**. Ground and surface water, because it has filtered through the Earth, solvates minerals from the rock and soil. Some of these minerals*—namely magnesium and calcium—have the ability to complex with anions in the water. The resulting ionic compounds form salty crusts around water fixtures and in sinks and tubs. This **calcification** can damage fixtures and impair water flow. There's plenty of Na^{+} (and other group 1 cations) in tap water as well, but calcium and magnesium salts display the interesting property of having low water solubility. Despite the fact that ionic compounds should theoretically be water-soluble, some of them are held together so tightly by electrostatic forces that they don't tend to dissolve. This phenomenon is commonly observed in salts containing group 2 and group 3 cations. As a result, sodium in water doesn't produce faucet calcification and isn't a factor in water hardness, but the group 2 minerals do, and are thus a factor.

Further, soap molecules are anionic,♦ meaning that they can also complex with calcium and magnesium. Sodium and soap form soluble salts, but calcium and magnesium make for insoluble soap salts. These calcium and magnesium soap salts don't behave the way soap generally does, which is why it's difficult to form a lather in hard water, and soap scum tends to **precipitate**—meaning collect—on tub walls in the familiar tub ring. Hard water is not a health hazard; it's merely a nuisance. However, it's possible to "soften" water by exchanging the undesirable calcium and magnesium ions for sodium ions (Image 4.18). Water passes through a softening device that performs an **ion exchange**. Ca^{2+} and Mg^{2+} ions displace Na^{+} ions from the exchange resin (a resin is a bit like a wax), and the sodium ions end up in the water, while calcium and magnesium ions stick to the resin. Since sodium salts are soluble in water, neither calcification of faucets nor precipitation of soap occurs.

Not all municipal water is hard—coastal regions of the United States that utilize treated rainwater (which contains no dissolved minerals) have very **soft water**. Soft water doesn't result in calcification of fixtures and soap lathers effectively. However, soft water doesn't rinse soap from the skin as easily

*NOTE

Minerals will be discussed further in Chapter 8, but in short, include the metal cations and halide (halogen) anions.

♦NOTE

Soaps will also be discussed further in Chapter 8, but it is sufficient for now to know that they are large, negatively charged molecules.

CONCEPT CHECK:

How are rainwater and chemically softened water the same? How are they different?

Answer: They are both soft (no calcium or magnesium ions), but chemically softened water contains sodium ions, while rainwater does not.

as hard water, which can leave skin feeling a little slimy when bathing or washing hands.

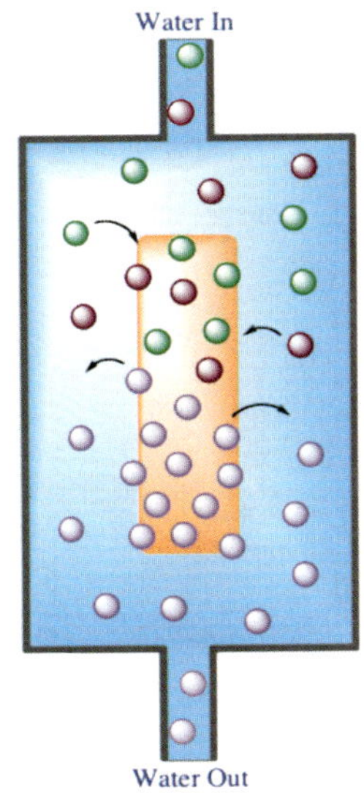

Image 4.18. An ion-exchange resin forms the heart of a water softener. Water full of Ca2+ and Mg2+ ions (red and maroon) enters the softener and the cations stick to the resin, displacing Na+ ions (purple). The water exits the softener enriched in sodium, and without its calcium and magnesium.

4.9 BOTTLED VERSUS TAP WATER—SOCIAL IMPLICATIONS

There are places in the world where clean, safe drinking water is hard to come by. The United States is not one of these places. Still, bottled-water manufacturers have managed to turn consumers' concern with the safety of their drinking water into a multibillion-dollar business (soda is the only beverage whose sales outstrip those of bottled water—beer, juice, and milk lag far behind). However, Americans' love affair with bottled water comes at a high social price: the energy associated with producing bottles, processing and bottling water, and transporting the finished product is immense.

Several studies have evaluated and reported the energy cost associated with bottling water, and have determined that the energy input is multifaceted. In the first place, petroleum and petroleum products are used in the manufacture of bottles—plastic bottles are literally made of oil. Second, energy is required to treat and process the water that is to be bottled. Bottled water falls into one of two categories, per United States Food and Drug Administration (FDA) regulations. *Spring water* is derived from an underground source, while *purified water* is merely municipal tap water that has been further distilled, de-ionized,

Image 4.19. The image that bottled water manufacturers try to evoke (left) is at odds with the source of most bottled water (right). (Waterfall © Scott Lefler, 2009.)

or otherwise treated. Regardless, this represents an increased energy cost above that of tap water production. Energy is then required to package bottled water and still more energy to transport it (actually, transportation represents one of the greatest energy inputs due to the weight of a truckload of water). Overall, the energy cost per bottle of water in terms of oil consumed has been estimated as equivalent to one-quarter of the volume of the bottle itself! This represents an energy expenditure of 2000 times that associated with producing municipal tap water.

Another cost associated with the consumption of bottled water is the price. The United States Department of Agriculture (USDA) advises us to drink eight 8-oz. glasses of water per day. If that consumption were filled entirely by bottled water, the cost could be upwards of $1400 annually (bottled water costs, in general, about 2–2.5 times the cost of gasoline by volume)! By comparison, the annual price tag on that same quantity of tap water would be about 50 cents. Presumably, given that the cost of water is astronomically high (both energetically and financially), consumers are enamored of it for good reason. Perhaps it's purer, healthier, or tastier than tap water? Unfortunately, none of these rationales for bottled water consumption stands up to scrutiny.

In terms of purity and healthfulness, municipal water is regulated by the EPA, which has strict processing and purity standards. Tap water also contains added fluoride, routinely demonstrated to have a positive effect on dental health in appropriate concentrations. Bottled water, on the other hand, is meant to be regulated by the FDA, though most of it falls outside their purview since very little crosses state lines, exempting it from regulation. Further, independent studies show no improved purity of bottled water over tap, and less than 5 percent of bottled water contains the concentration of fluoride recommended by the American Dental Association. In the end, most bottled water is nothing but low-fluoride tap water that's been subjected to additional processing, producing an infinitesimal improvement upon the impressive purity of municipal water (if any improvement at all).

Regarding taste, while no controlled scientific studies have addressed this issue (taste being a subjective matter), there is overwhelming nonscientific evidence that in blind taste tests, bottled water can't be distinguished from tap. In fact, many cities routinely hold water "taste-offs" to raise awareness of municipal water quality, with consistent success. Regardless, even those for whom tap water has an undesirable taste have the option of a home tap filter.

A final issue to consider is that plastic water bottles, while recyclable, are often used once and discarded. This increases community (and global) waste. Further, even those bottles that are recycled are reprocessed at an additional energy cost. It's somewhat interesting (and distressing) to consider the costs we're willing to accept to obtain bottled water, despite having instant access to incredibly pure, minimal-cost water available from every tap in our homes. Overall, the decision to avoid bottled water saves money, saves a tremendous amount of energy, and comes at no cost to health or wellness.

4.10 THE LAST WORD—SUPERWATERS

Some bottled waters go above and beyond simply evoking the image of a mountain spring. Market shelves are now filled with bottles upon bottles of so-called "superwaters," that contain flavorings, vitamins, and so forth. These superwaters promise to do more than just quench thirst—claims range from improving sports performance to promoting weight loss. In a particularly bizarre misconception of both chemistry and basic physiology, some waters are *oxygenated*—O_2 has been bubbled into the water for greater oxygen saturation than is normally found in drinking water. These products promise enhanced cellular oxygenation and suggest that oxygenated water boosts performance, cognition, energy level, and a host of other functions. Unfortunately, this claim simply doesn't, well, hold water! Humans, like other land animals, absorb oxygen through the lungs. We have no mechanism whatsoever for absorption of oxygen through other body surfaces or membranes. Supplementing the gut with oxygen from overly fancy water does nothing but increase the amount of gas in the stomach, which has no effect whatsoever except perhaps to contribute to eructation (an overly fancy word for belching).

SUMMARY OF MAIN POINTS

- Water is critical to life.
- olutions are mixtures of solvents and solutes—water is an excellent solvent.
- Electronegativity is a measure of an atom's ability to pull electrons in bonds.
- Pure covalent bonds form between atoms with identical electronegativity.
- Polar covalent bonds form between atoms with different electronegativity, and result in dipole moments.
- A bond dipole is unequal distribution of electrons in a bond, and results in partial charges on the atoms in the bond.
- A net dipole is the sum of a molecule's bond dipoles.
- Atoms with net dipoles are polar; atoms with no net dipole are nonpolar.
- Water is extremely polar.
- Polar things dissolve well in other polar things; nonpolar things dissolve well in other nonpolar things.
- Compounds with very large bond dipoles (along F–H, O–H, or N–H bonds) interact strongly with each other via hydrogen bonding—water is an excellent hydrogen bonder.
- Due to water's formation of hydrogen bonds, it has a high boiling point, melting point, and density. It is resistant to temperature change, is very cohesive, and forms a solid that is less dense than its liquid.
- Ionic bonds form between atoms with very different electronegativity—metals and nonmetals—and are the result of complete electron transfer.
- Cations and anions—the result of electron transfer reactions—are electrostatically attracted to one another and form ionic salts.
- The charge on an atom when it ionizes can be predicted from its group number on the periodic table.
- Ionic compounds always form so that the charges balance.
- Ionic compounds dissolve well in water.
- Ionic compounds dissociate in water, forming electrolytic solutions that can conduct electricity (molecular compounds are not electrolytes).
- Earth's water is primarily unavailable for drinking—only 1 percent of the planet's water is fresh AND accessible.
- To make water potable, it must be treated—surface water requires more extensive treatment than groundwater.

- The availability of safe drinking water is an important global health issue.
- Water with large amounts of dissolved Ca^{2+} and Mg^{2+} is called hard water and is a nuisance; it can be softened via ionexchange.
- Bottled water is incredibly expensive financially and energetically, and affords no real advantage over municipal water.

Questions and Topics for Discussion

1. Why would a chemist or biologist call water the single most critical chemical to life?
2. Explain the concept of electronegativity. How does it affect chemical bonding?
3. To which element is hydrogen's electronegativity considered to be nearly identical? What is the ramification of this upon bonds between hydrogen and this element?
4. Why do bond dipoles result in formation of partial charges?
5. What kinds of molecules have bond dipoles? Net dipoles? What kinds of molecules have bond dipoles but NO net dipole?
6. Consider the molecules CH_4, CH_3Cl, CH_2Cl_2, $CHCl_3$, and CCl_4. How are they similar? How are they different? Which have net dipoles?
7. What is the primary predictor of solubility? What sorts of compounds dissolve well in water?
8. What is hydrogen bonding? How does water's hydrogen bonding affect its physical properties, making it unusual as a molecule?
9. How are particles arranged in solids? In liquids? In gases? What generalizations regarding mobility of particles, compressibility, and density can be made for each?
10. Why does ice float on liquid water? Why is this odd behavior for a chemical?
11. Why do some elements form covalent compounds, while some form ionic compounds? Which types of element pairs will form covalent compounds? Which types of pairs will form ionic compounds?
12. Why is it unnecessary to use Greek prefixes in the naming of ionic compounds?
13. Why is it not strictly correct to think of an ionic compound as a "molecule?" What is different about a sample of an ionic compound in terms of delineating a single molecule?
14. What is an electrolyte? What can electrolytes do that nonelectrolytes can't?
15. Describe the behavior of an ionic compound when dissolved in water. How does it interact with water?
16. What types of water are found on Earth and what percent of total water falls into each category? Is most of Earth's water drinkable? Is most of Earth's drinkable water accessible?
17. Which parts of the world have the least access to safe drinking water?
18. How is drinking water processed in the United States? Is tap water safe and drinkable?
19. What is the difference between drinking water and 100 percent H_2O?
20. What is hard water? Is it dangerous? How does it form? How is it softened?
21. What are the major social, economic, and financial ramifications of drinking bottled water? Do you think bottled water is good for a community? For the world?

Problems

1. Define each of the following: solute, solvent, solution, aqueous solution, concentration.
2. For each of the following, identify the solute and the solvent: sugar water, vitamin A dissolved in fat, salt water.
3. Put the following in order of increasing electronegativity: F; H; C; Cl; Br; and Li.
4. Put the following in order of increasing electronegativity: S; N; Mg; I; H; and O.

5. Define each of the following: dipole; electronegativity; partial charge; periodic trend; pure covalent bond; and polar covalent bond.
6. Sketch each of the following molecules, indicating bond dipoles: HBr; Cl_2; COH_2; and CH_2I_2.
7. Sketch each of the following molecules, indicating bond dipoles: HCN; O_2; H_2S; and CF_2Cl_2.
8. Sketch each of the following molecules, indicating partial charges: HBr; Cl_2; COH_2, and CH_2I_2.
9. Sketch each of the following molecules, indicating partial charges: HCN; O_2; H_2S; and CF_2Cl_2.
10. Sketch each of the following molecules, indicating net dipole: HBr; Cl_2; COH_2; and CH_2I_2.
11. Sketch each of the following molecules, indicating net dipole: HCN; O_2; H_2S; and CF_2Cl_2.
12. Classify each of the following molecules as polar or nonpolar: HBr; Cl_2; COH_2; and CH_2I_2.
13. Classify each of the following molecules as polar or nonpolar: HCN; O_2; H_2S; and CF_2Cl_2.
14. Which of the molecules shown is capable of hydrogen bonding? Sketch the hydrogen bonding interaction.

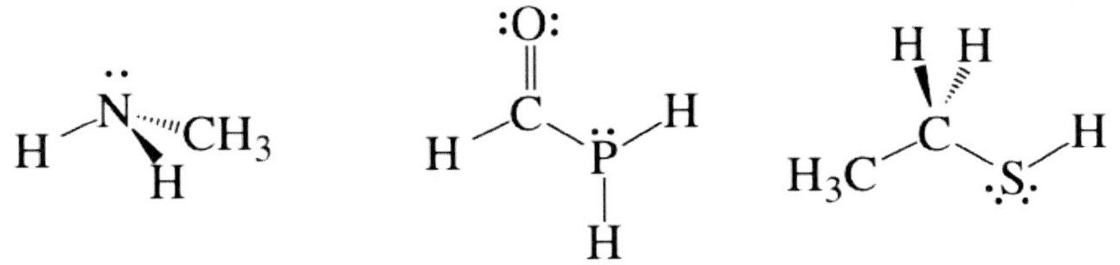

15. Which of the molecules shown is capable of hydrogen bonding? Sketch the hydrogen bonding interaction.

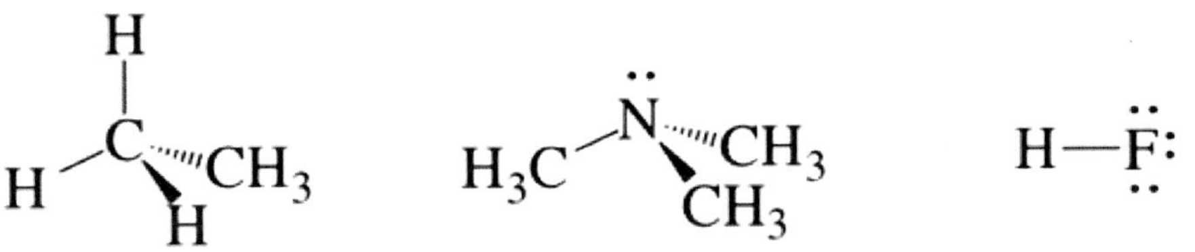

16. Predict the charges on ions of each of the following elements: K; Al; P; and I.
17. Predict the charges on ions of each of the following elements: Cs; Ba; Se; and Cl.
18. Draw Lewis Dot Structures showing the formation of an ionic compound from each of the following combinations of elements (show electron transfer and resulting charges): Cs and F; Cs and O; Ca and O; and Ca and Cl.
19. Draw Lewis dot structures showing the formation of an ionic compound from each of the following combinations of elements (show electron transfer and resulting charges): Rb and S; Rb and N; Sr and F; and Sr and S.
20. Draw Lewis dot structures showing the formation of an ionic compound from each of the following combinations of elements (show electron transfer and resulting charges): Mg and N; and Al and S.
21. Determine the chemical formula of each of the following: lithium oxide; potassium selenide; magnesium oxide; and calcium fluoride.
22. Determine the chemical formula of each of the following: rubidium fluoride; sodium oxide; sodium nitride; and beryllium chloride.
23. Indicate which of the following are electrolytes: H_2S; PF_3; Li_2O; CH_4; and NaF.
24. Indicate which of the following will dissociate in water: CH_2O; $COCl_2$; $MgCl_2$; Na_2S; and KCl.
25. In which states is the Ogallala Aquifer located?
26. What type of water represents the largest percentage of Earth's water? The largest percentage of Earth's freshwater?
27. Define the following: aquifer; groundwater; surface water; and potable water.
28. What ions are responsible for hard water? What ions are used to soften water?
29. What are the effects of hard water?

30. What are the four ways in which energy is required as an input to producing bottled water?

Answers to Problems

1. A solute is the chemical being dissolved; a solvent is the chemical doing the dissolving; a solution is a mixture made up of a solute and solvent; an aqueous solution has water as the solvent; concentration is a measure of the relative proportion of solute to solvent.
3. Li, H, C, Br, Cl, F.
5. A dipole is an unequal distribution of electrons in a bond (or a separation of charge); electronegativity is a measure of how tightly electrons are drawn toward an atom's nucleus; partial charge results from partial electron loss or gain through a dipole; a periodic trend is one that can be predicted from looking across rows or up/down columns on the periodic table; a pure covalent bond is one in which there is equity of electron sharing; a polar covalent bond is one in which there are differences in electronegativities of atoms in a bond, resulting in unequal electron sharing.
7.

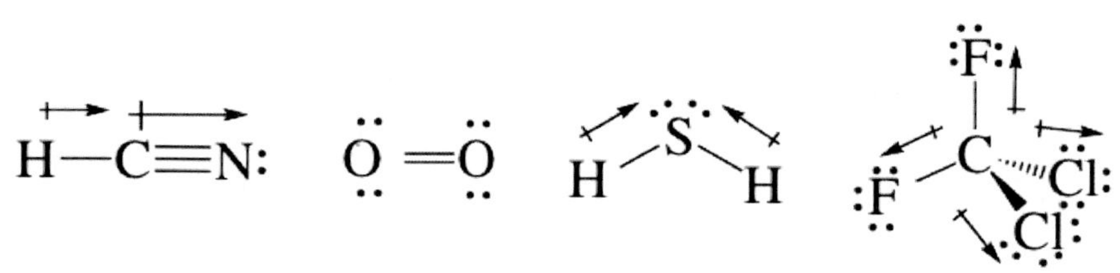

9. Note: Where effects are mixed (such as on C in CH_2I_2), partial charges have been omitted.

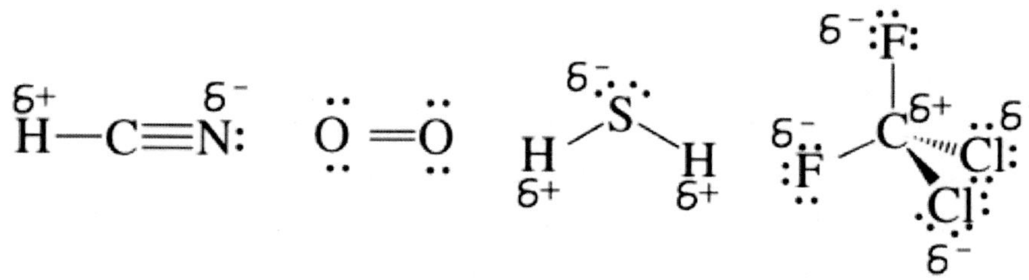

11.

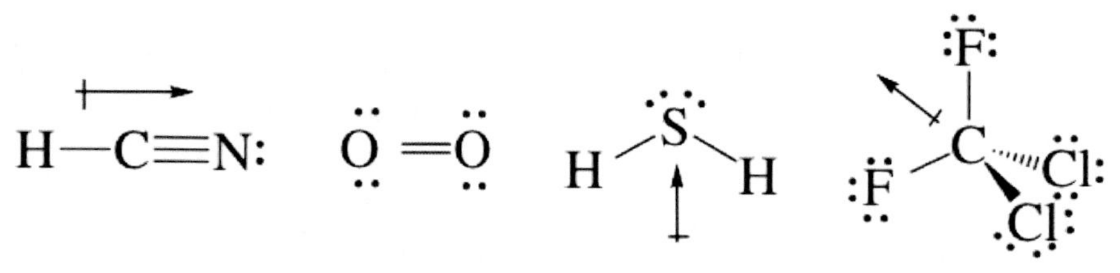

13. HCN is polar, O_2 is nonpolar, H_2S is polar, CF_2Cl_2 is polar.
15.

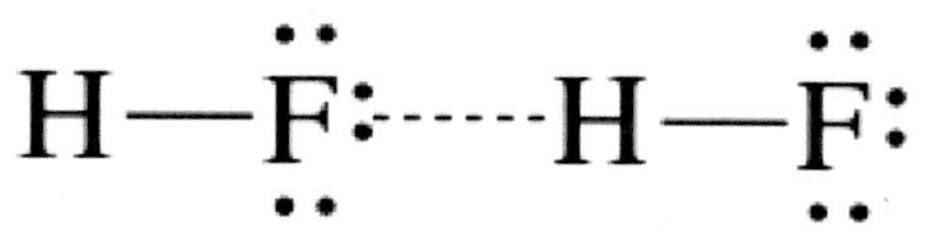

17. Cs^+, Ba^{2+}, Se^{2-}, Cl^-.

19.

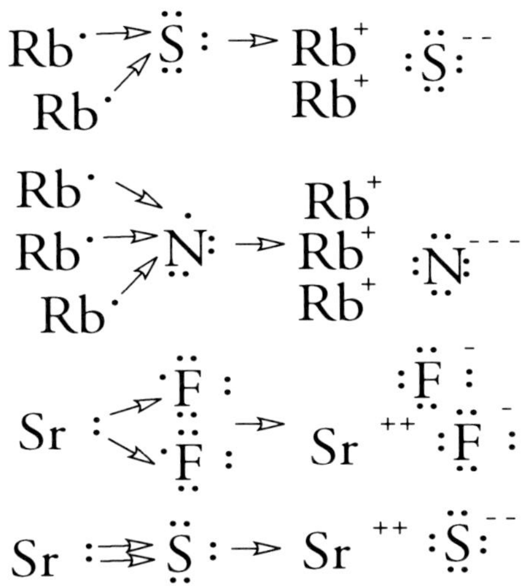

21. Li_2O, K_2Se, MgO, CaF_2.
23. Li_2O and NaF are electrolytes.
25. WY, SD, NE, CO, KS, OK, NM, TX.
27. An aquifer is an underground pool of water; ground water is located underground; surface water is on the surface of the Earth (rivers, lakes, etc); potable water is fit for human consumption.
29. Hard water causes calcification of pipes and fixtures and precipitation of soap.

Sources and Further Reading

Role of Water in Life

Alpert, P. (2005). Sharing the secrets of life without water. *Integrative and Comparative Biology*, 45(5), 683–84.

Rothschild, L.J., Mancinelli, R.L. (2001). Life in extreme environments. *Nature*, 409, 1092–101.

Water Distribution and Safety

Azoulay, A., Garzon, P., Eisenberg, M. (2001). Comparison of the mineral content of tap water and bottled waters. *Journal of General Internal Medicine*, 16(3), 168–75.

www.epa.gov/ow

www.un.org/waterforlifedecade

www.usgs.gov

www.who.int/topics/drinking_water/en

Bottled Water

Gleick, P.H., Cooley, H.S. (2009). Energy implications of bottled water. *Environmental Research Letters*, 4(1), 014009.

Lalumandier, J.A., Ayers, L.W. (2000). Fluoride and bacterial content of bottled water vs. tap water. *Archives of Family Medicine*, 9(3), 246–50.

5 The Chemistry of Acids and Bases

We encounter acids every day of our lives. Vinegar in salad dressing is acetic acid; vitamin C is ascorbic acid; we put muriatic acid (also called hydrochloric acid) in our pools; the burning sensation in hardworking muscles comes from lactic acid, a metabolic waste product. We even use acids in food preparation—carbonic acid causes bread to rise, while lactic acid turns milk into cheese. Some commonly encountered acids have negative effects on us and on the environment. For instance, pollution results in acid rain, which kills aquatic life and destroys architecture. Oxalic acid is found in many foods, including leafy green vegetables, and can cause kidney stones in some individuals. Spilled industrial or laboratory acid can cause burns and act as an environmental pollutant. Like all substances, acids can be either useful or detrimental, either nontoxic or harmful. What is an acid, though, from a chemical standpoint? What gives it the physical and chemical properties that distinguish it from a nonacid, or from its chemical opposite, a base? Most importantly, in what context are acids helpful and safe, as opposed to dangerous and toxic?

5.1 ACIDS

An **acid** is a chemical that behaves in a particular way in aqueous solution. The simplest definition of an acid is that it dissociates in water (just like an

Image 5.1. The vitamin C in a grapefruit and other citrus fruit, the chemical making this hardworking runner's legs burn, and the substance helping to prevent contamination of a swimming pool are all acids.

ionic compound) to produce the cation H^+ and ANY anion. Let's examine this statement a bit closer. First of all, it's perhaps a bit surprising to see hydrogen ionizing to form H^+; after all, thus far we've only seen it form covalent bonds. Remember, though, the nature of hydrogen is a bit dichotomous, hence its common representation in two different places on the periodic table! As a nonmetal (which it is, since as an element it does NOT have the physical properties of a metal), we expect to see hydrogen forming a single covalent bond to another nonmetal to fill its one valence shell hole. Of course, given its very low electronegativity, this bond is generally highly polarized (unless it is bonding to another atom of hydrogen or to carbon). As a group 1 element, however, hydrogen has properties in common with the other elements in its group. Like lithium, sodium, and so on, hydrogen has a single electron in its outermost shell. The group 1 metals ionize in the presence of sufficiently electronegative elements (nonmetals), emptying their valence shells and leaving the atoms stable (and cationic). Hydrogen, it turns out, can do the same under the right circumstances. While hydrogen's bonds to very electronegative elements act covalent (molecular) MOST of the time, the bonds can actually polarize completely in aqueous solution. This leaves hydrogen cationic, and the atom to which it was bonded (which is now holding both of the electrons from the former covalent bond) becomes anionic. It is this ionization of hydrogen in aqueous solution that defines an acid, and as such, it becomes a bit easier to predict what sorts of chemicals will be acids—by necessity, they must contain hydrogen bonded to an electronegative element. For instance, all of the compounds of hydrogen with group 7 nonmetals are acids, and there are many examples of acids in which hydrogen is bonded to oxygen. Unfortunately, this definition only works in one direction. While acids always contain hydrogen bonded to an electronegative element, not all chemicals in which hydrogen is bonded to an electronegative element are acids—water is an obvious example.* Conveniently, the acids we'll see in this text all have the word *acid* in their name, as in *hydrochloric acid*, *sulfuric acid*, and *lactic acid*. For now, it's sufficient to know that the acids we'll be discussing fall into one of three groups—binary acids, oxyacids, and carboxylic acids. **Binary acids** are molecules made up of hydrogen and only one other element: In the examples we'll see in this text, the second element will be a halogen, as in HCl. **Oxyacids** are combinations of one or more atoms of oxygen with another nonmetal, plus one or more atoms of hydrogen. Examples of oxyacids include nitric acid (HNO_3) and sulfuric acid (H_2SO_4). **Carboxylic acids** are an incredibly large and varied class of compounds, but all include the characteristic combination of atoms shown here:

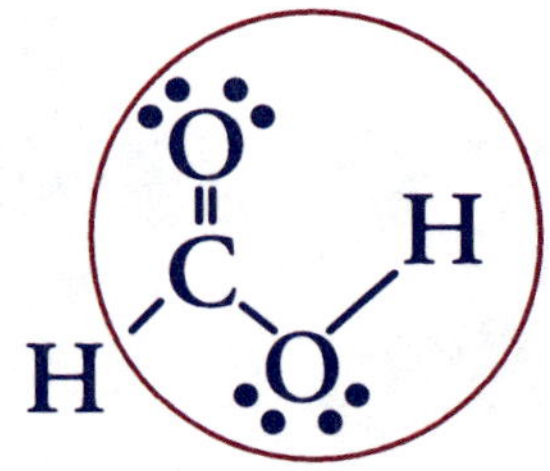

The circled area represents the arrangement of atoms typical of carboxylic acids. This is formic acid.

***NOTE**

It's worth noting that electronegativity is not the SOLE determinant of an atom's ability to ionize its bond with hydrogen in aqueous solution. For instance, despite oxygen's high electronegativity, water is not acidic. However, despite the fact that sulfur is not as electronegative as oxygen, H_2S is acidic. It turns out that a combination of factors dictate acidity (including atomic size and properties of the resulting anion after loss of hydrogen), and electronegativity is only one of these factors. It's quite safe to say, though, that elements with low electronegativity, such as carbon, do not ionize their bonds with hydrogen in aqueous solution!

In a chemical formula, the combination shown above is written *–COOH*. Formic acid's chemical formula, for instance, would be HCOOH. Acetic acid, shown in Image 5.2, has the chemical formula CH_3COOH. These formulas look a bit different than the ones we've seen thus far in the text. Why, for instance, is formic acid written HCOOH rather than CH_2O_2? The bigger molecules get, the harder it becomes to draw Lewis structures from formulas—just imagine trying to figure out how atoms are connected to one another in CH_2O_2! By writing the formulas in ways that give information about structure (the presence of *–COOH*, for instance, always indicates a particular atomic arrangement), it makes it easier for us to determine the structure of a molecule from its formula. Technically, CH_2O_2 is a perfectly accurate chemical formula for formic acid, but looking at that formula, you'd not be able to tell right away that the compound is acidic. HCOOH, however, makes the presence of the *–COOH* arrangement of atoms very apparent. Since this arrangement always indicates that a compound is an acid, you can instantly tell that the formula HCOOH refers to an acidic compound, without having to draw the structure or hear the name.

Acid Formula	Acid Name
HCl	Hydrochloric acid
H_2SO_4	Sulfuric acid
HNO_3	Nitric acid
H_2CO_3	Carbonic acid
HCOOH	Formic acid
CH_3COOH	Acetic acid (vinegar)

Table 5.1. Names worth knowing: some of the most commonly encountered acids.

To visualize an acid's behavior in water, we simply imagine BOTH electrons from the bond between hydrogen and whatever it's attached to being pulled so tightly toward the more electronegative element that ionization occurs. This leaves us with an H^+ cation, and an anion made up of the rest of the molecule:

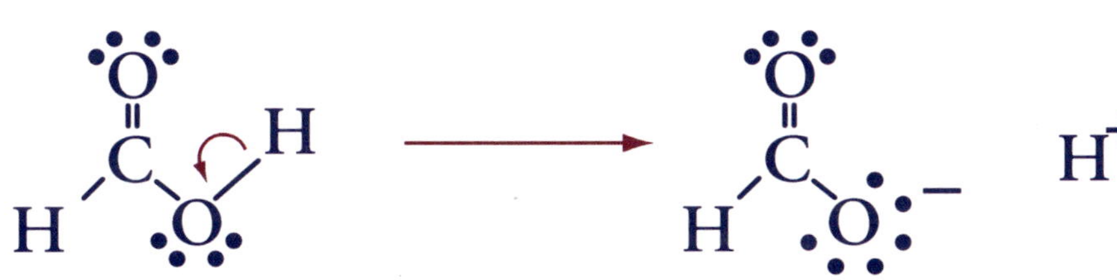

When acid dissolves in water, the electronegative element pulls both electrons from the covalent bond into its valence

The result is an H^+ cation, and an anion made up of the remaining parts of the acid molecule

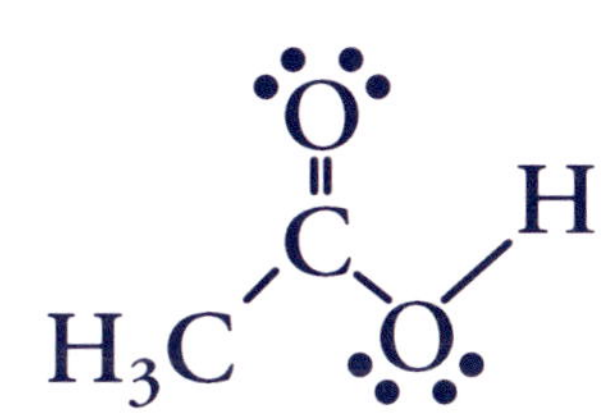

Image 5.2. Acetic acid (vinegar) is a carboxylic acid with the chemical formula CH3COOH.

CONCEPT CHECK:

In the context of an acid, how does hydrogen act like a nonmetal? How does it act like a metal?

Answer: It forms a covalent bond to another atom (like a nonmetal) that polarizes and leads to dissociation in water (like a metal).

5.2 POLYATOMIC IONS

In the dissociation of formic acid above, we see the formation of an interesting species: HCOO-, also called the *formate anion*. It's fairly elementary to see why hydrogen becomes cationic when dissociation occurs—it's lost its electron. Why, though, does the oxygen end up with a negative charge? Both in formic acid and in formate, oxygen has a full octet of eight electrons, so there's been no change in the valence. We can verify that by counting electrons—there are two bonds and two lone pairs around oxygen in formic acid, and there is one bond and three lone pairs around oxygen in formate. That's a full octet in both cases. Normally, ionization reactions involve a change in an atom's electron count, so what's going on? In reality, oxygen HAS experienced a change in its electron count! Remember that in formic acid, the electrons in the O-H bond don't both BELONG to oxygen—one of them is oxygen's and one is hydrogen's. Oxygen can count both of the electrons toward its valence (that's the nature of a covalent bond), but it doesn't have personal "ownership" of hydrogen's electron, so that electron doesn't throw off its atomic proton/electron balance, and oxygen is still neutral. When the O-H bond is pulled entirely away from hydrogen and toward oxygen, oxygen does take ownership of hydrogen's electron, leaving hydrogen cationic (as we see), and oxygen anionic, because it has acquired a seventh "personal" valence electron, as compared to the six it started with.

Ions such as formate, consisting of two or more atoms covalently bonded together, are called **polyatomic ions**, and are relatively common in chemistry (especially in the chemistry of acids and bases). Names, charges, and formulas of the most common polyatomic ions appear in Table 5.2.

Formula	Name
NH_4^+	Ammonium
OH^-	Hydroxide
NO_3^-	Nitrate
SO_4^{2-}	Sulfate
PO_4^{3-}	Phosphate
CO_3^{2-}	Carbonate
HCO_3^-	Bicarbonate

Table 5.2. Names and chemical formulas of common polyatomic ions.

When we discussed formation of ionic compounds in Chapter 4, we saw the reaction taking place in two parts. First, the more electronegative atom(s) took electrons from the less electronegative atom(s), forming anions and cations. Next, the resulting ions (with their electrostatic attraction to each other) formed an ionic bond. For now, we will not address the formation of polyatomic ions—we'll just assume they already exist. We'll see some instances of how they form in the real world later in the text (sulfate, for instance, is a by-product of volcanic eruption and combustion of fuel), but for now, we don't need to concern ourselves with these details. Suffice it to say, ions are not permanently "stuck" with the atoms with which they originally reacted—even if a sodium cation, for instance, formed when sodium lost an electron to fluorine, it's possible for the NaF to break apart (as we saw in our discussion of electrolytes). When this happens, new ionic bonds can form. This is exceedingly common in aqueous solution, and allows for the rearrangement of cations and anions into different compounds. In any case, assuming for now that the polyatomic ions have formed through some chemical means, we can certainly figure out how they might combine with other ions to form compounds.

Polyatomic ions can combine with other anions or cations to form ionic compounds the same way that monoatomic ions do. Just as in binary compounds of a metal and nonmetal, salts containing polyatomic ions form such that the compound is neutral. Hydroxide, for instance, with its -1 charge, can combine with one sodium cation to form NaOH. A magnesium cation, however, with a +2 charge, will need to combine with two hydroxide anions to create a neutral compound. Magnesium hydroxide therefore has the formula $Mg(OH)_2$. Note the use of parentheses around the OH; this is to indicate that the subscript refers to the entire OH group, and not simply the H. This compound, per the subscript, consists of an ion of magnesium and two units of the OH^- anion. $MgOH_2$, written without parentheses, would refer to a molecule made up of one magnesium atom, one oxygen atom, and two hydrogen atoms—and wouldn't be a valid chemical formula, incidentally!

Polyatomic ions also share nomenclature with binary ionic compounds: Just as in compounds of metals and nonmetals, when naming ionic compounds with polyatomic ions, we name the cation first (without changing its name), then the anion. Greek prefixes are not used.

Example: NH_4Cl would be called *ammonium chloride.*

Example: K_2SO_4 would be called *potassium sulfate.*

TRY THIS:

What is the name and chemical formula of a compound consisting of Ca^+ and CO_3^-?

Answer: $CaCO_3$, calcium carbonate.

TRY THIS:

What is the name and chemical formula of a compound consisting of NH_4^+ and SO_4^{2-}?

Answer: $(NH_4)_2SO_4$, ammonium sulfate.

5.3 BASES

Unlike acids, **bases** do not ionize in water to produce hydrogen cations. Instead, they "force" water to ionize, by taking one of its hydrogen atoms:

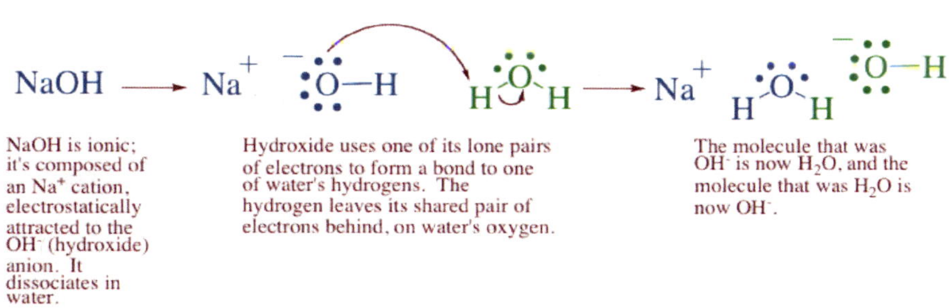

There are several important points to note regarding the reaction shown above. First, the compound NaOH is ionic, as it consists of sodium (which never forms covalent bonds) and what we now recognize as the hydroxide anion. NaOH, like all ionic compounds, will therefore dissociate in water. From that point on, the interesting chemistry is perpetrated entirely by OH^- and other species present—Na^+ is not an active participant in the chemical reaction. This is actually quite typical of Na^+ and other metal cations; once they've formed, they're quite stable, and while we'll see them electrostatically attracted to different anions (essentially, they're attracted to whatever negatively charged particles are present in the solution mixture), they don't participate in reactions. The Na^+ is called a **spectator ion**, meaning it "watches" the reaction rather than taking part in it. A second important point has to do with charge. For clarity, let's name the species in the reaction above "blue hydroxide" and "green water." When blue hydroxide "steals" a hydrogen from green water, the hydrogen leaves behind both of the electrons it shared with oxygen in green water. Because green water's oxygen has now acquired an additional "personal" electron, the oxygen becomes negatively charged. Note that its valence hasn't changed; the difference is merely that what was a shared electron has become an unshared electron. Further, because any chemical bond must consist of a pair of electrons, blue hydroxide must provide BOTH electrons that will go toward making the bond with green water's hydrogen (the hydrogen contributes no electrons toward the bond, because it left them behind). As such, the blue hydroxide's oxygen is essentially "giving" an electron to green water's hydrogen—it's taking a nonbonding pair and turning it into a shared pair. This means green hydrogen can now claim one of blue oxygen's electrons. Blue oxygen has therefore essentially lost an electron—not from its valence, because it still has an octet, but from its "personal" electron count—and it is now neutral rather than negatively charged.

Not all bases are ionic salts of hydroxide, however, as we see in the case of ammonia (NH_3) which is also basic:

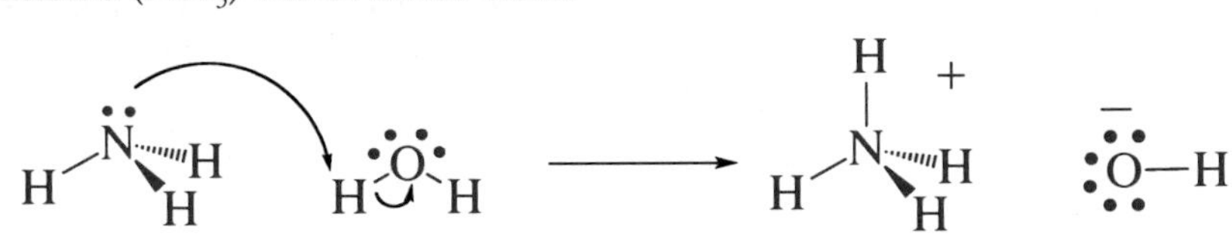

Ammonia uses its lone pair of electrons to form a bond to one of water's hydrogens. The hydrogen leaves its shared pair of electrons with oxygen.

Because ammonia has acquired an ionized hydrogen (an H^+) without acquiring any additional electrons, it is now positively charged (NH_4^+). What was water is now OH^-.

***NOTES**

There's a pattern here: Bases all react with water by taking one of water's hydrogens, so they must all have a lone pair of electrons available to form a bond to that hydrogen. Be careful, though—all bases have lone pairs, but not all molecules or atoms with lone pairs are bases!

In this case, we are not starting with a dissociation reaction—NH_3 is purely covalent, and the only way to predict its behavior in water is to know it's a base. Bases are quite a bit harder to recognize than acids, but it's worth knowing that the most common bases are salts of the OH^- anion and ammonia. Like any base, ammonia reacts with water by using a nonbonding pair of electrons* to steal hydrogen away from water. Hydrogen releases both electrons it shared with oxygen, resulting in the formation of the OH^- anion. Because nitrogen from NH_3 has used a nonbonding pair of electrons to make a bond to hydrogen, it has essentially "lost" an electron—it is now positively charged—and ammonia has become the NH_4^+ cation. Note that charge is always conserved in these reactions—if neutral starting materials react (as in the case above), the products might be charged, but the sum of the charges will be zero. Also note that when a base takes a hydrogen from water, the base will ALWAYS lose a unit of negative charge. If it was negative before the reaction (like OH^-), it will become neutral. If it was neutral before the reaction (like NH_3), it will become positive.

> **CONCEPT CHECK:**
>
> Why do bases end up producing OH^- in water, even if they're not hydroxide salts?
>
> Answer: When a base takes a hydrogen from water, hydroxide is left over.

5.4 ACIDS, BASES, AND TASTE

While acids and bases are chemically relevant to many applications, one of the places we experience them most frequently is in our sensation of flavor. What we think of as the *taste* of food is mostly smell. Nuances of flavor, such as the vanilla and cinnamon flavors in sugar cookies or the complex combination of vegetables, meat, and cheese in a lasagna, come from molecules in the food interacting with the *nasal epithelium*, which contains the smell-detecting receptors in our noses. We don't have the ability to literally TASTE many different flavors, as is well evidenced by the experience of eating food while suffering a bad head cold. When the nose is out of commission, we're reduced to five basic taste groups: sweet, sour, salty, bitter, and meaty. Different parts of the tongue contain taste buds specialized for different flavors: We taste sweet best with the tips of our tongues, sour is most noticeable along the sides, salty is strongest on the sides and toward the front, and the back of the tongue is best at sensing bitter flavors. Our ability to recognize a meaty flavor is relatively newly discovered, and this sensation

Image 5.3. The sweet flavor of cookies, the sour of oranges, saltiness of chips, bitter taste of broccoli, and meaty flavor of a nicely done steak are all chemical signals that help us instinctively eat the foods we need to, and avoid the foods that might harm us.

appears to be generalized across the entire tongue. Given that we use the nose for the majority of the flavor experience we get from food, what is the purpose of the tongue's ability to recognize these five basic flavors? It turns out that each of them provides us (unconsciously) with important information about what we're putting in our mouths!

Sweet foods, as we'll learn in Chapter 8, contain sugars, which provide energy. The flavor of sweet, therefore, lets us know that we're eating something capable of fueling our activities. In fact, we respond favorably to sweet foods on a deep-seated and instinctive level—cross-culturally, babies react positively to bottles of sugar water from the moment they're born. Salty flavors in food alert us to the presence of NaCl, which provides the important mineral, sodium. A typical modern Western diet is not lacking in sodium, but the mineral is somewhat difficult to come by in a hunter-gatherer's or forager's diet, so the ability to sense the flavor of salt (and the corresponding positive response to this flavor in low concentration) helped to guide early humans toward foods that would provide necessary sodium. Meaty flavors direct us toward consuming plenty of protein, which is a large molecule used to build muscles and structural components of cells. The remaining flavors—sour and bitter—are lent to foods by acid and base respectively. The mildly sour flavor of an orange is due to the presence of citric acid and vitamin C (ascorbic acid). The much less tolerable, more intense tartness of a lemon comes from higher concentrations of those acids. Our ability to sense acid (via the sensation of sour) directs us toward potential sources of important nutrients such as vitamin C, while directing us away from acids strong enough to do damage to our teeth and digestive tracts. Bitter flavors, such as those of some vegetables, coffee, and many herbs, are lent to foods by basic compounds (sometimes called *alkaloids*). While not all alkaloids are toxic, many toxic compounds are alkaloids. As a result, our ability to sense them through flavor helps us avoid ingesting toxins. Studies of newborns show that in addition to responding very favorably to sweet flavors, they find mildly sour and salty flavors to be interesting. They respond in a strongly unfavorable fashion, however, to bitter flavors, indicating an instinctive avoidance of potential toxins. The ability to tolerate and even enjoy the bitter flavors of some foods is developed over time—often late in childhood or into early adulthood—as we become capable of overruling our instincts.

5.5 MOLARITY AND CONCENTRATION

We saw in Chapter 4 that the proportion of solute to volume of solution is referred to as a solution's concentration. The most common quantification of concentration in chemistry is **molarity (M)**, which is defined as moles of solute per liter of solution:

$$molarity = \frac{moles\ of\ solute}{liters\ of\ solution}$$

Example: What is the molarity of 4.0 L of aqueous solution containing 1.75 moles of NaCl?

$$molarity = \frac{1.75\ mol\ NaCl}{4.0\ L\ solution} = 0.44\ \frac{mol}{L} = 0.44\ M$$

We can also use brackets to indicate a reference to a solution's concentration. For the solution above, we'd write: [NaCl] = 0.44M.

TRY THIS:

What is $[CaCl_2]$ if 2.5L of solution contain 1.3 mol $CaCl_2$?

Answer: [CaCl2] = 0.52M.

Solution:

$$molarity = \frac{1.3\ mol\ CaCl_2}{2.5\ L\ solution} = 0.52\ \frac{mol}{L} = 0.52\,M$$

We can use molarity as a conversion factor to determine how many moles of a solute are in a given volume of solution, or what volume of solution would contain a given amount of solute.

Example: How many moles of NH_3 are in 2.0L of a 0.50M solution?

$$2.0\ L\ solution\ x\ \frac{0.50\ mol\ NH_3}{1\ L\ solution} = 1.0\ mol\ NH_3$$

Example: What volume of a 0.85M NaCl solution contains 2.0 mol NaCl?

$$2.0\ mol\ NaCl\ x\ \frac{1\ L\ solution}{0.85\ mol\ NaCl} = 2.35\ L\ solution$$

TRY THIS:

1.5L of solution contains how many moles of KF, if [KF] = 0.75M?

Answer: 1.1 mol KF.

Solution:

$$1.5\ L\ solution\ x\ \frac{0.75\ mol\ KF}{1\ L\ solution} = 1.1\ mol\ KF$$

TRY THIS:

What volume of solution contains 0.50 mol NH_3, if $[NH_3]$ = 1.25M?

Answer: 0.40 L solution.

Solution:

$$0.50\ mol\ NH_3\ x\ \frac{1\ L\ solution}{1.25\ mol\ NH_3} = 0.40\ L\ solution$$

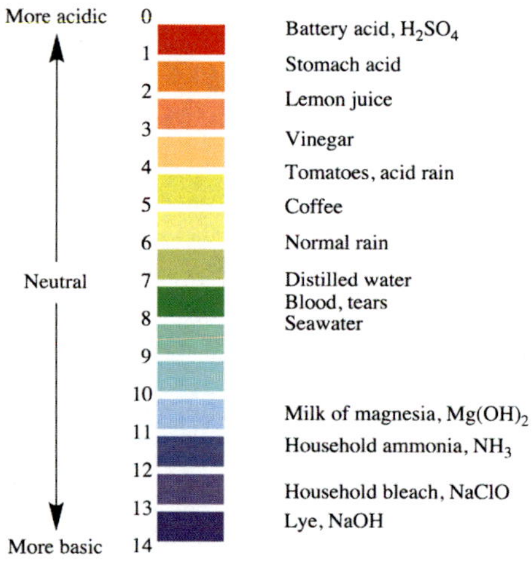

Image 5.4. The pH scale. Common household items have pHs from 0–14.

5.6 THE PH OF A SOLUTION

Concentration and molarity are very relevant to our discussion of acid. We can use the concentration of H^+ in an acidic solution to calculate the **pH** of the solution. Most of us have heard of pH in reference to a solution's acidity—pH values fall between 0 and 14 for most routinely encountered solutions (Image 5.4). If a solution has a pH of 7, the solution is referred to as **neutral**; that is, it's neither acidic nor basic. Solutions with a pH greater than 7 are **basic**, and solutions with a pH less than 7 are **acidic**.

Mathematically, pH is calculated as:

$$pH = -\log[H^+]$$

Example: What is the pH of a solution for which $[H^+]$ is 0.050M?

$$pH = -\log[H^+] = -\log(0.020) = 1.3$$

Example: What is $[H^+]$ in a solution of pH = 2?

$$pH = 2 = -\log[H^+],\ 10^{-2} = 0.01M$$

There are two important points to notice regarding the pH scale. First, a lower pH corresponds to a higher $[H^+]$, and a higher pH corresponds to a lower $[H^+]$. Second, because pH is a logarithmic function, a decrease of one pH point corresponds to a tenfold increase in $[H^+]$. A two-point decrease in pH corresponds to a hundredfold increase in $[H^+]$, and so forth.

Example: How many times more $[H^+]$ is in a solution of pH 4 vs. in a solution of pH 7?

For pH 4: $pH = 4 = -\log[H^+],\ 10^{-4} = [H^+] = 0.0001M$

For pH 7: $pH = 7 = -\log[H^+],\ 10^{-7} = [H^+] = 0.0000001M$

$[H^+]$ is 1000 times higher in the solution of pH 4.

TRY THIS:

What is the pH of a solution for which $[H^+] = 0.64M$?

Answer: pH = 0.19.

Solution:

$$pH = -\log[H^+] = -\log(0.64) = 0.19$$

TRY THIS:

What is $[H^+]$ for a solution with a pH of 12.2?

Answer: [H+] = 6.31 x 10-13.

Solution:

$$pH = 12.2 = -\log[H^+], 10^{-12.2} = [H^+]$$
$$= 0.000000000000631M = 6.31 * 10^{-13} M$$

5.7 NEUTRALIZATION

One important bit of chemistry involving acids and bases is the way they interact. When acids and bases react in solution, they **neutralize** each other. To understand this, remember that acids dissociate in water to produce H^+, while bases "steal" H^+ ions from water. Put together in solution, the base will simply react with the H^+ ions that have been liberated from acid rather than reacting with water—this is because it's far quicker and easier for a base to take a hydrogen ion that's simply floating in solution than having to take one off a water molecule. Note in the reactions below that left to its own devices, HCl dissociates (top reaction), and NaOH abstracts an H^+ from water (middle). Together in solution, HCl and NaOH do not involve water in their reaction. Instead, NaOH abstracts a hydrogen directly from HCl (bottom):

Acid dissociates to produce H^+

Base dissociates to produce OH^-

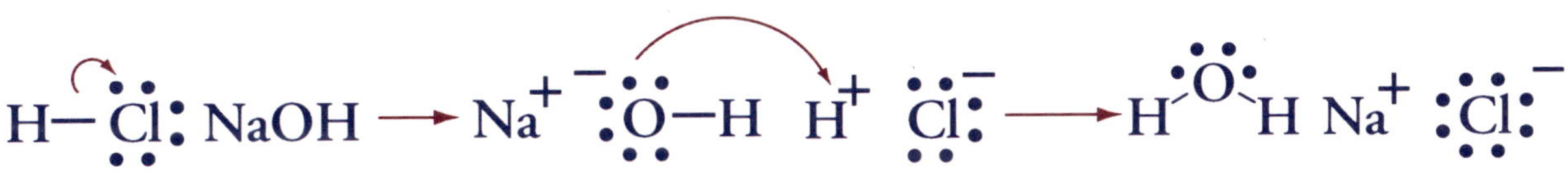

When HCl reacts with NaOH, the OH^- from NaOH bonds with the H^+ from HCl, producing water. The Na^+ and Cl^- are left over, and form NaCl in solution.

Neutralization reactions involving bases that contain the OH^- anion produce water and an ionic salt. The salt is always formed by spectator ions—the reaction above, for instance, forms NaCl. An interesting application of acid neutralization is the use of antacids to relieve the symptoms of acid stomach. Our stomachs produce an acidic secretion called *gastric acid*, which consists largely of HCl in solution at a pH of approximately 2, and which is used to chemically break down the food we eat. Occasionally, insults to the stomach (including eating greasy or very spicy foods, or excessive alcohol consumption) can result in overproduction of gastric acid, leading to a sensation of heartburn or indigestion. This is treated by neutralizing the excess acid with a base, often magnesium hydroxide $Mg(OH)_2$, commonly called *milk of magnesia*). When $Mg(OH)_2$ neutralizes HCl in stomach acid, water and an ionic salt are formed. In this case, because the spectator ions are Mg^{2+} and Cl^-, the ionic salt will be $MgCl_2$. Our balanced chemical reaction will look like this:

$$2\ HCl + Mg(OH)_2 \rightarrow 2\ H_2O + MgCl_2$$

Note that since $Mg(OH)_2$ has TWO hydroxide units (each of which need a hydrogen with which to react), TWO HCl molecules are required to complete the neutralization. This results in the formation of two molecules of water. Conveniently, it also provides the two chloride anions necessary to balance the charge on the magnesium cation so that a neutral salt can form.

Another common household remedy for acid indigestion is to make a solution of baking soda and water. Baking soda is the ionic salt $NaHCO_3$, or sodium bicarbonate. Bicarbonate (HCO_3^-) is a polyatomic anion. The neutralization reaction of HCl and $NaHCO_3$ is very similar to the reaction of HCl with a hydroxide-containing base. Just as it always does, HCl dissociates in aqueous solution to produce H^+ and Cl^-. $NaHCO_3$, like all ionic salts, dissociates in solution to produce Na^+ and HCO_3^-. HCO_3^- acts as a base, taking

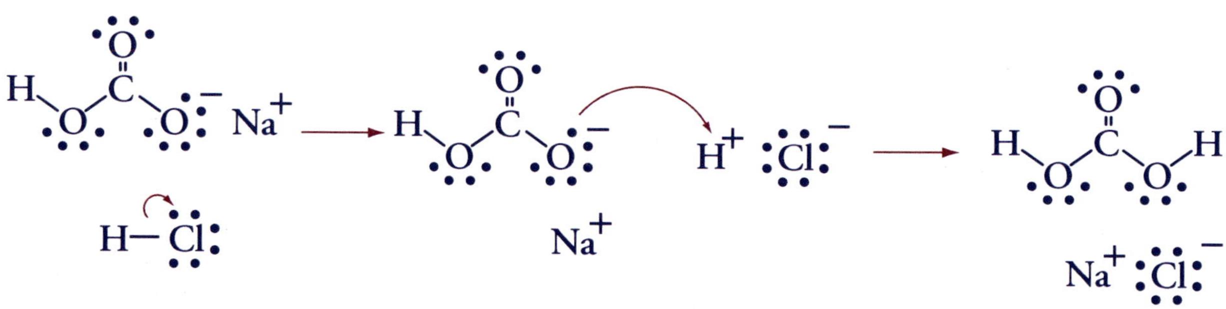

the H^+ liberated from HCl. Na^+ and Cl^- are spectator ions:

The products of this reaction are an ionic salt (NaCl), and the compound H_2CO_3, carbonic acid, which would be acidic except that it is quite unstable, and quickly reacts to form carbon dioxide and water:

$$H_2CO_3 \rightarrow CO_2 + H_2O$$

We saw this reaction going in the opposite direction in Chapter 2—carbon dioxide and water can react to produce carbonic acid. In fact, the direction in which the reaction proceeds is determined by the relative concentrations of the reactants and products. In a system in which there is lots of CO_2, the CO_2 will tend to combine with water to form carbonic acid. This is the reason that an individual who holds their breath and accumulates CO_2 in the bloodstream will produce carbonic acid, lowering the pH of the blood. If there is a lot of carbonic acid relative to the amount of CO_2, the carbonic acid will tend to fall apart. When an individual takes a bicarbonate antacid, as quickly as CO_2 is formed, it's expelled via burping. As a result, CO_2 concentrations in the stomach stay low, and carbonic acid keeps reacting to form carbon dioxide and water. Since carbon dioxide from the antacid is expelled, the neutralization products left in the stomach are simply NaCl and water. As an antacid, baking soda is just as effective as milk of magnesia, but they each have their own unique side effects. The large amount of CO_2 generated by

Concept Check:

Why will a base react with an acid instead of with water in aqueous solution?

Answer: It's easier to take a hydrogen ion from acid than from water.

baking soda can lead to burping, and milk of magnesia is not only an antacid, but also a laxative!

5.8 CHEMISTRY OF BAKING

The production of carbon dioxide by the reaction of sodium bicarbonate and acid makes baking soda useful for much more than neutralizing stomach acid. An important part of baking is the ability to **leaven** bread, or cause the bread to rise. This is accomplished by chemically generating gas within the dough matrix, resulting in air bubbles in the finished baked goods. Leavened baked goods like cakes, cookies, and many breads, have a lighter texture than unleavened baked goods such as tortillas and pita bread.

One technique used to produce leavening is to incorporate baking soda into the dough mixture. In the presence of an acidic ingredient such as buttermilk or yogurt (both of which contain lactic acid), the baking soda will neutralize the acid, producing carbonic acid:

$$CH_3CHOHCOOH \text{ (lactic acid)} + NaHCO_3 \rightarrow CH_3CHOHCOO^- \text{ (lactate)} + Na^+ + H_2CO_3$$

The carbonic acid then reacts further to produce water and carbon dioxide. As the dough bakes, the carbon dioxide gets trapped in the dough, expands, and causes the dough to rise:

$$H_2CO_3 \rightarrow CO_2 + H_2O$$

This reaction is quite rapid, and the dough can be baked immediately (in fact, breads made with baking soda are often called *quick breads*). Because of the rapid generation of carbon dioxide upon contact of baking soda with an acidic ingredient, however, it's important not to over-mix quick breads or cookies—to do so would prematurely release the CO_2, and it wouldn't get trapped in the dough.

A drawback to leavening baked goods with baking soda is that the leavening action requires acid. Since a common source of acid for cooking is lactic acid from partially fermented dairy products (which have a distinct sour flavor), this impacts the flavor of the final product. An alternative to baking soda is baking powder, which is baking soda with added starch and an acidic salt, often cream of tartar or calcium hydrogen phosphate. The acidic salt precludes the need to use a wet acid ingredient—the salt is activated (and becomes acidic) as soon as any moisture is added to the dry ingredient mixture.* The starch in baking powder helps to slow the reaction time. It does not affect overall CO_2 production, but slowing the rate at which the gas is formed helps to space out the air bubbles more evenly and prevents flattening of bread due to over-mixing.

In practical terms, then, knowing the chemistry of leavening allows us to make substitutions in recipes. Where a recipe calls for baking soda, it's ok to use baking powder instead (but because the baking powder consists of things other than $NaHCO_3$, it's generally necessary to add about three times as much

Image 5.5. The tortillas on the top are unleavened and don't contain air bubbles in the dough. The French bread below them, however, has been allowed to rise, resulting in its light texture. Gaps in the dough matrix formed by trapped carbon dioxide are clearly visible in the slice of bread.

***NOTE**

Kitchen hint: This is the reason many recipes instruct you to mix dry ingredients in one bowl, wet ingredients in a second bowl, and then combine the dry and wet right before baking—mixing separately helps to avoid activating the dry acid salt too soon!

baking powder as the recipe calls for soda). Where a recipe calls for baking powder, though, it won't work to substitute baking soda, since this substitution reduces the amount of acid in the mixture. Some chefs will add a bit of vinegar to a recipe—particularly if the taste will be covered up by other strong flavors or complemented by bitter flavors such as chocolate—to serve as an acid should they need to substitute baking soda for powder. Others will use lemon juice, and still others will replace part of the milk or water in a recipe with yogurt or buttermilk.

Incidentally, baking soda and baking powder are not the only ways to leaven baked goods. A common alternative used more frequently in breads and biscuits than in cookies or cakes is yeast. Yeast are fungal microorganisms (*saccharomyces cerevisiae*) that consume sugar for energy and release CO_2 as a waste product. As a result, if yeast are given a source of nutrition and sufficient time, they will leaven bread through their natural metabolic action. If you look at a bread recipe that's leavened with yeast, you'll notice a few key elements. First, the yeast (purchased dry) are reconstituted in water that has a bit of sugar stirred in. The sugar provides the yeast with a food source. Second, once the dough is mixed up, the yeast are given time to metabolize (digest) the

Image 5.6. Freshly mixed yeast dough (left), and the same dough after having been allowed to rise for nearly an hour (right).

sugar, and produce CO_2. Yeast breads often need to rise for an hour or more, and recipes recommend that the dough be kept covered in a warm location (in order to keep the yeast active) during the rising period.

5.9 ACID AND MICROBES

While yeast metabolize sugar to produce carbon dioxide, other microbes, including the bacterial species *Lactobacillus acidophilus*, metabolize sugar to produce lactic acid. A favorite sugar of the *L. acidophilus* bacterium is lactose, the sugar in milk. While any sample of milk that's not straight from the cow (or some other animal) has a small bacterial load, the population of *L. acidophilus* and other bacteria steadily increases as milk sits in a container. The bacteria metabolize the milk sugar and produce lactic acid, which eventually causes the milk to curdle, or spoil. If this spoiling process is carefully controlled (and if only certain bacterial species are allowed to participate), the result is any one of a number of desirable products: yogurt, buttermilk, and cheese are all products of milk fermentation by *L. acidophilus*. Acid causes one of the proteins in milk, called casein, to precipitate out into soft, semisolid clumps called *curds*. These can be eaten as they are (cottage cheese is a lightly fermented cheese curd mixture, rinsed of acid to reduce the sour taste) or further processed (sometimes with the help of additional bacteria or fungi) into other types of cheese.

While some microbes are incredibly beneficial to humans, we are susceptible to infection by others. A function of gastric acid in the stomach, beyond simply digesting food, is to help protect us from infection by ingested bacteria, many of which are acidsensitive. We harbor native bacterial populations in our digestive tracts, including several strains of *Escherichia coli*. Other strains of *E. coli*, however, are incredibly harmful to us and if ingested via consumption of contaminated food can cause infection. In general, the harmful strains (including *E. coli* O157:H7, responsible for numerous human deaths in recent years) produce toxins that can severely damage human organs. Thankfully, most *E. coli* are unable to survive in the stomach's low pH environment, and are killed shortly after they're swallowed. In an interesting interaction between chemistry and society, however, our desire to provide ourselves with a ready supply of meat has resulted in the selective breeding of acid-tolerant strains of *E. coli* and an increased incidence of human infection by the bacteria.

Cattle are one of a number of animals with multiple digestive compartments, in contrast to our single stomach. Unlike the very acidic stomachs of humans, the digestive tract of the cow is relatively neutral in pH (slightly above a pH of 7) under normal circumstances, when the cow is being fed its native diet of grasses and hay. In the advent of industrial agriculture, however, it was determined that high-calorie grains such as corn cause cattle to put on weight very quickly, thus bringing them up to appropriate slaughter size in record time. While the corn feeding of cattle has ramifications far beyond its effect upon their weight and digestive tracts, it is nevertheless the digestive effect of the corn that is most relevant here. Grain, it turns out, significantly acidifies the cows' digestive systems. Not only is this harmful to the cow (resulting in ulceration of the digestive organs, among other things), it also provides the perfect environment for acid-tolerant bacteria. When cattle are slaughtered, meat often becomes contaminated with colonic bacteria. Under ordinary circumstances (in the case of a grass-fed cow) this is not particularly dangerous to consumers of the meat. The bacteria that live comfortably in a cow's neutral digestive tract are quickly killed by the acidic environment of the human stomach. Corn-fed beef, however, becomes contaminated at slaughter with acid-tolerant bacteria that are NOT killed upon contact with stomach

Image 5.7. (Top to bottom) Cheddar is an aged, heavily salted cheese. The holes in Swiss cheese come from CO_2 produced by the bacteria of the family Propionibacter. The blue veins in Stilton and other blue cheeses are due to the fungus Penicillium roqueforti.

Image 5.9. Acid for industrial applications is transported in containment vehicles like this one.

acid, and if these bacteria are of a toxic strain (such as *E. coli* O157:H7) they will produce potentially lethal infections in humans.

5.10 ACID SPILLS AND INDUSTRIAL ACID DISASTERS

We've seen so far that acids are important and useful to us in many ways, including food preparation and chemical protection from microbe infection. Unfortunately, however, the very property of acid that makes them dangerous to microbial invaders also makes them damaging to structures and non-microbial living creatures. Acids, particularly if concentrated, have the ability to erode solid surfaces and burn living tissue. As a result, acid spills and accidents can have disastrous consequences on humans and on the environment.

HF, hydrofluoric acid, is a common industrial acid used in oil refining. It is also used in the production of stainless steel and to clean silicon for use in semiconductors. Due to some of its unique properties as an acid, however, HF also represents a unique toxicity liability in the case of an accidental spill.

Many acids, including HCl (hydrochloric acid) and H_2SO_4 (sulfuric acid), react at the surface of the skin, causing a painful burn on contact. Unlike most acids, however, HF is absorbed into deep tissues, where it interferes with nervous system function and often prevents the burn victim from feeling the extent of their injuries. This often delays burn treatment, and can result in exacerbation of tissue damage. Further, upon dissociation, the F^- ion has the ability to bind to Mg^{2+} and Ca^{2+} ions, which are found in the bloodstream. Ca^{2+}, in particular, is vital to normal biological function, and formation of CaF_2 (which has very low solubility in the blood, and precipitates out of solution) leads to lowered concentrations of Ca^{2+}. This can result in extreme

Image 5.8. Symbols commonly associated with hydrofluoric acid. The skull and crossbones represent toxicity, while the "corrosive" symbol indicates the acid's ability to damage surfaces and tissue. Four-colored "fire diamonds," like the diamond for HF shown here, are standard mechanisms for providing information about chemicalsand are found on chemical bottles and transport vehicles (as well as anywhere else industrial chemicals are routinely encountered). The number in red is an indication of flammability, blue represents health hazard, and yellow indicates reactivity. Additional important information is abbreviated in the white area. Chemicals are ranked on a scale of 0–4 in each of these categories. HF is not flammable, represents a maximum health hazard, is moderately reactive, and is corrosive.

tissue damage and pain, and can even lead to complications such as cardiac arrhythmias (irregular heartbeat) in severe cases. To prevent the formation of CaF_2 from blood calcium, HF burns are generally treated with the chemical calcium gluconate, which is an ionic salt of Ca^{2+}. F^- therefore forms ionic bonds with the Ca^{2+} from the calcium gluconate (simply because there's a lot of it), which prevents it from bonding to blood calcium. Of course, large-scale spills of hydrofluoric acid represent a tremendous danger to the surrounding community, as evidenced by the 2009 evacuation of 5000 local residents following the spill of 33000 pounds of hydrofluoric acid when a truck overturned near Plainfield, Pennsylvania.

5.11 THE LAST WORD—FERMENTATION

Earlier in this chapter, we discussed that yeast consume sugar and can be used to produce carbon dioxide for leavening baked goods. The other byproduct of sugar metabolism by yeast is ethanol (CH_3CH_2OH)—the alcohol in beer and wine. While ethanol concentrations in yeast bread are so low as to be inconsequential (the alcohol bakes out), the very same chemical reaction responsible for leavening bread also produces alcohol and carbonation (CO_2 dissolved in liquid) in alcoholic beverages. As in baking, the yeast require a source of sugar. Commonly used sources include malt (to make beer) and pressed grapes (to make wine). There are many different strains of yeast—specifically, the organism *Saccharomyces cerevisiae* is used in many food applications.

Why do yeast produce carbon dioxide and ethanol? With regard to carbon dioxide, yeast make it for the same reason we do—chemically breaking down sugars* for energy results in the production of CO_2. Frankly, most organisms that use sugar (or starch) for energy are going to end up producing carbon

***NOTE**

While the chemical nature of sugars and starches will be discussed in more detail in Chapter 8, it's worth knowing that starches are chemically very similar to sugars and are processed by organisms in much the same way.

Image 5.10. Beer and wine are both produced through the same reaction that leavens bread—yeast consume sugar and produce carbon dioxide and ethanol (alcohol) as by-products.

dioxide waste. Though yeast are not the only organism that produces CO_2, they are used for leavening bread because they are noninfectious to us, and because they are quite small and can easily be added to a dough mixture. The fact that they produce ethanol (which we, other animals, and bacteria do not) is a function of their particular metabolic mechanism, which is called fermentation. Early cultures discovered fermentation—and the interesting effects of consuming fermented beverages—when airborne yeast got into stored containers of food. Consumption of these yeast-fermented foods and beverages resulted in the feelings we associate with consumption of beer and wine. Just as many individuals now find these effects enjoyable, so did early peoples, and they worked to determine ways of purposely fermenting various substances. Any starch-containing substance can be fermented. While wine and beer are made by many different cultures, grapes and grain are by no means the only fermentable foods. Geographical and cultural regions have characteristic alcoholic drinks, representative of their most ready source of fermentable sugar or starch. For instance, vodka can be made from potatoes, rum from sugar cane, and tequila from the agave plant. Some Asian cultures even ferment mare's milk.

Cultures throughout the world find ethanol to have pleasant side effects, including social relaxation, euphoria, and a feeling of well-being. However, it's worth noting that chemically, ethanol is a toxin; for this reason, individuals under the influence of alcohol are referred to as *intoxicated.* Ethanol has a short-term effect on the nervous system, affects consciousness and thought, and changes the activity of certain brain chemicals. In high doses, it can cause acute alcohol toxicity, which includes decreased blood flow to the brain, depressed breathing, unconsciousness—even death. Because ethanol is toxic, the liver (our major toxin-processing organ) works hard to remove it from the body. As a result, excessive alcohol use over time severely damages the liver, leading to *cirrhosis* (scarring of the liver) in some cases. Interestingly, ethanol isn't just toxic to us—it's also toxic to yeast! While low concentrations of ethanol are tolerable to yeast, they die at higher concentrations. For this reason, it used to be rare to find wine containing more than 12–13 percent alcohol; that was the maximum concentration the yeast could survive. Now, however, wines commonly contain up to 15 percent alcohol, thanks to the efforts of biochemists and microbiologists who have been selectively producing strains of yeast with higher ethanol tolerance!

SUMMARY OF MAIN POINTS

- Acids dissociate in water to produce an anion and the H+ cation.
- Most acids are binary (usually hydrogen with a halogen), oxyacids, or carboxylic acids (-COOH).
- When an acid dissociates, the electrons from the bond to hydrogen remain with whatever atom hydrogen was bonded to. This produces an anion.
- Ions composed of two or more atoms covalently bonded to one another are called polyatomic ions.
- Polyatomic ions form ionic compounds with other ions—compounds can be predicted from the charges of the ions involved, and nomenclature is comparable to that of binary ionic compounds.
- Bases react with water, taking a hydrogen from it to produce the OH- anion.
- Our perception of flavor is primarily based on smell, but there are five main tastes; each one provides important information: sweet (sugar), sour (acid), salty (sodium), bitter (potential toxin), and meaty (protein).
- The concentration of a solute in a solvent is often measured in moles of solute per liter of solution and is called the solution's molarity (mol/L, or M).
- The acidity of a solution is related to the molarity of H+, and is measured as pH, where pH = -log[H+].
- Typical pH values range from 0–14, where lower pH corresponds to a more acidic solution, and pH of 7 is neutral (distilled water).
- Acids and bases react with each other through neutralization reactions.
- Common examples of neutralizations include antacids used for heartburn and the leavening action of baking soda (base) with an acid.
- Baked goods can also be leavened with yeast, which metabolize sugar to produce CO_2.
- Lactic acid is produced by some microorganisms and is used in the food industry to make cheese.
- Acid in the stomach protects us from microorganism infection, but conditions that select for acid-tolerant bacteria (such as the acidic digestive tract of corn-fed cattle) produce bacteria that can't be killed by our acidic stomach, which can cause epidemic infection.
- Acids have the potential to be very damaging, particularly in the case of large-scale spills or industrial accidents.
- Hydrofluoric acid is a commonly used industrial acid that, if spilled, has the potential to be highly toxic or deadly.

Questions and Topics for Discussion

1. In what ways does hydrogen act like a group 1 element? Like a group 7 element? How should it be grouped on the periodic table?
2. In general, how do acids behave in water?
3. Some acids are stronger than others. For instance, HCl is a stronger acid than H_2S. Based upon what you know about the periodic table, propose a reason.
4. Is the bond between H and Cl in HCl more properly characterized as covalent or ionic? Justify your answer.
5. Why do you think it becomes increasingly important to write out chemical formulas in atomic groupings (i.e., CH_3COOH as opposed to $C_2H_4O_2$) as molecules get larger?
6. What are some of the acids you've encountered in everyday life, and where have you come across them?

7. Why do polyatomic ions have charges? Use the example of nitrate, below, to compare the number of protons to the number of electrons (TOTAL!) in the molecule:

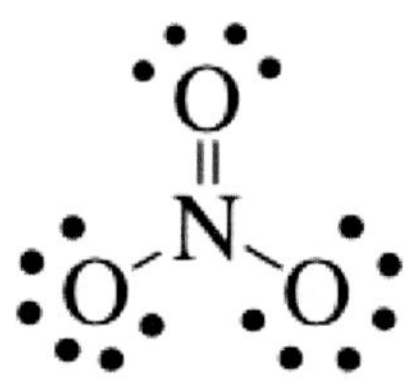

8. Why does the polyatomic ion ammonium (NH_4^+) form when ammonia reacts with water?
9. In general, how do bases behave in water?
10. In the reaction of KOH with water, is there a spectator ion? If so, what is it, and what makes it a spectator?
11. Why do you think it's important for us to taste the basic flavors of sweet, sour, salty, bitter, and meaty?
12. Why do you think it's important to use smell to get information from our food above and beyond the five basic flavors?
13. Describe the sensation of eating with a stuffy nose. If you can't remember the last time you did so, try eating while holding your nose. How does this compare to a typical eating experience?
14. Think of three bitter foods that children typically don't like, but that adults tend to enjoy. Do a little Internet research—are these foods alkaloid-containing?
15. Why is it necessary to have a measure of concentration for solutions?
16. Molarity is defined as moles of solute per liter of solution. Would a solution made by adding 1 mol of NaCl to 1 liter of water be 1 M? Why or why not?
17. What is the difference in hydrogen ion concentration between a solution of pH = 1 and pH = 2? Explain.
18. Which is more acidic: a solution of pH = 1 or a solution of pH = 2? Explain.
19. Why does pH decrease as acidity increases?
20. Why do you think the pH scale is logarithmic rather than linear? What purpose does this serve?
21. Explain the general mechanism through which a base reacts with an acid. What is this kind of reaction called?
22. How are acid-base neutralizations useful in everyday life? Think of an example besides those mentioned in the text.
23. Find a recipe that calls for baking soda, but not baking powder (you may need to use the Internet as a reference). Is there a source of acid in the recipe? If so, what is it? How could you modify the recipe if you had no baking soda?
24. Find a recipe that calls for baking powder, but not baking soda (you may need to use the Internet as a reference). Is there a source of acid in the recipe? If so, what is it? How could you modify the recipe if you had no baking powder?
25. How can yeast be used to make bread dough rise?
26. How can microorganisms be used to make cheese?
27. What is the pH of the human stomach, and what effect does this normally have on microorganisms?
28. Why does which food cattle eat affect the likelihood of human infection upon eating meat from the cattle?
29. What are some of the applications for which hydrofluoric acid is useful?
30. How does hydrofluoric acid act similarly to other acids in the case of a burn? How does it act differently?

Problems

1. Using arrows to indicate where electrons in bonds go, show the behavior of each of the following acids in water: CH_3COOH; HBr; and HI.
2. Using arrows to indicate where electrons in bonds go, show the behavior of each of the following acids in water: CH_3CH_2COOH; HF; and H_2S.
3. Define each of the following: Acid; polyatomic ion; binary acid; oxyacid; and carboxylic acid.
4. Categorize each of the following as a binary acid, oxyacid, or carboxylic acid: HF; H_3PO_4; CH_3CH_2COOH; and H_2S.
5. Categorize each of the following as a binary acid, oxyacid, or carboxylic acid: HI; HBr; $HClO_4$; and HCOOH.
6. Provide the chemical formula and name for a compound made of each of the following pairs: Ammonium and bromine; ammonium and carbonate; sulfate and cesium; and carbonate and magnesium.
7. Provide the chemical formula and name for a compound made of each of the following pairs: Lithium and hydroxide; phosphate and sodium; phosphate and ammonium; and nitrate and magnesium.
8. Provide the chemical formula for each compound: Rubidium phosphate; ammonium bicarbonate; calcium sulfate; and calcium phosphate.
9. Provide the chemical formula for each compound: Barium carbonate; barium nitrate; strontium phosphate; and sodium hydroxide.
10. Provide the name for each of the following: NH_4NO_3; $LiHCO_3$; Li_3PO_4, and $(NH_4)_3N$.
11. Provide the name for each of the following: NH_4Cl; $Ca(NO_3)_2$; $SrSO_4$; and $Sr(OH)_2$.
12. Using arrows to indicate where electrons in bonds go, show the behavior of LiOH in water.
13. Using arrows to indicate where electrons in bonds go, show the behavior of CH_3NH_2 in water (hint: the N acts just like the N on ammonia!).
14. Calculate the molarity of a solution containing 1.5 mol KF in 2.5 L.
15. Calculate [HCl] in a solution containing 0.25 mol HCl in 5.2 L.
16. Calculate [NaCl] in 5.0 L of a solution containing 3.5 mol NaCl.
17. Calculate $[NH_3]$ in 3.5 L of a solution containing 1.25 mol NH_3.
18. How many moles of $Ca(OH)_2$ are in 0.75 L of a 2.5 M solution of $Ca(OH)_2$?
19. How many moles of $NaHCO_3$ are in 1.4 L of solution, if $[NaHCO_3]$ = 1.2 M?
20. What volume of a 2.2 M solution of LiOH contains 0.65 mol LiOH?
21. What volume of a 0.45 M solution of $NaNO_3$ contains 1.5 mol $NaNO_3$?
22. What is the $[H^+]$ in a solution of pH = 2.25?
23. What is the hydrogen ion concentration in a solution of pH = 12.4?
24. What is the pH of a solution in which $[H^+]$ = 2.2×10^{-2} M?
25. What is the pH of a solution with a hydrogen ion concentration of 5.7×10^{-13}M?
26. What is the pH of a 1 M solution of HCl (assume all the HCl dissociates)?
27. How does $[H^+]$ in a solution of pH = 6 compare to that of pH = 9?
28. How does $[H^+]$ in a solution of pH = 2 compare to that of pH = 4?
29. Which pH values are acidic? Basic? Neutral?
30. Using arrows to show where electrons in bonds go, show the behavior of each of the following pairs in water: LiOH and HI; HBr and NH_3; and HCOOH and $NaHCO_3$.
31. Using arrows to show where electrons in bonds go, show the behavior of each of the following pairs in water: HCl and KOH; NH_3 and CH_3COOH; and HBr and $NaHCO_3$.
32. Write out a balanced chemical equation for the reaction of $Ca(OH)_2$ and HCl.
33. Write out a balanced chemical equation for the reaction of HNO_3 and NaOH.
34. Write out a balanced chemical equation for the reaction of NH_3 and HI.

35. Write out a balanced chemical equation for the reaction of H_2SO_4 and LiOH (hint: both H's will ionize in sulfuric acid).
36. Write out a chemical reaction showing how baking soda ($NaHCO_3$) produces carbonation for leavening bread. Use CH_3COOH as your acid.

Answers to Problems

1.

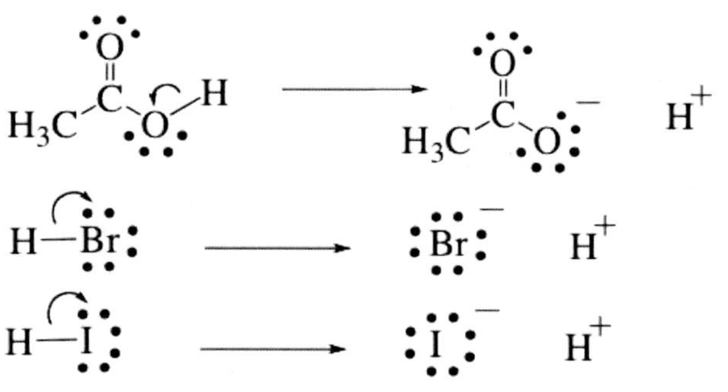

3. An acid is a chemical that dissociates to produce an H^+ and an anion in water; a polyatomic ion is a charged group of covalently bonded atoms; a binary acid is made up of a hydrogen and one other element; an oxyacid is made up of hydrogen, oxygen, and one other element; a carboxylic acid contains the characteristic arrangement of atoms –COOH.
5. HI is a binary acid, HBr is a binary acid, $HClO_4$ is an oxyacid, HCOOH is a carboxylic acid.
7. LiOH, lithium hydroxide; Na_3PO_4, sodium phosphate; $(NH_4)_3PO_4$, ammonium phosphate; $Mg(NO_3)_2$, magnesium nitrate.
9. $BaCO_3$, $Ba(NO_3)_2$, $Sr_3(PO_4)_2$, NaOH.
11. Ammonium chloride, calcium nitrate, strontium sulfate, strontium hydroxide.
13.

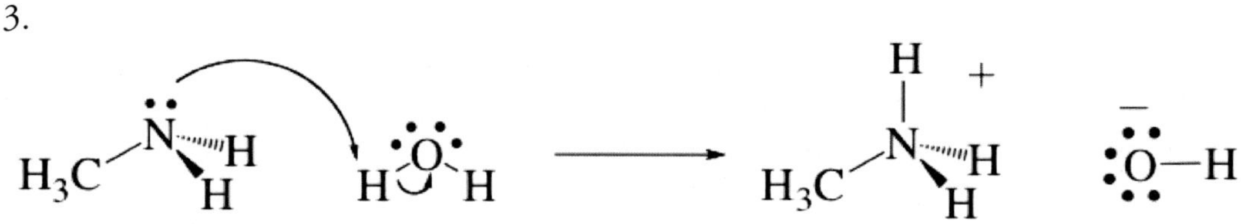

15. 0.048 M.
17. 0.36 M.
19. 1.7 mol.
21. 3.3 L.
23. 3.98×10^{-13} M.
25. pH = 12.2.
27. pH = 6 has 1000x higher $[H^+]$ than pH = 9.
29. pH values less than 7 are acidic, 7 is neutral, greater than 7 is basic.

31.

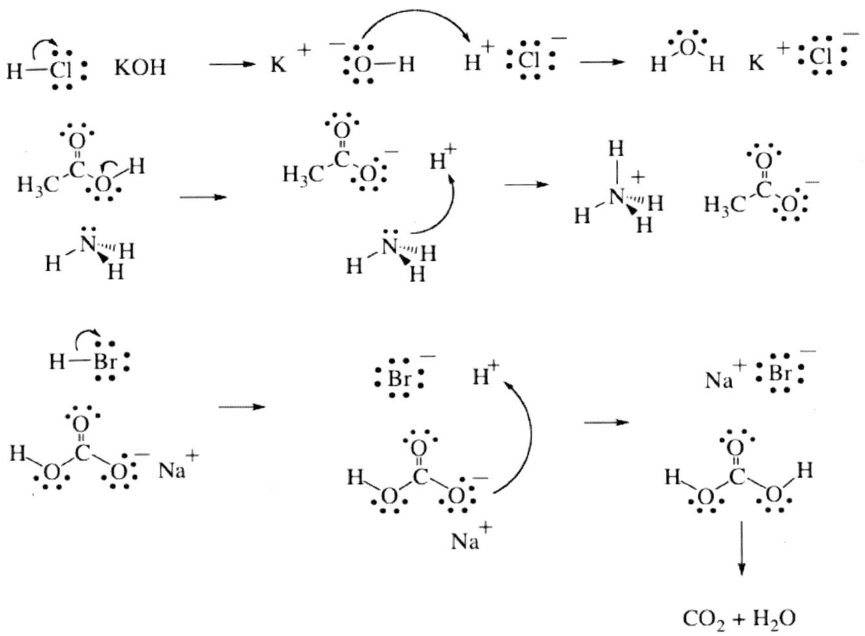

33. $HNO_3 + NaOH \rightarrow H_2O + NaNO_3$
35. $H_2SO_4 + 2\ LiOH \rightarrow 2\ H_2O + Li_2SO_4$

Sources and Further Reading

Taste

Steiner, J.E. (1974). Innate, discriminative human facial expressions to taste and smell stimulation. *Annals of the New York Academy of Sciences*, 237, 229–33.

Pathogens and Acidity

Callaway, T.R., Carr, M.A., Edrington, T.S., *et. al.* (2009). Diet, *Escherichia coli* O157:H7, and cattle: a review after 10 years. *Current Issues in Molecular Biology*, 11, 67–79.

Diez-Gonzales, F., Callaway, T.R., Kizoulis, M.G., *et. al.* (1998). Grain feeding and the dissemination of acid-resistant Escherichia coli from cattle. *Science*, 281(5383), 1666–68.

Pennsylvania Acid Spill

CNN. (2009, March 21). Acid spill evacuation ends for 5,000 Pennsylvania residents. Retrieved from http://www.cnn.com/2009/US/03/21/pennsylvania.spill/index.html.

Yeast and Alcohol

Moneke, A.N., Okolo, B.N., Nweke, A.I., *et al.* (2008). Selection and characterization of high ethanol tolerant Saccharomyces yeasts from orchard soil. *African Journal of Biotechnology*, 7(24), 4567–75.

Nevoight, E. (2008). Progress in metabolic engineering of Saccharomyces cerevisiae. *Microbiology and Molecular Biology Reviews*, 72(3), 379–412.

Pretorius, I.S., (2000). Tailoring wine yeast for the new millennium: novel approaches to the ancient art of winemaking. *Yeast*, 16(8), 675–729.

6 The Chemistry of Pollution

If you live in a major metropolitan area, no doubt you've looked out over the city and seen a dense, smoggy cloud obstructing your view. This cloud (which can vary in color from white to blue through gray to brown) is composed of various atmospheric pollutions that are produced as the result of human activity.

In addition to being aesthetically unappealing, pollution is harmful. Studies show that high levels of air pollution are associated with greater numbers of hospitalizations for headache, cardiovascular events, respiratory difficulty and respiratory distress—particularly among asthmatics and sufferers of chronic obstructive pulmonary disease (COPD). Pollution produced through the combustion of coal and diesel has also been identified as a carcinogen. In addition to direct health effects, pollution results in the formation of acid rain, leads to climate change and alteration of the environment, and depletes the ozone layer.* A discussion of pollution raises several important questions. From a chemical perspective, what is pollution? Where does it come from, and of what is it composed? How is it harmful to individuals and the environment? Most importantly, what can we do as a society to reduce the amount of pollution produced?

Image 6.1. Santiago, the capital of Chile, lies in a valley between the Chilean coastal mountain range on the west and the towering Andes on the east. Thermal inversions in the winter trap pollutants in the valley, making Santiago one of the most polluted cities in South America.

***Note**

There are many pollutant compounds, each with a different effect upon the atmosphere and environment. In this chapter, we'll focus only upon the toxicity of these substances. Climate change and ozone depletion will be addressed in later chapters.

6.1 COMMON POLLUTANTS—CARBON MONOXIDE

One theme we'll see as we begin our discussion of pollutants is that the vast majority of them are products of combustion reactions, either directly or indirectly. Recall from Chapter 1 that combustion is any chemical reaction in which something (often a hydrocarbon or hydrocarbon derivative) burns in the presence of oxygen.

We've already seen that methane (a gaseous hydrocarbon that is the major component of natural gas) burns in oxygen by the reaction:

$$CH_4 + 2\,O_2 \rightarrow CO_2 + 2\,H_2O$$

In an analogous reaction, octane, a hydrocarbon constituent of gasoline, also combusts:

$$2\,C_8H_{18} + 25\,O_2 \rightarrow 16\,CO_2 + 18\,H_2O$$

The octane combustion reaction shown above, however, is somewhat idealized. A significant amount of oxygen (25 molecules for every two molecules of octane) is required to completely transform octane to CO_2 and H_2O. Because combustion is an exceedingly rapid reaction, most fuel (including coal, natural gas, and gasoline) burns via an alternate reaction that requires less O_2. Octane, for instance, can react as follows:

$$2\ C_8H_{18} + 17\ O_2 \rightarrow 16\ CO + 18\ H_2O$$

Here, the major product aside from water is CO—**carbon monoxide**. This is a colorless gas with no discernible odor. As an environmental pollutant, however, it's quite toxic. The reason for carbon monoxide's toxicity arises directly from its structural similarity to oxygen.

Body cells require oxygen in order to function, and **hemoglobin** (a protein* in red blood cells) is responsible for carrying oxygen to the cells. In order to transport oxygen effectively, hemoglobin must chemically bind O_2 at the lungs, and release O_2 at the tissues. Due to its structural similarity to O_2, CO can also bind to hemoglobin. However, CO binds so strongly that, unlike oxygen, it is not released by the hemoglobin for a long period of time (many hours). Hemoglobin with CO bound is unavailable to transport O_2. If sufficient CO is inhaled into the lungs, a large percentage of the hemoglobin in the body will be incapable of delivering oxygen to the tissues, and carbon monoxide poisoning (which is a form of chemical suffocation) results. Mild carbon monoxide poisoning produces headache, nausea, and weakness, as well as exacerbation of existing cardiovascular or respiratory disease. More severe cases cause confusion, seizures, and death. The treatment for carbon monoxide poisoning requires that high-concentration oxygen be administered to the patient—this helps to saturate the lungs with O_2, and dislodges CO from hemoglobin approximately four times faster than it would otherwise be dislodged.

***Note**

Proteins will be discussed further in Chapter 8, but they're nothing more than large molecules.

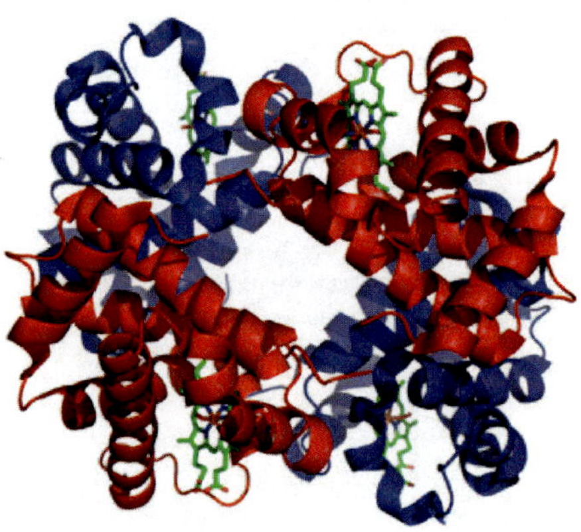

Image 6.2. Hemoglobin, a large molecule portrayed here, transports oxygen from the lungs to the tissues. It's made of protein (shown as red and blue ribbons) and four molecules called porphyrins (shown in green). Each porphyrin has an Fe^{2+} ion at the center that binds a molecule of O_2.

Concept Check:

Why do hydrocarbons produce CO when burned (rather than just carbon dioxide and water)?

Answer: It takes less oxygen to produce CO than to produce CO_2.

Carbon monoxide is both an indoor and an outdoor pollutant. Indoors, unvented heaters, leaking chimneys, wood-burning stoves and fireplaces, and gas-powered equipment can result in high levels of carbon monoxide. The EPA sets limits for acceptable levels of environmental pollution due to various contaminants. While there are no regulatory standards for indoor carbon monoxide, the EPA nevertheless recommends that measures be taken (particularly during the winter months and in colder climates) to reduce potential indoor exposure. Outdoors, carbon monoxide sources include smoke from fires, coal-fired power plants, and automobile exhaust. In particular, carbon monoxide builds up as a result of heavy and congested traffic. In order to reduce CO emissions from vehicles, the catalytic converter* was invented in the 1950s and refined in the 1970s. A catalytic converter functions by combusting carbon monoxide in more oxygen to produce CO_2:

$$2\,CO + O_2 \rightarrow 2\,CO_2$$

Catalytic converters serve the further purpose of reacting any un-combusted hydrocarbon fuel that makes its way into the exhaust system. It's worth mentioning that while catalytic converters are fairly efficient at converting carbon monoxide to carbon dioxide, we mustn't assume that CO_2 is a harmless solution to the problem. As we'll discuss in Chapter 13, CO_2 is a potent greenhouse gas, and has been identified by scientists as a cause of global warming.

> ***NOTE**
>
> The word *catalyst*, as in catalytic converter, refers to a substance that speeds up a chemical reaction without being consumed in the reaction. In the case of an automobile catalytic converter, the catalyst in question is generally a metal—often platinum, palladium, or rhodium. We'll see catalysts again in Chapter 10.

6.2 COMMON POLLUTANTS—NOx

NOx (pronounced "nocks") is the collective term for mononitrogen oxides—chemicals consisting of one nitrogen atom and one or more oxygen atoms. NOx are a subclass of the **nitrogen oxides**, which are binary compounds of nitrogen and oxygen, and include NO (nitrogen monoxide), NO_2 (**nitrogen dioxide**), N_2O♦ (dinitrogen monoxide), and several others. NOx are produced when atmospheric nitrogen combines with oxygen in various proportions. Despite its abundance (78 percent of the atmosphere), its triple bond makes N_2 quite **inert**, or chemically unreactive. Under very high temperature conditions, however, O_2 decomposes to form highly reactive atomic oxygen, which in turn can react with elemental nitrogen to form NOx. The required conditions for this reaction exist in a lightning storm (a natural source of NOx) or in the combustion chamber of an automobile engine or coal-fired power plant. In order to provide oxygen for the fuel reaction, combustion chambers have an air intake. Since air is a mixture of nitrogen and oxygen (as well as traces of other substances), and since the temperature of an internal combustion engine can easily exceed 2000K, all the required pieces for formation of NOx are in place:

$O_2 + \text{heat} \rightarrow 2\,O$ (this reaction begins the process)

$O + N_2 \rightarrow NO + N$ (this reaction couples with the next to form a continuous loop)

$N + O_2 \rightarrow NO + O$

> **♦NOTE**
>
> N_2O is also commonly called *nitrous oxide*. It's sometimes used by dentists during routine procedures, in part because it is an anesthetic and reduces the sensation of pain. Because many dental procedures are not so much painful as they are psychologically uncomfortable, however, nitrous oxide's popularity may be due mostly to its powerful relaxant effect. The chemical has the ability to produce feelings of euphoria and dissociation—essentially, a sense of not particularly caring what a medical professional is doing to one's teeth!

Image 6.3. Pollution threatens not only urban areas, but the beauty of the natural world, as demonstrated by this power plant near Grand Canyon in Arizona. (Photo © Scott Lefler, 2010)

NO and NO_2 can be interconverted by some relatively complex chemistry. Suffice it to say, while most NOx are emitted as NO, the rate of conversion to NO_2 is rapid, and within minutes of emission, both NOx species are well represented in exhaust. $NO_2$is quite toxic—exposure causes respiratory irritation, particularly in individuals with underlying respiratory conditions. Long-term exposure can result in permanent damage to the lungs. It also has an acrid odor and a yellowish-brown color that can be observed in many urban pollution clouds.

In an attempt to reduce the environmental impact of the NOx pollutants (which as we'll see later in this chapter are also implicated in acid rain formation), modern catalytic converters decompose NO and NO_2 back to N_2 and O_2:

$$2\ NO \rightarrow N_2 + O_2$$

$$2\ NO_2 \rightarrow N_2 + 2\ O_2$$

Unfortunately, the catalytic material used to effect the reactions above can also result in the formation of several undesirable products, including NH_3, which is toxic upon inhalation, and N_2O, a potent greenhouse gas.

6.3 COMMON POLLUTANTS—VOLATILE ORGANIC COMPOUNDS AND POLYCYCLIC AROMATIC HYDROCARBONS

Volatile organic compounds (VOCs) are small carbon-based molecules (organic means carbon based) that vaporize easily and enter the atmosphere. **Polycyclic aromatic hydrocarbons (PAHs)** are largecarbon-based molecules. VOCs and PAHs are large classes of pollutant compounds, as opposed to pure substances, but are classifiedas groups by the Environmental Protection Agency (EPA) and other regulatory agencies due to the similarities in their sources and in the hazards they present.

VOCs are both indoor and outdoor pollutants. Indoors, common pollutant culprits include formaldehyde (CH_2O), an industrial solvent that leeches into the air from building materials and irritates mucous membranes.* Outdoors, VOCs come largely from un-combusted hydrocarbon fuel released in the exhaust of engines and coal-fired power plants. The EPA has found that outdoor VOC concentrations rarely reach levels of concern with regard to human exposure. Indoor VOC concentrations, by contrast, are often many times higher than outdoor concentrations, and activities that result in increased shedding of organic compounds from building materials—such as paint stripping—can take the indoor VOC concentration to 1000 times those found outdoors. Regardless of the low outdoor concentration of these compounds, they nevertheless lead to formation of another atmospheric pollutant—ozone (see below).

***NOTE**

Need further proof that formaldehyde is irritating to mucous membranes? It's a preservative, disinfectant, and tissue-firming agent. For these reasons, it's a common ingredient in embalming solutions, whereby it temporarily stalls softening of tissue and bacterial decay in corpses.

As a class, PAHs are much larger molecules than VOCs. Like VOCs, they are products of incomplete fuel combustion, the primary sources of which are motor vehicle engines, coal-fired power plants, and residential wood and coal-burning. PAHs are of concern as airborne pollutants because of their documented capacity as **mutagens** (substances that cause mutation) and **carcinogens** (substances that cause cancer).

6.4 COMMON POLLUTANTS—OZONE

In Chapter 10, we'll see that ozone plays a vital role in protecting our planet and its living occupants from overexposure to damaging radiation from the Sun. The ozone we depend upon to protect us, however, is not at ground level. Instead, our protective ozone layer is sequestered in the stratosphere, well away from Earth's occupants. Tropospheric ozone, however, is an environmental pollutant.

Ozone, O_3, is an **allotrope** of oxygen, meaning that it's an alternate form of the element. It's a bluish-gray gas (the color of which can sometimes be observed in summer smog clouds over sunny, hot metropolitan areas) with a sharp, almost metallic odor.* Ozone is far less stable than O_2, and in addition to being highly reactive, it's a toxic respiratory irritant. Its effects are particularly hazardous to asthmatics and others with chronic respiratory disorders. In sufficient concentrations, however, respiratory irritation from ozone is felt even by those with completely healthy lungs—increased risk of infection, shortness of breath (particularly during exertion), and coughing or sore throat are common symptoms of ozone exposure.

Ozone, unlike NOx and CO, is not a direct product of combustion. Instead, it forms when VOCs and NOx react with the Sun's ultraviolet radiation (indicated in chemical reactions by the shorthand *hv*♦). NO_2 reacts as below:

$$NO_2 + hv \rightarrow NO + O$$

$$O + O_2 \rightarrow O_3$$

The reactions by which VOCs produce ozone are quite complex and involve chemicals called **free radicals**,** which are highly reactive chemicals with unpaired electrons. In short, VOCs accelerate conversion of NO (a pollutant in its own right, but not directly implicated in ozone production) to NO_2, thus increasing concentrations of NO_2 available to react with sunlight and produce ozone.

Concept Check:

What does it mean to say ozone is a "secondary pollutant"?

Answer: It is not produced by combustion directly, but is instead produced when combustion products (NOx) react with oxygen in sunlight.

*Note

Ozone is produced naturally during lightning storms. If you've ever had the misfortune to be uncomfortably close to a lightning strike and have smelled or tasted an acrid, metallic note in the air afterward, that was ozone!

♦Note

Why is hv used to represent ultraviolet radiation in chemical equations? Actually, it's used to represent ANY electromagnetic radiation. It is interpreted to mean *radiation of the appropriate wavelength*, and it's assumed that the individual reading the chemical equation will know (or find out) what that radiation is. This concept will be discussed further in Chapter 10, but for now, hv in a chemical equation can be taken to mean *light energy*.

**Note

You've likely come across the term *free radical* in the context of nutrition and wellness; many products advertise that they reduce the activity of free radicals in the body, and thus diminish the effects of aging, risk of cancer, and so forth. We'll discuss this concept further in later chapters.

Image 6.4. PM10 and PM2.5 contribute to the opaque nature of this pollution from a processing plant in China.

Image 6.5. The dramatic difference in visibility in Bryce Canyon (southern Utah) on a clear day versus a hazy day. PM is the major cause of visibility reduction. The EPA tracks visibility in the national parks as part of an effort to control and reduce air pollution.

6.5 COMMON POLLUTANTS—PARTICULATE MATTER (PM_{10} AND $PM_{2.5}$)

> **CONCEPT CHECK:**
>
> $PM_{2.5}$ is generally considered more detrimental to health than PM_{10}. Why?
>
> Answer: Smaller particles penetrate deeper into the lungs.

Particulate matter (PM) is the collective term for solid particles and liquid droplets that may become suspended in air. There are a variety of materials that may be categorized as particulate matter: soot, dust, aerosolized metal, smoke, organic chemicals, and even small molecules.

Particulate matter is divided into two classes. *Inhalable coarse particles,* also called **PM_{10}**, are particles of diameters greater than 2.5 μm (μm is the symbol for a *micrometer*, which is 10^{-6} m, or 1/1000 of a millimeter), but less than 10 μm. These particles are able to pass into the lungs, where they can severely affect health. There are also *fine particles*, or **$PM_{2.5}$**, which have a diameter less than 2.5 μm. As these particles are smaller than PM_{10}, they penetrate more deeply into the lungs. Health effects of both PM classes include exacerbation of respiratory symptoms in those with preexisting conditions, development of respiratory symptoms (including difficulty breathing, coughing, and susceptibility to infection) in those without preexisting conditions, and irregular heartbeat or other cardiovascular symptoms. In addition, $PM_{2.5}$ pollution is the major cause of haze and visibility reduction in cities and the surrounding countryside.

6.6 COMMON POLLUTANTS—SULFUR DIOXIDE

> ***NOTE**
>
> The nitrogen in petroleum and coal can react with oxygen when burned, providing yet another mechanism for NOx formation.

All the common fuels combusted for energy are hydrocarbon in nature—natural gas is mainly CH_4, while petroleum and coal are mixtures of much larger and more complex hydrocarbons. As we'll see in Chapter 13, however, the source of petroleum and coal means they also contain organic compounds that include the elements oxygen, nitrogen,* and sulfur. Coal, in particular, can be quite high in sulfur (up to 7 percent, with sulfur content varying greatly

Image 6.6. SO2 is released in a plume of volcanic gas (left) from Halema'uma'u Crater in Hawaii. Pingyao, China (right), is often covered with a dense haze of smog due to its many coal-fired power plants.

by region). This sulfur can combine with oxygen during combustion, resulting in the formation of SO_2, **sulfur dioxide**:

$$S + O_2 \rightarrow SO_2$$

Sulfur dioxide gas is a respiratory irritant, and causes symptoms including coughing and difficulty breathing. It's released naturally (in small amounts, relatively speaking) by volcanic activity. The primary **anthropogenic** (derived from human activities) source of SO_2 is coal combustion. The gas is colorless (though the high water content of volcanic plumes gives them their characteristic white, cloudy look), and has the odor of burnt matches.

6.7 COMMON POLLUTANTS—MERCURY AND LEAD

In addition to containing sulfur, coal also contains trace amounts of the element **mercury**. When coal is burned, the mercury is **aerosolized**, or released into the atmosphere, and eventually settles into the ground and water. Aerosolized mercury can travel quite some distance before settling; the EPA has estimated that upwards of 75 percent of mercury emissions from any given source enters global atmospheric circulation and becomes widely distributed. Mercury is quite toxic—it can cause birth defects, has been linked to cancer, and is a potent **neurotoxin** (a compound that damages the brain and nervous system). Exposure is particularly dangerous to developing embryos.

Environmental mercury exposure occurs primarily through breathing contaminated air and eating contaminated food products. Mercury does not pass through an organism the way some toxins do. Instead, it accumulates in the tissues and skeletal system. This means it also **bioaccumulates** in the food chain. That is to say, when mercury pollution enters an aquatic system, it is taken up first by microorganisms such as bacteria and plankton. These

Image 6.7. The EPA recommends avoiding king mackerel and other large predator fish due to their high levels of mercury.

organisms are consumed by small invertebrates, which are in turn eaten by fish. Predatory fish prey on smaller fish, and humans consume predatory fish. Through each step in the chain, the higher organism acquires the mercury from all the prey species it consumes, meaning that mercury concentrations in predatory fish can reach startling and dangerous levels. The EPA currently recommends that consumption of shark, king mackerel, tilefish, and swordfish be avoided, as mercury levels are too high for safety. Other EPA recommendations limit amounts of various fish and seafood that should be consumed within a given period of time—these recommendations are more stringent for pregnant and nursing women, and for children.* While fish and seafood pose a fairly well recognized threat with regard to mercury exposure, there are lesser-known sources of mercury in food. Researchers have recently determined that measurable levels of mercury can be found in many samples of high-fructose corn syrup, a ubiquitous sweetener in processed foods and beverages. The significance of high-fructose corn syrup as a source of mercury in the diet depends upon the manufacturer of the sweetener as well as the dietary habits of the individual: Those who consume large amounts of soda, fast food, and processed foods are at greater risk.

*Note

The EPA's recommendations re-garding the mercury in certain fish species are not the only recommended limitations on seafood consumption. Several organizations (including the Monterey Bay Aquarium) have published lists of seafood that should be avoided due to endangerment of the species, non-sustainable harvesting practices, and non-sustainable farming (see Sources and Further Reading).

Concept Check:

Why would it be more dangerous to consume lots of tuna (a large predator fish) as opposed to consuming a lot of herring (a smaller fish)?

Answer: Mercury bioaccumulates and concentrates in organisms at the top of the food chain.

Lead, unlike mercury, is no longer a major airborne pollutant in the United States. Until the 1980s, lead was a common gasoline additive. In recent decades, however, leaded gasoline is no longer marketed to the public, and airborne lead pollution is restricted mainly to the areas surrounding lead processing plants (including manufacturers of lead batteries and other lead-containing products). These days, the most common mechanism of exposure to lead in the United States is through paint. Lead-based paint was commonly used on both interior and exterior walls of houses until 1978, at which time the U.S. government banned its use. Paint on walls of older houses, however,

Image 6.8. Lead paint on the walls of older homes (such as the paint shown in this home flooded by Hurricane Katrina) presents a health hazard to occupants, particularly children.

commonly contains lead and represents a hazard (even if the older paint has been covered over). Old lead paint frequently peels, and the lead imparts a sweet taste to the paint chips. Children, particularly very young ones, have a tendency to pick up the paint chips and consume them, owing to a natural inclination to put things in their mouths and the pleasant taste of the chips. Further, if paint is scraped or sanded down, lead contaminates the soil around the house. This provides further potential for exposure to lead if soil is aerosolized and inhaled (as it would be by children playing in the yard), or if soil is consumed (owing again to the predilection of young children to explore their environments orally). Lead, like mercury, is a potent neurotoxin.

Image 6.9. A forest in the "Black Triangle" area of the Czech Republic near the borders with Germany and Poland. Heavy post-World War II industrialization led to pollution and significant precipitation of acid rain. Spruce trees in the former forest have been decimated.

6.8 POLLUTANT RAMIFICATIONS—ACID RAIN

Apart from their intrinsic toxicities, airborne pollutants are implicated in two additional environmental concerns. The first, that most are potent greenhouse gases, will be addressed in Chapter 13. The second concern is that they are directly involved in the production of **acid rain**, a phenomenon that presents a danger to aquatic and terrestrial environments, as well as to man-made structures.

It turns out that while distilled water is neutral, with a pH of 7, normal rain is actually slightly acidic. This is due to the slight solubility of carbon dioxide (which occurs naturally in the atmosphere) in water, and the subsequent reaction of the two to form carbonic acid:*

$$CO_2 + H_2O \rightarrow H_2CO_3 \rightarrow H^+ + HCO_3^-$$

***NOTE**

Recall that we've seen this reaction several times already: The concentration of CO_2 in the bloodstream affects the pH of the blood, and the production of carbonic acid in baked goods (from the reaction of baking soda with acid) results in formation of CO_2 bubbles.

The hydrogen ion concentration, therefore, is higher in rainwater than in pure water, giving rainwater a pH of around 5.6. Other natural processes may slightly affect the pH of rain, but it's generally accepted that normal rain has a pH between 5 and 7.

Over-acidification of rain (pH < 5) cannot be accomplished by CO_2. Instead, **SOx** (monosulfur oxides including SO_2, pronounced "socks") and NOx emissions are responsible for acid rain formation. The chemistry of this process revolves around formation of the highly reactive free radical species OH·, the hydroxyl radical. We're used to seeing OH^-, the hydroxide anion. In this case, the species in question is not an ion, and has no charge. Instead, it is an oxygen atom covalently bonded to a hydrogen atom, with an unpaired electron represented as a dot. This is an example of a free radical species. The chemistry of free radicals is somewhat complex, but they all have a **radical**, or unpaired electron—that is to say, an electron that is neither part of a bond NOR part of a nonbonding pair. As you might imagine, this unpaired electron means there is a hole in the atom's valence (an octet can only be filled if all electrons are paired), and since atoms are most stable with full octets, free radicals are highly reactive. Formation of the hydroxyl radical occurs in the troposphere as a by-product of the reaction of pollutant ozone with sunlight:

$$O_3 + h\nu \rightarrow O_2 + O$$

$$O + H_2O \rightarrow 2\ OH\cdot$$

The hydroxyl radical goes on to react with NO_2, resulting in the production of the highly acidic species nitric acid (HNO_3) which, when formed in the troposphere, can dissolve in the water of forming raindrops. Once dissolved, nitric acid (like all acids) dissociates in water to produce hydrogen ions:

$$OH\cdot + NO_2 \rightarrow HNO_3 \rightarrow H^+ + NO_3$$

SO_2 also reacts in the atmosphere with water and hydroxyl radicals to form the acidic species H_2SO_3 (sulfurous acid) and H_2SO_4 (sulfuric acid):

$$SO_2 + H_2O \rightarrow H_2SO_3 \rightarrow 2\ H^+ + SO_3^{2}$$

$$H_2SO_3 + 2\ OH\cdot \rightarrow H_2SO_4 + H_2O$$

$$H_2SO_4 \rightarrow 2\ H^+ + SO_4^{2-}$$

Note that in the cases above, sulfurous acid and sulfuric acid have two ionizable hydrogens, both of which contribute to the pH of the resulting rain. The other common SOx species, SO_3, is formed from SO_2, and reacts with water to form H_2SO_4:

$$2\ SO_2 + O_2 \rightarrow 2\ SO_3$$

$$SO_3 + H_2O \rightarrow H_2SO_4$$

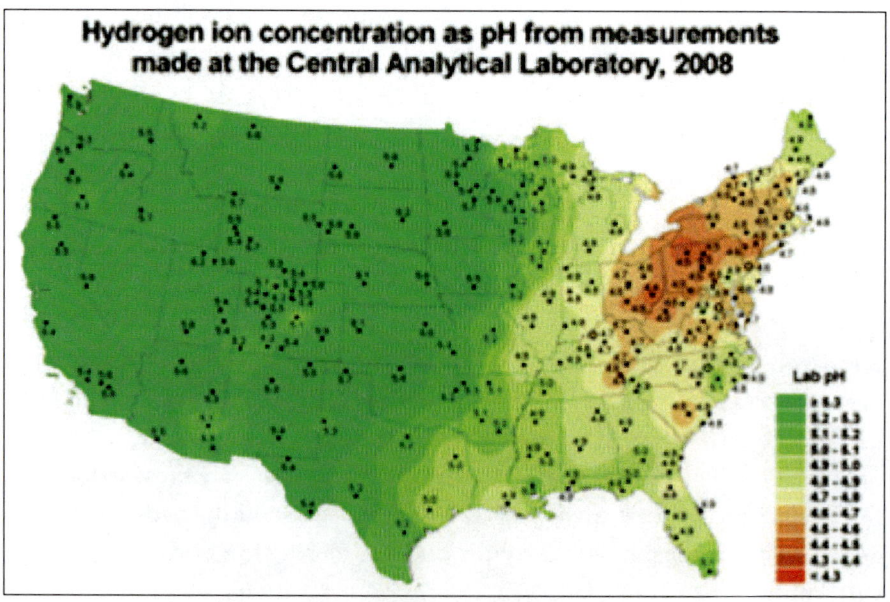

Image 6.10. Acidity of rain in the United States as measured by the National Atmospheric Deposition Program, supported by the EPA.

While there are natural sources of SOx and NOx, anthropogenic sources are far more significant, and are the major contributors to global acid rain. The EPA estimates that up to 67 percent of SOx emissions in the United States are the result of burning coal for electricity.* Burning of biomass, such as in slash-and-burn tropical farming, is a second major global anthropogenic source of SOx and NOx emissions.

The ramifications of acid rain are significant and far-reaching. Because of global atmospheric cycles, pollutant emissions have the ability to travel hundreds of miles or more, meaning that acid rain is not limited to heavily industrialized areas (though it's more common and more concentrated in these areas). When it falls on fields or wooded areas, acid rain damages plant leaves. This limits the ability of the plant to obtain nutrients (leaves are a tree's sunlight-gathering and food-producing unit), and can weaken or kill them. The decreased pH of soil upon which acid rain has fallen further affects trees and smaller plants by reacting with nutrients and leeching them from the soil. Acidification of surface water is a serious concern—fish gills can be eroded by acidified lakes and rivers, and acid damages fish skeletons. Shelled invertebrates and other aquatic species are also adversely affected. The compound $CaCO_3$, calcium carbonate, is a major component offish skeletons and aquatic invertebrate shells. While not water-soluble ($CaCO_3$ is also the major component of classroom chalk, which does not dissolve in water), the compound reacts with acid to form a familiar product—carbonic acid:

$$CaCO_3 + 2\,H^+ \rightarrow Ca^{2+} + H_2CO_3$$

$$H_2CO_3 \rightarrow CO_2 + H_2O$$

In general, aquatic species begin to suffer the effects of acidity below a pH of 6, and most fish eggs can't hatch below a pH of 5. Low-pH lakes eventually die, meaning they become incapable of supporting life.

In addition to the effect of acid rain upon the natural world, human-made structures are adversely affected. Automobile paint and clear coating can be etched by acid rain. Metals, such as the bronze used in many statues, are corroded. Stone, including the limestone and marble used in many older buildings and statues, literally dissolves under the assault of acid rain (Image 6.11). $CaCO_3$ is a major component of both marble and limestone, rendering these substances susceptible to acid, just as aquatic shells are.

While the destruction of man-made structures by acid rain does not affect the health of humans or other organisms, structures, statues, and art are nevertheless culturally important and represent a further impact of acid rain on society. Further, these

Concept Check:

If rain has a pH of 6, is it acid rain? Why or why not?

Answer: Rain with a pH of 5–7 is normal; the acidity comes from dissolved carbon dioxide.

*Note

It's worth noting briefly that electricity is sometimes promoted as a "clean" energy source—electric cars, for instance, are advertised as non-polluting. However, this sort of advertising fails to acknowledge the source of the electricity used to power the vehicles. In the United States, most electricity is generated by coal combustion, which, as we've seen, is highly polluting.

Image 6.11. A gargoyle from a European cathedral. The limestone (largely $CaCO_3$) from which this carving was made is not water-soluble, but dissolves in acidic solution, hence the "melting" of the gargoyle's once distinctive features.

Image 6.12. This cartoon, drawn in 1831, depicts steam coaches in England. The artist's depiction of the large plumes of black smoke issuing from these engines is quite representative of the uncontrolled emissions that marked the early industrial era.

visual manifestations of acid rainfall act as reminders of its presence and the prevalence of pollutant emissions.

CONCEPT CHECK:

What are the applications in which we've seen carbonic acid so far in the text?

Answer: Regulation of blood pH, bicarbonate antacids, leavening baked goods, the normal acidity of rain, and the product formed when acid rain reacts with aquatic shells or limestone.

The effects of acid rain upon the ecosystem and upon human establishments are made particularly clear in areas such as the "Black Triangle" in eastern Europe, an area of the northern Czech Republic that was quickly industrialized following World War II. The industrialization resulted in a dramatic increase in SOx and NOx emissions and heavy acidification of the local rain. Forests died, lakes and streams became too acidic to support aquatic life, and structures began to deteriorate.

6.9 THE INDUSTRIAL REVOLUTION

In addressing the question of when pollution began to affect society, we must look to the **Industrial Revolution**. This period of history, which began in the mid- to late-eighteenth century and was in full swing by the mid-1800s, represents a major transition in human society in many ways. Foremost among these, the period marked the changeover from human- and animal-powered labor and manufacturing to machine-powered. The hallmark of the Industrial Revolution was the development of coal-fired steam generation, which was used to power machines and engines. This technology was followed closely by the invention of the internal combustion engine. In short, this period of history marked the beginning of humankind's major dependence upon fossil fuels.* Prior to this time, the environmental problem of air pollution simply didn't exist—any combustion that took place, such as burning of wood to heat buildings, was on too small a scale to result in the production of significant quantities of pollutant material. As the Industrial Revolution

***NOTE**

We'll discuss fossil fuels in more detail in Chapter 13, but for now, these include natural gas, petroleum, and coal. Early in the Industrial Revolution, coal was the major accessible fossil fuel.

Image 6.13. This historic photograph shows the English town of Widnes under a pall of heavy coal smoke in the late nineteenth century.

CONCEPT CHECK:

If people were burning things (like wood) long before the Industrial Revolution, why was this period the first time that air pollution became an issue?

Answer: Wood was burned in much smaller amounts than coal; since work was done using human or animal power, wood was burned for heat only.

began to change the shape of society in Europe and worldwide, however, its effects became apparent.

Early industrialized societies, including England and the United States, quickly began to recognize the beginnings of a pollution problem. At first, this awareness was limited to the observation of the unsightly black smoke that poured from factory and steam engine smokestacks—the soot tended to blacken the sides of buildings and was observed to worsen the number of respiratory problems in the city. Several unsuccessful acts and orders were passed both in the United States and in Europe—the city of Cincinnati, for instance, ordered for smoke reduction measures in 1881, though the order was not enforced.* An attempt by the city of St. Louis to pass an ordinance outlawing the emission of thick black or gray smoke was overturned by the State Supreme Court, which called the legislation *unreasonable*.

Pollution awareness came to a head in 1952, when a **smog**—a combination of smoke and fog—settled over the city of London, England. While this event wasn't the first of its kind (many similar events were observed in the years between the beginning of the Industrial Revolution and the London event), it was well publicized and had a shocking death toll. The London Fog, also called the Great Smog of 1952, was the result of a temperature inversion that held fuel emissions over the city rather than allowing them to dissipate into the higher atmosphere and around the countryside. Pollutants, including sulfur dioxide, nitrogen dioxide, particulate matter, and other toxic emissions reached deadly concentrations♦ within a matter of hours, with concentrations remaining high for four days. Reportedly, London became so darkened and obscured by the smog that even in the middle of the day, men had to walk in front of city buses with lanterns to light the street for the driver. Conservative estimates hold that 4,000 individuals lost their lives as a result of the London Fog, but recent reexamination of the event suggests the total number of deaths may have been closer to 12,000.

Image 6.14. The Great Smog of 1952 in London, England.

***NOTE**

Despite the fact that the leading causes of death in Cincinnati at that time were tuberculosis, pneumonia, and bronchitis—all respiratory in nature.

♦NOTE

The sulfur dioxide concentration during the London Fog is estimated to have reached 10 times the EPA's current acceptable standard for U.S. air, while the particulate matter concentration is estimated to have topped out at nearly 30 times the current acceptable standard.

6.10 POLLUTION REGULATION

In response to increasing awareness of the dangers of pollution, the U.S. Congress passed a series of **Clean Air Acts** in the 1960s and 1970s (also creating the EPA in 1970), with a further Clean Air Act amendment in 1990. One of the major provisions of the Clean Air Act was that the EPA would develop programs to regulate and reduce the common air pollutants (identified as PM, SO_2, NOx, ozone, CO, and lead). A second major requirement was for a phase-out of leaded gasoline (see below). Further provisions dealt with emissions from motor vehicles (responsible for over half of NOx emissions, and approximately half of VOC and other pollutant emissions) and emissions from "non-road" motorized equipment such as recreational vehicles and lawnmowers.

For 20 years, the Clean Air Act was arguably one of the most successful pieces of environmental legislation in U.S. history, with impressive reductions in the six common pollutants identified for regulation and reduction by the

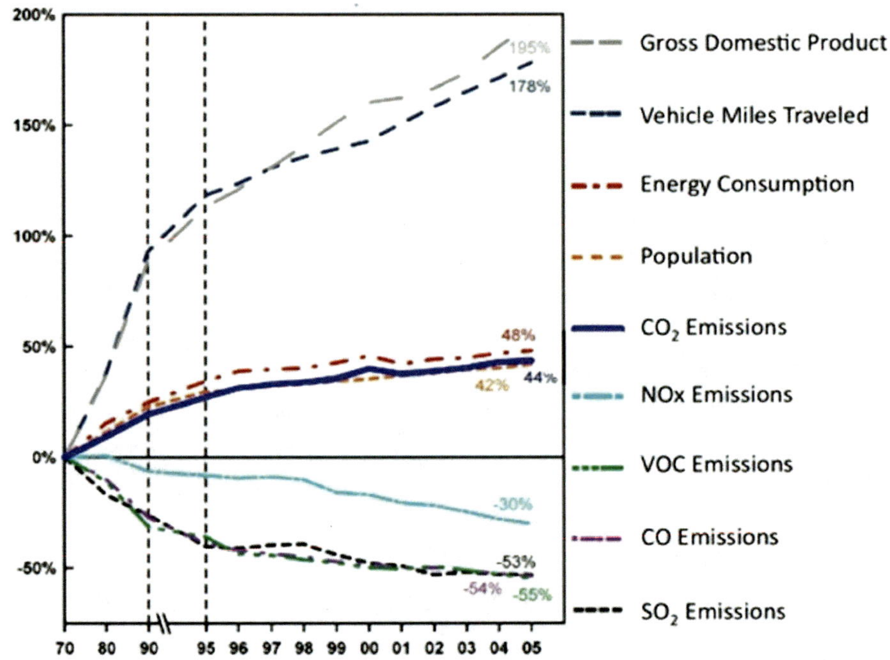

Image 6.15. The Clean Air Act's effect on major airborne pollutant emissions per EPA data, 1970–2005. Pollution values are given as percent change above or below 1970 values.

EPA (Image 6.15). Unfortunately, the initial legislation failed to address another key airborne emission and potent greenhouse gas: carbon dioxide. This led to a 2007 U.S. Supreme Court case in which a number of environmental organizations, twelve states and several cities sued the EPA for failure to regulate carbon dioxide and other known greenhouse gases.* A further weakness of the act is that the 1990 amendment allowed for emissions trading, a practice whereby heavy polluters can buy unused pollution allowances from light polluters, rather than coming into compliance of their own accord. This, combined with recent lax enforcement of the act by the EPA, has somewhat reduced the effectiveness of the legislation in recent years.

> ***NOTE**
>
> The EPA had determined in 2003 that they lacked the authority to regulate greenhouse gases, and that even if they had the authority to do so, they would not use it. It was determined by the court that the regulation of CO_2 does indeed fall under the auspices of the Clean Air Act, and as such, under the EPA, which has since begun the process of exploring greenhouse emissions regulation.

As a result of Clean Air Act stipulations regulating SOx emissions, coal plants have begun to preferentially process lower-sulfur coal. This reduces, but does not eliminate, SOx emissions. Furthermore, coal-fired power plants now employ one or more **scrubbing** (chemical washing) or filtering techniques to reduce SOx emissions. The most common technique relies upon wet limestone slurry—largely composed of $CaCO_3$—which absorbs some of the sulfur from combustion gases. While nearly 90 percent efficient (when operating correctly), smaller sulfur-containing particles are still able to pass through scrubbers and coal pollution still contains measurable SOx. Further, the waste absorbed by the scrubber is neither eliminated nor rendered chemically inert—rather, it is isolated within the scrubbing sludge instead of being aerosolized, rendering the sludge into a land and water pollutant. All in all, scrubbing is a tremendously resource-intensive process. It is estimated that a typical 500-megawatt coal plant (the size necessary to power a city of about 140,000 people) burns approximately 1,430,000 tons of coal in a year, and uses 2.2 billion gallons of water and 146,000 tons of limestone for scrubbing. Despite the scrubbing, 10,000 tons of SOx are still emitted from each plant on an annual basis, along with 10,200 tons of NOx, 700 tons of carbon monoxide, and 500 tons of PM_{10} and $PM_{2.5}$. Further, each plant produces nearly 200,000 tons of scrubber sludge annually. The sludge is disposed of in landfills or is converted into concrete and drywall, but is high in mercury and lead. There is concern that these heavy metal pollutants leach out of the sludge, ending up in the ground and in water.

CONCEPT CHECK:

Why is coal considered such a dirty fuel?

Answer: It contains nitrogen (source of NOx) and sulfur (source of SOx) in higher proportions than other fuels. It also contains mercury and lead; sludge from scrubbing is an environmental pollutant.

6.11 AIR QUALITY INDEX

One problem faced by the EPA in quantifying the quality of air on a given day is that each airborne pollutant is dangerous in different concentrations. The EPA uses two factors to determine acceptable concentrations of any given pollutant: toxicity and exposure. **Toxicity** relates to the inherent harmful potential of a given pollutant—that is, how likely the pollutant is to do damage. **Exposure** is a measure of the amount of substance likely to be encountered; it is a combination of the concentration of a pollutant and the amount of time the pollutant is in the atmosphere. As a means of providing citizens with a simple schema for judging air quality, the EPA developed the **Air Quality Index (AQI)**. Pollutant concentrations are normalized on a scale of 0–500, which is divided into subcategories corresponding to the level of health concern (Image 6.16).

Air Quality Index (AQI) Values	Levels of Health Concern	Colors
When the AQI is in this range:	*...air quality conditions are:*	*...as symbolized by this color:*
0–50	Good	Green
51–100	Moderate	Yellow
101–150	Unhealthy for sensitive groups	Orange
151–200	Unhealthy	Red
201–300	Very unhealthy	Purple
301–500	Hazardous	Maroon

Image 6.16. AQI values, health concerns, and corresponding colors.

Each range of AQI values is symbolized by a particular color. Additionally, each value range and color correspond to a level of health concern. *Good* air quality conditions are not reasonably expected by the EPA to cause health problems in most individuals, though *moderate* conditions may cause extremely sensitive members of the population some difficulty. Air quality conditions falling into the category of *unhealthy for sensitive groups* may place certain groups at risk, including the very young and elderly and those with underlying cardiovascular and respiratory conditions. *Unhealthy* air is likely to adversely affect nearly everyone, while *very unhealthy* air would cause relatively severe symptoms in all members of a population. *Hazardous* air represents an air quality emergency. It is important to note that a separate AQI value is assigned to each airborne pollutant being monitored, but that unhealthy levels of even one pollutant create a health concern—not all pollutants are necessarily at the same AQI value at all times.

Using the normalized AQI (available for cities with air monitoring programs through the EPA-run Website www.airnow.gov), residents of the United States can easily assess air quality without having to keep track of the disparate concentrations of each pollutant associated with each air quality category. For instance, in order to fall into the "good" range as defined by the EPA, eight-hour average ozone concentrations must be no higher than 0.064 **ppm** (meaning parts per million, or ozone molecules per million molecules of air), while eight-hour average carbon monoxide concentrations can rise as high as 4.4 ppm (Image 6.17).

Even without committing these concentrations to heart, though, a concerned individual need only glance at the AirNow Website to determine that if, for instance, their city is "green" for carbon monoxide, but "orange"for $PM_{2.5}$, carbon monoxide is in the acceptable range, but particulate matter levels may cause problems for sensitive groups. This helps individuals choose outdoor activities appropriate to the air quality and their personal health—an asthmatic might not engage in strenuous outdoor activity on an "orange"pollutant day, for instance.

CONCEPT CHECK:

Why is the AQI a helpful tool for getting information about air quality?

Answer: It normalizes the acceptable concentrations of various pollutants to a single, easy-to-read scale.

Air Quality Index (AQI) Values	$PM_{2.5}$ 24-hour Average (micrograms/m^3)	8-hour Ozone Concentration (ppm)	Carbon Monoxide 8-hour average (ppm)
0–50	0-15.4	0-0.064	0.0-4.4
51–100	15.5-40.4	0.065-0.084	4.5-9.4
101–150	40.5-65.4	0.085-0.104	9.5-12.4
151–200	65.5-150.4	0.105-0.124	12.5-15.4
201–300	150.5-250.4	0.125-0.155	15.5-30.4
301–500	>250.4	>0.155	>30.4

Image 6.17. The relationship between AQI value and concentration for three separate pollutants.

6.12 CHANGING PERCEPTIONS OF RISK

One fascinating area of interplay between chemistry and society is in the public perception of risk associated with an activity or substance and how that perception varies with time and circumstance. The advent (and subsequent phase-out) of leaded gasoline in the United States is an excellent example.

Early in the development and popularization of the automobile, the four-stroke internal combustion engine (Image 6.18) used to power the vehicle was subject to a phenomenon known as *knocking*, which occurred when fuel combusted prematurely.

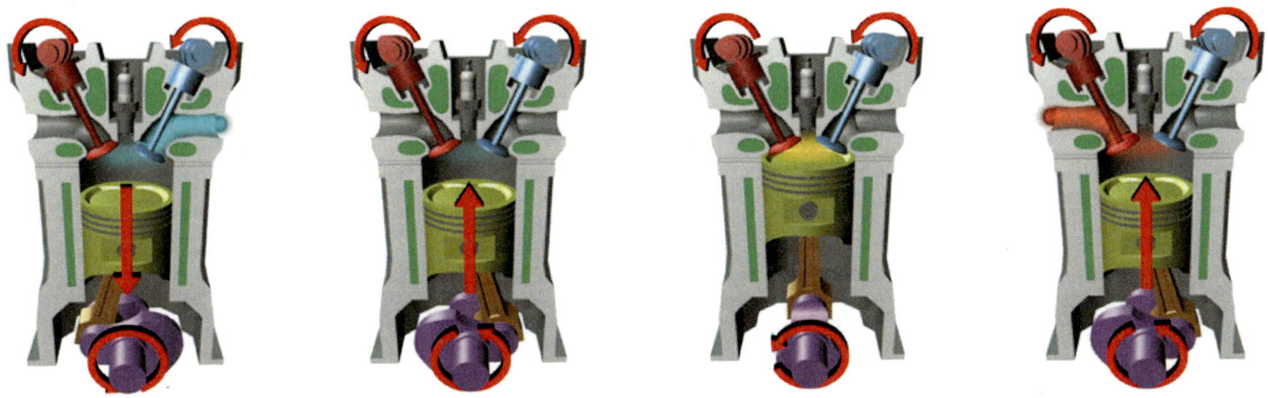

Image 6.18. The four parts of a four-stroke engine cycle: Intake (fuel is brought into combustion chamber), compression (fuel is compressed for increased generation of power from combustion), ignition (fuel is ignited with a spark), and exhaust (products of combustion are released).

Knocking resulted in an engine that, rather than supplying power continuously, did so in a series of power bursts (similar to a set of explosions). This reduced vehicle power and overall efficiency. **Thomas Midgley**, an inventor and mechanical engineer, was working for General Motors when he determined that the knocking phenomenon was due to a problem with the fuel used in engines rather than the engines themselves. In 1921 he added **tetraethyl lead**, $(CH_3CH_2)_4Pb$, to gasoline in order to smooth combustion and prevent engine knocking. For a variety of reasons, including the patentability and profit potential of the additive, leaded gasoline was heavily promoted and became pervasive in the marketplace, despite concern expressed by several scientists and health experts.* The dangers of lead were not in the forefront of public awareness at that time, though they became more apparent as the manufacture, distribution, and sale of leaded gasoline began to take a toll on individuals and the environment. Midgley himself, after working with lead in the lab, found he was developing lead poisoning and took time off to recover. Plant workers were among the first to succumb to lead poisoning, experiencing a variety of symptoms culminating in insanity and death, in numbers that raised the concerns of authorities. A conference was convened in 1925 by the surgeon general to address these concerns and to make a determination on the wisdom of proceeding with production and sale of lead additives. Several of the scientists present at the conference presented strong arguments against the use of lead. Of particular note, the director of a large metropolitan hospital made a statement well in line with what is now known as the **precautionary principle**, which has its roots in maxims such as *better safe than sorry*. The idea is that in the absence of proof and with the possibility of harm, it's far better to take a more conservative approach

> ### *NOTE
>
> While the dangers of lead poisoning were known to the industry at this time, they were not yet widely recognized by the public. Still, GM took the marketing precaution of referring to the fuel additive as *ethyl* to avoid any mention of lead during its introduction to the marketplace.

than wait for evidence of danger before initiating change or regulation. While this principle has a multitude of origins and no "conception" as such, it is generally accepted that formalization of the principle took place in 1998 during a meeting of scientists and policy makers, and states:

> *"When an activity raises threats of harm to human health or the environment, precautionary measures should be taken even if some cause and effect relationships are not fully established scientifically. In this context the proponent of an activity, rather than the public, should bear the burden of proof. The process of applying the precautionary principle must be open, informed and democratic and must include potentially affected parties. It must also involve an examination of the full range of alternatives, including no action."*

The hospital director suggested that the use of lead additives ought to be suspended until research demonstrating its safety was available. In contrast, however, a doctor named Robert Kehoe suggested that the ruling on lead use should be made only upon the basis of research available at that time. Since there was little research available, he suggested that lead should not be barred from manufacture or use. In the end, the surgeon general adopted Kehoe's perspective, stating that until there was evidence of harm, tetraethyl lead would be allowed. It's well worth remarking that Dr. Kehoe was initially a consultant of the lead industry and was later director of the Kettering Laboratory—a research group funded by the lead industry to produce lead safety studies. It is easy to conjecture that the economic and professional pressure on Dr. Kehoe to avoid producing any studies demonstrating the toxicity of lead was considerable.

In a further effort to demonstrate the innocuous nature of the additive (and with questionable wisdom, given his previous bout with lead poisoning), Midgley famously washed his hands with tetraethyl lead during a press conference. Despite the best efforts of the industry, however, public health concerns over exposure to the lead additive through misuse, mishandling, spills, and accidents increased. Further, lead oxides from fuel exhaust contaminated the air, ground, and water, leading to widespread and rampant lead pollution. One study in particular examined lead concentrations in New York City dust, finding concentrations 50 percent higher in 1934 as compared with ten years prior. This study and others like it initially received little attention, and U.S. consumption of leaded gasoline continued to increase. By the 1970s, though, growing concern over lead pollution and toxicity resulted in the aforementioned EPA-mandated phase-out of leaded gasoline. While industries vigorously opposed the notion that lead from the gasoline was leaching into the environment and argued that the phase-out was unwarranted and unnecessary, the second U.S. National Health and Nutrition Examination Survey, published in 1980, extinguished any doubt of a connection. The survey showed that as use of leaded gasoline decreased from 1976 to 1980, blood lead levels closely paralleled that decrease. This led the EPA to more closely examine the dangers of leaded gasoline, and eventually resulted in the much more aggressive phase-out of leaded gasoline mandated in 1982. Lead is no longer a major airborne pollutant in the United States, thanks to these measures, and lead pollution and risk of exposure are now isolated to manufacture sites and sources of old paint. Countries in which lead is still used in gasoline, however, continue to have problems with lead toxicity and pollution.

Of note in this discussion is that despite concerns expressed by environmental researchers and experts, production and marketing of tetraethyl lead as a gasoline additive was allowed to proceed. Only after many preventable deaths and 60 years' worth of unnecessary lead pollution of the environment was legislation finally put in place to curtail the use of lead fuel additives. During this period of time, public opinion regarding the dangers of lead changed markedly. It's worth mentioning that in a battle for control of public opinion between independent researchers and corporations or corporation-sponsored researchers, those with access to the advertising dollars will most likely win public favor. Lessons from lead can be applied as we face modern environmental challenges and questions: It's important to remember that manufacturers of products that may be environmentally harmful are highly motivated to produce results that support their claims of safety and to encourage suppression of results that don't. In the end, there are two questions worth asking whenever evaluating scientific information. First, is there data supporting the claim of risk or safety? Second, who funded the study?

6.13 THE LAST WORD—SMOKING

Very few psychologically healthy individuals, particularly if armed with information about the health dangers of pollutants, would purposely concentrate and inhale the smoke from a coal-fired power plant or the tailpipe of a car. However, to compare these clearly off-putting inhalants to the smoke from a cigarette is not a scientific stretch—in fact, in may ways, cigarette smoke is even worse. While the cardiovascular and respiratory consequences of smoking are well known at this point and while smoking is acknowledged to cause cancer, it's interesting to consider cigarettes from another, less articulated angle—that of their chemistry.

From a chemical standpoint, cigarette addiction is due in large part to the compound nicotine. There are additional (significant) psychosocial influences on smoking behavior and addiction, but a discussion of these is beyond the scope of this text. Nicotine is a powerful **neurochemical**, inducing a feeling of alertness that is followed some time later by relaxation. It also suppresses appetite, which proves to be a motivating factor for smoking—and a deterrent to smoking cessation—for some weight-conscious individuals. Regardless of the myriad chemical effects of nicotine, however, the molecule itself does not cause cancer. The carcinogenic properties of cigarettes are due to the hundreds of other chemicals present in both the raw materials and the products of combustion reactions taking place within a lit cigarette.

Cigarettes consist of tobacco leaf (which contains the nicotine) and as many as 599 additives, a list of which was finally made available to the Centers for

Image 6.19. Cigarette companies strive to conjure images of nature and health (such as that of this tobacco leaf) rather than images of soot, tar, and products of chemical combustion.

Disease Control and Prevention (CDC) in 1994. While the additives are all federally approved for use in food, they have largely not been tested under combustion conditions, which in many cases result in the formation of carcinogenic products. Among disturbing additives from the list are vanilla, honey, and chocolate—included in cigarettes for their sweetness and appealing aroma and suspected by anti-tobacco lobbies to be a mechanism for luring children into smoking. Ammonia is added to enhance nicotine absorption and increase addiction. Other additives serve various purposes, including preserving the tobacco, influencing taste, and masking symptoms of smoke inhalation and illness. These include benzene, formaldehyde, methanol, and acetylene torch fuel (all VOCs), as well as cyanide (a highly toxic chemical used in gas chamber executions in some states), arsenic (a potent inhibitor of cellular energy production), and lead.

The list above addresses only the raw ingredients in cigarettes; a further litany of compounds is produced during combustion. These include all the major airborne pollutants—NOx, SOx, CO, PAHs, VOCs, and so forth—and additional carcinogens not generally found in exhaust. In addition pollutants are found in cigarette smoke at concentrations much higher than those routinely encountered even in highly polluted air. The CO concentration of undiluted cigarette smoke, for instance, is about 4.5 times higher than that of undiluted, concentrated car exhaust.

Clearly the chemistry of an unlit cigarette is complex enough, to say nothing of the chemistry that takes place upon lighting. This is further complicated by the chemistry (much of it still unknown or incompletely understood) of these compounds upon absorption into the body. In many ways, smoking can be compared to lead: for a long time, large corporations either prevented public dissemination of studies or funded their own in order to control knowledge available to the public. Public perception of the risk of cigarette smoking was very different in the early- and mid-1900s than it is today; even now, the transition toward perceiving cigarettes as something that should be regulated is incomplete. This is due jointly to the desire of individuals to maintain freedom of choice and the desire of manufacturers to maintain their profit margins. It's interesting to consider, however, whether one day a textbook might use cigarettes (just as this text uses lead) as a mechanism for introducing the precautionary principle.

SUMMARY OF MAIN POINTS

- CO is a product of incomplete carbon combustion and reduces oxygen delivery to tissues.
- Catalytic converters help reduce CO emissions, but produce CO_2, a potent greenhouse gas.
- NOx are the products of high-temperature reactions of N_2 with O_2.
- NO_2 is a respiratory irritant.
- VOCs are largely indoor pollutants and are toxic.
- PAHs are products of incomplete fuel combustion and are carcinogenic.
- Ozone (O_3) is a pollutant in the troposphere and a secondary pollutant formed via the reaction of NO_2 with sunlight.
- O_3 is a respiratory irritant.
- $PM_{2.5}$ and PM_{10} are aerosolized particulate matter from various sources and are respiratory irritants.
- SO_2 is produced naturally (in small quantities) by volcanoes and anthropogenically by coal combustion.
- SO_2 is a respiratory irritant.
- Mercury pollution is due in large part to coal combustion, and is an air, land, and water contaminant.
- Mercury concentrates in the food chain, particularly through consumption of aquatic organisms. The EPA recommends limiting consumption of predator fish due to high mercury levels.
- Lead is mostly found in and around old buildings (due to lead paint) in the United States. It is a potent neurotoxin.
- Rain is naturally slightly acidic due to dissolved CO_2.
- Rain with a pH of less than 5 is considered to be acid rain.
- Acid rain is due to dissolved SOx and NOx in raindrops, producing H_2SO_4 and HNO_3 (both strong acids).
- Acid rain damages terrestrial and aquatic environments, enhances rust formation, and damages man-made structures.
- The Industrial Revolution was the point in history during which anthropogenic pollution levels became significant.
- Regulation of pollution in the United States began in earnest with the Clean Air Act, creation of the EPA, and amendments to the Act.
- Coal scrubbing results in the reduction of SOx emissions from coal burning, but produces contaminated sludge.
- The EPA determines acceptable levels of airborne pollutants by considering two important factors for each: toxicity and exposure.
- To simplify air-quality reporting, the EPA created the Air Quality Index (AQI).
- Perceptions of risk change over time: Lead was a common additive to gasoline through the 1970s but is now highly regulated by the EPA.
- Corporations producing products may be highly motivated to prevent the publicizing of studies demonstrating toxicity of their product.
- The precautionary principle can be summarized to state that in the absence of evidence, it's better to be safe than sorry. This principle is appropriate to take into consideration in making major decisions as a society.

Questions and Topics for Discussion

1. Why do you suppose most tents have a label printed inside that warns occupants not to use a stove in the tent?
2. What are the benefits of a catalytic converter? What are the drawbacks?

3. Why do you suppose that breathing O_2-enriched air helps reduce the symptoms of carbon monoxide poisoning?
4. Natural gas produces low levels of CO compared to larger hydrocarbons. Compare the combustion reactions of CH_4 and C_8H_{18}, and propose a reason for this.
5. Why are VOCs generally considered an indoor rather than an outdoor pollutant? Which other pollutants might reach concentrations of concern indoors, and why?
6. The lungs consist of airways that branch and grow smaller as they get deeper in the lung tissue. Based upon this information, suggest why $PM_{2.5}$ might be more harmful than PM_{10}.
7. Why is the EPA concerned with visibility in national parks?
8. Natural gas is considered the cleanest burning of the common fuels. Why?
9. Despite claims (by the natural gas industry) that natural gas is the cleanest burning of the common fuels, its combustion still produces many pollutants. Which are most likely? Which are least likely?
10. Why might natural gas combustion result in less NOx emission than coal or petroleum combustion?
11. Why is it accurate to say that mercury concentrates up a food chain? Why is this effect more significant in aquatic food chains as opposed to terrestrial food chains? (Hint: think about the number of "links" in each chain.)
12. Why is a shark likely to have higher mercury levels than, for instance, a sardine?
13. What are the effects of acid rain on terrestrial, aquatic, and man-made environments?
14. Destruction of sculptures and buildings doesn't really constitute an "environmental" or "toxicity" effect of acid rain, yet these effects are generally of great concern. Why is destruction of man-made structures so worrisome?
15. What are SOx and NOx, and why are they of environmental concern?
16. Where is acid rain most prevalent in the United States? Why do you think this is so?
17. Why do you think $CaCO_3$ dissolves in acid, even though it doesn't dissolve in water?
18. What are the major ways in which the Industrial Revolution changed society? Think about this both in terms of positive and negative effects.
19. Why do you think initial regulation of pollution was so difficult to achieve?
20. In what ways were the Clean Air Act and its amendments successful? In what ways were they unsuccessful?
21. Do you feel the practice of emissions trading should have been allowed for by the Clean Air Act Amendment of 1990? Why or why not?
22. Coal is a highly controversial fuel in the United States—despite the fact that it is plentiful, it is a very dirty fuel. How is coal dirtier than other common fuels?
23. How does coal scrubbing help reduce airborne pollution? What problem does it create? Which do you think is worse?
24. Look up the AQI for your city on www.airnow.gov. What are your air quality values today? What sorts of pollution does your area struggle with?
25. Why was the AQI an important communication tool created by the EPA?
26. Why was lead considered a major airborne pollutant by the EPA in the 1970s, while it's not commonly encountered in the environment in the United States now?
27. How did the fuel industry affect public opinion of lead exposure risk in the 1920s?
28. What is the precautionary principle? What are the benefits and drawbacks of abiding by it? (Hint: think about both industrial and environmental perspectives.)
29. Why are independently-funded or government-funded (as opposed to industry-funded) scientific studies sometimes ignored by the public (as in the case of the lead studies in the early and mid-1900s)?
30. Why is it important to consider the funding source for a study in evaluating its efficacy? Where did funding contribute to public opinion of lead safety?

Problems

1. Write a balanced combustion reaction (idealized—CO_2 and H_2O products) for propane (C_3H_8), a common gas grill fuel.
2. Write an alternate combustion reaction (CO and H_2O products) for propane (C_3H_8), a common gas grill fuel. Compare to the reaction from question 1; explain why this reaction requires less oxygen.
3. What is the anthropogenic source of CO?
4. What are the effects of exposure to CO?
5. What is a catalytic converter?
6. Why is N_2 so chemically unreactive?
7. What is the anthropogenic source of NO_2?
8. What are the effects of exposure to NO_2?
9. What are VOCs and where are they commonly encountered?
10. What are the effects of exposure to VOCs?
11. What are PAHs and where do they come from?
12. What are the effects of exposure to PAHs?
13. Write a Lewis dot structure for ozone.
14. What is an allotrope? Give an example.
15. In what portion of the atmosphere does ozone occur naturally? In what portion is it a pollutant?
16. What is the anthropogenic source of ozone?
17. What are the effects of exposure to ozone?
18. Why is ozone referred to as a "secondary" pollutant? Which other pollutants are involved in ozone production?
19. What is the current source of most lead exposure in the United States?
20. What are the effects of exposure to lead?
21. What are the effects of exposure to mercury?
22. Why is rain naturally acidic?
23. Write out the chemical reaction explaining the natural acidity of rain.
24. What is the approximate pH of normal rain?
25. What pH constitutes acid rain?
26. What is the chemical source of acid rain? How does it form?
27. Why are aquatic invertebrate shells and fish skeletons susceptible to acidification of water?
28. Why are marble and limestone damaged by acid rain?
29. What is the AQI and how is it used?
30. Which two factors are used by the EPA to determine acceptable pollutant concentrations in air?
31. Define toxicity and exposure.
32. What is tetraethyl lead and what was it used for? Is it still used?

Answers to Problems

1. $C_3H_8 + 5\ O_2 \rightarrow 3\ CO_2 + 4\ H_2O$
3. CO comes from combustion of hydrocarbons in insufficient oxygen.
5. A catalytic converter helps convert CO to CO_2, and decomposes NO and NO_2 to N_2 and O_2.
7. NO_2 comes from the reaction of N_2 with O_2 at high temperature.
9. VOCs are volatile organic compounds and are found most commonly indoors, where they leach from building materials.
11. PAHs are polycyclic aromatic hydrocarbons and come largely from uncombusted fuel.

13.

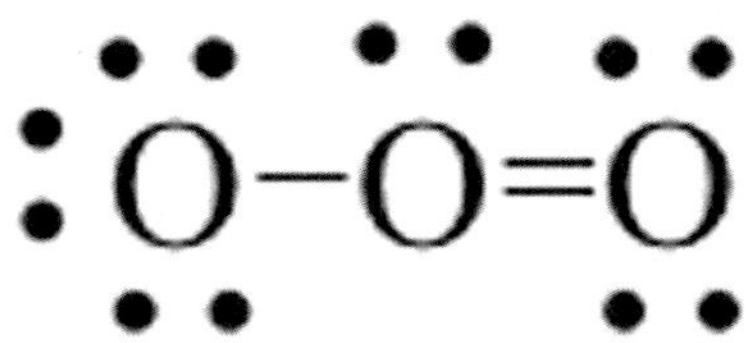

15. Ozone occurs naturally in the stratosphere and is a pollutant in the troposphere.
17. Ozone is a respiratory irritant.
19. Most lead exposure in the United States is through old (pre-1978) paint in homes or in the dust around older homes.
21. Mercury is a neurotoxin.
23. $CO_2 + H_2O \rightarrow H_2CO_3 \rightarrow H+ + HCO_3-$
25. Acid rain has a pH less than 5.
27. Aquatic invertebrate shells and fish skeletons are largely composed of $CaCO_3$, which reacts with acid to form H_2CO_3, which dissociates to form H_2O and CO_2. The shells and skeletons dissolve.
29. The AQI normalizes pollutant concentrations onto a scale from 0–500. AQI values are divided into several categories based upon health hazard (good to hazardous) so that citizens can easily assess the quality of air without having to read and interpret individual pollutant concentrations.
31. Toxicity is the intrinsic health threat of a pollutant. Exposure takes into consideration both the concentration of a pollutant and the amount of time it is in the atmosphere.

Sources and Further Reading

Industrial Revolution

Jacobson, M. (2002). *Atmospheric Pollution: History, Science, and Regulation.* Cambridge University Press.

Pollution and Health

Bell, M. L., Peng, R. D., Dominici, F., *et. al.* (2009). Emergency hospital admissions for cardiovascular diseases and ambient levels of carbon monoxide results for 126 United States urban counties, 1999–2005. *Circulation*, 120(11), 949–69.

Chabot, F., Gomez, E., Guillaumot, A., *et. al.* (2009). Acute exacerbations of chronic obstructive pulmonary disease. *LaPresse Médicale*, 38(3), 485–95.

Chiu, H. F., Yang, C. Y., (2009). Air pollution and emergency room visits for arrhythmias: are there potentially sensitive groups? *Journal of Toxicology and Environmental Health*, 72(13), 817–23.

Dales, R. E., Cakmak, S., and Vidal, C. B. (2009). Air pollution and hospitalization for headache in Chile. *American Journal of Epidemiology*, 170(8), 1057–66.

Grant, W. B. (2009). Air pollution in relation to U.S. cancer mortality rates: an ecological study; likely role of carbonaceous aerosols and polycyclic aromatic hydrocarbons. *Anticancer Research*, 29(9), 3537–45.

Peng, R. D., Bell, M. L., Geyh, A. S., *et. al.* (2009). Emergency admissions for cardiovascular and respiratory diseases and the chemical composition of fine particle air pollution. *Environmental Health Perspectives*, 117(6), 957–63.

Srebot, V., Gianicolo, E. A. L., Rainaldi, G., *et. al.* (2009). Ozone and cardiovascular injury. *Cardiovascular Ultrasound*, 7, Article 30.

Common Air Pollutants

www.epa.gov

VOCs and PAHs

Finlayson-Pitts, B. J., and Pitts, J. N. (1997). Tropospheric air pollution: ozone, airborne toxics, polycyclic aromatic hydrocarbons, and particles. *Science*, 276(5315), 1045–51.

Gulyurtlu, I., Abelha, P., Gregorio, A., *et. al.* (2004). The emissions of VOCs during co-combustion of coal with different waste materials in a fluidized bed. *Energy and Fuels*, 18(3), 605–10.

Hannigan, M. P., Cass, G. R., Penman, B. W., *et. al.* (1997). Human cell mutagens in Los Angeles air. *Environmental Science and Technology*, 31(2), 438–47.

Leiter, J., Shimkin, M. B., and Shear, M. J. (1942). Production of subcutaneous carcinomas in mice with tars extracted from atmospheric dusts. *Journal of the National Cancer Institute*, 155(3), 167–74.

NOx, VOCs, and Ozone

Ryerson, T. B., Trainer, M., and Holloway, J. S. (2001). Observations of ozone formation in power plant plumes and implications for ozone control strategies. *Science*, 292(5517), 719–23.

Mercury

Dufault, R., LeBlanc, B., Schnoll, R., *et. al.* (2009). Mercury from chlor-alkali plants: measured concentrations in food product sugar. *Environmental Health*, 8(1).

http://www.montereybayaquarium.org/cr/seafoodwatch.aspx

Acid Rain

Jacob, D. J., (1999). *Introduction to Atmospheric Chemistry.* Princeton University Press.

Coal Scrubbing

Dunham, J. T., Rampacek, C., and Henrie, T. A. (1974). High-sulfur coal for generating electricity. *Science*, 184(4134), 346–51.

www.ucsusa.org/clean_energy

Air Quality Index

www.airnow.gov

Lead

Kaye, S., and Reznikoff, P.A comparative study of the lead content of street dirt in New York City in 1924 and 1934. *Journal of Industrial Hygiene and Technology*, 29(3), 178–79.

Kovarik, Bill. (1994, rev. 1999). *Charles F. Kettering and the 1921 Discovery of Tetraethyl Lead in the Context of Technological Alternatives.* Presented at the Society of Automotive Engineers Fuel and Lubricants Conference, Baltimore, Maryland.

Moore, Colleen. (2003). *Silent Scourge: Children, Pollution, and Why Scientists Disagree.* Oxford University Press.

www.cdc.gov/nchs/nhanes

Smoking

Rabinoff, M., Caskey, N., Rissling, A., et. al. (2007). Pharmacological and chemical effects of cigarette additives. *American Journal of Public Health*, 97(11), 1981–91.

http://www.pmintl-technical-product-information.com

Unit 3: Personal Chemistry

Pharmaceuticals and Poisons 7

Almost everyone has benefited at times from the healing power of pharmaceutical compounds. We don't often stop to ponder the chemical nature of the medications we're taking, how they were discovered, or what exactly they do. We simply take the pills we're prescribed, and we generally get better. There's a fine chemical line, however, between a miracle cure and a deadly poison—a line we generally try not to cross, occasionally flirt with, and sometimes completely disregard. Pondering the nature of this line leads us to consider some important questions. Chemically, what are pharmaceuticals? How are they discovered? What differentiates pharmaceuticals from natural compounds, or from toxins? In this chapter, we will explore the chemistry of compounds that heal and those that kill, and the very fine line that separates the two.

Image 7.1. The bark from the white willow tree (*salix alba*) is used to make analgesic tea.

7.1 FROM TEA TO TABLETS

Throughout history, humans have sought out ways to reduce our pain and treat our illnesses. Until the middle of the nineteenth century, many treatments came in the form of herbs known by reputation to have a healing effect. For instance, the ancient Greeks used a tea made from white willow (*salix alba*) bark to reduce headache, inflammation, pain, and fever.

Because making tea from willow diluted whatever ingredient in the bark was responsible for its effect, chemists began to wonder whether there might be a way to concentrate the active ingredient in order to yield a more powerful and effective compound. In the mid-1800s, a chemist named Johannes Buchner discovered that it was possible to isolate yellow crystals* of a substance that he called salicin, since it was extracted from *salix alba*. Further investigation of the compound, both by Buchner and others, revealed that salicin was actually a glucose (sugar) unit bonded to a hydrocarbon ring to make one large molecule. Chemical removal of the sugar yielded a second pure substance, named salicylic acid for its parent compound and its acidity:

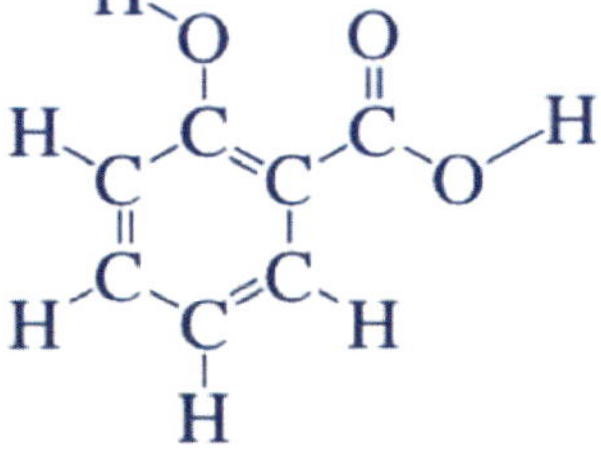

***NOTE**

Chemists often use the physical properties of a solid to help them determine whether what they've isolated or synthesized is a pure substance or a mixture. Pure substances tend to form crystals rather than amorphous, muddy-looking masses. Also, pure substances have discrete and identifiable melting points, while mixtures tend to melt over a wide temperature range.

Salicylic acid is a very large molecule, and its Lewis structure is somewhat cluttered and unwieldy. This is a common problem when portraying large molecules—in particular, **organic compounds** tend to be quite large. **Organic chemistry** is the study of organic compounds, which are carbon-based molecules made up of the major and minor elements of life. The fact that these molecules are always carbon based allows us to simplify the manner in which they are drawn. In organic molecules, carbon always makes four bonds. Remembering this, we can actually leave the hydrogen atoms that are bonded to carbon out of our organic structures—we simply mentally fill in the requisite number of hydrogens when looking at a molecule. Because our discussion of acids in Chapter 5 showed us that the number of bonds formed by oxygen, nitrogen, and other elements may vary somewhat depending upon the circumstances, we can't make the same simplification with regard to hydrogen atoms bonded to elements other than carbon—those hydrogens must be explicit. Using our hydrogen atom simplification, salicylic acid would look like this:

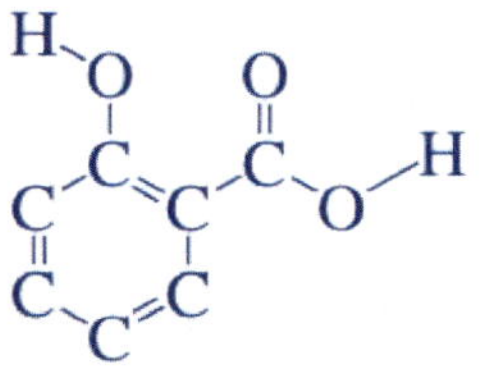

Note that the hydrogen atom attached to each oxygen atom is still shown. The hydrogens attached to carbon are not shown, but are still part of the molecule—we're merely rendering the molecule quicker and easier to draw by making the hydrogen implicit rather than explicit. In order to determine how many hydrogen atoms are attached to a particular carbon, we count the explicit bonds and then mentally fill in enough hydrogen atoms to complete carbon's set of four bonds:

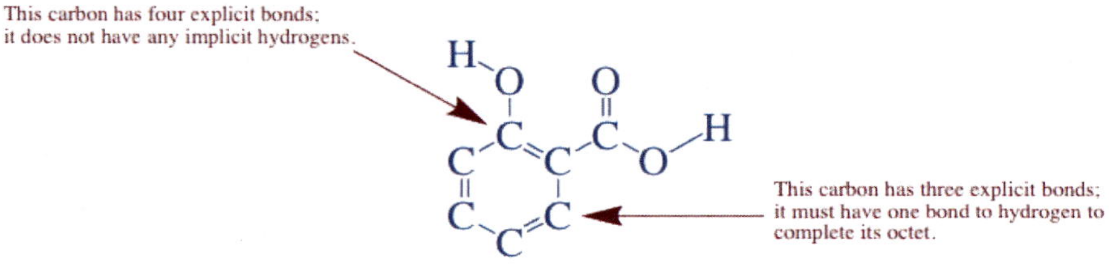

There is one further simplification we can make in drawing organic molecules. Knowing that organic compounds are carbon based, we can eliminate the atomic symbol for carbon from our structure. Atoms of elements other than carbon are denoted by their atomic symbols, and anywhere in the molecule that an atomic symbol is omitted, we assume the element in question is carbon. Portrayed thus, our salicylic acid looks much less cluttered:

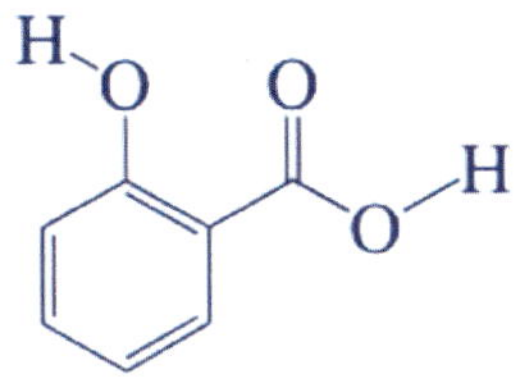

Note that the bonds make it quite clear where the carbon atoms are located, and that despite our simplifications, the molecular structure above (called a **line-angle** diagram) allows a significant amount of information to be conveyed by a relatively uncluttered visual. Any compound with at least one bond from carbon to a non-hydrogen atom* can be drawn in line-angle. In order to help make sense of the line-angle diagrams, it's important to use correct bond angles. In a small molecule like ethane (CH_3CH_3), the line-angle diagram requires no portrayal of bond angles; ethane simplifies to a mere line:

> ### *Note
>
> It's not possible to draw very small molecules in line-angle. Methane (CH_4), for instance, would disappear entirely!

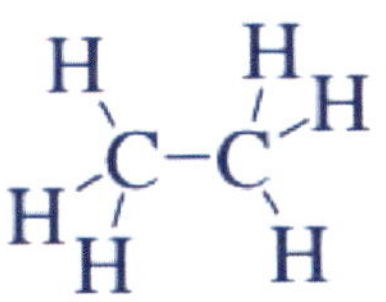

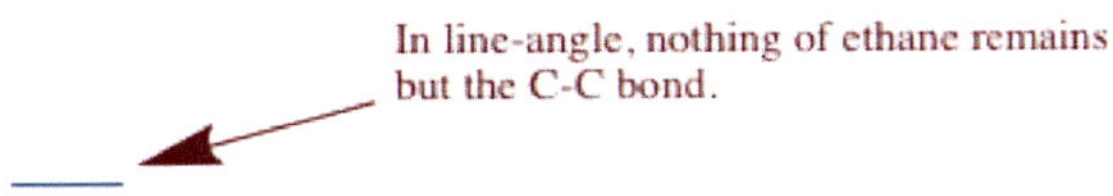

Notice that in this example, the full Lewis structure of ethane is drawn without the wedge-dash notation we've been using to indicate tetrahedral geometry. Instead, plain bonds are drawn where the wedge and dash lines would have gone, and are taken to imply wedge and dash. Even without explicit wedge-dash notation, we are still meant to understand that a tetrahedral structure, with one hydrogen from each carbon coming out of the page toward us, and one hydrogen from each carbon headed away from us, is being represented. This is another common way to portray organic structures. It's not nearly as concise as line-angle, but it provides some simplification and is useful in drawing very small molecules such as ethane that don't lend themselves particularly well to line-angle.

> ### Concept Check:
>
> Why is line-angle useful?
>
> Answer: It allows us to draw large molecules without them being cluttered or difficult to understand.

A larger molecule, like pentane ($CH_3CH_2CH_2CH_2CH_3$), would be impossible to make sense of if we tried to simplify it without attending to bond angles. Imagine the (incorrect) line-angle diagram that would result if we simplified a flat Lewis structure of pentane:

Without attending to bond angles, our pentane would be nothing more than a line. This would incorrectly suggest that, rather than five carbons, the molecule in question consisted of only two (like ethane).

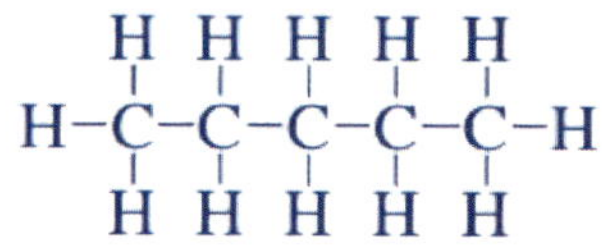

Instead, if we are careful to portray pentane with appropriate 109.5° bond angles, we get a sensible, readable line-angle simplification:

In an appropriately drawn line-angle diagram, pentane's five carbons are clearly evident.

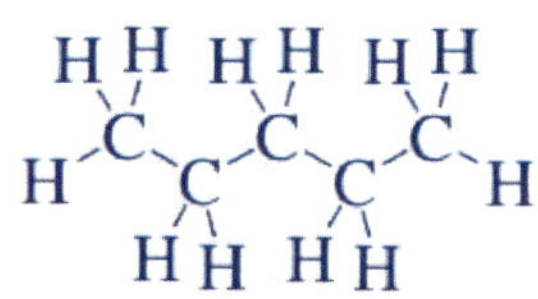

TRY THIS:

Draw line-angle diagrams for the following molecules:

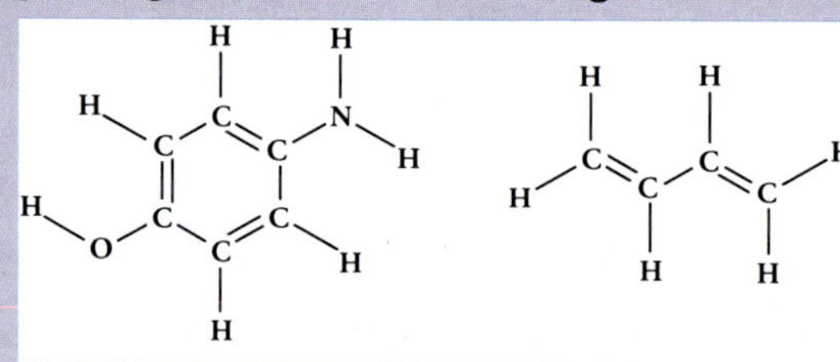

Answers:

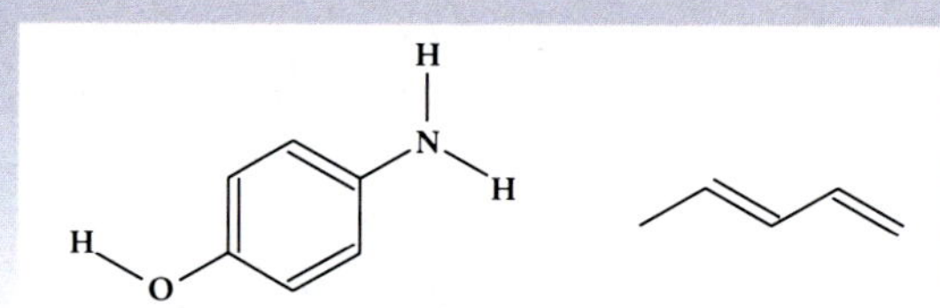

TRY THIS:

Redraw each of the following as full Lewis structures (including C's and H's):

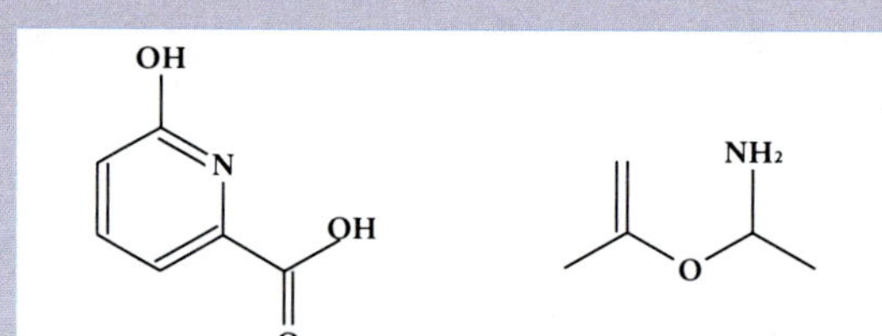

Answers:

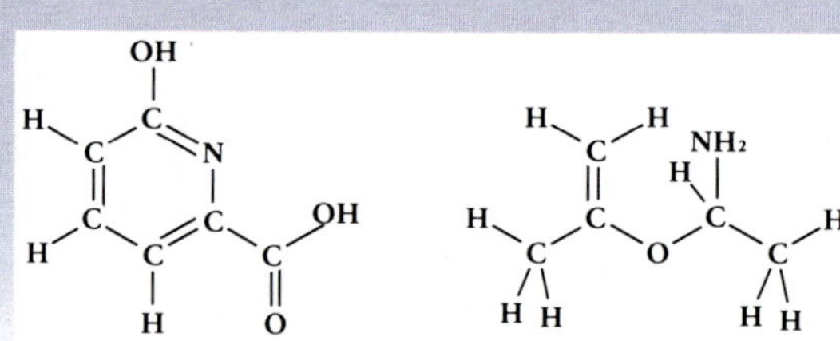

Returning to our discussion of willow bark extract, it was determined through experimentation that salicylic acid, like salicin, was an anti-inflammatory and analgesic. Unfortunately, it was also terribly irritating to the stomach. Felix Hoffmann, a chemist working for the Bayer Company, chemically modified the molecule. His modified molecule was called acetylsalicylic acid:

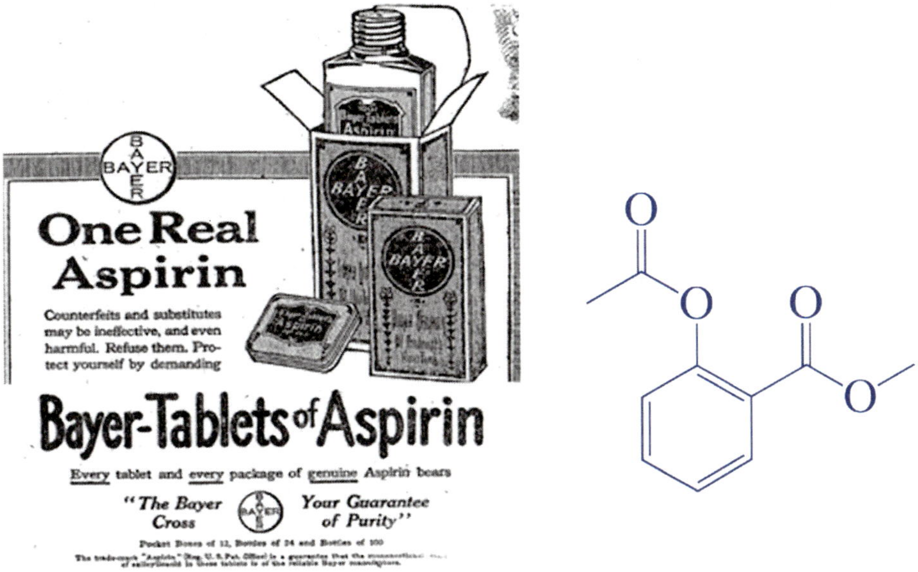

Image 7.2. An early advertisement for aspirin (left), and the acetylsalicylic acid molecule (right).

Hoffmann's molecule, when tested on human patients, proved to be both effective and far less irritating to the stomach than salicylic acid. Bayer named and marketed the drug as aspirin. Aspirin's initial popularity and subsequent ubiquity have worked together to make it the best-selling pharmaceutical of all time.

In the years since the synthesis of aspirin, other pharmaceuticals have been developed in this same way—a natural product is observed to have a desirable effect, but side effects are often noticed as well. In an attempt to amplify the positive while reducing or eliminating the negative, pharmaceutical companies

Image 7.3. (From left to right) Cinchona, foxglove, and Pacific yew—some of the many plants from which pharmaceuticals have been extracted or derived. (Cinchona photo by Forest & Kim Starr.)

purify out the active ingredient and often modify it. Examples of drugs obtained or developed in this way include quinine from cinchona tree bark (used to treat malaria), digitalis from foxglove (used to treat heart conditions), and paclitaxel from the Pacific yew (a potent chemotherapeutic).

7.2 PHARMACEUTICAL DESIGN

When Felix Hoffmann modified the willow derivative salicylic acid to the less irritating compound known as aspirin, how did he know what to do? Was he simply playing "mad scientist," mixing chemicals in his lab and hoping for the best? Or was he completely sure of what needed to be done in order to produce an effective pharmaceutical? The reality lies somewhere between these extremes. Pharmaceutical chemists rarely work without any sense of direction whatsoever. In Hoffmann's case, the structure of salicylic acid was known. Further, enough was known about the **functional groups** on the molecule to reasonably conjecture which of them was responsible for the side effects. A functional group is a characteristic arrangement of atoms in a molecule that lends a predictable set of properties to that molecule. We've already come across one functional group—the carboxylic acid. Note that salicylic acid is a carboxylic acid, with the characteristic *–COOH* atomic arrangement. Not only does the presence of the carboxylic acid provide us with information about the molecule (it's acidic), it also allows us to condense these atoms further in the line-angle diagram of salicylic acid, as shown below:*

> ***NOTE**
>
> In the case of a condensation of a grouping of atoms into a meaningful cluster like this, we do not leave the atomic symbol for carbon out of that portion of the line-angle diagram.

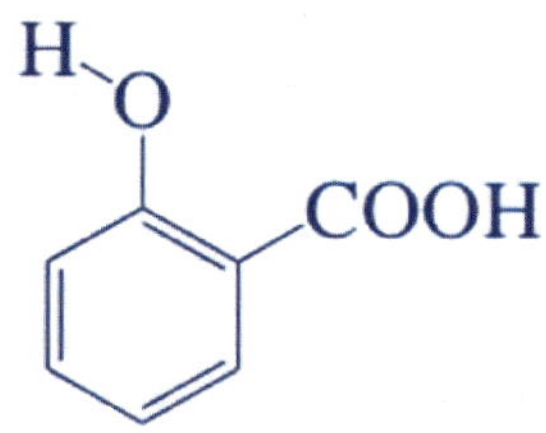

In addition to being acidic, carboxylic acids typically smell unpleasant or bracing. Vinegar, an aqueous solution of acetic acid (CH_3COOH) is one example of the characteristic, biting odor of these molecules. Another carboxylic acid, butyric acid ($CH_3CH_2CH_2COOH$), contributes to the fragrances of rotting butter and human body odor. There are many functional groups besides that of the carboxylic acid. A **benzene ring** is one such example. This functional group consists of six carbons arranged into a hexagon, with alternating single and double bonds between them. Benzene rings (also called **aromatic rings** because the molecules containing them are often highly fragranced—sometimes for better, sometimes for worse) are very common in pharmaceuticals. The salicylic acid molecule contains a benzene ring. In addition, salicylic acid contains a functional group that was of particular interest to Hoffmann. An *–OH* functional group is generally referred to as an **alcohol**. Common household alcohols include methanol (wood alcohol), ethanol (grain alcohol, the kind in beer, wine, and liquor), and isopropanol (rubbing alcohol).

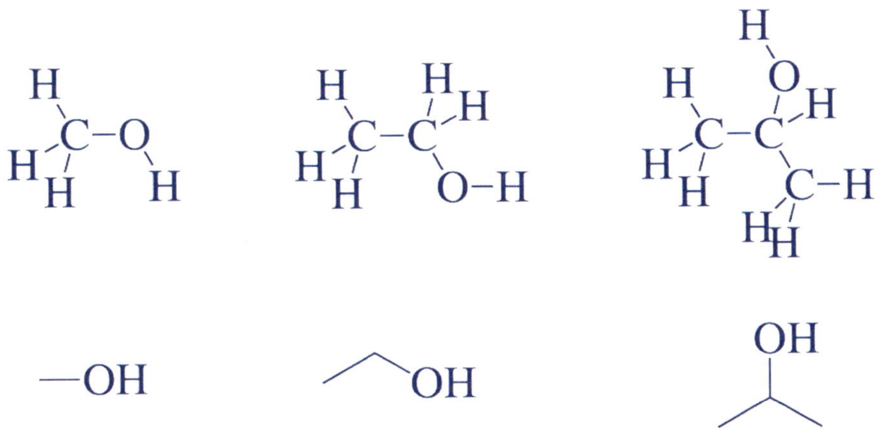

Image 7.4. (From left to right) Methanol, ethanol, and isopropanol structures and line-angle diagrams.

In the particular case of an alcohol that is attached directly to a benzene ring, a new compound functional group is formed. This group is called a **phenol**, and phenol groups were known by Hoffmann and his contemporaries to be very irritating to the skin and mucous membranes.* Hoffmann hypothesized that it might be the phenol functional group that was responsible for irritation of the stomach lining. By modifying rather than simply removing it, he was able to maintain much of the molecule's original shape (an important consideration, as we'll see later in this chapter). As you can see, knowledge of functional groups and organic chemistry was the driving force behind Hoffmann's ability to manipulate willow extract to produce the far more effective pharmaceutical, aspirin.

***Note**

Phenol-containing compounds were among the first antiseptics, used to wash and cover wounds in order to prevent infection. Unfortunately, they also had a tendency to burn the skin chemically.

Organic molecules sometimes contain a single functional group, but far more commonly they contain many. Methanol, as we saw above, has only the alcohol functional group (carbons singly bonded to other carbons and carbons bonded to hydrogens are not considered functional groups—rather, these are backbone elements). For this reason, we can say that methanol is an alcohol. Salicylic acid, though, is a much larger and more complicated molecule. It's in possession of two separate functional groups: a phenol and a carboxylic acid. As such, we wouldn't say *salicylic acid IS a phenol*; rather, it is correct to say *salicylic acid contains a phenol.*

In addition to the functional groups relevant to salicylic acid, there are several others worth discussing. **Amines** are nitrogen-containing groups. They are common in pharmaceuticals—we'll see them later in this chapter as one of the functional groups in morphine, for instance. Amines often have dead, unpleasant, or fishy odors. Ammonia is one of the more pleasant amines, and still, few people would care to smell it for any length of time. Putrescine and cadaverine, also amines, lend rotting cadavers their putrid odor.

You may be sensing that organic nomenclature is quite a bit more complex than the binary compound nomenclature discussed in Chapter 1. A complete discussion of the organic naming system is beyond the scope of this text. In general, though, organic compounds often have two names—one systematic (as in *methanol*), and one common (as in *wood alcohol*). Sometimes,

Image 7.5. (From left to right) Ammonia, putrescine (1,4-diaminobutane), and cadaverine (1,5-diaminopentane) exhibit the characteristic unpleasant odor of amines to varying degrees.

the common names sound like systematic names. Putrescine, for instance, sounds like a chemical name, but is in fact the common name for a compound systematically called 1,4-diaminobutane. The name putrescine shares the twin benefits of being less of a mouthful to pronounce than the systematic name and being representative of a notable source of the compound (putrid flesh). Common names are encountered frequently in organic chemistry and are often informative (β-*carotene*, for instance, is found in and named for carrots) and sometimes amusing (a large organic molecule with a shape similar tothat of the geodesic dome designed by architect Buckminster Fuller is called *buckminsterfullerene*).

Esters, like phenols, are compound functional groups. These combine a carboxylic acid with an alcohol. Unlike phenols, in which the benzene ring and alcohol can't easily be separated chemically, esters are literally synthesized from (and can be decomposed into) their constituent parts:

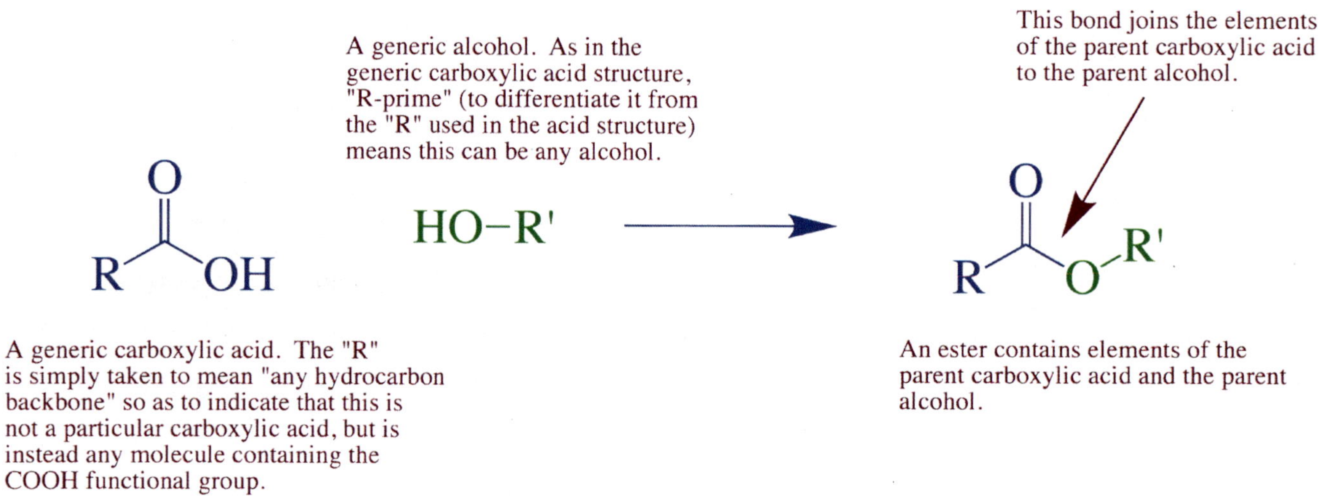

The resulting ester is neither a carboxylic acid (it is not acidic) nor an alcohol, but has its own unique set of properties. Esters are highly fragrant molecules, but unlike the unpleasant carboxylic acids and amines, these compounds generally have very pleasing odors. In fact, esters are responsible for many fruit and floral fragrances.

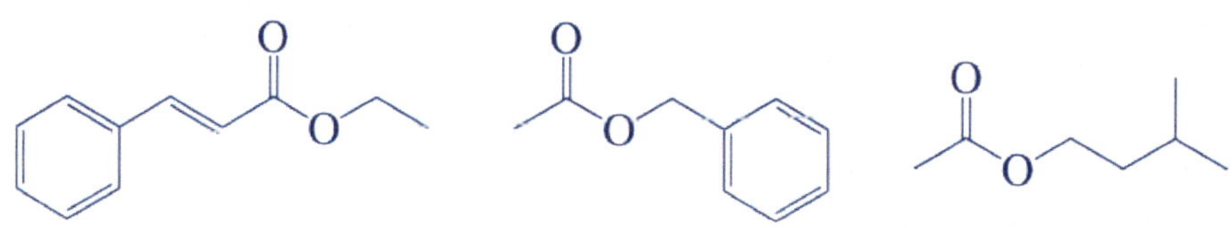

Image 7.6. (From left to right) Ethyl cinnamate (cinnamon), benzyl acetate (in the fragrance of pear and strawberry), and isopentyl acetate (a component of banana) are esters.

It's worth remembering that the ester is, at its chemical heart, the combination of a carboxylic acid and an alcohol, and under the right circumstances the ester can decompose into its constituent parts. In fact, the decomposition of an ester takes place readily in the presence of water and a very small amount of acid or base. The amount of acid or base required for esters to decompose is so small that when exposed to surfaces or air (which are always contaminated with small quantities of mildly acidic or basic compounds), esters will generally react fairly quickly. This explains why expensive perfume fragrances tend to wear true throughout the day,* while inexpensive perfume fragrances change (often becoming unpleasant) over time. Perfumes are made using scented molecules, many of which are esters. More expensive perfumes use oil as the solvent in which fragrance molecules are dissolved. The oil protects the delicate esters from water and keeps them from reacting chemically. Less expensive than oil, however, is the alcohol and water solvent used in cheaper scents. In the bottle, protected from acids and bases, the esters do not react. Sprayed on the skin, however, esters are exposed to the environment, and react in short order. The resulting carboxylic acids tend to smell far less pleasant than the original esters.

***NOTE**

All perfumes smell differently on a person than they do in the bottle, and in fact, smell differently on different people. This is due to individual body chemistries and odors interacting with and mixing with perfume molecules.

TRY THIS:

Identify the functional groups in the molecule below:

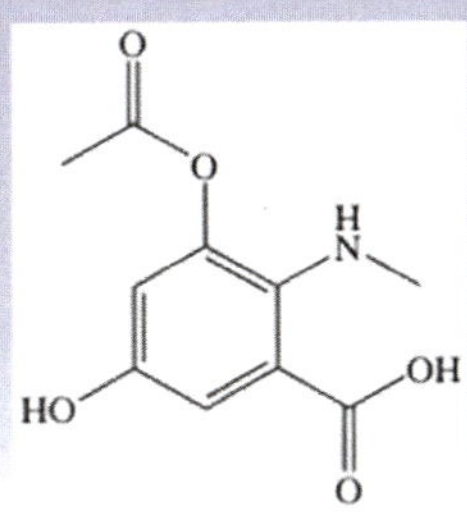

Answer:

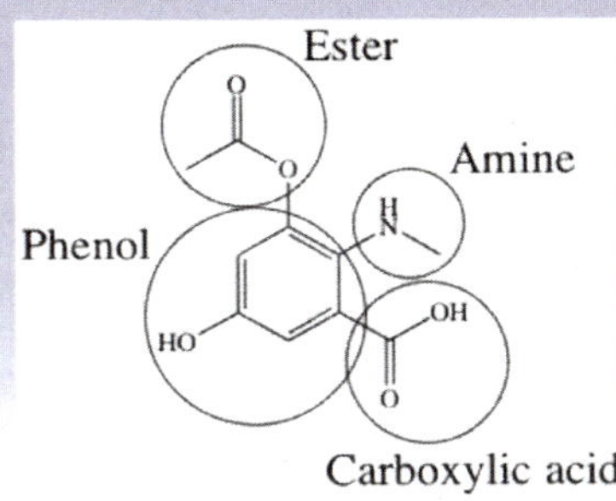

7.3 WONDER DRUGS

An **ether** is yet another functional group. Unlike the highly reactive esters, ethers are fairly inert. As a group, they consist of two hydrocarbon backbones—these can be two of the same hydrocarbon or they can be different from one another—joined by a common oxygen. As in the ester example above, R and R' are taken to mean *hydrocarbon of some sort*, and are used in portraying a generic ether:

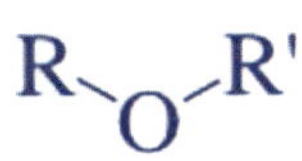

Most of us have heard the word ether before—it's commonly used to refer to a particular molecule more appropriately called by its chemical name, *diethyl ether.*♦

♦NOTE

It's not unusual for the most common (or first discovered) member of a group of molecules to be referred to in popular parlance by the functional group name. *Ethanol*, grain alcohol, is usually simply called *alcohol*. The simplest of the phenols, an otherwise unsubstituted benzene ring with an OH group attached, is just *phenol*.

Diethyl ether gained popularity as an early anesthetic. Before the middle of the nineteenth century, no anesthetics were known. Surgical procedures were completed on fully (or at least mostly) conscious patients. Prior to surgery, patients were often encouraged to consume large quantities of alcohol in the hope that this would reduce their sensation of pain somewhat, or at least render them less aware of their milieu. Regardless, surgical assistants were sometimes required to forcibly restrain patients.* It goes without saying that the horror of surgery in the days before anesthetic made it a last resort medically—surgeons had to work quickly on their suffering patients, making procedures sloppier than those practiced as a near art form today, and patients often succumbed to shock from the pain (not to mention the prevalence of postsurgical infection). Diethyl ether, first used by the dentist William Morton in 1846, was found to reduce consciousness so that surgery could be performed on a still (and mercifully pain-free) patient. Not only did this increase the popularity of surgery as a therapy, it also allowed for the honing of surgical skills and development of techniques that increased the effectiveness of procedures. Despite the fact that ether was abandoned by the medical world rather quickly (it had the potential to be explosive and left patients feeling very ill upon waking), it was replaced with a series of increasingly advanced anesthetics. A modern surgeon has at his or her disposal a variety of anesthetics (so many, in fact, that anesthesiology has become a medical specialization in its own right) that can be selected based upon the procedure, patient allergies, or a mere matter of preference.

***NOTE**

Old surgical accounts suggest that the surgeons detested inflicting these surgeries as much (or nearly so) as their patients detested experiencing them—it was not unknown for surgeons to brace for an operation with a bit of alcohol themselves!

CONCEPT CHECK:

What functional group does aspirin have that salicylic acid lacked?

Answer: Aspirin has an ester group.

7.4 DRUG SOLUBILITY

As noted in Chapter 4, the polarity of a compound informs its solubility. Highly polar compounds (particularly ionic and hydrogen bonding molecules) dissolve well in water, while nonpolar compounds are hydrophobic and dissolve well in other nonpolar molecules, including fats. As we turn our discussion to organic molecules, most of which are quite large, the issue of polarity becomes somewhat more complex. Take, for instance, the aspirin molecule. This molecule has some areas that are highly polar, and other areas in which the hydrocarbon backbone predominates:

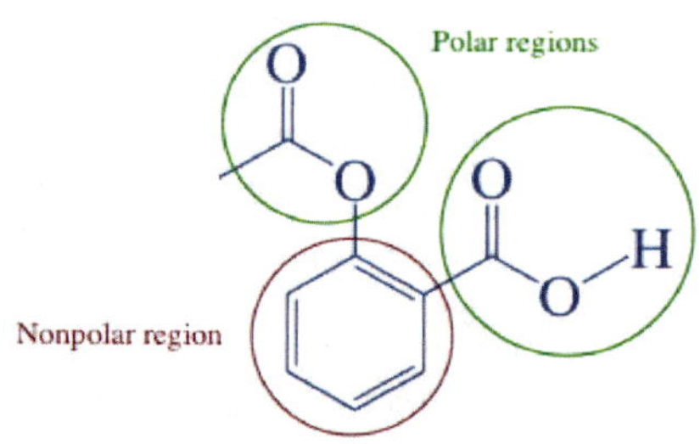

In a case such as this, how do we ultimately define the polarity of the molecule? It turns out that solubility is not as rigorously defined as implied earlier. Molecules need not be either entirely **hydrophilic** (which means water-loving) or entirely hydrophobic. Rather, solubility is a continuum, with polar functional regions lending hydrophilicity, and nonpolar regions lending hydrophobicity. To determine solubility, we compare the predominance of polar and nonpolar regions and come to a judgment. Because the continuum makes it difficult to judge in absolute terms the polarity of a compound, in most cases we'll simply use relative terms to compare two or more compounds' solubilities. For instance, a compound with more OH groups relative to its number of carbons is likely to be more water-soluble than a compound with fewer OH groups relative to its number of carbons (Image 7.7).

Image 7.7. Ethylene glycol (left) is likely to have greater water solubility than ethanol (center), which is likely to be more water-soluble than hexanol (right).

TRY THIS:

Determine which areas of the molecule below are polar, and which are nonpolar.

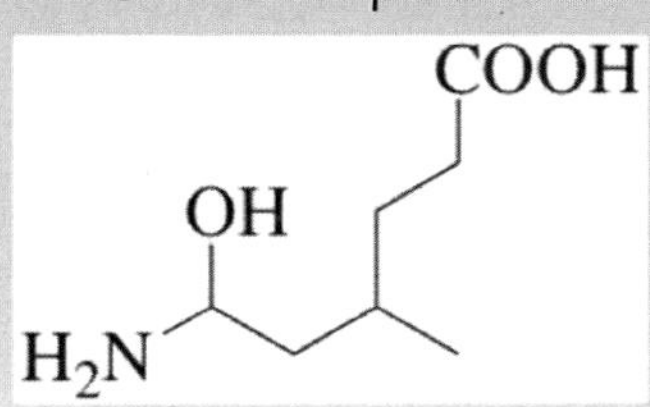

Answer:

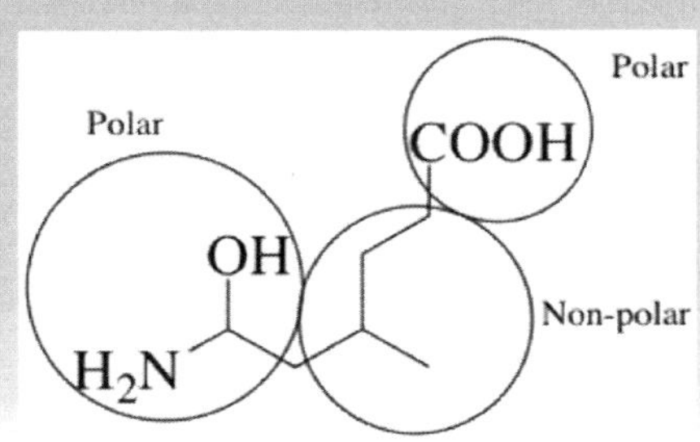

TRY THIS:

Which is likely to be more water-soluble and why?

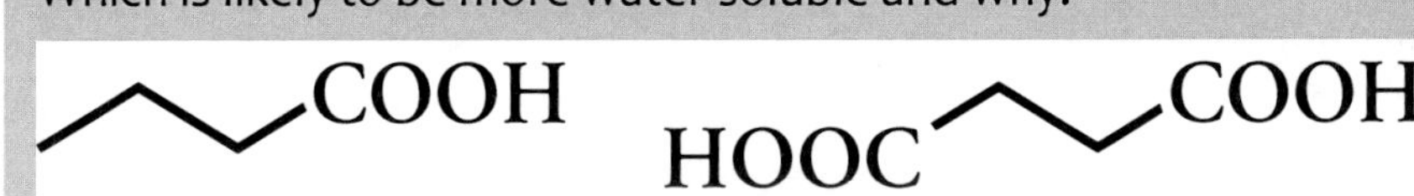

Answer: The molecule on the right will be more water-soluble—it has more polar functional areas (-COOH is polar) compared to the number of carbons.

In designing pharmaceuticals, water solubility is of great importance—a drug must dissolve in water in order to be absorbed from the digestive tract or dissolve in the aqueous media of the body. While some drugs are appropriately hydrophilic on their own, others (including many of the large amines) are only slightly soluble in water. Morphine, a potent analgesic shown below, has very limited water solubility, making it difficult to administer in its pure form:

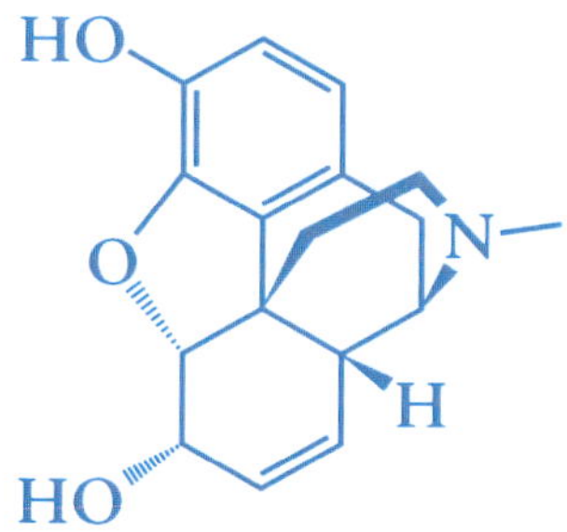

In order to make morphine water -soluble, it's possible to take advantage of the chemical properties of one of its functional groups. Ammonia, as we saw in Chapter 5, is basic. It turns out amines are basic as well. It really doesn't matter whether a nitrogen atom has three hydrogen atoms, three hydrocarbon groups, or a combination of the two attached to it—the chemical property of basicity is consistent across the amines. It's possible to take advantage of this by neutralizing the amine group on morphine with an acid (often HCl). The resulting ionic salt (written *morphine HCl*) has greatly enhanced water solubility:

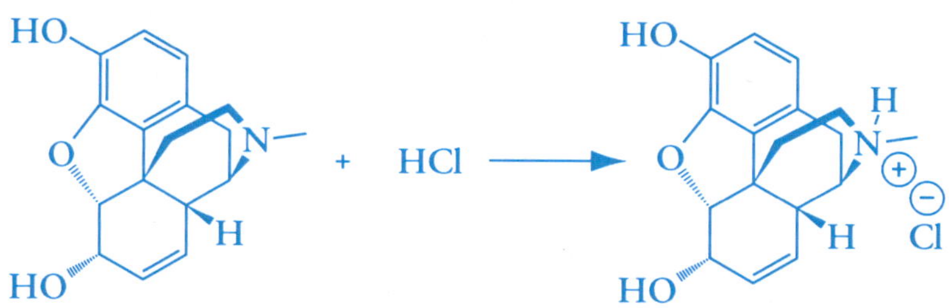

It's quite common to see acidic salts of pharmaceuticals; a few common examples from the medicine cabinet include diphenhydramine HCl (an antihistimine used to reduce allergy symptoms), pseudoephedrine HCl (a decongestant), and dextromethorphan HBr (a cough suppressant). In each case, the parent pharmaceutical had low hydrophilicity, while the amine salt is significantly more water-soluble.

The primary goal of salt formation from pharmaceuticals is to enhance water solubility, but there are additional benefits. Shelf lives of salts are longer than those of the parent compounds, and odor is reduced—a major consideration for many amine-containing pharmaceuticals!

7.5 RECEPTORS AND PHARMACOPHORES

In addition to being an interesting example of a free-based pharmaceutical, morphine has a long and intriguing history. Produced naturally by the opium poppy (*Papaversomniferum*), morphine's analgesic, sleep-inducing, and hallucinogenic properties have been appreciated for hundreds, perhaps thousands, of years. Unfortunately, morphine is highly addictive, which limits its pharmacological utility. As a result, pharmaceutical chemists have long sought to produce alternative compounds that would replicate the activity of morphine without the addictive side effects. Two compounds that produce similar analgesia (with greatly reduced hallucination and potential for addiction) are codeine and meperidine (marketed under the trade name Demerol). While the chemical similarity between codeine and morphine is clear from the structures, at first glance, meperidine looks quite different:

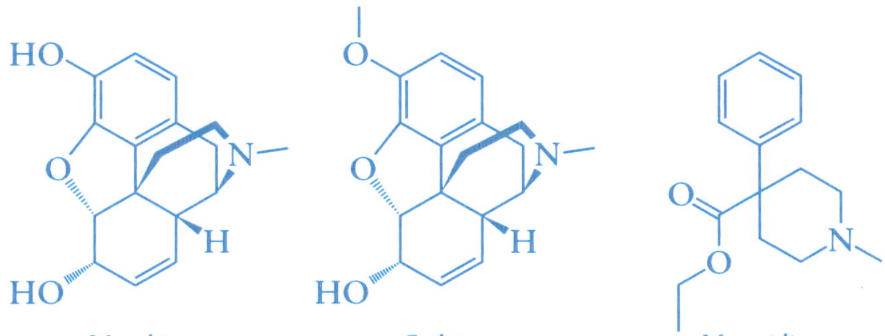

Morphine Codeine Meperidine

Drawn from a slightly different three-dimensional perspective, however, the similarities between meperidine and the other two compounds become more clear:

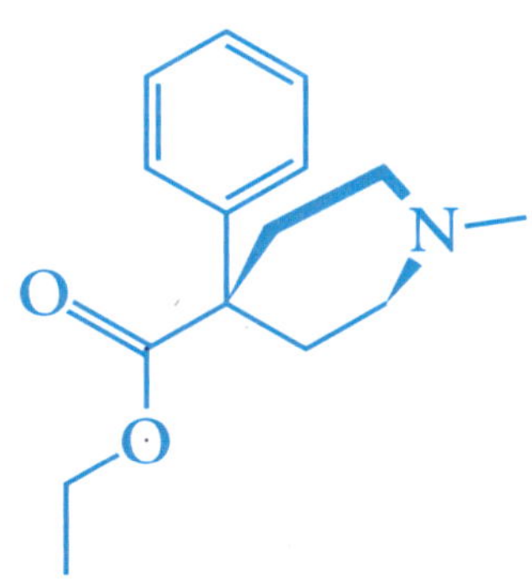

It turns out that the similarity in shape among these molecules is no coincidence—pharmaceuticals operate in the body by mimicking, enhancing, or repressing the action of the body's own products. In essence, pharmaceuticals act as artificial "keys" for the body's own "locks," where the "locks" are more properly called **receptors**, and are large molecules on body cells. A **substrate**, or compound that fits into a receptor like a key, binds to the receptor and produces an effect. Morphine, incidentally, fits into a receptor whose native substrate is an *endorphin*, a pain-reducing compound produced by the body's cells. Receptors may have one or more natural substrates, but any molecule with a shape similar to that of the receptor's natural substrate will be able to bind. With regard to morphine and compounds of similar

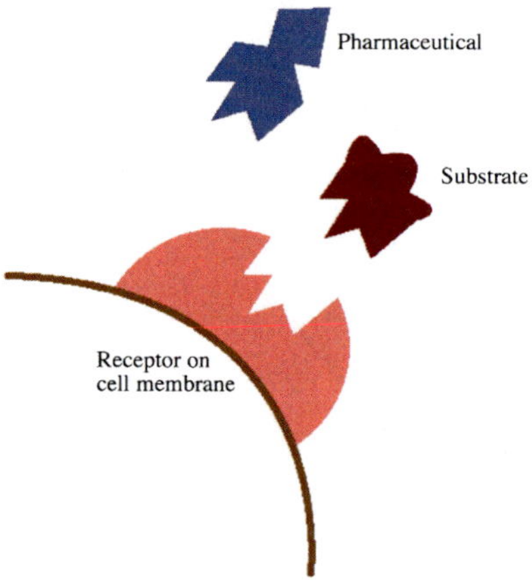

Image 7.7. A natural substrate fits into the receptor site. Any pharmaceutical binding to the same receptor would be expected to have a functional area (pharmacophore) similar to that of the natural substrate. The rest of the molecule, however, could look quite different.

activity, meperidine shows us that it isn't necessary for a molecule to be nearly identical to morphine (as codeine is) to produce the morphine effect—in fact, only a few key functional areas appear to be important. This is quite typical of pharmaceuticals. Generally, a core functional area (called a **pharmacophore**) is responsible for binding to the receptor. This pharmacophore would be expected to be common to all substrates binding to a given receptor. The rest of the molecule, however, can have any of a number of shapes associated with it. Areas apart from the pharmacophore are responsible for various pharmaceutical properties, but do not bind to the receptor. It is through manipulation of non-pharmacophore molecular structure that chemists achieve a reduction in side effects and other differences between similar drugs.

While the morphine-like molecules codeine and meperidine are frequently used today (as is morphine, though with great caution), a very early attempt to synthesize a lessaddictive morphine analog produced unexpected results. Shortly after Felix Hoffmann acetylated salicylic acid to produce aspirin, he was asked by the Bayer Company to work on producing a less addictive form of morphine. Probably suspecting he was going to end up with codeine (which was known at that time to be less addictive but which had not yet been synthesized), Hoffmann acetylated morphine. The result was diacetylmorphine:

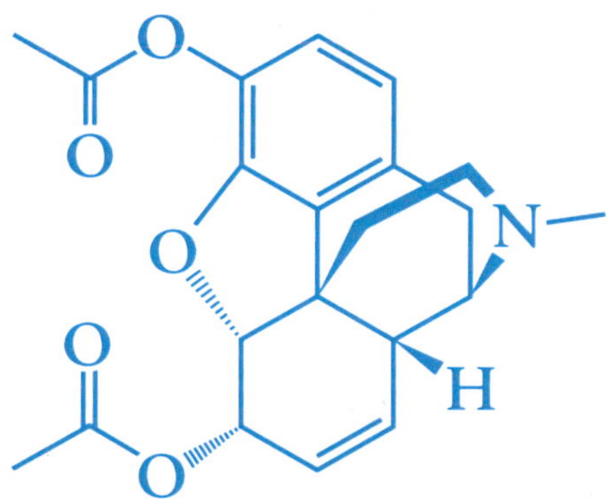

Bayer marketed the new drug extensively as a nonaddictive morphine substitute and a cough suppressant. Unfortunately, it was soon discovered that diacetylmorphine was quickly and readily converted to morphine in the body and was, in fact, even more active than its parent compound. It was also more addictive. While most of us probably aren't familiar with the chemical name *diacetylmorphine*, almost everyone has heard of Bayer's trademark name for the drug—Heroin. The heroin episode was quite embarrassing for Bayer, for obvious reasons.

7.6 EARLY PHARMACEUTICAL REGULATION

With the rapid advancements in pharmacological chemistry taking place in the late nineteenth and early twentieth centuries, clearly some regulation was in order. In 1906 President Theodore Roosevelt signed the **Food and Drug**

Act, the primary stipulations of which prevented the transport of goods across state lines unless they were correctly labeled and free from contaminants. There were no provisions for safety or efficacy; the proverbial "snake oil" salesman was free to sell any substance he liked—provided that it was pure and correctly labeled—whether it was toxic, ineffective, or both. The weaknesses of the 1906 Food and Drug Act are readily apparent—Coca Cola, for instance, won a lawsuit (*United States vs. Forty Barrels and Twenty Kegs of Coca-Cola*, 1909) in which they were charged with putting excessive caffeine (as well as cocaine) into their product. Thanks to accurate labeling and lack of contamination, however, Coca-Cola was spared legal sanctions.*

A major improvement in food and drug safety legislation came in 1938, but only after a tragic incident demonstrated beyond argument the need for further control. In 1937 a pharmaceutical company found what appeared to be an excellent solvent for the antibiotic sulfanilamide. While quite effective, sulfanilamide has exceedingly low solubility in water. It's also very bitter, making it difficult to administer to children. The company determined that the compound dissolved readily in diethylene glycol. The solvent had the added advantage of being sweet, increasing palatability to children. Unfortunately, diethylene glycol is also quite toxic (a fact that was known to most chemists, but was somehow overlooked by the pharmaceutical company), and no animal testing preceded the marketing of the new elixir.

Of a group of 353 children to whom the drug was administered, 100 died of kidney failure. This created a firestorm of public, scientific, and political response, the result of which was the **1938 Food, Drug, and Cosmetic Act.**♦ This act required truthful labeling and proven safety of all drugs. Manufacturers of pharmaceuticals had to request a new drug application, to be reviewed by the FDA (which had been created about a decade prior). The FDA required that the new drug application include submission of toxicity studies.

It is worthy of mention that while safety standards were made more stringent via the 1938 FD&C Act, proof of efficacy was still not required. A major concern associated with a legislative act that allows for pharmaceuticals to be produced and marketed with some proof of safety but none whatsoever of efficacy is that the consumer is unable to formulate a risk-to-benefit ratio for purposes of decisionmaking. If a drug is slightly risky but highly effective, such that the potential for benefit far outweighs the potential for harm, a consumer might choose to use the medication. If a drug is risky and minimally or rarely effective, a consumer might decide against using the drug. Because no drug is completely free from safety concerns, it is only when armed with information about the potential benefit of using a substance that a consumer is able to use risk-benefit analysis in determining their course of action. The lack of an efficacy requirement was a major weakness of the 1938 legislation. And then came thalidomide, and everything changed.

***NOTE**

The cost of the legal battle was significant, however. Coca-Cola ended up settling out of court, and did eventually reduce the caffeine in their product. Cocaine, however, remained an ingredient for some time.

♦NOTE

We see evidence of this regulatory measure every time we read the ingredient label on a food or drug containing artificial coloring. FD&C Red 40, for instance, is a red color approved under the Food, Drug, and Cosmetic Act.

Image 7.8. Sulfanilamide elixir, with its toxic diethylene glycol solvent, was the driving force behind more stringent safety regulations.

7.7 THROUGH THE LOOKING GLASS—HEALER OR KILLER?

Beginning in the late 1950s, thalidomide was marketed as a mild sedative and anti-nausea drug. It was particularly appreciated by women in the early months of pregnancy, for whom it helped relieve the symptoms of morning sickness. The drug was prescribed extensively in Germany and England, but was refused an application for release in the United States by FDA inspector Frances Oldham Kelsey, who felt that more safety testing was necessary before the drug should be brought to market. Shortly after thalidomide made its debut in Europe, physicians became aware of an increased incidence of what had been a historically rare birth defect called phocomelia, meaning "flipper limb." During certain key periods of gestation, thalidomide inhibits the ability of the embryo to develop normally, resulting in severe physical deformities. Hundreds of thousands of children were born with these defects during the 1950s and 1960s, mostly in Europe. Kelsey's refusal to allow marketing of the drug spared the United States from widespread effects.

There are two interesting issues associated with thalidomide. The first is that despite Kelsey's feeling that the drug's safety was not adequately established, it had, in fact, undergone testing. Adult studies of thalidomide revealed it to be safe and low in side effects. Unfortunately, developing embryos are often much more sensitive to drugs than adults—the *teratogenic*, or development-related, effects of the drug were unknown because drug studies are not traditionally done on pregnant women—the risk is felt to be too

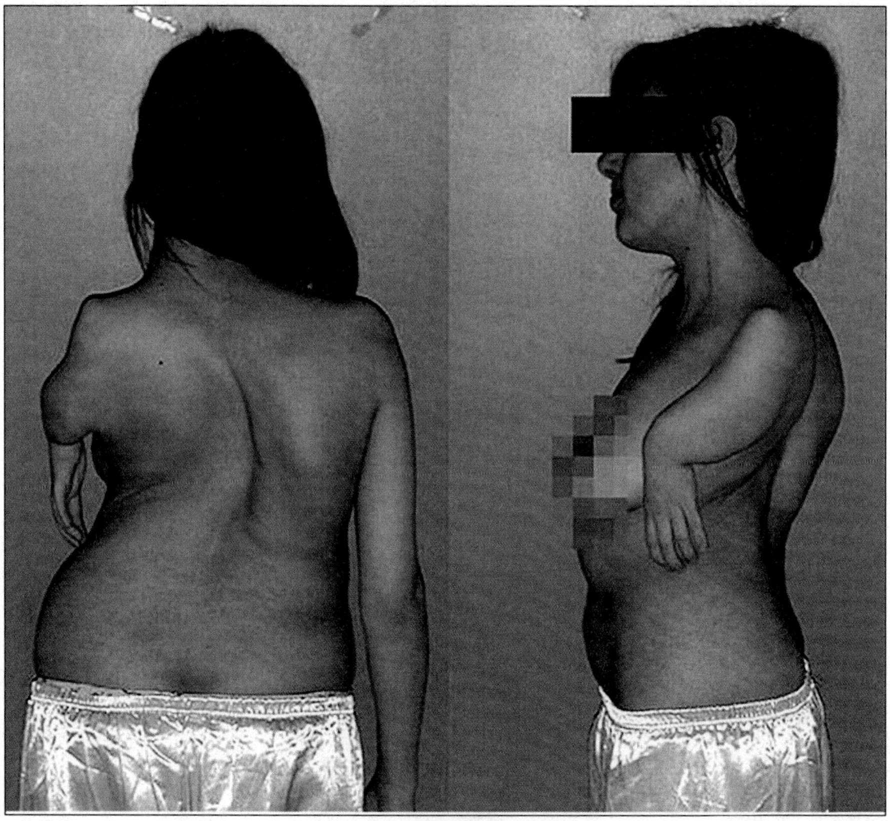

Image 7.9. An adult with phocomelia.

great. A second interesting bit of information about thalidomide is that it exists in two forms that are mirror images of each other. One form is a safe sedative, the other is a **teratogen** and causes birth defects.

How is it possible for a molecule to have a mirror image? To answer this question requires some three-dimensional visualization. Imagine, first, a symmetrical object like a baseball. If a baseball is held up to a mirror, the mirror image of the baseball is identical to the original. An asymmetrical object, however, such as a hand, is not identical to its mirror image (Image 7.10). The mirror image of a right hand is a left hand—they have many similarities, and can do many of the same things, but they are not **superimposable**, meaning that where the left hand has a thumb, the right has a pinkie finger and vice versa. If we flip one hand over so that thumbs and pinkies are together, now the left hand has a back where the right has a palm, and so forth. A hand is not superimposable upon its mirror image.

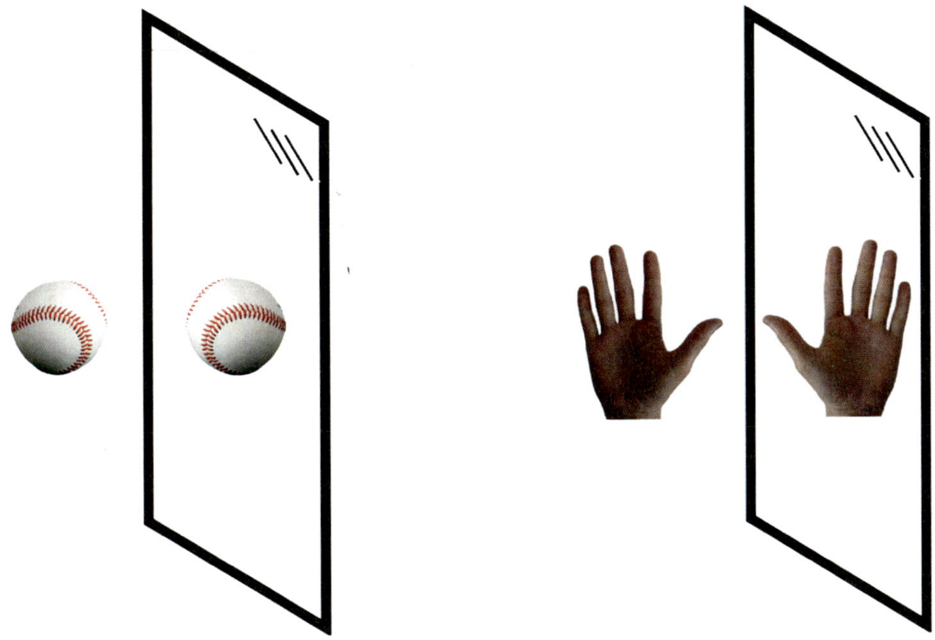

Image 7.10. Mirror images of symmetrical objects are identical to (superimposable upon) the original object. This is not so with asymmetrical objects.

Chemicals, depending upon their symmetry, can be superimposable upon their mirror images (like baseballs), or non-superimposable (like hands). Chloroform, an early anesthetic, is an example of a symmetrical molecule. If a chloroform molecule is reflected in a mirror and the mirror image is rotated, we can see that the mirror image is identical to the original (Image 7.11). Remember that molecules, like all physical objects, can be turned over and rotated; this does not change the molecule chemically. We can turn a mirror image of a molecule any way we need to in order to superimpose it upon the original. We mustn't break bonds or rearrange the molecule, however—to do that would be to make a new, different molecule.

Symmetrical molecules like chloroform, which are superimposable upon their mirror images, are called **achiral** molecules. Alternately, a molecule may be asymmetrical. The compound in Image 7.12 is an example of such a molecule:

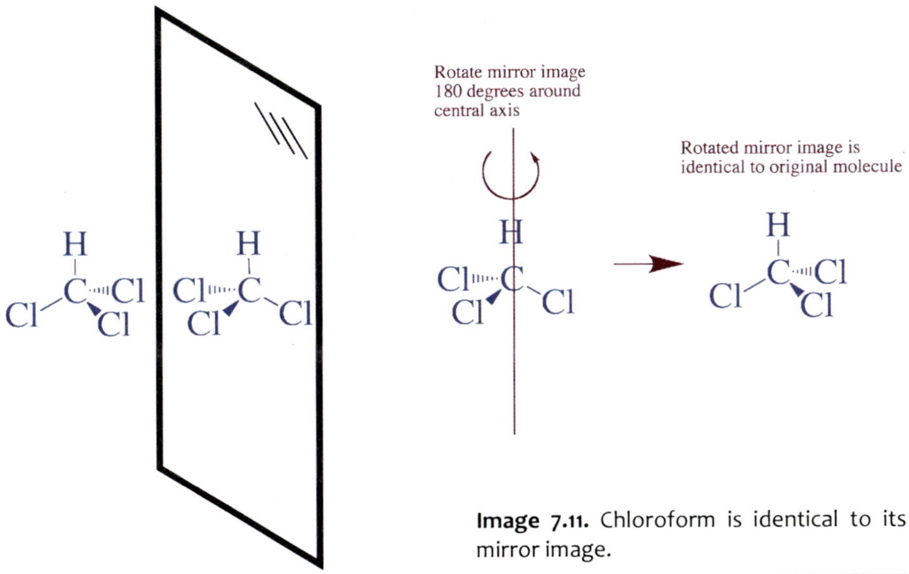

Image 7.11. Chloroform is identical to its mirror image.

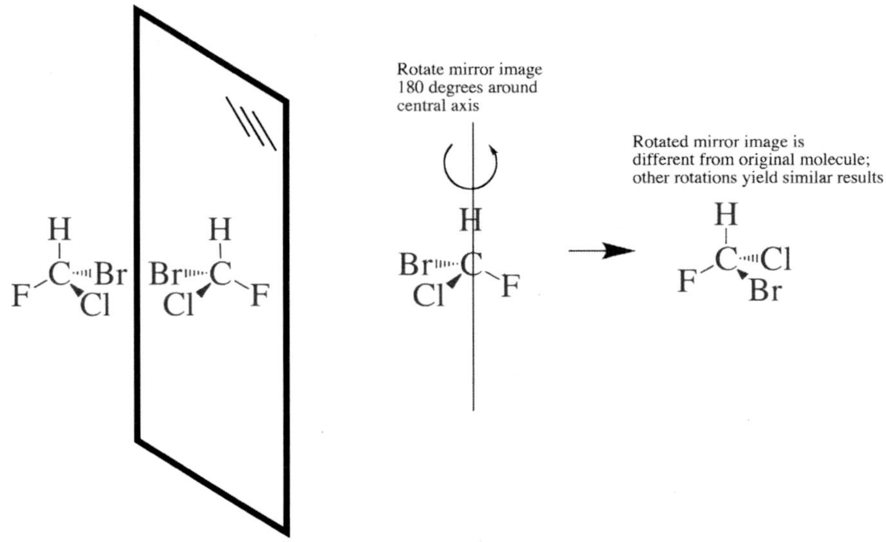

Image 7.12. This molecule is not identical to its mirror image.

There is no way, short of breaking bonds and rearranging the molecule, to superimpose the molecule in Image 7.12 on its mirror image. Molecules like this, which are non-superimposable on their mirror images, are called **chiral** molecules. The original molecule and its mirror image are referred to as a pair of **enantiomers**. Enantiomers have identical physical properties—they have the same melting point, boiling point, color, and so forth. However, they react very differently in the body. Let's return to thinking about the substrate-receptor interaction that takes place in the body—the substrate fits into the receptor like a key into a lock. If you look closely at your house key, you'll notice it's asymmetrical. That is, its mirror image would not be identical to, or superimposable on, the original key. If you had the key's enantiomer made, you wouldn't be able to fit it into the locks on your front door. The same is true of substrates: If a substrate is chiral (and most are), only the correct enantiomer will be able to fit into the receptor and produce an effect.

Where this becomes a problem for pharmaceutical chemists is that, owing to the incredible similarity of a pair of enantiomers, it's very time consuming and difficult to produce only one of the two. One of the amazing things about living cells is that they can do quite easily what is very difficult for chemists—cells generally only make ONE enantiomer of a pair as a natural product. As a result, the receptors for that product are specialized to fit only the proper enantiomer. A way to think about this is in terms of hands (substrate) and gloves (receptor). If the body makes "right hand" molecules as a natural product, then the body will only produce "right glove" receptors for that substrate. If a pharmaceutical chemist wants to make an artificial substrate, he or she generally synthesizes a drug containing BOTH the "right hand" and the "left hand" molecules. A mixture like this, composed of 50 percent of one enantiomer and 50 percent of the other, is called a **racemic mixture**. The "right hand" molecules from the racemic mixture will bind to receptors in the body. The "left hand" molecules will not. They might be harmlessly excreted, they might cause a side effect, or they might be highly toxic: it depends upon the molecule, and it's important for a chemist to be aware of this when designing a drug.

It can be difficult to glance at a molecule and ascertain whether it is chiral or achiral. An easy shortcut is to search the molecule for a carbon atom with four different **substituents**. (A substituent is any group bound to the carbon.) Regardless of the size of the molecule, even ONE carbon atom with four different groups bonded to it indicates asymmetry, and means the molecule is chiral (and has a mirror-image enantiomer). Each of the molecules below is chiral; the carbon with four different substituents is indicated in each case:

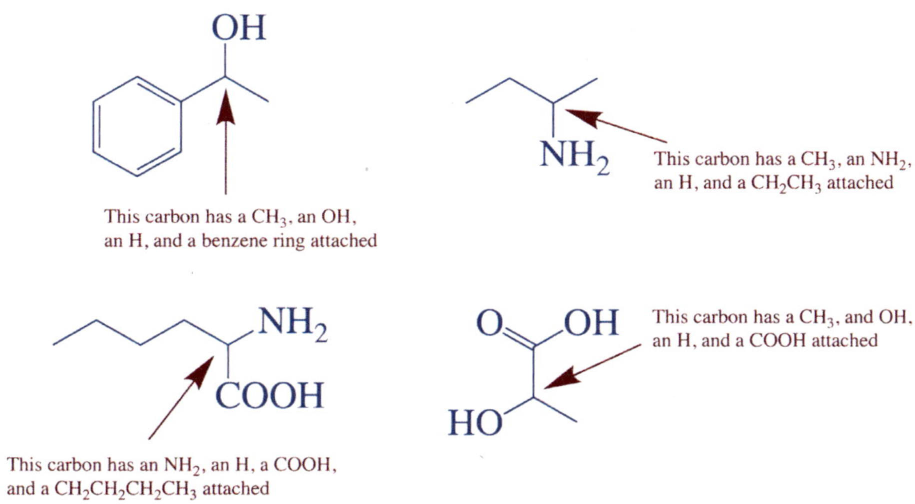

TRY THIS:

Indicate the carbon(s) with four different substituents on the following molecule:

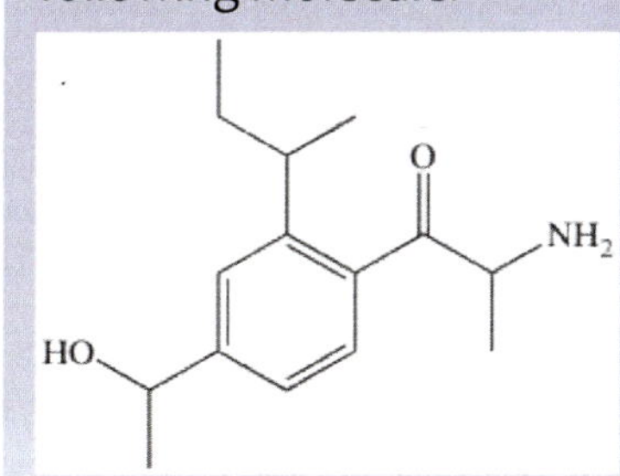

Answer:

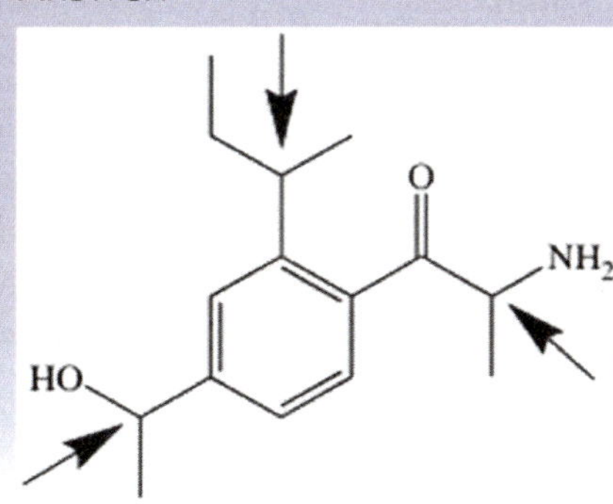

CONCEPT CHECK:

If a wristwatch were a molecule, would it be chiral or achiral?

Answer: Chiral—it's asymmetrical.

As mentioned earlier, where chirality becomes a major consideration for the pharmaceutical industry is in the disposition of the non-substrate enantiomer; it's critical to know whether this enantiomer will have any physiological activity, and if so, what that activity is. For example, take a chemical called *methorphan.* This compound is chiral—the two enantiomers are distinguished from one another in nomenclature by the addition of prefixes to their names—one form is called *levomethorphan*, while the other is *dextromethorphan.**

***NOTE**

There are many ways to indicate through nomenclature that a compound is chiral. One is the use of the prefixes *levo-* and *dextro-*, another is through using the prefixes *R-* and *S-*. Yet another mechanism puts the symbol (+) in front of one enantiomer, and (-) in front of the other. Why one mechanism is used over another depends upon many factors, not the least of which is simply the preference of the pharmaceutical company. A further frustration is that many compounds (like morphine) are chiral, but carry no telltale prefix in front of their names. (In actuality, there is a levomorphine and a dextromorphine, but these names are rarely used.) As a result, a discussion of enantiomer nomenclature is well beyond the scope of this text!

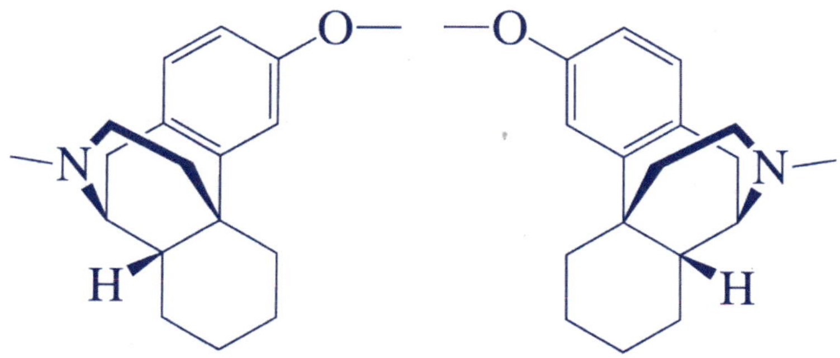

Dextromethorphan Levomethorphan

Dextromethorphan is a nonaddictive, very effective cough suppressant. Levomethorphan, however, is an addictive opiate—if you compare the structure to that of morphine from earlier in the chapter, you'll notice many similarities between the shapes! Dextromethorphan, while appearing to have many key chemical similarities to morphine, contains a mirror image of the morphine pharmacophore and can't bind to the morphine receptor. When a pharmaceutical company synthesizes methorphan, both the dextro- and levo- forms are produced. Because it's known that the levo- form is not safe, the enantiomers must be separated from one another and the levomethorphan discarded. This process is time consuming and expensive, but in this particular case, it's quite necessary.

A much more favorable situation is presented by ibuprofen. Like methorphan, ibuprofen is synthesized as a combination of two enantiomers. Dexibuprofen is an analgesic and anti-inflammatory. Levibuprofen has no activity in the body. Luckily for the pharmaceutical company, however, a liver enzyme converts levibuprofen to the active form. This means the drug can be produced inexpensively as a racemic mixture.

Levibuprofen Dexibuprofen

Thalidomide presents a third scenario—like methorphan and ibuprofen, thalidomide is chiral. As mentioned previously, one enantiomer is a safe, effective sedative. The other is teratogenic. Unlike methorphan, however, pharmaceutical companies can't simply purify the drug and give only the safe form—the safe enantiomer is actually converted into the teratogenic enantiomer by the liver. As such, there is no safe way to administer thalidomide—at least, not to relieve symptoms of morning sickness!

In 1962, following the thalidomide tragedy, an amendment to the Food, Drug, and Cosmetic Act made the safety restrictions on pharmaceuticals even tighter than they already were. Provisions included that before human trials of a drug could begin, an Investigational New Drug application (IND) had to be submitted, replete with results of animal testing. A further stipulation provided that, for all drugs introduced after 1962, both proof of efficacy and documentation of safety were required. While new, stricter safety standards were not retroactively applied to drugs submitted under the Food, Drug, and Cosmetic act between 1938 and 1962, the amendment nevertheless required that manufacturers of drugs introduced during this period of time supply proof of efficacy in order to maintain their pharmaceutical on the market. Of course, a major result of the new legislation was to increase the cost of pharmaceutical development into the hundreds of millions of dollars per prescription drug (including the cost of mandatory post-marketing surveillance required by the FDA). This has had a direct effect upon the cost of prescription drugs to consumers, as pharmaceutical companies have to recoup their development costs. A second result of the more stringent regulation is that the time necessary to take a drug from the lab bench to the market is, on average, eight to ten years. Patients and families of patients with illnesses for which there are currently no appropriate pharmaceuticals find the wait intensely frustrating, and understandably so. However, the FDA must balance public safety with pressure from both the pharmaceutical companies and the public to bring new drugs to market quickly.

In the end, thalidomide has found an application as a useful drug for treating symptoms of Hansen's disease (leprosy), some forms of cancer, and a small variety of other ailments. It is a highly controlled substance, and is never administered if there is any possibility that the patient might be or become pregnant. The thalidomide case, however, highlights the utility of having not only safety information for a pharmaceutical, but also efficacy information. In two separate applications, the risk-to-benefit ratio of using thalidomide turns out very differently. For a pregnant woman (high risk), taking thalidomide for morning sickness (beneficial, but symptoms are not life-threatening) represents an unacceptable risk-to-benefit ratio. For a non-pregnant (minimal risk) patient with a painful and disfiguring disease (high benefit), the risk-to-benefit ratio clearly points in the direction of using the drug. Since the 1962 amendment, we have both safety and efficacy information at our disposal, allowing for this type of risk-benefit analysis in decision making.

Image 7.13. (Clockwise from top left) The death cap mushroom, poison dart frog, and castor beans are just a few examples of organisms producing completely natural chemicals that are deadly poisons.

7.8 WHAT DOES "NATURAL" MEAN, ANYWAY?

To many individuals concerned with health, welfare, and well-being, it's important to avoid "chemicals" or the "artificial," and to seek out the "natural." The avoidance of chemicals, as we saw in Chapter 1, is simply not possible. All matter is composed of elements, compounds, or mixtures—in short, chemicals. Our bodies are made entirely of chemicals, water is a chemical, the air we breathe is chemical, the food we eat is entirely chemical. Dispensing with the notion that we can in any way avoid chemicals, let's instead examine the difference between "natural" and "artificial," or man-made, chemicals.

A good place to start is with flavoring agents—some of us, in choosing which foods to eat and which to avoid, tend to steer clear of products containing artificial flavors. Perhaps as we read those words we picture scientists mixing chemicals in a lab, producing completely synthetic and falsified (and possibly dangerous!) versions of the naturally scented compounds found in nature. By contrast, when we read the words *natural flavor*, perhaps images of fruits, vegetables, and spice plants come to mind. Unfortunately, the dichotomy is not nearly so clear. Natural and artificial flavors are often identical chemicals; the natural flavors are extracted from foods, and the artificial are made in the lab. While that definition seems to support the typical notions of flavors and their origins, let's look a bit deeper. Natural flavors don't have to—and in fact, often don't—come from the food they taste like. While some of the chemicals used in natural banana flavor may be extracted from banana, natural almond flavor comes from the pits of stone fruits (like peaches). Furthermore, the artificial almond flavor made in the lab is the SAME chemical as the natural almond flavor extracted from peach pits (an organic molecule called benzaldehyde). Surprisingly, artificial flavors are often safer and more pure than natural flavors, since the chemicals have been controlled for purity in the lab and are much more likely to be free from contaminants. Natural almond flavor, for instance, may be laced with cyanide, a toxic chemical that occurs naturally in peach pits—artificial almond flavor never is. This is not to say that artificial flavor is "better for you" than natural, or that since natural flavor contains the same chemicals as artificial flavor, which is made in a lab, they should both be avoided—remember, chemicals CAN'T be avoided, by definition. It's simply to say that upon even mild scrutiny, the idea of equating *natural* with *good* and *artificial* with *bad* (or even trying to discriminate between the two) falls apart.

To take another perspective on this, there are many completely natural chemicals synthesized by plants and animals (often as a form of self-defense) that are deadly poisons for humans. *Amanita phalloides*, commonly called the "death cap," is probably the most poisonous mushroom in the world. Ingestion leads to nausea, vomiting, diarrhea, and often to liver and kidney failure.

Other examples of deadly natural substances include a variety of toxic alkaloids from the poison dart frog, ricin from castor beans, and atropine and scopolamine from deadly nightshade, to name just a few.

Occasionally, our desire for the "natural" leads us to eschew man-made pharmaceuticals in favor of herbal or traditional medicines. We should remember that many pharmaceuticals—including aspirin, digitalis, and anti-malaria drugs—had their roots (so to speak!) in herbal extracts, so to draw a line between pharmaceuticals and herbals in the first place is exceedingly difficult. Additionally, in much the same way that artificial flavor is frequently of higher purity than natural flavor, so too are man-made pharmaceuticals subjected to much greater scrutiny and regulation than are herbals.

While it's difficult to classify a substance as purely toxic (just as it's difficult to classify a substance as purely safe—the same digitalis that is a life-saving cure in one situation can be deadly if misused) some herbals are so inherently toxic as to have no safe dose. For example, the traditional Chinese herb *aristolochia*, used for weight loss, causes kidney failure. With regard to pharmaceuticals, we count on the FDA to regulate drug safety, and expect that any preparation prescribed to us by a physician or available on drugstore shelves will not do us harm. Unfortunately, we don't have the FDA's assurance (nor that of any other regulatory agency) with regard to the safety or efficacy of herbals. The lax government regulation of herbs means that important safety information, such as the toxicity of aristolochia, is not necessarily disclosed to consumers, and potentially toxic herbs are sold legally. In the particular case of aristolochia, many unwitting U.S. customers who purchased the herb ended up requiring dialysis or kidney transplant. The difference in the way pharmaceuticals and herbals are regulated has its roots in the 1962 amendment to the Food, Drug, and Cosmetic Act. The amendment classified herbals as dietary supplements and provided that together, supplements and herbals were to be held to lower safety (and no efficacy) standards. The **Dietary Supplement Health and Education Act (1994)** further defined the role of the FDA in regulating herbals and dietary supplements. Provisions of this act included exempting dietary supplements and herbals from pre-marketing evaluation by the FDA, requiring that new supplements/herbals be "reasonably" expected to be safe, and grandfathering supplements/herbals marketed prior to 1994, meaning they were not required to have a reasonable expectation of safety. A further legislative order passed in 2007 required dietary supplement and herbal manufacturers to ensure—but not prove—lack of product contamination and correct labeling. A cursory glance back at the history of FDA regulation of pharmaceuticals reveals that current supplement and herbal regulation is no more stringent than the pharmaceutical regulation mandated by the original Food and Drug Act of 1906—before the diethylene glycol and thalidomide tragedies forced further legislation. The FDA requests that supplement manufacturers track and report adverse reactions and evidence of toxicity, but reporting is still on a voluntary basis at this time.

CONCEPT CHECK:

Does the current herbal supplement legislation provide the information necessary for making a risk-to-benefit analysis?

Answer: No. Efficacy information is not provided, and safety information is not required.

It appears that in the end, any attempt to equate *natural* with *safe* (or to equate *artificial* with *dangerous*) falls apart—even when it comes to government regulation. The FDA regulations of pharmaceutical drugs help provide us with the information we need (with the help of our physicians) to assess the risk-to-benefit ratio of using a drug. There is by no means assurance that simply because a pharmaceutical is available for purchase or prescription, it is 100 percent safe. Many pharmaceuticals are quite toxic even in small doses, and all are toxic in sufficient dose. Still, safety and efficacy information allow for the determination of an appropriate dose, such that the potential for harm from the drug is far outweighed by the potential that it will do us good. With regard to "natural" substances (whether a tea made from tree bark, a flavoring agent extracted from a seed, or a field herb), we rarely have safety and efficacy information. Our ability to weigh the potential risk against the potential benefit of, say, an herb, is nonexistent in the absence of scientific data. Some man-made chemicals are, as we've already seen, highly toxic. It's important to remember, though, that the deadliest toxins aren't the ones we produce, but the ones carefully crafted by nature—which has proven itself to be the most accomplished of synthetic chemists.

7.9 THE LAST WORD—THE PARACELSUS PRINCIPLE

Paracelsus, a Renaissance botanist, said of chemicals, "All substances are poisons; there is none that is not a poison. The right dose differentiates a poison from a remedy." Throughout this chapter, we've seen examples of pharmaceuticals that are safe in one application, and harmful in another, like thalidomide. Further, almost all pharmaceuticals are safe in the correct circumstances and dosage, but highly toxic if misused. This doesn't come as any particular revelation to most people—in general, we approach pharmaceuticals and other "man-made" chemicals with caution (though hopefully this chapter has at least partially unseated the myth that natural means safe and man-made means harmful). Perhaps the most striking exhibition of the Paracelsus Principle's veracity, though, comes through examining two very different substances—one widely regarded as the safest, most life-giving chemical there is, and the other infamous as the deadliest poison known to man.

Clostridium botulinum is a bacterial species that produces the botulinum toxin, considered the deadliest poison on Earth. Exceedingly small doses of the substance are lethal.* While *C. botulinum* can infect wounds (leading to a condition known as gas gangrene, which was far more common in the days before the advent of antiseptics), it's also possible to contract botulism through the ingestion of contaminated foods, particularly improperly canned goods. Symptoms include muscle weakness and paralysis, which eventually spread to the respiratory muscles and preclude the ability to breathe. Left untreated, most botulism is fatal. Despite the incredibly deadly nature of botulinum toxin, however, it has found medical utility in a number of applications. Perhaps the bestknown, medical botulinum toxin (under the trade name Botox) is used in cosmetic procedures. Injected into facial muscles, the toxin acts as a paralytic, preventing contraction and reducing the appearance of existing—and development of new—wrinkles. Further, Botox is used to reduce excessive sweating, ease the muscle spasms and associated pain of temporomandibular joint syndrome (TMJ) and other spasmodic muscular disorders, and ease symptoms of diabetic neuropathy (damage to the peripheral nervous system, particularly common in the feet of diabetic individuals), among a host of other applications. A clear demonstration of the efficacy of the Paracelsus Principle, in the right dose and for the right reason, deadly botulinum toxin becomes a valuable tool in the physician's arsenal.

***NOTE**

The LD50 (the dose that will kill 50 percent of the people exposed) for botulinum toxin is 5 ng/kg. An ng (nanogram) is 10^{-9} g; in other words, the LD50 is 5 billionths of a gram per kilogram of body weight.

On the flip side of the coin, water (as we saw in Chapter 4) is indispensable, unavoidable, and inextricably necessary for survival. We consume it daily, surround ourselves in it regularly, and consider it so safe a substance that the general population doesn't think of it as a "chemical" at all. Of course, like all matter, water is a chemical—and like all chemicals, water is a poison. While water intoxication and water toxicity occur rarely (since the dose required to render water toxic is quite large), they do occur. Over-consumption of water can lead to imbalances in electrolytes that cause a host of symptoms, culminating in brain function abnormalities that can be harmful or fatal. Water intoxication is observed most commonly among athletes,* occasionally among the mentally ill (who may consume large amounts of water without regard to thirst), and in at least one bizarre incident, in a highly competitive individual out to win a contest. In 2007 a radio talk show offered a video game prize to the contestant who could drink the most water in the shortest period of time. Jennifer Strange, a California woman, died of water toxicity after drinking approximately two gallons in about an hour. While tragic, this incident and others like it serve to remind us that just as toxic substances can find safe medicinal applications, safe substances can be toxic. As Paracelsus said, "The right dose differentiates a poison from a remedy."

***Note**

Sweating results in a loss of salts (electrolytes) AND water; replacing the water without the salts, which happens when athletes quench thirst over long periods of time with water rather than sports drink, can lead to hyponatremia, meaning low body salt. While the mechanism for low electrolyte levels in this case is different than in the case of an individual who drinks a vast quantity of water in a short period of time, the end result is the same.

SUMMARY OF MAIN POINTS

- Salicin, isolated from white willow, was chemically modified to amplify its effectiveness while reducing its side effects.
- Organic chemistry is the study of organic molecules, which are (often large) compounds with carbon skeletons.
- Organic molecules are simplified into line-angle diagrams, in which hydrogen atoms bonded to carbon are implicit, as are the atomic symbols for carbon.
- Functional groups are characteristic arrangements of atoms with predictable chemical properties, and are of importance to synthetic chemists
- Important pharmaceutical functional groups include carboxylic acids, benzene rings, alcohols, phenols, amines, esters, and ethers.
- The solubility of large compounds is complicated—compounds often have areas of hydrophobicity and hydrophilicity, where solubility is the additive combination of different functional areas.
- Amines are common in physiologically active compounds, but often have poor water solubility, so are reacted with acid to make a free-base, or ionic salt.
- Pharmaceuticals interact with receptors for natural substrates in the body.
- Pharmaceuticals (and natural substrates) fit into a receptor like a key into a lock. As a result, compounds (natural or man-made) that fit into the same receptor must have areas of similar shape.
- The major functional area of a molecule that contributes to its activity in the body is called the pharmacophore.
- Early drug regulation included the Food and Drug Act (1906), which required correct labeling and lack of contamination of substances transported across state lines.
- The 1938 Food, Drug, and Cosmetic Act added that proof of safety was required for new drugs.
- Thalidomide, used in Europe in the 1950s and 1960s for morning sickness, was highly teratogenic, but was not approved for use in the United States.
- Symmetrical molecules are superimposable on their mirror images and are called "achiral."
- Asymmetric molecules (identified as molecules containing one or more carbons with four different substituents) are non-superimposable on their mirror images and are called "chiral."
- A molecule and its mirror image are called "enantiomers"; a 50:50 mix of a pair of enantiomers is a racemic mixture.
- Thalidomide is chiral; one form is safe, the enantiomer is teratogenic.
- Drug manufacturers must be aware of enantiomers when synthesizing drugs; generally only one of the enantiomers is active as a pharmaceutical, but the other may have any one of a number of effects in the body.
- It's cheaper to synthesize a drug as a racemic mixture, but it cannot be marketed as such if one enantiomer is toxic.
- The 1962 amendment to the Food, Drug, and Cosmetic Act added proof of efficacy to the FDA's requirements of a drug. This amendment made it possible to use risk-to-benefit analysis in decision-making.
- Natural does not always mean "good" or "safe," and "man-made" does not always mean "bad" or "toxic"; many natural compounds are quite toxic even in low doses.
- Herbal compounds are nowhere near as stringently regulated as pharmaceuticals; legislation requires only that they be "reasonably" expected to be safe, correctly labeled, and free from contamination.

Questions and Topics for Discussion

1. Explain the two simplifications made in converting a Lewis structure into a line-angle diagram. Why do these simplifications make sense—that is, why is it still easy to read line-angle diagrams despite all the "missing" pieces?
2. What kinds of molecules can and can't be drawn in line-angle?
3. Why are bond angles so important when drawing in line-angle?
4. Why do so many pharmaceuticals have their roots in herbals or natural products? Why would it be hard for a pharmaceutical chemist to invent a drug *de novo*, meaning without anything to use as a template?
5. What is the purpose of keeping track of functional groups in organic chemistry?
6. List the key characteristics of each of the following functional groups: carboxylic acids, benzene rings, alcohols, phenols, amines, esters, and ethers.
7. Why do perfumes start to stink if exposed to water and acid or base? What about them, chemically, causes this?
8. Why is it more difficult to determine the solubility or polarity of a large organic compound than of a small compound? What does it mean to refer to areas of hydrophilicity or areas of hydrophobicity on an organic molecule?
9. Which functional groups are likely to contribute to water solubility? Which are likely to contribute to fat solubility?
10. Why are some amine-containing pharmaceuticals reacted with HCl?
11. What is a pharmacophore and why is it important to pharmacological chemistry?
12. Can you determine the morphine pharmacophore simply from looking at the structure of morphine? What additional information do you get if you also look at the structure of codeine? Of morphine, codeine, and meperidine together?
13. Why are receptors and substrates sometimes explained in terms of locks and keys?
14. Describe the 1906 Food and Drug Act. What did it require of pharmaceuticals?
15. What led to the passing of the 1938 Food, Drug, and Cosmetic Act? What did this act require of pharmaceuticals?
16. What led to the passing of the 1962 amendment to the Food, Drug, and Cosmetic Act? What did this amendment require of pharmaceuticals?
17. What is a risk-to-benefit ratio? Why were consumers unable to use risk-benefit analysis prior to 1962?
18. Why do some molecules have nonidentical mirror images, while others don't? Think of a non-molecular (macroscopic) example of a "chiral" object and an "achiral" object (don't use baseballs or hands, since those were used in the text).
19. Draw a simple molecule consisting of only a carbon with four different atoms attached (you can use W, X, Y, and Z if you like). Draw the mirror image of the molecule. Are they superimposable? Is your molecule chiral?
20. Why do pharmaceutical chemists prefer to market a drug as a racemic mixture? Under what circumstances can they do so? Under what circumstances must a drug be marketed as a single enantiomer?
21. Why is a drug's enantiomer not expected to behave in the same manner as the drug in the body?
22. Why are stereoselective receptors (receptors for only one enantiomer of a pair) so common in the body, even though it's hard for chemists to produce a single enantiomer of a pair?
23. Sketch dextromethorphan and levomethorphan. Sketch morphine. What do you notice about your sketches? Would you be able to predict (even without being told) which methorphan is the addictive opiate? How can you tell?
24. Describe the risk-to-benefit logic that might be used to decide whether to take a drug in each of the following cases: a drug has a 50 percent chance of treating a fatal illness, but a 10 percent chance of

causing a fatal toxic reaction in the patient; a different drug has a 50 percent chance of treating acne, but a 10 percent chance of causing a fatal toxic reaction in the patient.

25. Why can't we make the assumption that natural things are safe?
26. An herbalist tells you that an herb has the ability to treat headaches, but has no side effects or toxicity. Do you believe the herbalist? Why or why not?
27. Compare and contrast herbal and supplement regulation with pharmaceutical regulation. What are the similarities and differences in the law?
28. Is it possible to use risk-benefit decision-making in determining whether to use an herbal in the United States? Why or why not?

Problems

1. Convert the following to line-angle:

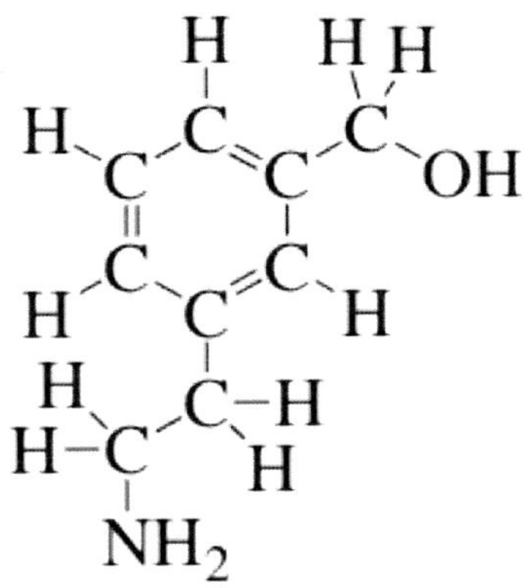

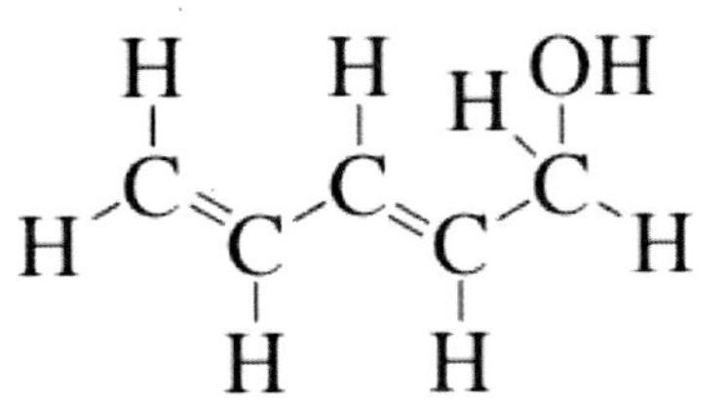

2. Convert the following to line-angle:

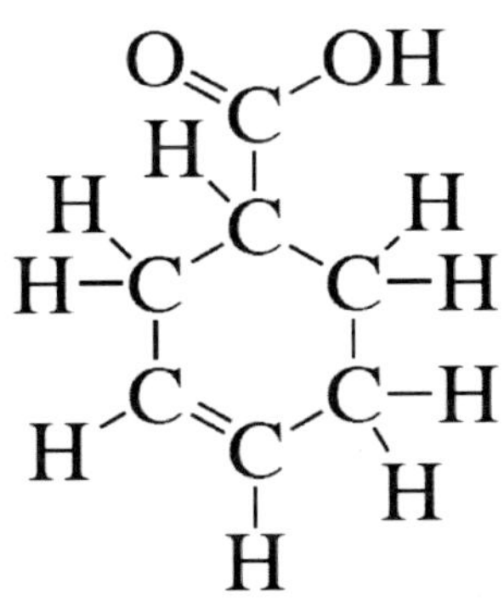

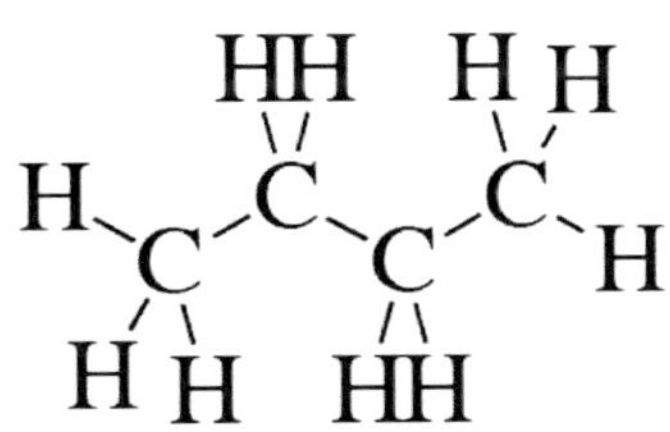

3. Convert the following to full Lewis structures:

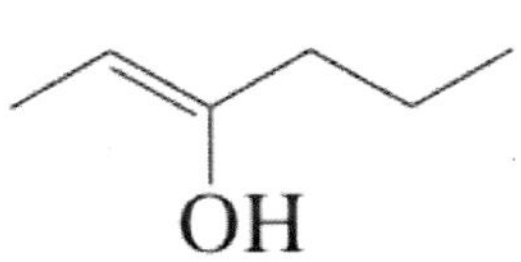

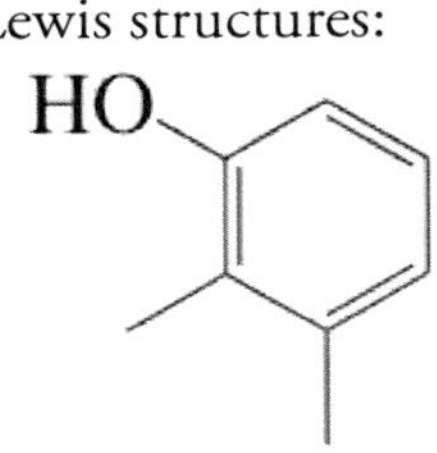

4. Convert the following to full Lewis structures:

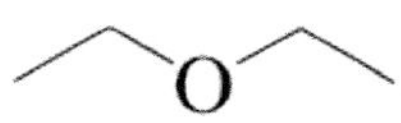

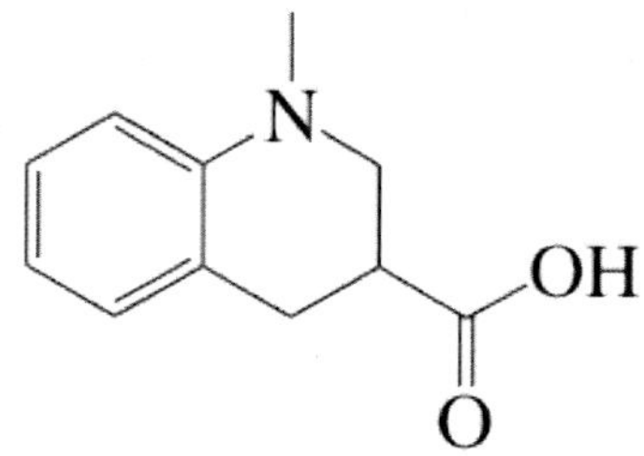

5. Count the hydrogens and carbons on the following:

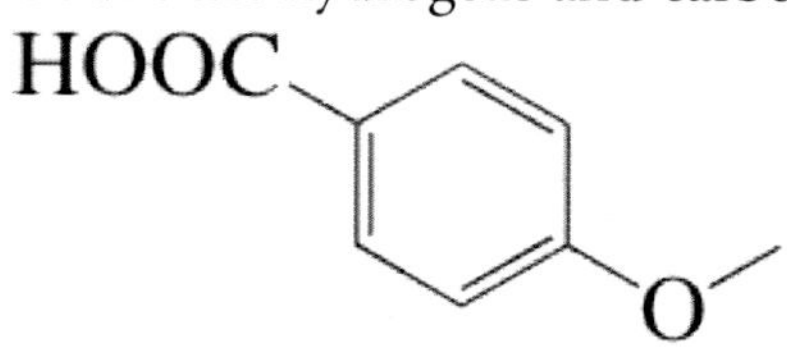

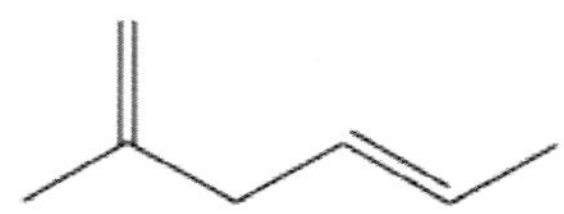

6. Count the hydrogens and carbons on the following:

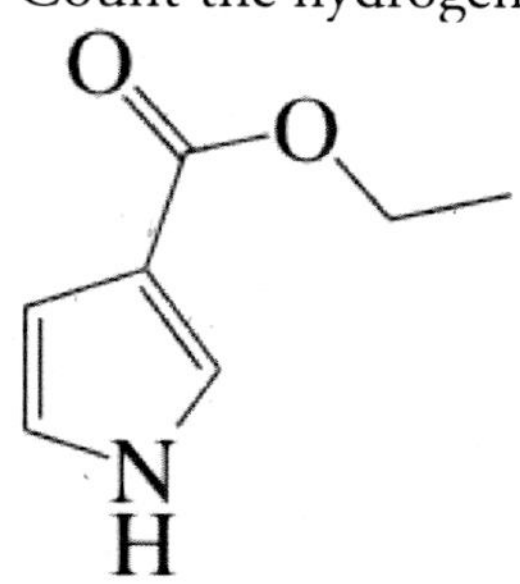

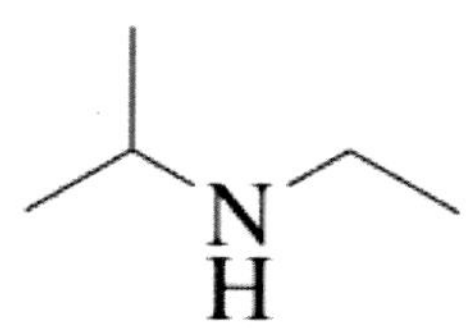

7. Circle and label all the functional groups in the molecule below:

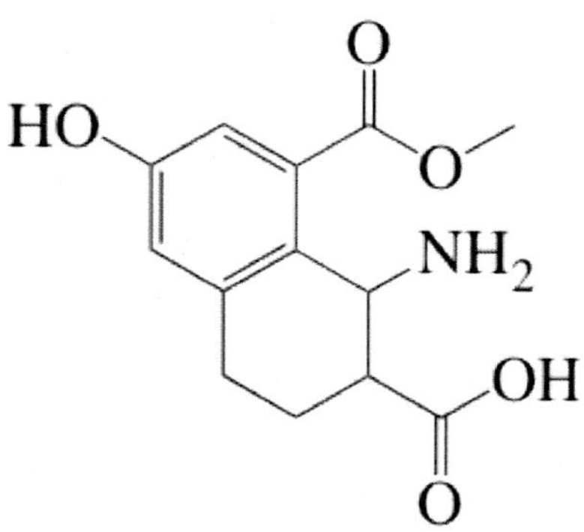

8. Circle and label all the functional groups in the molecule below:

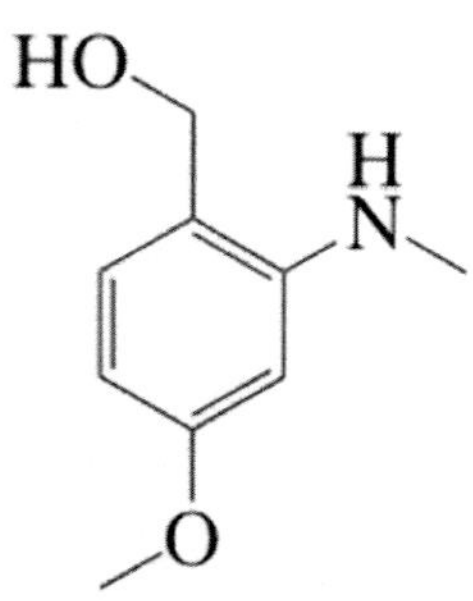

9. Describe each of the following functional groups in terms of their characteristic arrangement of atoms: carboxylic acids, benzene rings, alcohols, phenols, amines, esters, and ethers.

10. Circle and label regions of polarity and nonpolarity on the molecule below:

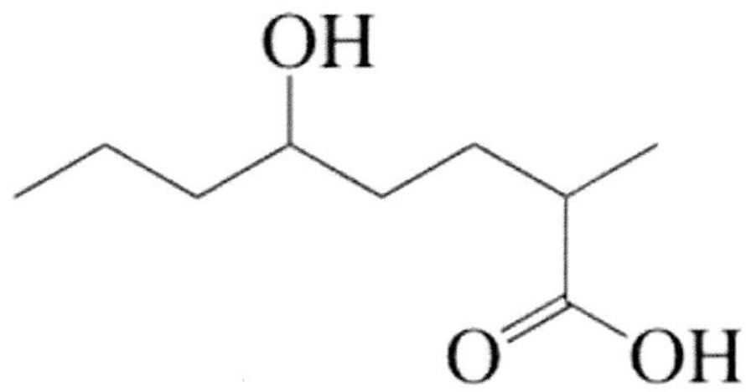

11. Circle and label regions of polarity and nonpolarity on the molecule below:

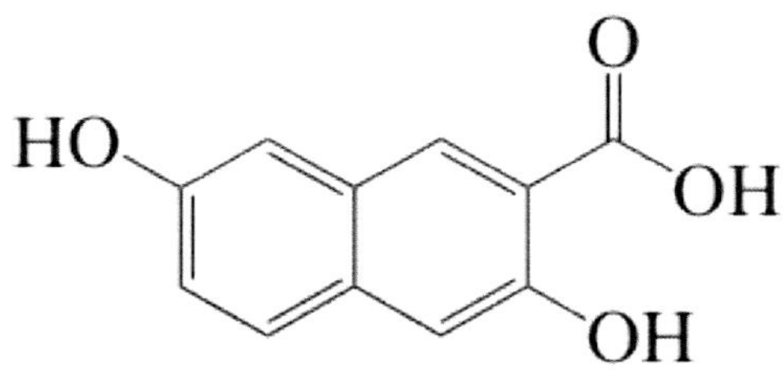

12. Arrange the following molecules in order of increasing solubility:

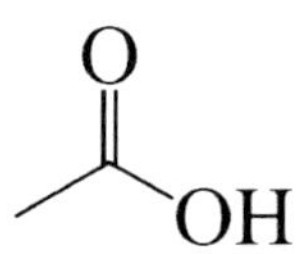 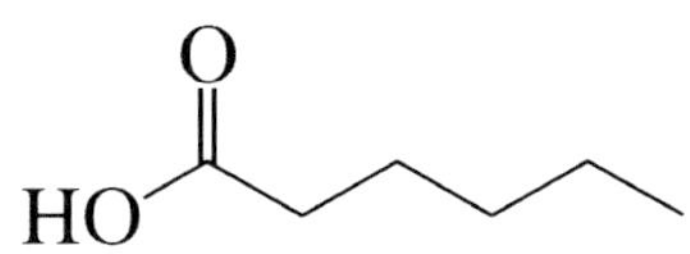 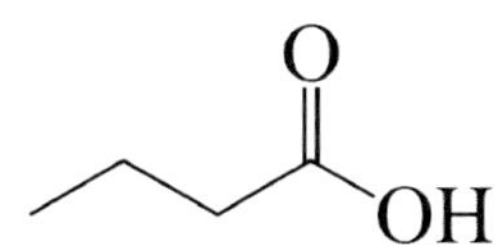

13. Arrange the following molecules in order of increasing solubility:

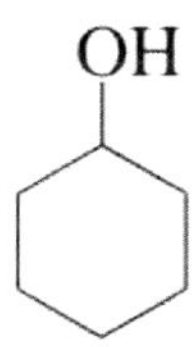 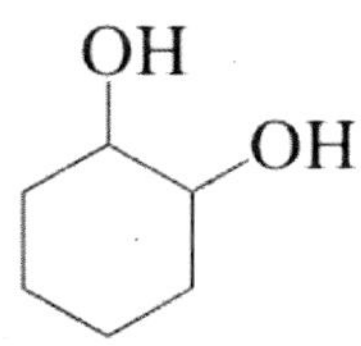 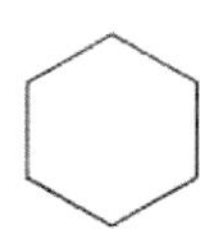

14. Define: pharmacophore and substrate.
15. Define: chiral; achiral; enantiomer; and racemic mixture.
16. If the following items were molecules, determine whether they would be chiral or achiral: Foot; basketball; die (singular for dice); human; and an ear.
17. Indicate the carbon(s) with four different substituents on the following molecule:

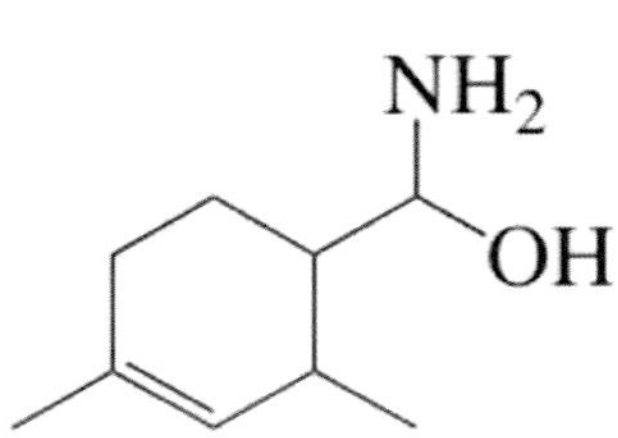

18. Indicate the carbon(s) with four different substituents on the following molecule:

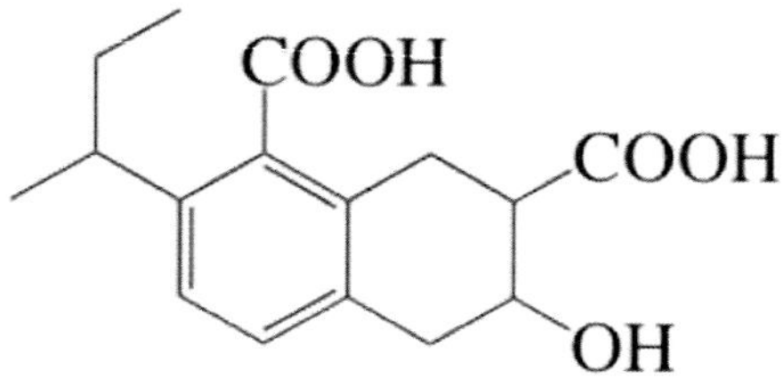

19. Determine whether the following molecules are chiral or achiral:

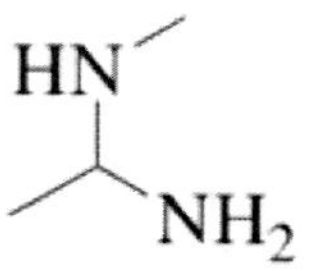 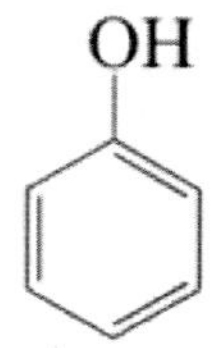 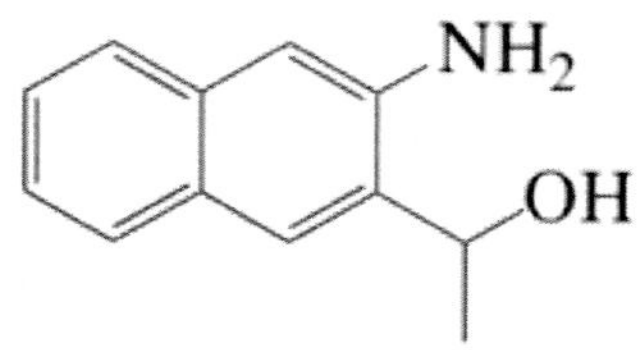

20. Determine whether the following molecules are chiral or achiral:

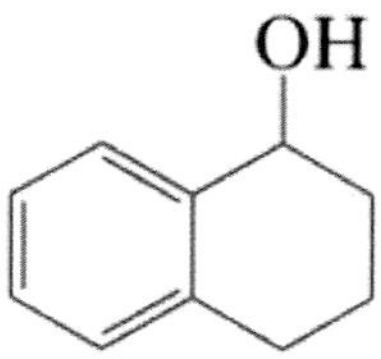 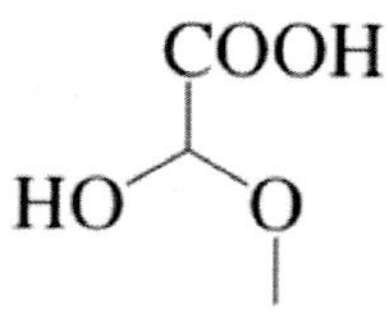 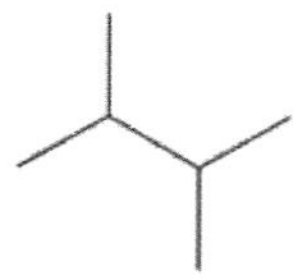

21. Why must methorphan be sold as a single enantiomer, while ibuprofen can be sold as a racemic mixture?
22. Given that one enantiomer of thalidomide is safe and non-teratogenic, why can't pregnant women take the safe enantiomer for morning sickness?
23. Why can thalidomide be used as a pharmaceutical in some cases, even though it's highly teratogenic?

Answers to Problems

1.

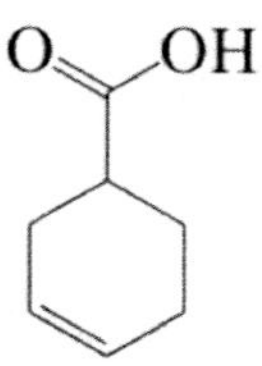 

3.

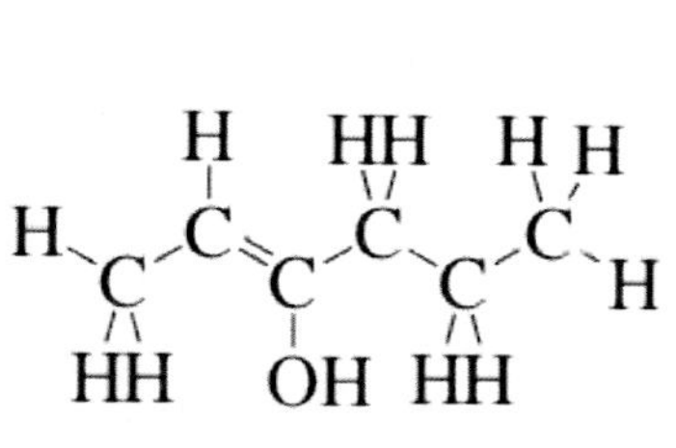 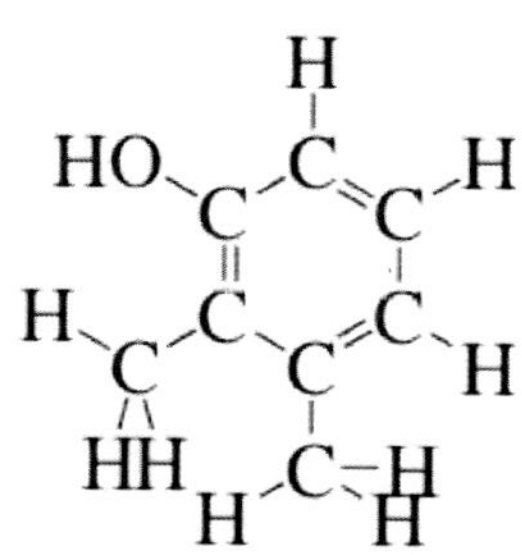

5. 8C, 8H. 7C, 12H.
7.

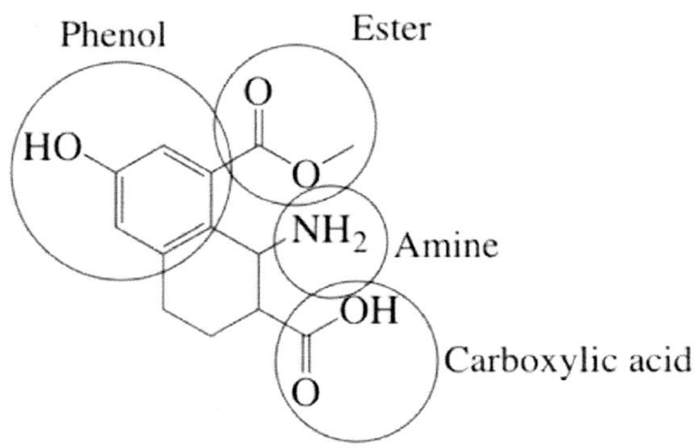

9. Carboxylic acids contain a COOH group; benzene rings are six-carbon rings with alternating single and double bonds; alcohols contain an OH group; phenols are alcohols attached directly to a benzene ring; amines are nitrogen-containing; esters are a combination of a carboxylic acid and an alcohol, or a C double bonded to an O and single bonded to another O that has an R group attached; ethers are two R groups connected by a common O.

11.

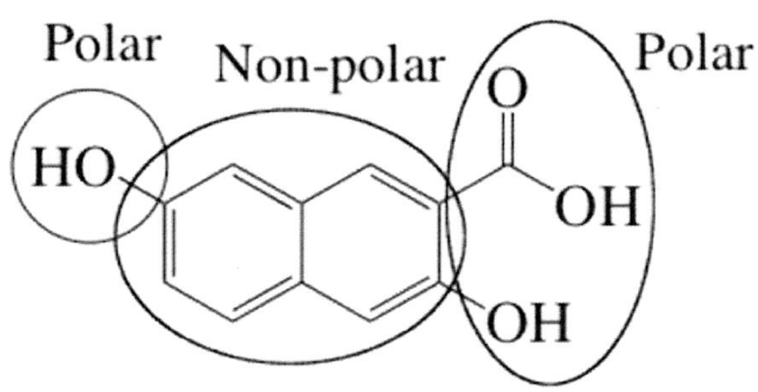

13.

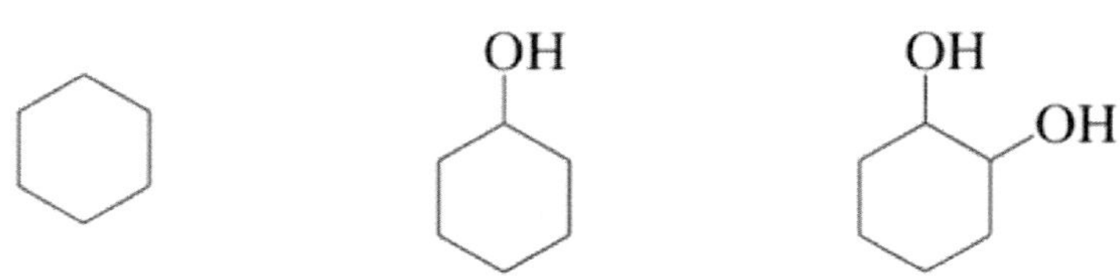

15. A chiral molecule is asymmetrical; an achiral molecule is symmetrical; enantiomers are a pair of non-superimposable mirror images; a racemic mixture is a 50:50 mixture of two enantiomers.

17.

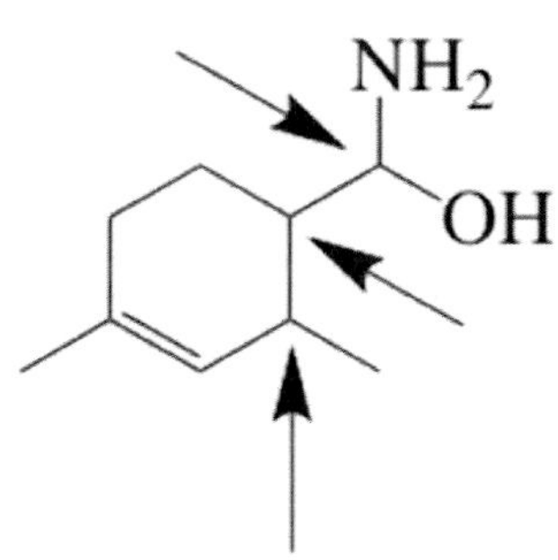

19. Chiral; achiral; chiral.

21. Methorphan has one safe enantiomer and one enantiomer that is an addictive opiate, so only the safe one can be taken as a pharmaceutical. Ibuprofen has one active (safe) enantiomer, and one inactive enantiomer, but the inactive enantiomer is converted into the active form by the body.

23. Thalidomide has medical applications in certain diseases, but can only be used if there is no chance the patient is or will become pregnant.

Sources and Further Reading

Le Couteur, Penny, and Burreson, Jay. (2003). *Napoleon's Buttons.* Penguin, New York.

Schlosser, Eric. (2001). *Fast Food Nation.* HarperCollins, New York.

Timbrell, John. (2005). *The Poison Paradox.* Oxford University Press.

www.fda.gov

Paracelsus Principle

www.cdc.gov

Associated Press. (2007, January 13). *Woman dies after water-drinking contest.* Retrieved from MSNBC website: http://www.msnbc.msn.com/id/16614865/

The Chemistry of What We Eat 8

Every living thing needs to get energy from its environment. Some organisms get energy from the Sun, while others consume food. Humans, of course, fall into the second category: we need to eat. Unlike pandas, which are able to extract all the nutrition they need from bamboo, or marmots, which eat nothing but grass and still manage to store enough fat to support eight months of hibernation, we require a much more complex and varied diet. What is it that we need from our food in order to stay alive and healthy? Why must we concern ourselves not simply with the energy content of what we consume, but the vitamin and mineral content as well? Most importantly, what do we actually do with our food, from a chemical standpoint, once it's chewed and swallowed?

Image 8.1. A yellow-bellied marmot (*Marmota flaviventris*) in Great Basin National Park builds up its energy stores for hibernation by eating nothing but plenty of grass (Photo © Scott Lefler, 2009). Humans need—and crave—a variety of foods.

8.1 ENERGY AND CHEMICAL BONDS

We've learned that atoms form molecules because in so doing, they are stabilized in a number of ways. Not only does bond-formation allow atoms to fill their valence shells, but additionally, electrons in bonds are stabilized by the presence of two attractive nuclei rather than one. In this way, we have a qualitative sense of the stabilization that occurs when molecules form. We can also look at the process of bond formation (and the opposite process of **bond dissociation**) in a quantitative sense—it's possible to use information about the types of bonds forming as a molecule is assembled to determine exactly how much stabilization of the atoms takes place. Examining the energetic processes that molecules undergo as they form, dissociate, and rearrange is the study of **thermodynamics**. While there are several principles and laws of thermodynamics, only a few are critically important to our discussion. First, we must understand that stabilization is simply a reference to energy. High-energy molecules (or atoms) are less stable than lower-energy molecules or atoms. Atoms with incomplete valence shells are less stable (that is to say, have more energy) than atoms in molecules with full octets. As atoms form bonds, they become more stable and lower in energies. The extra energy—or the energy difference between the higher- and lower-energy states—is released, and can be quantified. There's a characteristic amount of stabilization (i.e., energy release) for each bond that forms—energy

*NOTE

A kcal is 1000 calories, where a calorie is the amount of heat (energy) required to increase the temperature of 1 gram (1 mL) of water by 1°C.

released depends both upon the identities of the atoms and the type of bond—and the total amount of stabilization (energy released) when a molecule forms is simply the sum of the individual bond energies. The table below shows the amount of energy (in kcal/mol*) released for each type of bond that forms (these are in molar quantities).

Bond Type	Energy Released (kcal/mol)
H–H	104.2
C–H	99
O–H	111
N–H	93
C–C	83
C–O	85.5
C–N	73
C=C	146
C=O	192
O=O	119
N≡N	226
C≡C	200

Table 8.1. Bond energies for common bonds in organic molecules.

We can use this information to tabulate how much energy, for instance, would be released if 2 moles of H atoms and one mole of O atoms formed one mole of H_2O. When water forms, each hydrogen atom forms a bond to oxygen, such that two new O–H bonds form:

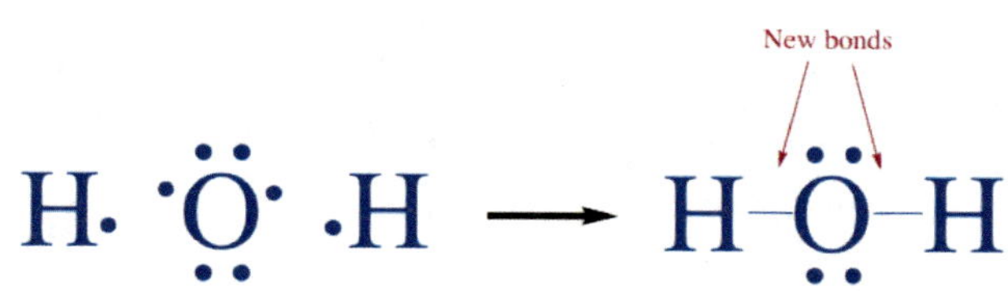

This represents a release of energy of 111 kcal/mol for each O–H bond, so the total energy released is calculated:

2 O–H bonds = 2 (111 kcal/mol) = 222 kcal/mol

In other words, each mole of water that forms from elements will release 222 kcal of energy.

What happens to the released energy? Well, that depends upon the circumstances. Let's take a nonchemical example that we can visualize easily. If I lift a box into the air, I use energy. Very formally, energy is defined as the potential to do **work**, where work moves an object against a force. So I have moved

Try This:

How much energy is released if three moles of H atoms and one mole of N atoms formed one mole of NH_3?

Answer: 279 kcal/mol would be released.

Solution:

3 N–H bonds = 3 (93 kcal/mol) = 279 kcal/mol

an object (the box) against a force (gravity)—therefore, I've done work, and it must have taken energy. What's happened to the energy I expended? I've given it to the box, which now has more energy (more potential to do work) than it had before. In particular, the box has **gravitational potential energy**, meaning it has the potential to fall to the ground. How can we tell that the box has energy? If I drop the box it can do work—like perhaps break my toe if I don't get out of the way quickly enough. This process involves several energetic transformations. The gravitational potential energy the box has while in the air is transformed into **kinetic energy**—the energy of motion—as the box falls. That kinetic energy then exerts a force on the bones of my toe, breaking them. Lifting a box and winding up with a broken toe is an example of a way in which energy can be transferred from me to one object (the box) and then another (my toe). Another possibility is that after lifting (and releasing) the box, I manage to move my foot so that the box falls harmlessly to the floor. In this case, I'd simply hear a "thud" sound as the box landed, and nothing further would happen. Here, it appears at first as though the energy has disappeared, but in fact, that is not possible. The **first law of thermodynamics** says that the energy of the universe is constant: that is, energy can never be created or destroyed. When I lifted the box, the energy it gained came from me; after lifting the box, it had more energy, and I had less. When the box landed on the floor, the energy it lost* must have gone somewhere. In the second scenario, the box made a noise ("thud"), which means molecules in the air were made to vibrate (they were given energy), and their vibrating caused my eardrums to vibrate (my eardrums were given energy), which is how I heard the sound. The floor probably shook a little (also a vibration, so the floor got some energy), and all the vibrating molecules in the air and the floor created a little bit of friction, meaning that the molecules heated up. Heat is a form of energy—warmer molecules have more energy than cooler ones—so we can see with all the vibrating and warming that our box's energy hasn't been lost at all; it's merely been transferred, as energy always is.

***NOTE**

How do we know the box has lost energy? Because a box sitting on the ground has no potential to fall toward the ground—no gravitational potential energy. Additionally, if it's motionless, it has no kinetic energy either.

8.2 ENTROPY

It's important to note that while *total energy* is always conserved, *usable energy* isn't. In the second box-falling scenario, we saw that gravitational potential energy was transformed into kinetic energy, which was then transformed into the

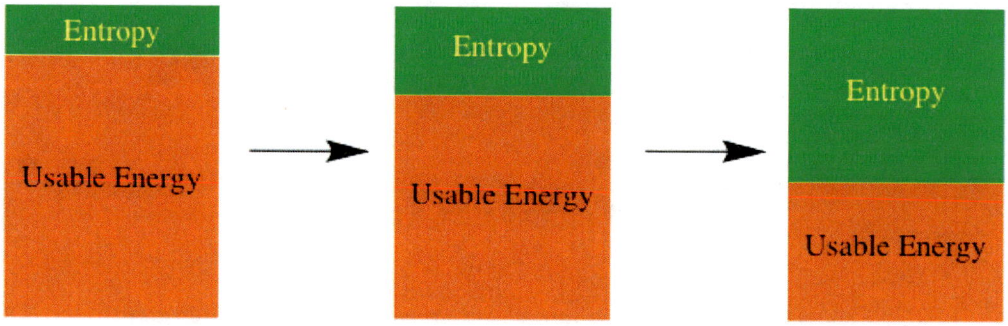

Image 8.2. While the energy of the universe stays constant, entropy increases over time (at the expense of usable energy).

energy of vibration and heat. However, since we can't harness the vibration or heat to do any work in this case, the energy from the box has been transformed into an unusable form. This disordered energy is called **entropy**, and it's an important concept in thermodynamics.* While the first law holds true—the energy of the universe is constant—the **second law of thermodynamics** states that for any process, the entropy of the universe increases, meaning that the usable energy in the universe decreases (Image 8.2).

Let's imagine one more box-dropping scenario. This time, instead of letting the box hit the ground (or break my toe), I drop it onto a lever system (Image 8.3) with a rock sitting on the other end. Now, as the box falls, it transfers its energy (via the lever system) to the rock, which is launched into the air (gaining gravitational potential energy), and which can then transfer its energy to something else.

Here, it's tempting to imagine that we've avoided losing any energy to entropy, but unfortunately, this is not so. The rock launched by the falling box won't have as much energy as the box did. When the box landed on the lever, disordered energy was produced in the form of molecular vibration, and not all the box's energy was transferred to the rock. In the end, we're left with a fundamental truth—every transfer of energy results in a "loss" of some usable energy as entropy, so no energy transfer can ever be 100 percent efficient. As our box examples demonstrate, however, different systems have different efficiencies of energy transfer. Our cleverly designed lever system, for instance,

***NOTE**

Heat is not always "lost" energy (entropy). In the case of a coal-fired power plant, for instance, heat released by combustion is used to produce steam, which is used to generate electricity. In the human body, however, no heat-utilization system exists, so heat released through combustion reactions is no longer usable energy. Regardless of the form it takes, energy that can no longer be harnessed for use is entropy.

Image 8.3. A lever system allows the transfer of energy from a falling box to a rock (clearly, there was something really heavy in that box!) (Image © Kirstin Hendrickson, 2010).

was more efficient at capturing the box's energy than the case in which the box simply landed on the floor.

To return to the original question from several pages back (we wanted to know what happens to the energy released when a bond forms), there are several possibilities, depending upon the circumstances. The energy could be released as heat energy, or it could be stored as bond energy in another molecular system. We'll see both of these take place in our discussion of nutritional chemistry.

8.3 ENERGY OF CHEMICAL REACTIONS

The last principle of thermodynamics important to this discussion is that processes (and chemical reactions) are reversible—I can pick up a box, or I can let a box fall. In one case, energy is required. In the reverse process, energy is released. This always holds true. For any process or chemical reaction, if the reaction releases energy in one direction, it requires energy to go in the other direction. Furthermore, the amount of energy required is always equal to the amount of energy released. To return to our water example, what happens, thermodynamically speaking, when water is broken into its constituent atoms? First of all, since energy was released when two moles of hydrogen and a mole of oxygen formed O–H bonds to make a mole of water, breaking water into atoms must require energy. Second, the energy required to break bonds is equal to the energy released when bonds are made. For a mole of water, the energy required would be:

$$2 \text{ O–H bonds} = 2\ (111 \text{ kcal/mol}) = 222 \text{ kcal/mol}$$

In other words, each mole of water that is broken into atoms requires 222 kcal of energy.

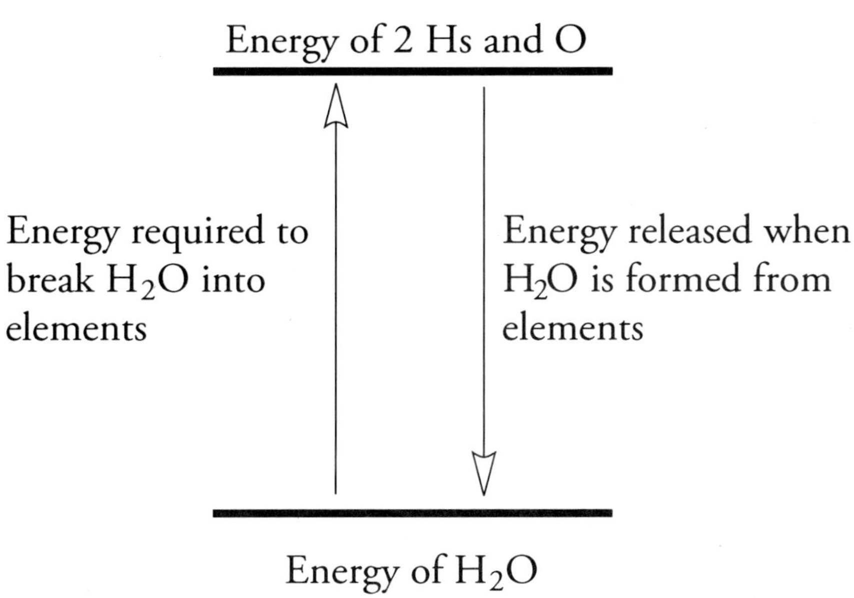

TRY THIS:

How much energy is involved (and is it released or required) when 1 mole of CH_4 is broken into atoms?

Answer: 396 kcal/mol required.

Solution:

4 C–H bonds = 4 (99 kcal/mol) = 396 kcal/mol; breaking bonds always requires energy.

Since forming a bond is ALWAYS stabilizing, it ALWAYS releases energy. Since breaking (dissociating) a bond is the reverse of forming a bond, it is ALWAYS destabilizing, and ALWAYS requires energy.

We very rarely need to think about forming molecules from atoms (or breaking molecules into atoms) in real life. Generally, we'll be looking at reactions in which molecules are rearranged into new ones. Reactions like these require that some bonds be broken and some new bonds be formed. By looking at which bonds break (and how much energy is required) and which bonds form (and how much energy is released), we can figure out whether a reaction releases or requires energy overall. For example, let's determine whether the combustion of methane in oxygen requires or releases energy. It's easiest to make such a determination by following a series of steps:

1) Balance the chemical reaction to accurately determine how many bonds of each type are involved. Since our bond energies are in kcal/mol, we'll think of our coefficients as molar quantities rather than molecular/atomic quantities.

$$CH_4 + 2\ O_2 \rightarrow CO_2 + 2\ H_2O$$

2) Drawing Lewis structures if necessary, determine what kinds of bonds (and how many of each) are broken in the reaction. Molecular shapes aren't really relevant to this discussion, so to keep things simple, we'll use plain Lewis structures:

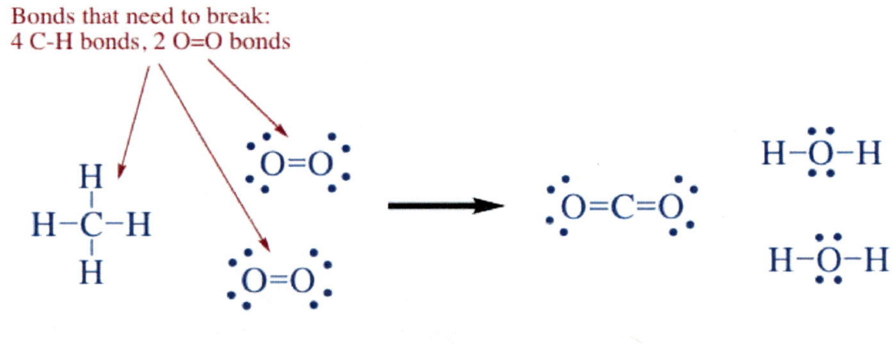

4 C-H bonds and 2 O=O bonds are broken

3) Using Table 8.1, determine how much total energy is required to break the bonds. By convention, these values are written as positive to indicate that the energy is entering the molecule(s).

4 (99 kcal/mol) + 2 (119 kcal/mol) = 634 kcal/mol required

4) Drawing Lewis structures if necessary, determine what kinds of bonds (and how many of each) are formed in the reaction.

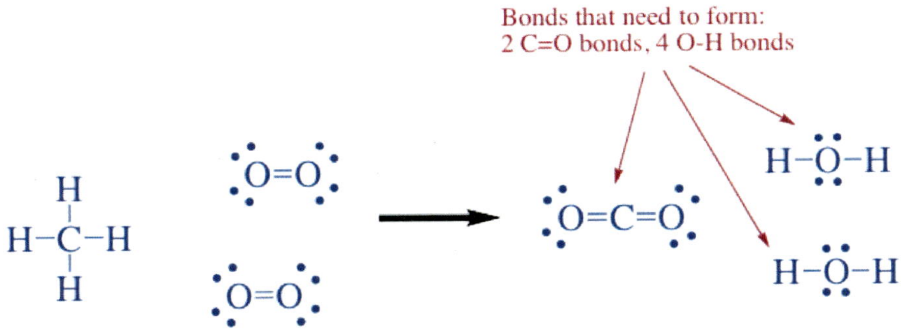

2 C=O bonds and 4 O-H bonds are formed

5) Using Table 8.1, determine how much total energy is released when bonds form. By convention, these values are written as negative to indicate that the energy is leaving the molecule(s).

4 (–111 kcal/mol) + 2 (–192 kcal/mol) = –828 kcal/mol released

6) Add the energy required (always a positive value) to the energy released (always a negative value). If the sum is positive, that's how much energy is REQUIRED by the reaction. If the sum is negative, that's how much energy is RELEASED by the reaction.

634 kcal/mol + –828 kcal/mol = –194 kcal/mol
Since more energy was released by forming bonds than required for breaking bonds, combustion of methane RELEASES 194 kcal of energy for every mole of methane burned.

Remember, if energy is released by a reaction, it can be released as heat, it can do work, or some combination thereof—the disposition will depend upon the circumstances. However, by convention, we refer to a reaction that releases energy as **exothermic** (from the Greek for *heat out*), and a reaction that requires energy as **endothermic** (from the Greek for *heat in*).

TRY THIS:

Determine whether the reaction of H_2 with N_2 to form NH_3 is exothermic or endothermic, and how much energy (kcal/mol) is released or required.

Answer: Exothermic; 19.4 kcal/mol released (−19.4 kcal/mol)

Solution:

$3\ H_2 + N_2 \rightarrow 2\ NH_3$

Energy to break bonds = 3 (104.2 kcal/mol) + 226 kcal/mol = 538.6 kcal/mol

Energy released when bonds form = 6 (−93 kcal/mol) = −558 kcal/mol

Total = 538.6 kcal/mol + −558 kcal/mol = −19.4 kcal/mol

8.4 HOW MOLECULES PROVIDE ENERGY

All of this helps us understand how molecules can provide us with energy. We eat molecules and transform them, through chemical reactions, into different molecules. As long as these reactions are exothermic—which, for food molecules, they always are—the reactions release energy. Some of that energy is released as heat (and helps us maintain our body temperature), but we also use some of the energy to do things like build structural molecules that make up our cells, run chemical reactions that require energy, and so forth.

The nutrients we take in as food fall into three major **macronutrient** categories: carbohydrate, protein, and fat. The word macronutrient is used to indicate that these three nutrients are used in large amounts by the body—that is, they are the energy-providing molecules. Depending upon the type of nutrient, we get a characteristic amount of energy (kcal) per quantity of food. Carbohydrates, for instance, provide us with about 4 kcal per gram. Protein, similarly, provides 4 kcal/g. Fat is much more nutritionally dense, at 9 kcal/g. This is one of the reasons we crave the taste of fatty foods—evolutionarily, they represented a great source of energy. Even though it's completely appropriate to refer to the energy supplied by a food in terms of kcal, it's a bit more common in the United States to see the unit **Calorie** (always capitalized) instead. A Calorie (Cal) is equal to a kcal, which is 1000 cal. The nutritional information on food packages always provides energy values in Cal.* It's important to recognize that not all of a food's weight comes from energy-yielding nutrients. For instance, lean ham is almost entirely protein—it contains very little fat and almost no carbohydrate. However, a 100g slice of ham does not contain 400 Cal, because in addition to protein, there's quite a bit of water and indigestible matter in the ham. Determining the caloric (or nutritional) content of a food can be quite difficult, since most foods contain a combination of nutrients, and weight alone doesn't allow an

***NOTE**

In other countries, different units of energy are more common. Europeans, for instance, typically report the energy content of food in kJ (kilojoules), where 1 kJ = 1000J, and 1J = 4.184 cal.

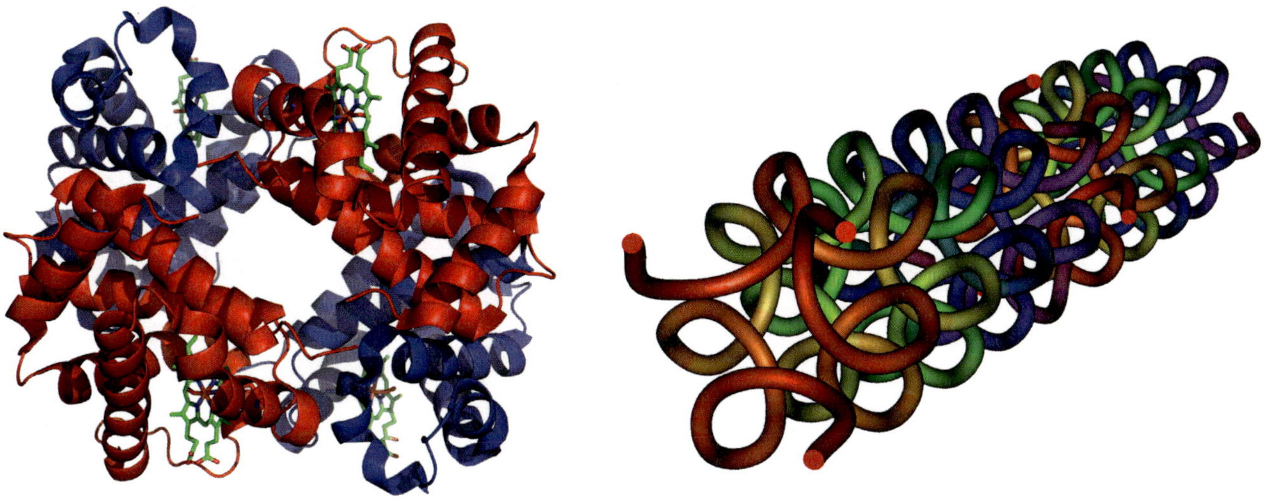

Image 8.4. Structural representations of two common proteins. In each case, the protein chain is represented as a "ribbon" (individual atoms are not visible at this resolution). Hemoglobin (left) in red blood cells carries oxygen from the lungs to the tissue, while collagen (right) adds elasticity to skin and epithelial tissue.

accurate measurement of the amount of nutritive material in the food. Thankfully, packaged foods are required to be labeled with nutritional information so that interested individuals can determine exactly what it is they're eating.

8.5 PROTEIN

The word **protein** comes from a Greek word meaning "of primary importance," because protein is more than just a source of energy. Large protein molecules form the structural components of cells and extracellular material. Proteins produced by the body help us digest our food, interact with our environments, and allow our cells to communicate with one another.

Proteins are made up of **amino acids**, which are small organic molecules. Amino acids consist of a short carbon skeleton, attached to which are two functional groups—one amine and one carboxylic acid. Additionally, each amino acid has a third characteristic group, called a side chain, attached to one of the carbons. The composition of the side chain determines the identity of the amino acid. There are 20 naturally occurring amino acids, each with a distinctive and unique side chain. The generic structure for an amino acid (with the side chain simply denoted as "R") is shown below:

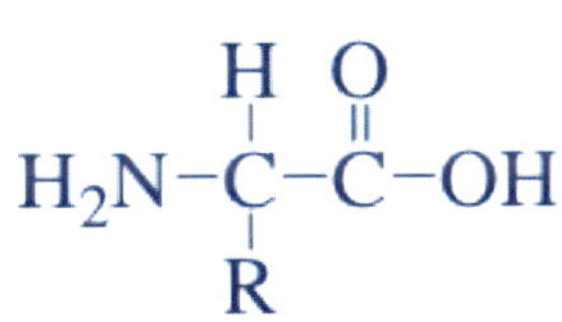

Amino acid structure

Side chains can be made up of a wide variety of atoms and functional groups. The smallest amino acid, glycine, has a very simple side chain consisting of nothing but an atom of hydrogen. Tryptophan, the largest amino acid, has a much more complex side chain, formed of two carbon rings (Image 8.5). Some side chains are purely hydrocarbon, while other amino acids have side chains that contain N, O, and S.

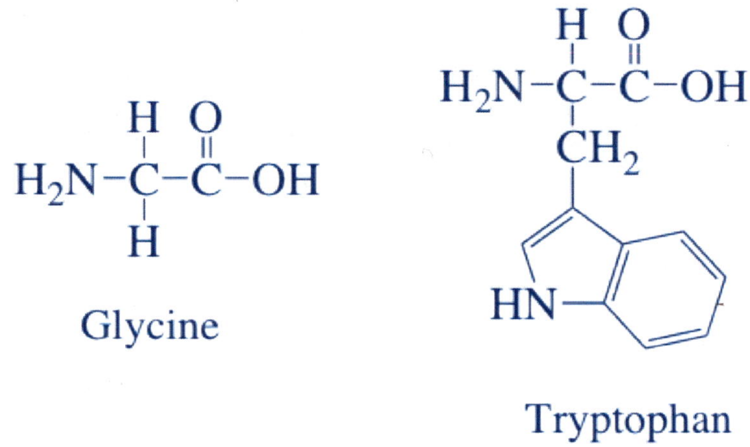

Image 8.5. Two very different amino acids. There are 20 total amino acids, each with a unique side chain.

Amino acids are protein building blocks—they are chemically linked together to produce long molecular chains called **polypeptides** (*poly* meaning "many" and *peptide* meaning "protein unit"). A polypeptide is a subcategory of the large class of organic molecules called **polymers** (*mer* means "unit," so a polymer is any large organic molecule made up of repeating units). Amino acids link together through the chemical removal of water, which is lost in pieces—H from the amine end of one amino acid, OH from the carboxylic acid end of the other amino acid—and the formation of a new bond. A chemical reaction like this, whether it takes place in the lab or in nature, is called a **dehydration synthesis**, since the atoms comprising a molecule of water are removed:

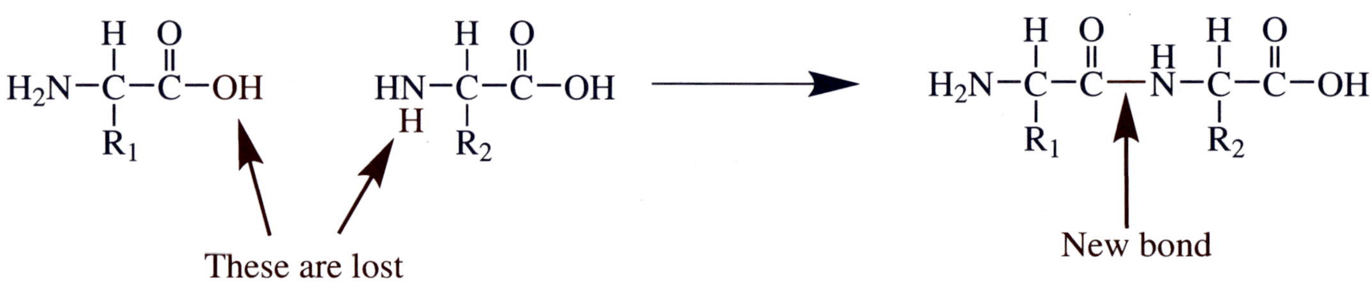

The new bond is called a **peptide bond**, because it joins the units of a polypeptide. For a polypeptide to become a fully functional protein, it must be processed further. This includes some chemical modification and folding, a full discussion of which is beyond the scope of this text. Regardless, at its heart, a protein is a chain of amino acids linked by peptide bonds. Proteins can vary in length from just a few amino acids to several thousand: The amino acids used to produce the protein, the order in which they are placed, and the length of the protein chain collectively determine the structure and function of the resulting protein. The short peptide shown in Image 8.6, called *met-enkephalin*, is an *endogenous opioid*—one of the body's own natural painkillers. This particular peptide consists of five total amino acids: tyrosine, two units of glycine, phenylalanine, and the sulfur-containing amino acid, methionine.

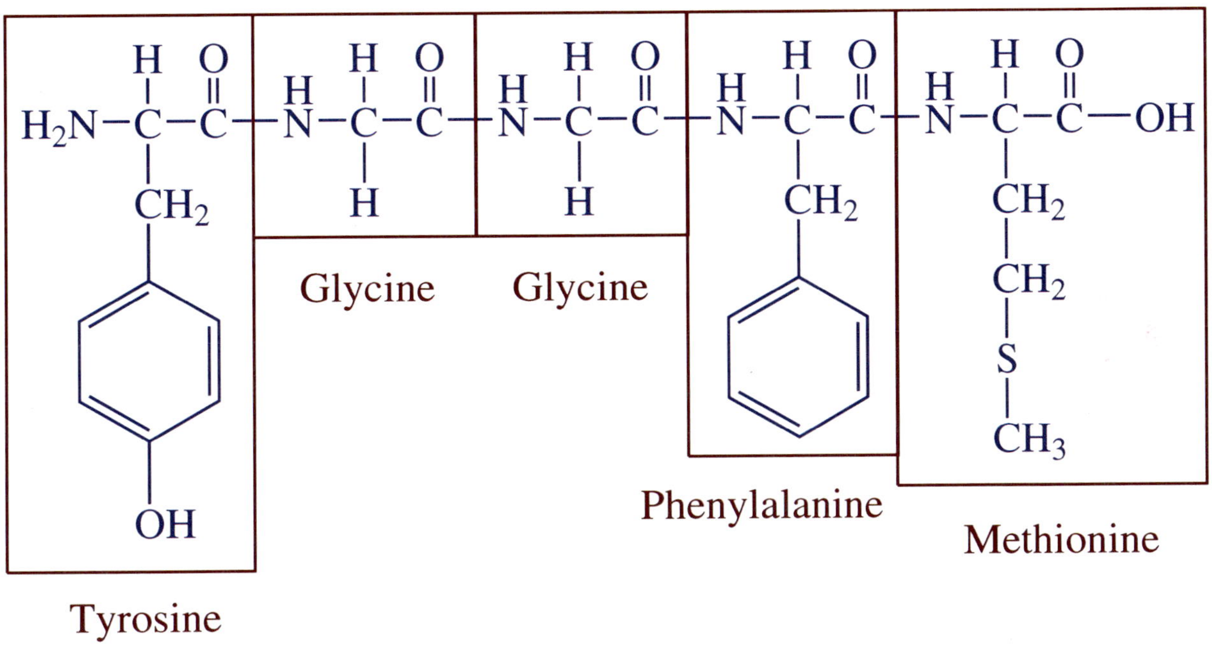

Image 8.6. Met-enkephalin, a painkiller made by the body, is a very short polypeptide.

Due to the myriad possible combinations of amino acids, proteins have the most structural and functional variability of any class of biomolecules. Receptors, such as those discussed in Chapter 7, are made of protein. Some proteins are **enzymes**, which speed up otherwise slow chemical reactions. Proteins can also be structural—our hair and nails are made of the tough, fibrous protein *keratin*, while the elasticity of our skin is maintained by the springy protein *collagen*. Muscle is made up of protein specialized for movement. Even certain **hormones** are made of protein. A hormone is a chemical (some, but not all of which are peptides) used for cell-to-cell communication in the body. Examples of peptide hormones include *insulin*, which helps us to regulate blood sugar, and *oxytocin*, which is important to childbirth and the emotional experience of love and attachment. Clearly, the variety of function that can be achieved by linking amino acids together is immense!

8.6 PROTEIN IN FOOD

Image 8.7. Fish and other lean meats, beans, and eggs are good sources of dietary protein.

Proteins are digested through a reversal of dehydration synthesis—water is added in pieces across the peptide bond, restoring the missing H and OH. This process, called **hydrolysis** (which literally means *to split with water*), breaks the bond and releases the amino acids. When the protein we consume is digested, the amino acids are taken up by cells, and can be broken down further to provide us with energy or used as building blocks for new body proteins.

CONCEPT CHECK:

If dehydration synthesis requires energy, what should we expect of hydrolysis?

Answer: It should release energy.

Some amino acids from the food we digest are neither used for energy nor to build new protein. Rather, they are chemically modified into related molecules, which serve various purposes in the body. One example is that of the **neurotransmitter** *serotonin* (Image 8.8). Our brains use neurotransmitters to gather information about the outside world and relay that information to the brain, to produce movement, and to experience emotion. Serotonin is a neurotransmitter involved in emotion—when released in the appropriate amount, it produces feelings of calm and well-being. It's synthesized in the body from the amino acid tryptophan, and some neurochemists speculate that consuming foods high in tryptophan may lead to relaxation or sleepiness.*

***NOTE**

It has been hypothesized that the sleepy feeling many people experience after eating large quantities of turkey (which contains tryptophan) on Thanksgiving is due to serotonin. However, in all likelihood, the real explanation is a bit simpler. Overeating, as many people tend to do on holidays, leads to diversion of blood away from the brain and toward the gut. The reduced blood flow to the brain produces a sensation of sleepiness. Further, turkey (like all animal protein) contains tryptophan, but not in particularly impressive amounts.

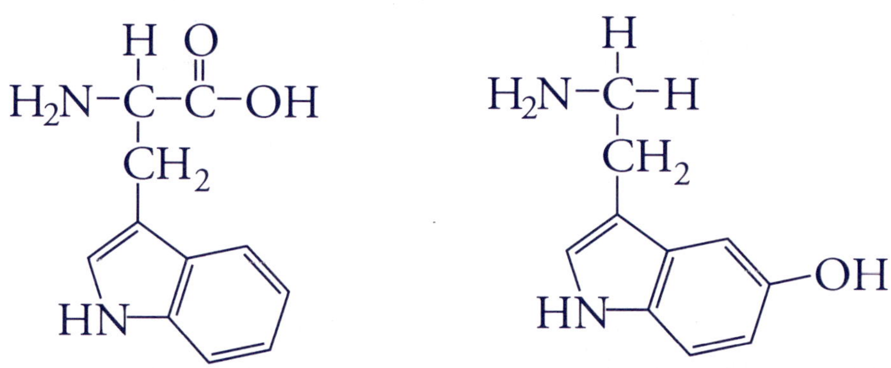

Tryptophan Serotonin

Image 8.8. In this side-by-side comparison, the structural similarities between serotonin and its synthetic parent compound, tryptophan, are apparent.

8.7 CARBOHYDRATES

Carbohydrates include sugars, starches, and indigestible dietary fiber. Even though sugars taste sweet to us while starch does not, they are very similar to one another chemically, and in fact are processed similarly by the body. Carbohydrates contain the elements C, H, and O, generally arranged into a ring, with many alcohol functional groups as substituents to the ring. A ring unit is called a **monosaccharide**, *mono* meaning "one" and *saccharide* meaning "sugar." Monosaccharides taste sweet, and are the building blocks of larger carbohydrates. A ubiquitous monosaccharide, **glucose**, appears in Image 8.10.

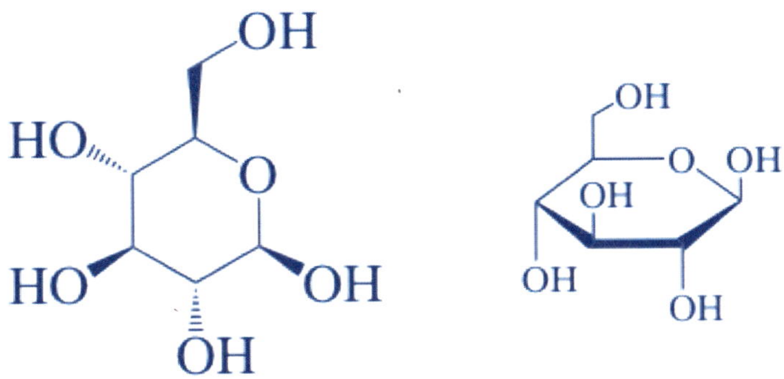

Image 8.10. A line-angle diagram of glucose (left), and a Haworth projection (right). The dark line at the bottom of the Haworth projection is taken to mean that edge of the ring is closer to the reader.

Image 8.9. Carbohydrates are found in honey and other sweets, in fruits and many vegetables, and in breads and other grain products.

While it's certainly OK to draw glucose in line-angle, sugars are often shown from a slightly different perspective, which makes it much easier to visualize the joining of monosaccharide units to form more complex carbohydrates. Tipping the sugar (imagine shoving the line-angle diagram so that it falls over backward) results in what is called a *Haworth projection*, and is a common way to portray carbohydrates. Glucose is not the only monosaccharide—others include **fructose** (fruit sugar) and galactose (a constituent of milk sugar) (Image 8.11).

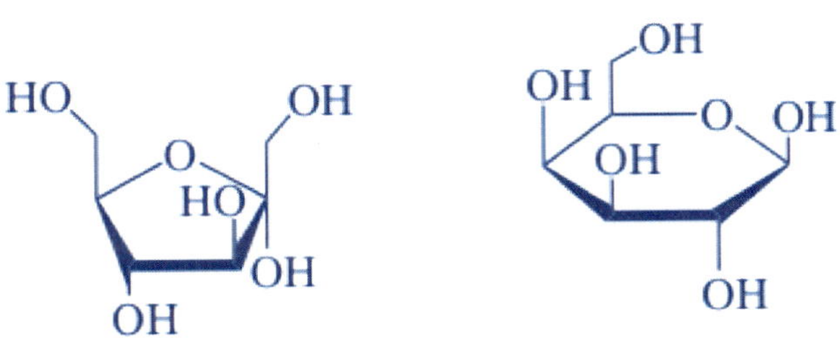

Fructose

Galactose

Image 8.11. Two monosaccharides. Both taste sweet and are chemically very similar to glucose.

To form slightly larger carbohydrates, two monosaccharides are chemically linked together. As with the peptide bond that links amino acids, sugar bonds result from the removal of the atomic constituents of water from two monosaccharides—H from one, OH from the other—and the formation of a new bond in their place. This produces a **disaccharide** ("two sugars") via dehydration synthesis:

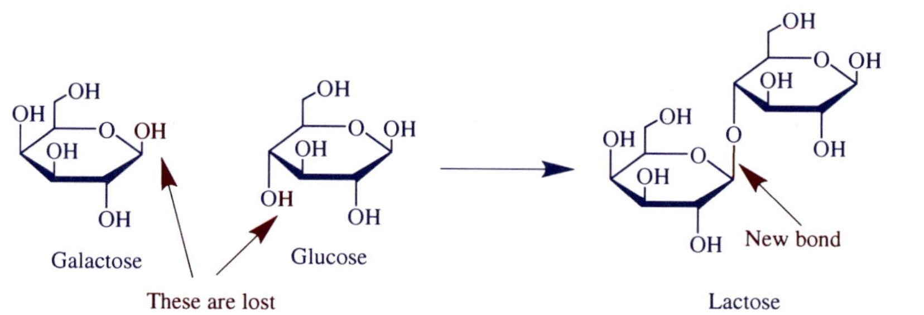

Common disaccharides include **sucrose** (table sugar), **lactose** (milk sugar), and maltose (malt sugar) (Image 8.12).* These molecules, like monosaccharides, have a sweet taste. In common parlance, both mono- and disaccharides are called *sugars*.

***NOTE**

Notice that like monosaccharides, disaccharides have names ending in -ose. This pattern will hold for larger carbohydrates, as well.

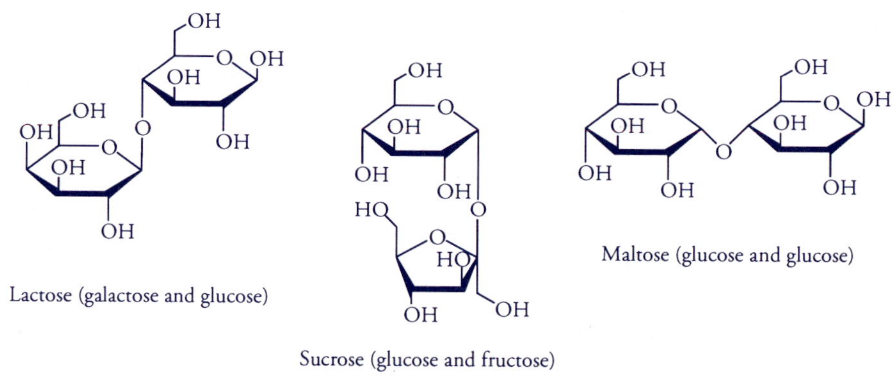

Image 8.12. Some disaccharides. Like monosaccharides, they taste sweet.

CONCEPT CHECK:

How is the sugar bond like a peptide bond?

Answer: Both are formed by dehydration synthesis.

When we eat starchy foods like bread, corn, or potatoes, we don't taste the sweet flavor of sugar. Regardless, we're consuming carbohydrate, this time in the form of long chains of monosaccharides. Like disaccharides, these chains (called **polysaccharides**, for "many sugars") are formed by dehydration synthesis. They may consist of hundreds to thousands of monosaccharide units (generally glucose). Depending upon the orientation of the bond linking the glucose units, the resulting polysaccharides have slightly different shapes, as we observe with **cellulose** (plant fiber) and **amylose** (plant starch):

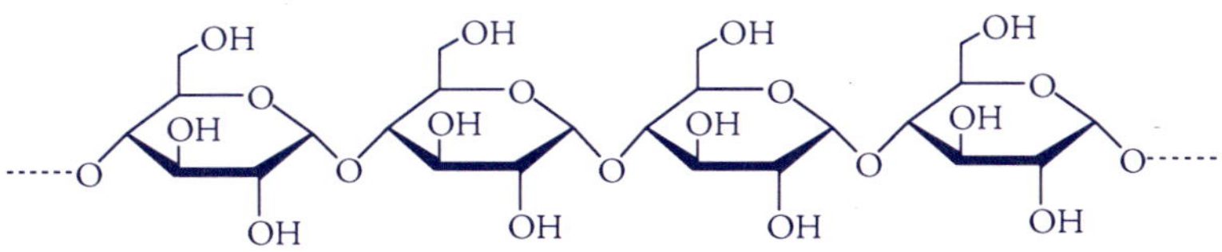

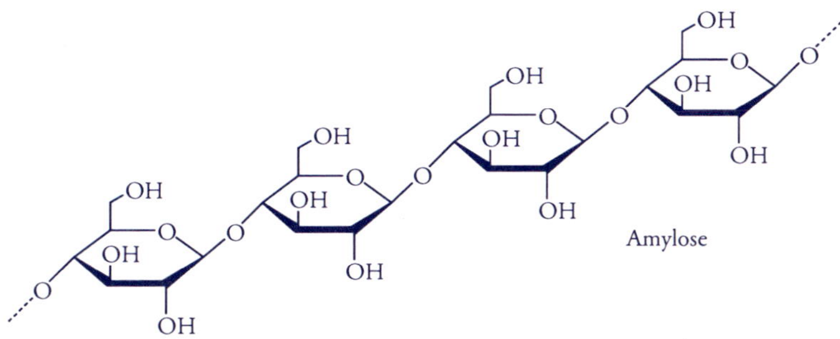

8.8 CARBOHYDRATE DIGESTION

The sugar bond, like the peptide bond, can be broken via hydrolysis. The elements of water are added across the sugar bond, breaking it and releasing the constituent monosaccharides. This is important, because the digestive tract is capable of absorbing only monosaccharides; larger carbohydrates must be broken down before absorption can occur. While this process will occur slowly on its own if carbohydrates are exposed to water, digestive enzymes are used to speed the reaction up significantly. Enzymes interact with their substrates in the same "lock and key" manner that we saw exhibited by receptors in Chapter 7. As such, the enzyme responsible for hydrolysis of lactose (called lactase) is not the same as the enzyme for sucrose (called sucrase) or amylose (called amylase). This helps us understand phenomena like *lactose intolerance*, a condition in which consumption of dairy leads to uncomfortable digestive symptoms. Lactose-intolerant individuals do not make lactase, and are unable to break lactose into its monosaccharides. As such, the lactose passes through the small intestine without being digested or absorbed. Lactose is easily digested by native bacteria in the large intestine, however, and as these bacteria process the lactose, they produce large quantities of gas. The gas buildup in the intestine leads to the bloating and cramping symptoms of lactose intolerance. Lactose-intolerant people still produce all the other normal carbohydrate-digesting enzymes, so they can eat other sugars and starches without difficulty.

The only difference between cellulose and amylose is the orientation of the linkages, but this is significant enough that different enzymes are required to digest the two polysaccharides. Humans and other plant-eating animals produce amylase, but lack the cellulose-digesting enzyme (cellulase). As a result, animals are unable to digest plant fiber, so it passes through the digestive tract without being absorbed, and provides no energy.* Despite the fact that we can't digest

***NOTE**

Animals that depend solely on fibrous plants for energy, like cattle, use symbiotic bacteria to help them digest their food. The bacteria produce cellulase, so are capable of breaking the sugar linkages in fiber.

fiber, however, it serves several important dietary purposes. Fiber provides our mechanical digestive tract with bulk—it's easier for the intestines to move food (and eventually food waste) if there's a fair amount of it. Low-fiber diets tend to produce less waste, but the waste spends more time in the lower digestive tract as a result of the difficulty associated with moving small amounts of material. In this case, toxins can build up as undigested food begins to decompose before being eliminated. Furthermore, fiber can physically bind to toxins (and cholesterol too, as we'll soon see!), helping to prevent them from being absorbed by the body. Finally, fiber serves the important role of helping to regulate the rate of nutrient absorption.

8.9 CARBOHYDRATE ABSORPTION

> ***NOTE**
>
> As already discussed, glucose is not the only monosaccharide. However, it's the most commonly occurring and, for the sake of simplicity, it is the one to which this text will refer in discussions of monosaccharide absorption and processing. Other monosaccharides are processed in an analogous manner.

As we've seen, monosaccharides* require no chemical digestion prior to absorption from the gut, while disaccharides and amylose must be hydrolyzed. This process, however, does not significantly slow the rate at which sugar from starch can be absorbed relative to sugar from a monosaccharide source. The hydrolysis of carbohydrate is quite rapid, such that even the amount of time required to hydrolyze amylose into pure glucose is trivial. This, it turns out, is very significant when it comes to looking at food from a chemical perspective. For instance, amylose is visually more similar to cellulose than it is to sucrose. Both amylose and cellulose are long chains of glucose, and neither has a sweet flavor. Sucrose, on the other hand, is a small, sweet-tasting disaccharide. From a sensory perspective, sucrose and amylose seem very different indeed. Chemically, however, amylose acts nearly identical to sucrose once swallowed. It, like sucrose, is quickly and efficiently hydrolyzed into its constituent monosaccharide units, which are then absorbed. From a cellular and chemical perspective, amylose is nothing more than sugar that doesn't happen to taste sweet. On the contrary, for all the visible similarity between cellulose and amylose, to us they could not be more different. We can digest amylose into glucose, but can't digest cellulose. Amylose provides us with energy, while cellulose does not.

The similarity between amylose and table sugar has major nutritional ramifications. To make sense of this, we must first understand that the human body works hard to maintain **homeostasis,** or a steady set of physiological

Image 8.13. Chemically, white rice (center, composed largely of amylose) has much more in common with table sugar (left) than it does with brown rice (right), which contains lots of cellulose.

conditions. We have mechanisms to keep our temperature, blood pH, salt levels, hydration level, and many other parameters constant over time. This is because our cells can only survive in a very specific set of conditions: If we get too hot or cold, too acidic or basic, too low on water or too high, the cells quickly begin to lose function and can die. One important parameter maintained by the body is the concentration of **blood glucose**, often just called blood sugar. A relatively constant blood sugar level provides plenty of available energy for the cells—too little blood sugar, called **hypoglycemia**, can lead to feelings of weakness and even unconsciousness. On the other hand, too much blood sugar, called **hyperglycemia**, is not safe for the cells. Hyperglycemia, if it goes on for quite some time, can cause kidney and heart disease, damage to the eyes, and nerve cell death. For this reason, a healthy body maintains a relatively constant level of blood sugar. This requires the use of several hormones, including **insulin**, which lowers blood sugar by causing the cells to take up glucose from the bloodstream, and **glucagon**, which raises blood sugar by causing the release of some glucose from stores in the liver. The activity of the hormone insulin, in particular, is relevant to our discussion of carbohydrate absorption. It turns out that cells are very sensitive to the amount and timing of insulin released in response to food—the so-called "strength" of an insulin signal. Sugar from foods that contain lots of fiber, like brown rice, whole grain, and most fruit, is absorbed slowly. This is because the cellulose, which is indigestible, literally gets in the way of digestive enzymes in the stomach as they attempt to hydrolyze starches and disaccharides. Further, the cellulose gets in the way of the proteins that are responsible for the absorption of monosaccharides into the bloodstream. As a result, the monosaccharides from these foods are absorbed very slowly, meaning that the blood sugar rises very slowly. In response to a slow increase in blood sugar, an organ called the **pancreas** slowly begins to release insulin. This "weak" insulin signal tells cells to take up blood sugar in accordance with their energy needs. If body cells have all the sugar they need to fulfill their requirements, extra glucose is stored. Glucose storage can occur through either of two different mechanisms—the sugar can be converted to fat, or it can be polymerized (made into a polymer) in the liver and muscles. The storage polysaccharide, called **glycogen**, bears a strong resemblance to amylose—it's made of glucose units connected in the same way that the amylose units are connected, meaning that it can later be hydrolyzed to release glucose if blood sugar starts to fall. The only significant difference between glycogen and amylose is that the former is highly branched, which allows for rapid mobilization of glucose in cases of need:

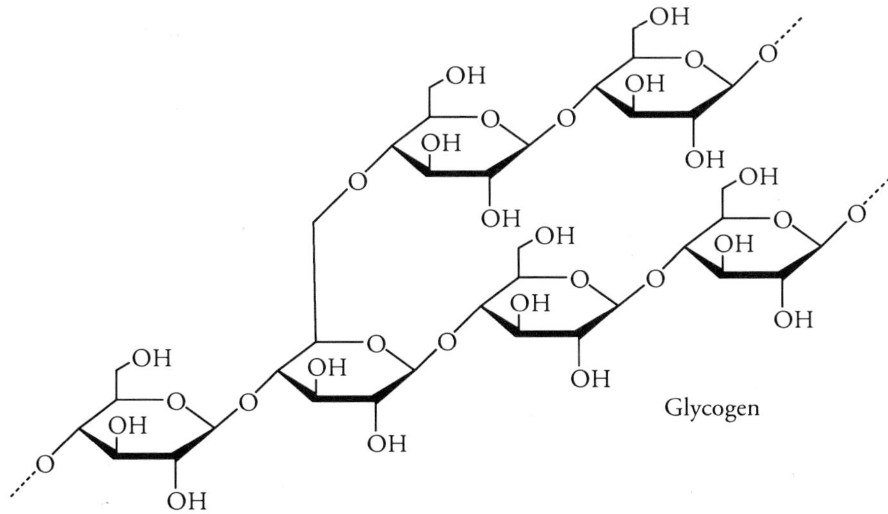

Glycogen

While muscle glycogen is an important source of energy for movement (as we'll see a bit later in the chapter), it's liver glycogen that is really critical to the organism as a whole. In the instance of hunger or fasting, cells begin to run out of energy quickly. The hormone glucagon (which, like insulin, comes from the pancreas) signals the liver to hydrolyze glycogen and release the glucose units into the bloodstream, raising blood sugar. This means that liver glycogen is a surrogate source of food for the body in the absence of a meal. The liver only stores about 2000 Cal of glucose in the form of glycogen, so it can't provide too many meals-worth of sugar to the body cells, but glycogen is nevertheless an important and rapid source of energy in a pinch—particularly in comparison with fat, which may be used for energy, but which is slower to access, making it less than ideal for energy emergencies. Further, some cells (including those in the brain) consume glucose preferentially, making it very important for the body to keep stored glucose on hand at all times. It's worth noting that the "weak" insulin signal associated with slow-to-digest sources of carbohydrate that are rich in fiber encourages the storage of extra glucose as glycogen rather than fat. This helps to ensure that the liver is full of glycogen and that fat is not stored unnecessarily.

The chemical process described above is in sharp contrast with that which prevails if a fast-digesting source of carbohydrate is consumed. Sugars and amylose are hydrolyzed (if necessary) and absorbed very quickly in the absence of cellulose. As a result, blood sugar rises sharply and dramatically. This rapid increase in blood sugar can lead to feelings of over-excitation, or "sugar high." The pancreas responds with a rapid and significant insulin release—a "strong" insulin signal. This particular type of insulin release appears to signal cells to store sugar as fat, rather than burning it to satisfy their energy requirements. Further, the strong and excessive cellular response causes blood sugar to be depleted far more rapidly (and to a greater extent) than normal, causing hypoglycemia. This leads to feelings of weakness, fatigue, depression, and hunger—a "sugar crash." Usually, individuals experiencing a sugar crash are inclined to eat, which raises blood sugar again and relieves the symptoms. We should note, though, that if the crash is treated with more fast-digesting carbohydrate, the entire cycle starts again! In addition, because cells are removing sugar from the bloodstream and storing it as fat (rather than using it for energy), it takes many meals of fast-digesting carbohydrate to sustain the cells for the same amount of time as they would be sustained by one meal of slow-digesting carbohydrate. This means that more total calories (most of which are stored as fat) are consumed in a given period of time. Meals consisting of significant amounts of amylose and sugar in the absence of cellulose or other digestion-slowing molecules* generally lead to unstable blood sugar and over-storage of body fat.

> ***NOTE**
>
> Protein and fat also slow digestion. This phenomenon will be discussed in further detail in Chapter 9.

Which carbohydrates tend to be slow-digesting versus fast-digesting? In general, it's rare to find fast-digesting foods in nature; natural sources of pure sugar are few and far between. Table sugar, for instance, can be obtained from plants such as sugar cane or beet, but must be purified out of the rest of the plant material. Fructose is found in fruit, but most fruit is also very high in cellulose, which leads to slow absorption of the sugar. Honey is essentially the only naturally occurring source of pure sugar that would have been available to humans before the advent of agriculture. Amylose apart from cellulose, similarly, is quite rare in nature. Most sources of starch, including grains, beans, and fruits, also contain significant amounts of cellulose. **Refined grains** such as white rice and white flour have had the cellulose (found in the **bran** casing around the grain) physically removed. The reason for this has nothing to do with nutrition and everything to do with sensation—people found that removing the bran from grain produced stickier rice and finer flour. Additionally, because the bran-removing process was time consuming and expensive, white flours were expensive and became associated with wealth, increasing their desirability. Unfortunately, however, to remove the bran from wheat or other grain leaves behind a core of nearly pure amylose, which, as we've learned, is chemically nothing more than so much sugar. It's interesting to think about the effects our gustatory desires have upon our body chemistry—the human system isn't really designed to handle significant quantities of plain sugar or refined grain, as apparent from the paradoxical insulin response to the consumption of such foods. While our individual and cultural preferences lead us to manipulate foods in order to produce particular flavors or sensations, altering foods from their natural forms can affect not only the taste and texture, but the chemistry within the body as well!

> **CONCEPT CHECK:**
>
> An apple tastes sweet, but contains lots of fiber. Is it likely to act more like sugar or like whole wheat in the body?
>
> Answer: Whole wheat.

8.10 FATS

Fat, unfortunately, has a very bad reputation as a nutrient class: many people consider fat as something to be avoided. It's important to understand, however, as we begin our discussion of the third macronutrient class, that fat in food and on the body is critical to survival. With regard to the food we eat, fat is a dense source of energy—remember that it provides 9 Cal/g consumed, as opposed to 4 Cal/g from protein and carbohydrate. Prehistorically, when food was scarce and had to be gathered or hunted, fat was valuable and difficult to come by; most plant foods are quite low in fat. A strong desire to

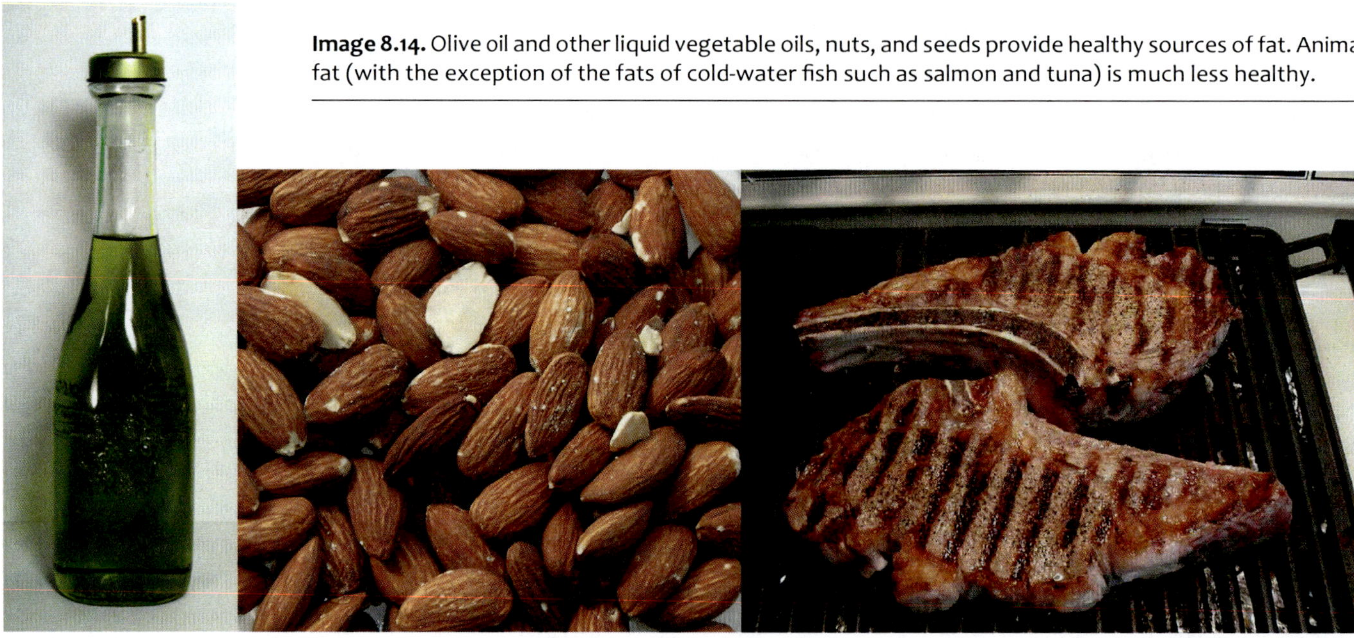

Image 8.14. Olive oil and other liquid vegetable oils, nuts, and seeds provide healthy sources of fat. Animal fat (with the exception of the fats of cold-water fish such as salmon and tuna) is much less healthy.

consume fatty foods and the sensation of pleasure we experience when eating them are both holdovers from a powerful evolutionary drive to seek out these rich sources of calories. In modern society, with food (and fat) plentiful and readily available, our drive to consume fat can lead us to take in too much of a good thing, leading to weight and health concerns (which will be addressed in Chapter 9). Regardless, fat remains an important part of the modern diet when consumed in appropriate amounts. It allows for maintenance of normal weight and body composition, is necessary to maintain immune and other body system functions, and allows for the uptake of fat-soluble vitamins from the digestive tract. It's important to note that just because a nutrient is consumed doesn't necessarily mean it is absorbed. While a mixed-vegetable salad is chock-full of vitamins, some of these are water, and others are fat-soluble (this will be discussed in detail later in the chapter). The vitamins that dissolve in fat can't be absorbed unless fat is being absorbed from the digestive tract at the same time.

Dietary fats fall into the class of molecules called **lipids**, which are large molecules, but which (unlike protein and carbohydrate) are not polymers. Furthermore, while the structures of proteins and carbohydrates are quite predictable—proteins are chains of amino acids, while carbohydrates are one or more sugar units linked together—lipids include a large variety of molecules. Some lipids consist of hydrocarbon rings, while others are open chains of atoms. Some are simple, and others are much more complex. Regardless, lipids share one definitive trait—they are all primarily nonpolar, or hydrophobic. The simplest lipid is called a **fatty acid**, and is nothing more than a very large carboxylic acid:

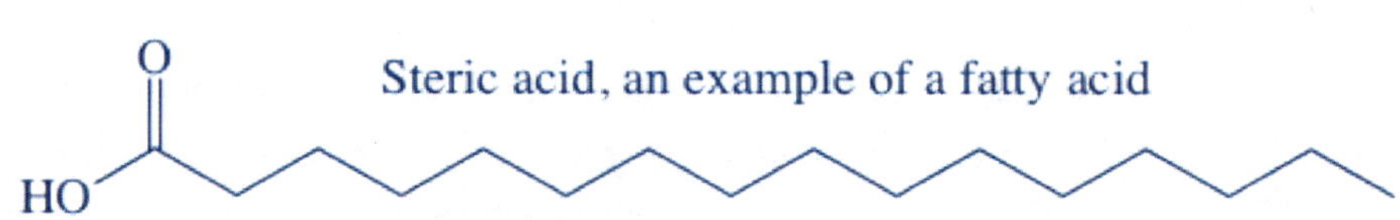

Steric acid, an example of a fatty acid

Note that while the carboxylic acid functional group—called the fatty acid's *polar head*—lends some polarity to the molecule, the majority of the structure is hydrocarbon in nature. The hydrocarbon chain is referred to as the molecule's *nonpolar tail*; due to its length in comparison to the polar head, the molecule is largely hydrophobic. Free fatty acids are not commonly found in nature. They are, however, often combined with an alcohol molecule called **glycerol** to make a complex lipid. Stoichiometrically, three fatty acids combine with each glycerol molecule, and at each junction, H from the glycerol and OH from the fatty acid's polar head is lost—H_2O in each case—making this reaction a dehydration synthesis:

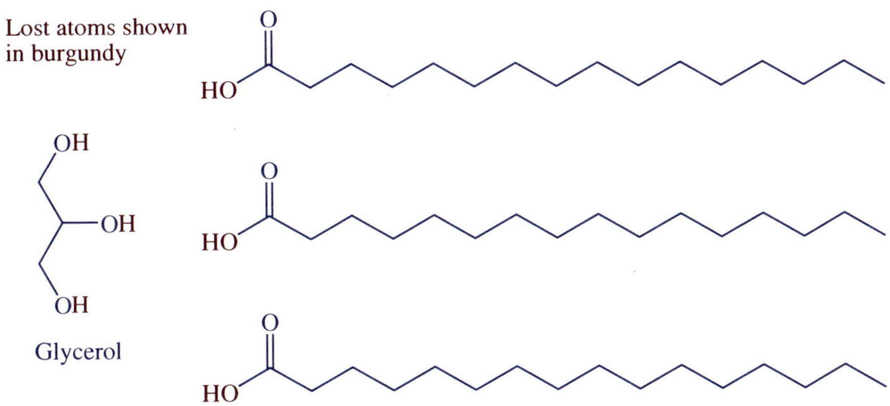

The resulting molecule, called a **triglyceride**, is even less polar than the original fatty acids:

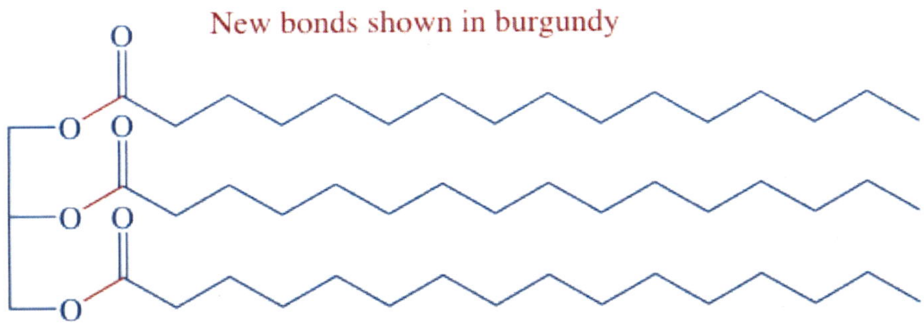

A triglyceride

Triglycerides are the molecules of which animal and plant fats and oils are composed. All living organisms store fat as a source of energy; when we consume an organism, we can digest its stored energy. In addition to digesting fat for energy, we can store it ourselves—our personal fat stores are also composed of triglycerides. There are two major reasons that fat (as opposed to protein or carbohydrate) is the primary energy storage molecule.* First, fat contains more than twice as much energy for the weight as do either of the other macronutrients—storing fat allows us to pack the same amount of energy into a much smaller and lighter package, which is convenient for an active animal. Second, like all lipids, triglycerides are nonpolar. This means they neither dissolve in water nor do they attract it. Carbohydrate, such as

***NOTE**

We saw earlier in the chapter that the liver stores glycogen, a carbohydrate, for consumption by body cells, but compared to the average human's fat stores, glycogen represents a small percentage of total stored energy. An average male human liver stores approximately 2000 Calories (enough for about one day of average activity) of glycogen, while an average male human body stores about 35,000 Calories worth of fat.

the glycogen stored by the liver and muscles, is highly polar (because of all the OH groups), and attracts tremendous amounts of water, which is co-stored with the glycogen. This increases both the weight and the bulk of the stored carbohydrate. Fat is the lightest, most compact way to store a lot of energy for times of need. It's noteworthy that unlike carbohydrate, which can be converted into fat, fat can't be converted into carbohydrate.

8.11 TYPES OF FATTY ACIDS

In the previous section of the chapter we saw stearic acid, a common component of animal triglycerides. Stearic acid is an example of a saturated fatty acid, meaning that it has no double bonds in its hydrocarbon tail. This is an important consideration, because it affects the molecule's melting point and physical state. Since fats are nonpolar, the primary attractions between them are van der Waals forces. While these attractive forces are weak in small molecules, they can be quite strong in larger molecules like fatty acids. Generally, saturated fatty acids and triglycerides composed of saturated fatty acids (called **saturated fats**) are soft solids at body temperature. The reason for this is that the long fatty acid tails can stack quite effectively against one another, almost like sticks in a woodpile (Image 8.15), making molecular packing very efficient. Recall from Chapter 2 that the more closely packed molecules are, the more solid their physical state.

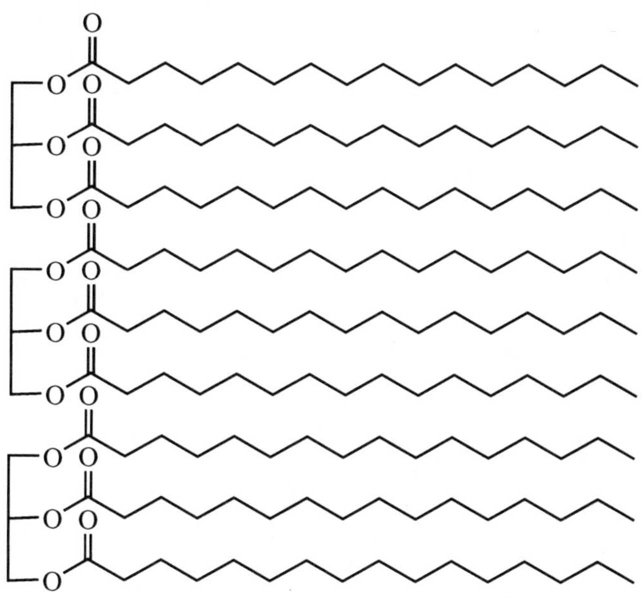

Image 8.15. Saturated fatty acids form triglycerides that can pack together tightly, producing soft solids at room temperature.

Some fatty acids have one or more double bonds in the tail. These are called unsaturated fatty acids, which combine with glycerol to form **unsaturated fats**; they have melting points significantly lower than those of saturated fats. Table 8.2 compares the structures and melting points of four fatty acids. Note that increasing the number of double bonds in a fatty acid decreases the melting point. The same is true of unsaturated fatty acids in triglycerides.

Name Melting Point	Structure
Stearic acid 69.9°C	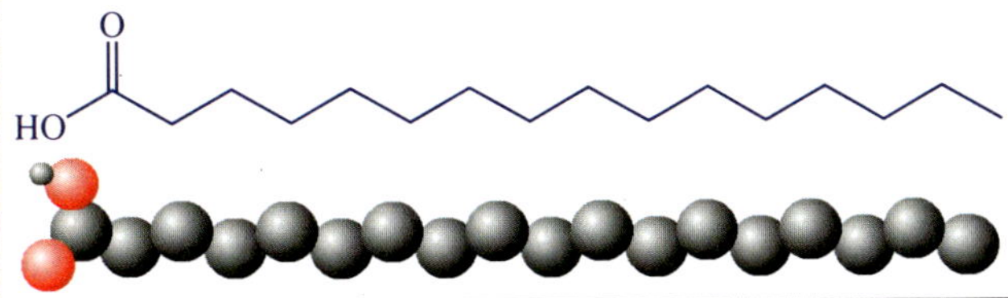
Oleic acid 13°C	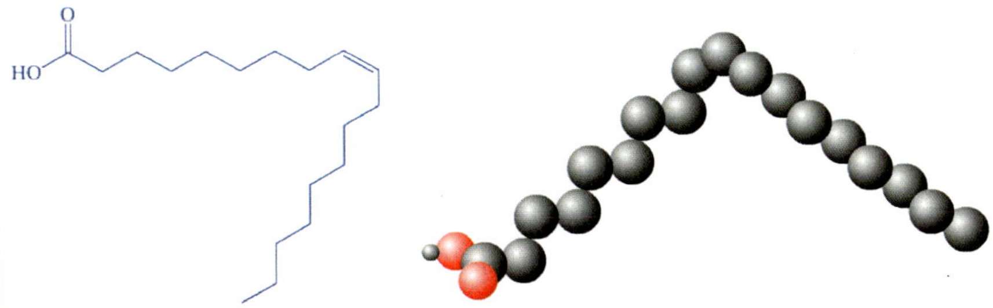
Linolenic acid –11°C	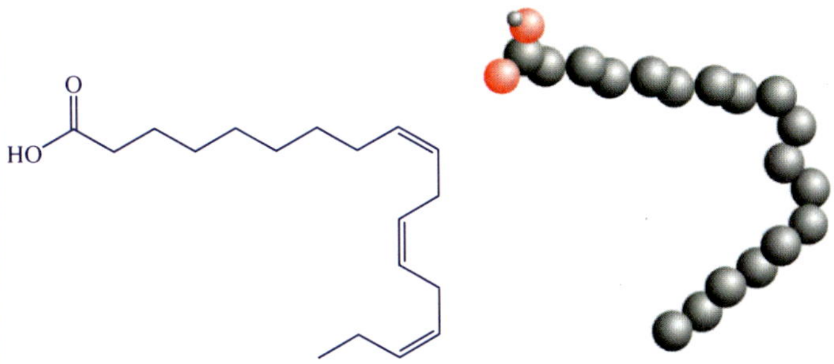
Arachidonic acid –49.5°C	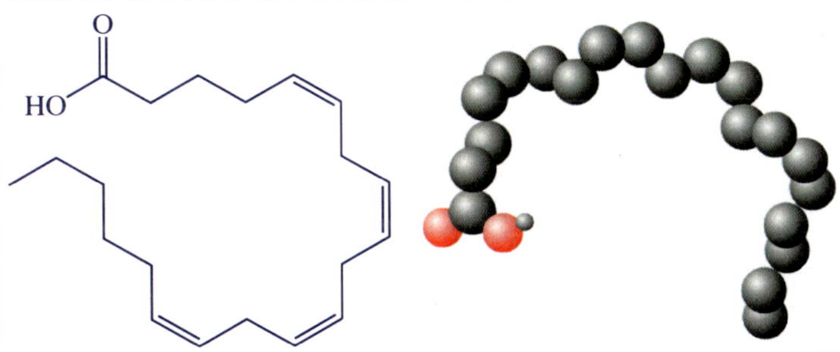

Table 8.2. Fatty acid structures and melting points.

The melting point of unsaturated fats is lower than that of saturated fats simply because the triglycerides can't pack as closely together in space (Image 8.16). This produces a physical state less dense than that observed in saturated fats—most unsaturated fats are liquid at body (and even at room) temperature.

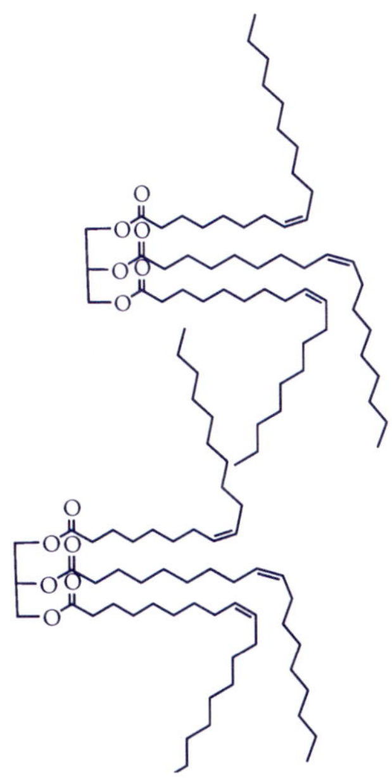

Image 8.16. Unsaturated fats pack much less densely and are therefore liquid at room temperature and human body temperature.

The double bonds in unsaturated fatty acids all put kinks in the tails—this is quite important. There are two possible types of double bonds in organic molecules: one called a ***cis*** double bond and one called a ***trans*** double bond:

A *trans* double bond

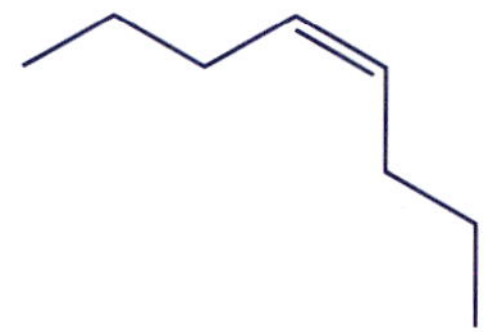

A *cis* double bond

While the *cis* double bond creates a kink in a hydrocarbon chain, a *trans* double bond does not. Almost without exception, naturally occurring unsaturated fatty acids have *cis* rather than *trans* double bonds; the reason has to do with the kink. Remember that the reason plants and animals store fat is for energy. In order to be an effective energy store, fat needs to be light and compact, which we've already learned it is. It must also be solid without being rigid—it would be quite hard for an animal to move around efficiently with either liquid or completely rigid energy stores all over its body. In addition to being incorporated into triglycerides as energy storage, fatty acids are also part of each cell's surrounding membrane. To function properly, cell membranes must be solid but malleable. For this reason, an animal that maintains a body temperature in the range of 37°C (like humans and other mammals) needs a fat that will be a soft solid at that temperature. Stearic acid, with its melting point of 69.9°C, and other saturated fatty acids work well. From this, we can surmise that most of our triglycerides are composed of saturated fatty acids. Further, the fat we consume from warm-blooded animal sources is mostly saturated fat. Tropical plants, such as the coconut, are also likely to contain saturated fat, owing to the very warm environment in which they live.

Consider, however, a cold-water fish such as the salmon. These animals are not able to regulate their body temperature like mammals. In the near-freezing water in which a salmon lives, saturated fat stores would solidify to the point that swimming would become impossible! As a result, salmon evolved the ability to produce and store unsaturated fat, which is liquid at much lower temperatures. This evolutionary adaptation allows salmon and other cold-bodied animals to survive in environments unsuited to saturated fatty acids comprising triglycerides and cell membranes. Besides cold-water fish, dietary sources of unsaturated fats are temperate-weather plants such as corn and soybean. The easiest way to visually determine whether a fat is saturated or unsaturated is its physical state at room temperature: saturated fats are solids, while unsaturated fats are oils.

Note that no such adaptive advantage would be conferred by the development of fatty acids with *trans* double bonds. These double bonds do not bend the fatty acid tails, and therefore triglycerides incorporating them would pack identically to (and have melting points identical to) saturated fats (Image 8.17). An animal with *trans* double bonds in its fatty acids would not be able to survive in temperatures any differently than those available to animals with

CONCEPT CHECK:

Which is more likely to contain saturated fat: an avocado (grows in tropical regions of South America) or grape seeds (grapes grow in temperate regions)?

Answer: Avocado.

Image 8.17. The packing of triglycerides with trans double bonds in them is identical to that of saturated fats.

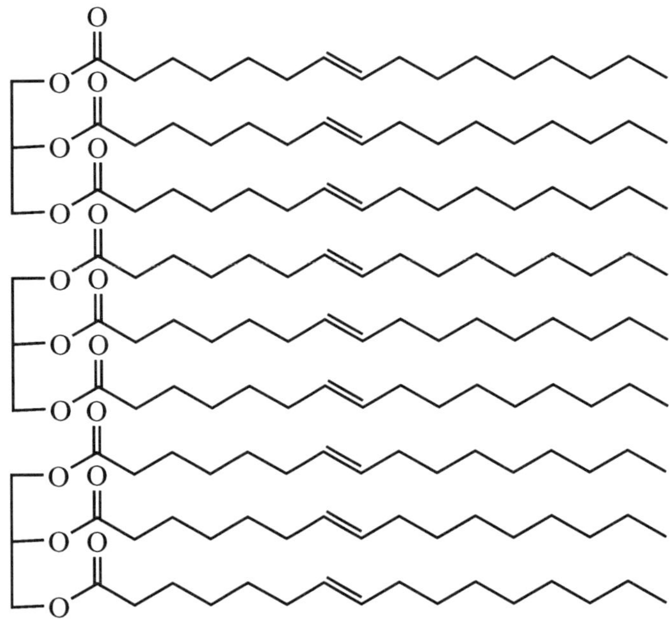

saturated fat. Because the *trans* fatty acid does not confer an adaptive advantage, it simply did not evolve in nature.

Despite the fact that *trans* fatty acids didn't evolve in nature, they do sometimes appear in food. Particularly common in highly processed foods, triglycerides containing *trans* fatty acids (called **trans fats**) come from unsaturated plant oils that have been chemically reacted so as to remove some of the double bonds in a reaction called *partial hydrogenation*. A side effect of this reaction is that some of the double bonds are not removed, but instead are *isomerized*, or changed from *cis* to *trans*. The resulting triglycerides pack much more tightly—and therefore have much higher melting points—than the parent oils used to produce them. This fat-modifying process was developed in order to solidify vegetable fats for cooking—shortening and margarine are examples of **partially hydrogenated oils**. While the health effects of trans fats will be discussed later in the chapter, it is telling to note that they are not utilized efficiently by bacteria, which reveals the reason for their ubiquity in processed foods—trans fats have long shelf-lives.

Two final important types of fatty acids are **omega-3** and **omega-6**. In fatty acid nomenclature, *omega* is used to refer to the carbon furthest from the carboxylic acid end. An omega-3 fatty acid has one or more double bonds—like other naturally occurring fatty acids, the double bonds will be *cis*—the furthest of which is three carbons from the omega end. An omega-6 fatty acid has its most distant double bond six carbons from the omega end. From Table 8.1, linolenic acid is an example of an omega-3 fatty acid, while arachidonic acid is an omega-6 fatty acid. Both omega-3 and omega-6 fatty acids, in addition to being healthy unsaturated fats, are necessary components of diet. They appear to play important roles in normal development, brain and immune system function, **metabolism** (the sum of all chemical reactions in the body), and several other functions. While both are essential to health, research suggests that an optimal diet should consist of about two to four times more omega-6 than omega-3 fatty acid. Typical Western diets, however, have up to 25 times more omega-6—common in cereals, grains, and most vegetable oils—than omega-3, which is found mostly in fish and some nut oils. Research suggests that a more balanced intake of omega-3 and omega-6 fatty acids may help to reduce the incidence and severity of chronic inflammatory conditions, as omega-6 fatty acids are used (among other things) as chemical precursors to *prostaglandins*, which are inflammatory molecules. Other therapeutic applications of omega-3 fatty acids well supported by research include reduction in symptoms or prevention of cardiovascular disease and high blood pressure.

Still further applications, including relief from symptoms of asthma and cancer prevention, are under investigation.

8.12 CHOLESTEROL AND STEROID HORMONES

Cholesterol, a lipid on the basis of its polarity, is not composed of fatty acids. Instead, the molecule consists of four connected rings (Image 8.18). An important biomolecule, cholesterol is the starting material for a number of synthesis reactions that take place in the body, with products including vitamin D and the steroid hormones which, like peptide hormones, are important in cell-to-cell communication. Some of the steroids synthesized from cholesterol are included in Image 8.18 and include cortisol (regulates metabolism), testosterone (male sex hormone), estradiol (female sex hormone), and progesterone (a pregnancy hormone). Note that cholesterol and the steroid hormones, while containing different functional groups, all share the same ring structure.

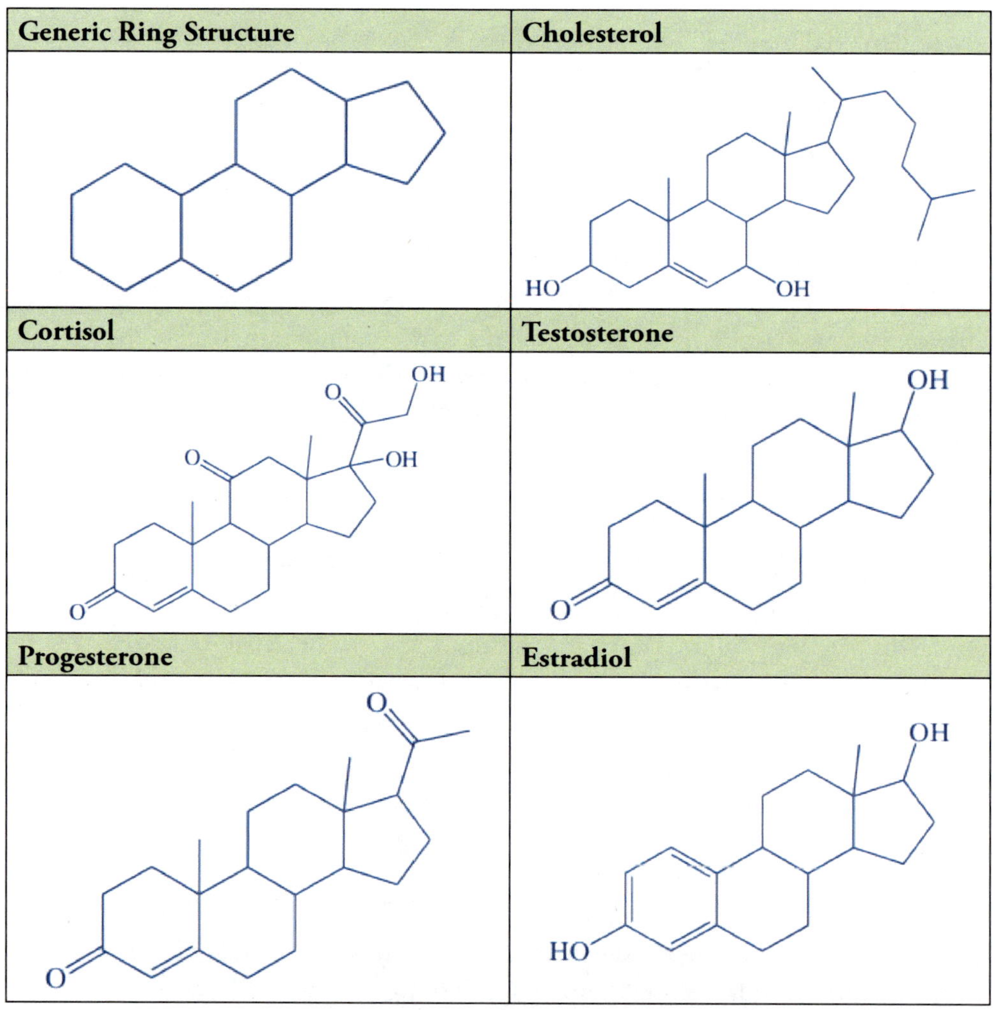

Image 8.18. Cholesterol and some important derivative molecules.

In addition to its role as a biosynthetic starting material, cholesterol serves the crucial function of helping to maintain appropriate fluidity in cell membranes. As discussed earlier, cell membranes must be malleable without being either too rigid or too fluid. In part, appropriate rigidity is maintained by incorporation of fatty acids with the appropriate degree of saturation. However, cholesterol is also incorporated into membranes and acts as a stiffening agent. Cells take up cholesterol from the bloodstream in accordance with their needs, but are able to make cholesterol if none is available to them. In other words, while we need cholesterol in order to survive, we needn't consume it; we can synthesize it. A normal, healthy individual really can't be cholesterol deficient, but it's quite easy to over-consume or overproduce cholesterol. High cholesterol has severe ramifications on cardiovascular health and leads to **atherosclerosis**, or hardening of the arteries. In order to understand why cholesterol can cause heart disease, it's necessary to discuss how it travels around the body.

It is a common misconception that there are two kinds of cholesterol, one of which is "good" and one of which is "bad." In fact, cholesterol (like water and glucose) is the name of a single molecule—there is only one kind of cholesterol. There are, however, two types of cholesterol transporters: **high-density lipoprotein (HDL)**, which helps remove excess cholesterol from the body, and **low-density lipoprotein (LDL)**, which transports cholesterol from the digestive tract to the cells. The key to this system is that HDL and LDL, both containing identical cholesterol and both traveling through the bloodstream, move cholesterol in different directions because of the location of their receptors. LDL is made in the liver and is packed with cholesterol absorbed from food. Its receptors are located on body cells. Cells carefully control what they take in and what they excrete. Therefore, a cell with plenty of cholesterol doesn't need to take in any more. In order to regulate cholesterol intake, only cells with a need for cholesterol produce LDL receptors. Cells with sufficient cholesterol stores do not produce the receptors and therefore can't accept cholesterol from the bloodstream. This protects the cells, but can lead to the cardiovascular effects mentioned earlier. If an individual consumes a diet high in cholesterol, LDL production is high. The abundant cholesterol quickly fulfills cellular needs and in response, they greatly reduce production of LDL receptors. As a result, LDL is not removed from the bloodstream, and the cholesterol from the LDL can collect in the arteries, forming **plaques** and stiffening the arterial walls. This increases the possibility of heart attack and stroke due to narrowed arteries—plaques can break off and block small vessels. HDL, on the other hand, lowers the risk of cardiovascular disease. It is produced in body cells, and is packed full of cholesterol that a cell doesn't need. The HDL receptors are on the liver, which takes up the cholesterol, modifies it, and excretes it via the digestive tract. By packaging and removing excess cholesterol from cells, HDL encourages them to produce LDL receptors so that LDL-contained cholesterol is removed more efficiently from the bloodstream.

The American Heart Association sets forth recommendations regarding cholesterol levels: In general, LDL (the so-called "bad" cholesterol) levels should be low and HDL (the "good" cholesterol) levels should be high. Not only does cholesterol intake influence the amount of LDL in the bloodstream, overall fat intake and the type of fat consumed also have an effect. Saturated

*NOTE

Alcohol consumption also increases HDL, which researchers are considering as one of the reasons for the "French Paradox"—the observation that some cultures that eat fatty foods but consume wine regularly have lower instances of heart disease than Americans. It's important to note, though, that anything above moderate alcohol consumption quickly leads to significant negative health effects that preclude any benefit.

Concept Check:

Which has the most negative effect upon cardiovascular health: saturated, unsaturated, or trans fat?

Answer: Trans fat.

fats tend to increase LDL levels. Unsaturated fats have long been thought to have a neutral effect on both LDL and HDL, though some recent evidence suggests that they may improve the LDL/HDL ratio. Trans fats appear not only to increase LDL, but also to decrease HDL. For this reason, most nutritionists suggest avoiding them entirely. The most efficient route to increased HDL appears to be regular, vigorous exercise.*

8.13 SOAP

In addition to the nutritional relevance of fatty acids and other lipids, they are also the molecules used in the production of soap. A possibly apocryphal tale tells of the discovery of soap in ancient times, when women would take dirty clothes to the river to rinse them in the water and rub dirt off on the rocks. Above the river was a hill, and on the hill there was a temple where animals were sacrificed and burned. Women of the lower classes had to use the undesirable portion of the riverbank where the remains of the sacrifices, including blood, fat, and ash from the fires, ran into the water. However, it was soon discovered that these women got their clothes a bit cleaner than those of more privileged classes. The reason for this phenomenon is that animal fat, if reacted with ash or lye (both of which contain NaOH), produces a foamy mixture that removes dirt and grease from clothing and skin far more effectively than water alone. By itself, water is far too polar to dissolve grease and oil (both of which are hydrophobic) and the dirt that clings to them.

Recall that animal fat is composed of triglycerides, which consist of glycerol and three fatty acid molecules. Since triglycerides are formed through dehydration synthesis, they can be taken apart again through hydrolysis. However, this reaction is quite slow. NaOH makes the hydrolysis much faster and more efficient, easily separating the three fatty acid molecules from glycerol. Because the fatty acids are acidic, however, and NaOH is basic, a further reaction occurs—the neutralization of the fatty acids to produce their sodium salts:

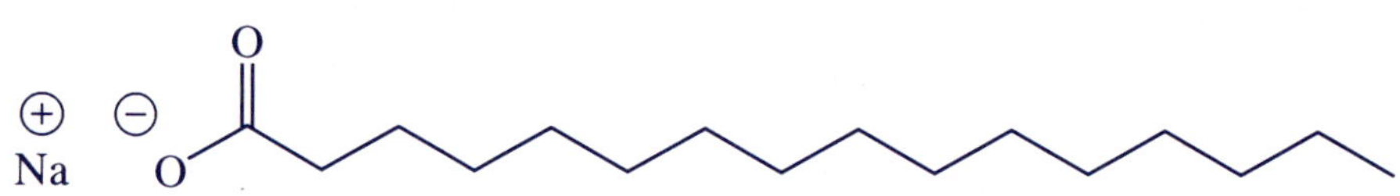

The molecule above is even more dichotomous with regard to its polarity than an ordinary fatty acid: in the case of the salt, the ionic head is highly water-soluble, while the tail remains quite hydrophobic. Because the molecule has distinct regions of polarity, it neither dissolves entirely in water nor in fat; instead, the fatty acid salt dissolves best when a combination of water and fat are present in a mixture. Molecules displaying behavior like this are called **amphiphilic**, meaning that they are both water- and fat-loving. They are also perfectly suited to removing grease from dirty surfaces. When a ball of grease, some soap, and some water are mixed together, the fatty acid salts arrange themselves in a sphere called a **micelle** around the grease. The tails cluster together, while the polar heads remain on the outside of the micelle, exposed to water (Image 8.19). The grease-containing micelle, because of the polar exterior, is very water-soluble and dissolves with ease, carrying the grease away.

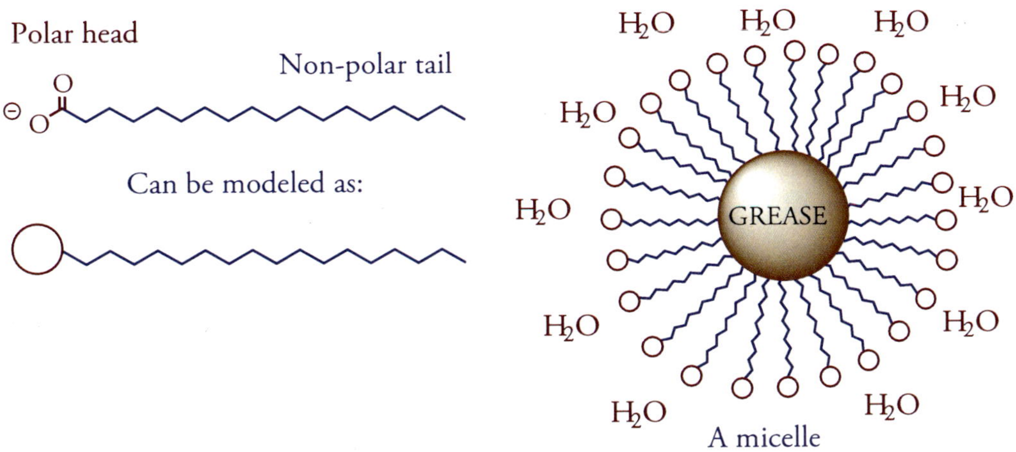

Image 8.19. A micelle forms naturally when fatty acid salts are dissolved in water (shown here in cutaway; a micelle is spherical in reality).

Today's soaps are a bit more chemically complex than the ancient variety—animal fat soaps made with lye tend to be quite drying to the skin and don't often smell very pleasant. Current soap technology incorporates pleasant fragrance, moisturizers for the skin, and other advancements. At its heart, though, the theory remains the same—amphiphilic molecules use areas of different polarity to make fat soluble in water.

8.14 VITAMINS

In addition to the macronutrients, we require some **micronutrients**. These are chemicals necessary in very small amounts that perform vital functions, but do not yield energy. **Vitamins** are organic molecules, and fall into two subclasses: water-soluble and fat-soluble. The solubility of a vitamin has significant ramifications upon the frequency with which it must be consumed and its potential for toxicity. The majority of the water-soluble vitamins aren't stored to any significant extent in the body, so they must be eaten on a regular basis. Since water-soluble vitamins consumed in excess of the body's needs are excreted, it is much more difficult to consume a toxic dose of these molecules* than it would be to overdose on a fat-soluble vitamin.

***NOTE**

Within reason—the Paracelsus Principle holds true here!

Vitamin C

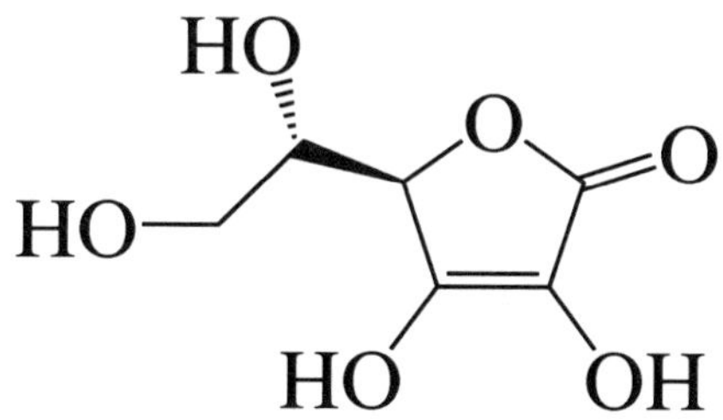

Vitamin C, or ascorbic acid, is a water-soluble **antioxidant**. Antioxidants help to protect the cells from environmental toxins, exposure to radiation (including the Sun's damaging ultraviolet light), and even damage from exposure to oxygen.* Vitamin C also has a role in healthy immune system function (though the popular myth that very large doses of the vitamin will prevent or cure a cold has never been supported by scientific studies) and is critical to the body's ability to produce collagen. Vitamin C deficiency is rare in modern cultures, because it's so easily obtained from fresh fruits and vegetables—especially citrus fruit, berries, and leafy greens. However, a disease called scurvy, which at one time was epidemic among sailors, caused a variety of symptoms, including susceptibility to infections, bleeding gums, and generalized weakness and malaise. The sailors' diets consisted almost entirely of dried meat and bread; these items resisted spoilage and were high in calories relative to their weight. Fruits and vegetables on ships were almost unheard of. Eventually, however, it was discovered that scurvy could be cured (and more importantly, prevented!) by the regular consumption of fruit, which from that point on was included in the sailors' rations. In fact, the term *limey*—slang for a British person—was derived from the British practice of consuming lime juice while at sea to hold scurvy at bay!

***NOTE**

While our cells require oxygen, it's nevertheless a very reactive molecule. O_2 is capable of engaging in reactions in the body that damage structural proteins and our genetic material, DNA. Vitamin C and other antioxidants react preemptively with the reactive oxygen species, rendering them harmless.

Vitamin B

There are actually many B-vitamins; they are referred to individually either by name or number (B1 is called thiamine, B2 is called riboflavin, and so forth). As a class, the B-vitamins play supportive but critical roles in metabolism. Many are **coenzymes**, meaning that they help metabolic enzymes function appropriately. The B-vitamins are all water-soluble, and as a result, they are not stored in the body. The only exception to this is B12, called cyanocobalamin, which can be stored in the liver for several years. While B-vitamins are found in many foods, including grains, fruit, and vegetables, B12 is found only in animal products (and one very esoteric species of yeast). This presents a challenge to vegans—without meat, eggs, or milk, the liver's stores of the vitamin become depleted. Since B12 is critical to synthesis of red blood cells, depletion leads to a condition called *pernicious anemia*. For this reason, life-long vegans supplement their diet with B12, either in the form of vitamin pills or vitamin-enriched food products.

Vitamin A

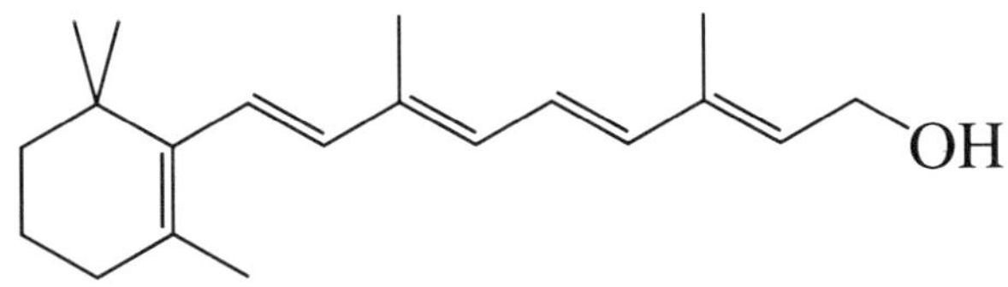

Also called retinol, vitamin A can be obtained directly from food or biosynthesized from the precursor compound, beta-carotene. Among its other functions in the body, retinol is crucial to vision; deficiencies can lead to reduced visual acuity, particularly in low-light conditions. Because vitamin A is fat-soluble, it does not need to be consumed daily—it can be stored in the body. In particular, it tends to concentrate in the liver. While this is beneficial in times of vitamin shortage in the diet, the ramification is that toxicity can result from an acute or chronic overdose. Retinol can be obtained from food in two different ways. Fruits and vegetables containing beta-carotene (particularly common in yellow-orange produce) allow for biosynthesis of the vitamin. Alternately, animal sources contain retinoids—partially synthesized vitamin A made by the animal. Common animal sources include eggs and dairy (as retinoids are required for the developing chick or growing calf), as well as organ meats like kidney and liver—the very same organs in which we store vitamin A!

Vitamin D

This vitamin is fat-soluble and crucial to the absorption of calcium (a mineral necessary for bone production and maintenance) from the digestive tract. It can be obtained in two ways: either from food sources or via biosynthetic modification of cholesterol in the skin upon exposure to sunlight. In most cases, about fifteen minutes of sunshine a few times a week is sufficient, though sunlight in extreme northern or southern regions of the planet may not be strong enough to cause the reaction. Compared to the other vitamins, there are few food sources of vitamin D. Some fish, including salmon, tuna, and cod, as well as the liver oils of these fish, are good sources. Beef liver, eggs, and some cheeses also contain the vitamin. Furthermore, many dairy products and cereals are supplemented with vitamin D in order to prevent deficiencies, which can be quite serious, especially in children. Childhood vitamin D deficiency can produce severe skeletal abnormalities and deformities due to incompletely hardened bones; this condition is called rickets, and was relatively common in the United States before vitamin D fortification of dairy and cereal products became common in the 1930s. Adult deficiency of the vitamin is associated with osteoporosis (weakening of the bones) and autoimmune disease.

Vitamin E

Like vitamin C, vitamin E is an antioxidant responsible for protecting cells from reactive oxygen species. Vitamin E, however, is fat-soluble, meaning that between the two antioxidants, all cellular environments are protected. The best sources of the vitamin include nuts, seeds, and vegetable oil. Leafy greens, like spinach and broccoli, are also good sources. Mild deficiency in the vitamin rarely causes symptoms, and severe deficiency is quite rare, since vitamin-E-rich foods are so common—generally, only those with malabsorption syndromes (inability to absorb nutrients from the digestive tract) are likely to be vitamin E deficient.

Vitamin K

Primarily known as the vitamin necessary to produce some of the proteins required for blood clotting, vitamin K also has a role in building bone. Deficiencies result in inability to clot blood, which can lead to bleeding disorders. Vitamin K can be obtained from vegetable sources—particularly leafy greens. In addition, native bacteria in the digestive tract, including *E. coli*, produce the vitamin, which we can then absorb from the intestine.

CONCEPT CHECK:

Why is it important to be careful not to overdose on fat-soluble vitamins?

Answer: They are stored in the body and can be toxic in large amounts.

8.15 MINERALS

Like vitamins, **minerals** are micronutrients. They do not supply us with energy, and we need only consume small amounts of them. As with vitamins, though, mineral deficiencies can be very serious. The minerals are trace elements, many of them metal or halide ions, and serve various purposes in the body. Iron, for instance, must be incorporated into hemoglobin for the protein to be capable of delivering oxygen to tissues. Calcium is critical to production and maintenance of the skeletal system, and plays key roles in muscle function and heartbeat. Sodium and potassium ions have important metabolic and neurological function—they are critical to maintaining fluid balance, allowing for transmission of nerve signals and muscle movement, and production of energy. As we saw in Chapter 7, severe sodium deficiencies can cause weakness and even death. The chloride ion is necessary to production of stomach acid (HCl), and is also involved in maintaining fluid and electrolyte balance. Copper is required for synthesis of hemoglobin and many of the body's structural proteins. Iodine maintains normal thyroid gland function, which helps to regulate metabolism. Minerals are generally obtained from food. A balanced diet composed of all the food groups, as will be discussed in Chapter 9, satisfies mineral requirements.

Throughout this chapter, we've examined the chemistry of both macro- and micronutrients in the human body: why they're needed, how they're utilized, and how they affect health. One theme that begins to develop through a discussion of nutritional chemistry—and which will be expanded upon in Chapter 9—is that while the chemistry that takes place in the body is quite complex, it is also quite predictable. When reactants are supplied and conditions are favorable, reactions proceed. When reactants are not supplied, reactions can't proceed. As complex as human nutritional requirements seem—the marmots and koalas of the world, with their very simple diets, have it much easier—everything we take in is toward the single purpose of providing reactants and energy for cellular reactions. In the end, we're really just big thermodynamic machines, running chemical reactions and harvesting the energy they release.

8.16 THE LAST WORD—VITAMIN ENRICHED!

A cursory glance at drugstore shelves stocked with personal care products reveals scores of bottles touting special ingredients—amino-acid-enriched conditioners, shampoos containing vitamins, and even shampoos containing lotion or conditioner. While it's tempting to believe television and magazine ads that encourage us to buy the newest and most "scientifically advanced" products, reevaluating these items in light of the principles of chemistry provides a different perspective.

With regard to hair, while shampoo and conditioner manufacturers tirelessly attempt to persuade us that tresses need to be provided with nutrition, the simple fact is that hair is composed primarily of the protein keratin—the same protein that forms the fingernails. In a manner highly analogous to the formation of skin, living cells from deep within the hair follicle rapidly divide and move upward, losing internal structures and functions as they fill with keratin until they are no longer living cells, but simply membrane sacs full of protein. Because the cells composing hair are neither alive nor metabolically active, they require no nutrition. We've seen in this chapter that vitamins allow living cells to run chemical reactions, an activity in which a dead cell does not engage. Live and actively dividing follicular cells, located within the scalp, do require both macro- and micronutrients in order to produce desirable quantities of strong, well-formed hair. However, these nutrients must be supplied in the same manner as they are to any cell—from the digestive tract via the bloodstream. In the end, there are only two things that can be accomplished by rubbing products on hair—keeping it clean (a function of soap), and keeping it from getting too dry and breaking (a function of some sort of moisturizer or conditioner).

On a similar topic, many products claim to provide vitamins to the skin if applied topically. While it is not a stretch to imagine that vitamin E, for instance, could penetrate the skin's outer layers, a chemist struggles to envision how the highly hydrophilic vitamin C might do any good at all. The membranes of all cells are composed of lipids; fatty acid complexes, cholesterol, and other nonpolar molecules surround the cell, separating its aqueous interior from the aqueous extracellular environment. As a result, while nonpolar molecules (particularly small ones) cross the membrane with ease, polar molecules are excluded. In the case of the skin, this function is highly protective—it keeps us from absorbing extraneous chemicals (particularly toxins) from our environments. Very few compounds have the ability to penetrate into the body through the skin. Some research has suggested that if chemically modified (by attaching a series of very long hydrocarbon tails through ester linkages), vitamin C can be made to pass through the skin. Other studies indicate that at low pH (around 3.5), vitamin C can be absorbed transdermally. Neither of these conditions, however, are met in the average hair or skin-care product, meaning that vitamin C in shampoo ends up where it is most soluble—dissolved in the shower water, running down the drain!

SUMMARY OF MAIN POINTS

- Thermodynamics is the study of the energetics of chemical reactions.
- The formation of bonds between atoms always releases energy, while breaking bonds always requires energy.
- Reactions are rearrangements of bonds—the formation of bonds that are lower in energy (more stable) results in a release of energy (and vice versa).
- The first law of thermodynamics states that energy must be conserved in all processes.
- Entropy is disordered (unusable) energy.
- The second law of thermodynamics says that entropy must increase for all processes.
- The energy released (or required) by a chemical reaction is the sum of energy required to break bonds minus the sum of energy released to make new bonds.
- Exothermic reactions release energy, while endothermic reactions require energy. The macronutrients we eat are broken down in exothermic reactions, and we harvest that energy to do work.
- Macronutrients are nutrients we need in large quantities, and include protein, carbohydrate, and fat.
- Food energy is measured in Calories (Cal), where 1 Cal = 1 kcal.
- Proteins are polymers of amino acids, which can be metabolized for energy, used to make new protein, or converted into other molecules.
- Protein is a major component of every cell: Enzymes, receptors, and fibrous structural materials are all protein.
- Carbohydrates include sugars (mono- and disaccharides), starch polysaccharides, and indigestible fiber.
- Different enzymes break disaccharides and starches into their monosaccharide units, but we lack the enzyme to digest cellulose.
- Amylose starch breaks down quickly into pure glucose, so from a chemical perspective, it's sugar.
- Sugar consumption elevates blood sugar, which is regulated by the hormones insulin (causes cells to take up glucose) and glucagon (causes release of glucose).
- Sugar and amylose cause rapid elevation of blood sugar and rapid, strong insulin responses; this leads to low blood sugar and enhanced fat storage.
- We store excess sugar as glycogen, or convert it into fat.
- Fat is an important dietary component. It's nutritionally dense, maintains immune and other system functions, and allows for absorption of fat-soluble vitamins.
- Fats are consumed (and stored) as triglycerides, which are made of a glycerol molecule and three fatty acids.
- We store most of our energy as fat because it's energetically dense and doesn't co-store with water, since it's nonpolar.
- Fatty acids with no double bonds are called saturated fats and are malleable solids at human body temperature.
- Naturally-occurring fatty acids with double bonds are called unsaturated fats, and the double bonds are all *cis*.
- Fatty acids with *trans* double bonds don't occur naturally—they are synthesized by partial hydrogenation and are common in processed foods.

- Omega-3 and omega-6 fatty acids have many double bonds (all *cis*), the last of which is three (or six) carbons from the tail of the fatty acid.
- An appropriate dietary ratio of omega-3 : omega-6 (about 1:2) is associated with good health.
- Cholesterol is a lipid made of four connected rings. It is important for biosynthesis of steroid hormones and vitamin D, and also appears in cell membranes.
- Too much systemic cholesterol can result in atherosclerosis and cardiovascular disease.
- LDL is a cholesterol transporter that takes cholesterol from the digestive tract to the cells; high levels are associated with cardiovascular disease.
- HDL is a cholesterol transporter that takes cholesterol from the cells to the liver for excretion; high levels are associated with good cardiovascular health.
- Saturated and trans fat consumption increase LDL, trans fat lowers HDL, and unsaturated fats have no effect on HDL or LDL levels. Most consumed fat should be unsaturated for these reasons.
- In the presence of base, triglycerides break down into fatty acid anions. These are soaps.
- Soaps are amphiphilic, since they have a polar region and a nonpolar region. They spontaneously form micelles in water, and can make grease water-soluble.
- Vitamins and minerals are micronutrients—we don't need to consume large quantities of them.
- Water-soluble vitamins are C (antioxidant) and the B-vitamins (mostly coenzymes).
- Fat-soluble vitamins are A (involved in vision), D (involved in calcium absorption and regulation), E (antioxidant), and K (blood clotting).
- There are many important minerals, including iron (for hemoglobin), calcium (for bones), sodium and potassium (for fluid balance and nerve/muscle function), and iodine (for thyroid).

Questions and Topics for Discussion

1. Explain why forming bonds releases energy, while breaking bonds requires energy. Does this seem reasonable from an atomic perspective? (Think in terms of subatomic particles and stability.)
2. Why do double (and triple) bonds have higher bond energies than single bonds?
3. Explain why breaking a given bond requires exactly the same amount of energy that would be released if that same bond formed.
4. In terms of total energy and entropy, how is burning sugar in a beaker and letting the heat escape different from burning sugar in a beaker and heating water?
5. If forming bonds required energy, how would the universe be different?
6. Why is food energy counted in Cal rather than cal? What is the relationship between Cal and cal?
7. Does it make sense that polymers made by dehydration synthesis are digested by hydrolysis? What do each of those mean chemically?
8. Why might a nutritionist say that starch is really just sugar? Is that reasonable?
9. What is the difference (with regard to digestion and blood sugar) between whole grain and refined grain? Which is more like sugar and why?
10. Why is a rapid increase in blood sugar detrimental from a chemical perspective? What does it lead to?
11. What generally happens to glucose when it's absorbed into the blood stream slowly? Quickly?
12. Why do you think trans fat didn't evolve as a storage fat in nature?
13. Why are most of our stored calories in the form of fat? What advantages does it have? Why is our storage fat mostly saturated?
14. A disease called familial hypercholesterolemia results from a genetic inability to produce LDL receptors. What sorts of symptoms do you think disease sufferers experience?

15. In popular culture, HDL is sometimes called "good cholesterol" and LDL is called "bad cholesterol." Are these accurate descriptions? Why or why not?
16. After a doctor's appointment including blood work, a friend tells you that he has high cholesterol. You find out that his LDL is in the normal range, but his HDL is very high. What might you say to him? Are you concerned for his health?
17. Early soaps were made with animal fat and lye. In what ways are modern soaps superior? In what ways must they be the same?
18. Bile salts are amphiphilic molecules released into the intestine by the gall bladder. They are used to aid in the digestion of fats. How do you think they achieve this, given that they do not chemically react with fats at all? (Hint: digestive enzymes are all water-soluble!)
19. Why do you think organ meat (like liver) is very high in vitamins A and D?

Problems

1. Calculate the energy released or required to form 1 mole of O_2 from atoms.
2. Calculate the energy released or required to form 1 mole of N_2 from atoms.
3. Calculate the energy released or required to form 1 mole of C_2H_4 from atoms (hint: draw a Lewis structure!).
4. Calculate the energy released or required to form 1 mole of COH_2 from atoms (hint: C is in the middle—draw a Lewis structure!).
5. Calculate the energy released or required to form 2 moles of CO_2 from atoms (hint: remember that Table 8.1 lists energies in kcal/mol).
6. Calculate the energy released or required to form 2 moles of CH_3OH from atoms (hints: draw a Lewis structure, and remember that Table 8.1 lists energies in kcal/mol).
7. Calculate the energy released or required to break 1 mole of H_2 into atoms.
8. Calculate the energy released or required to break 1 mole of O_2 into atoms.
9. Calculate the energy released or required to break 1 mole of C_2H_6 into atoms (hint: draw a Lewis structure!).
10. Calculate the energy released or required to break 1 mole of CH_3NH_2 into atoms (hint: draw a Lewis structure!).
11. Calculate the energy released or required to break 2 moles of H_2O into atoms (hint: remember that Table 8.1 lists energies in kcal/mol).
12. Calculate the energy released or required to break 2 moles of C_2H_2 into atoms (hints: draw a Lewis structure, and remember that Table 8.1 lists energies in kcal/mol).
13. Determine how much energy is released or required for the reaction shown. Is this reaction exothermic or endothermic?

    $2\ H_2 + O_2 \rightarrow 2\ H_2O$

14. Determine how much energy is released or required for the reaction shown. Is this reaction exothermic or endothermic?

    $CO_2 + 2\ H_2O \rightarrow CH_4 + 2\ O_2$

15. Determine how much energy is released or required for the reaction shown. Is this reaction exothermic or endothermic?

    $C_2H_4 + 3\ O_2 \rightarrow 2\ CO_2 + 2\ H_2O$

16. Determine how much energy is released or required for the reaction shown. Is this reaction exothermic or endothermic?

$C + O_2 \rightarrow CO_2$

17. Fill in the blanks in the following table with equivalent values across rows:

40 kcal	________Cal	________cal
________kcal	550 Cal	________cal
________kcal	________Cal	10,700 cal

18. Fill in the blanks in the following table with equivalent values across rows:

________kcal	487 Cal	________cal
________kcal	________Cal	13,600 cal
320 kcal	________Cal	________cal

19. What's a polymer? Which of the macronutrients are polymers and which are not?
20. Define: protein; amino acid; peptide bond; enzyme.
21. How can the digestion products of proteins be used in the body?
22. What are the general categories of carbohydrates, and how do they differ from one another?
23. Why can't we digest fiber?
24. Why is fiber an important component of diet?
25. What are the roles of the hormones insulin and glucagon? How does insulin affect body chemistry if it's released slowly? Quickly?
26. How is excess glucose stored in the body?
27. Why are fats important to the diet?
28. What is the difference between saturated and unsaturated fat?
29. Why is unsaturated fat considered much healthier than saturated or trans fat?
30. Are omega-3 and omega-6 fatty acids saturated, unsaturated, or trans fats? Are they considered healthy?
31. Why is cholesterol an important biomolecule? What is it used for?
32. What are HDL and LDL, and what do they do in the body?
33. What factors affect HDL and LDL levels, and how?
34. What is a micelle? What does it do?
35. Which vitamins are water-soluble? Fat-soluble? Which can be stored in the body?
36. Fill in the following table for vitamins:

Vitamin	**Good Source**	**Function in the Body**
A		
B's		
C		
D		
E		
K		

37. Fill in the following table for minerals:

Mineral	**Function in the Body**
Iron	
Calcium	
Sodium and potassium	
Chloride	
Copper	
Iodine	

38. Why are protein, carbohydrate, and fat called macronutrients, but vitamins and minerals are called micronutrients?
39. Do vitamins and minerals provide energy? If not, why do we need them?

Answers to Problems

1. 119 kcal released.
3. 542 kcal released.
5. 768 kcal released.
7. 104.2 kcal required.
9. 677 kcal required.
11. 444 kcal required.
13. 116.6 kcal/mol released; exothermic.
15. 313 kcal/mol released; exothermic.
17. Fill in the blanks in the following table with equivalent values across rows:

40 kcal	40 Cal	40,000 cal
550 kcal	550 Cal	550,000 cal
10.7 kcal	10.7 Cal	10,700 cal

19. A polymer is a large molecule made up of repeating units. Starches and fiber are polymers, as are proteins. Fat is not a polymer.
21. We can break down amino acids for energy or use them to make proteins. We can also make neurotransmitters and other molecules from them.
23. We lack the enzyme capable of breaking the sugar linkages in cellulose.
25. Insulin causes cells to take up blood sugar, while glucagon causes the release of glucose into the bloodstream. A slow insulin release causes sugar to be taken up by cells, used for energy, and much of the excess stored as glycogen. A quick insulin release causes the cells to take up large quantities of glucose from the blood (causing blood sugar to fall rapidly), and the glucose is converted to and stored as fat.
27. Fats are dense sources of energy, allowing for absorption of fat-soluble vitamins from food, and maintaining normal body composition.
29. Unsaturated fat provides the benefits of fat in the diet without the risk of cardiovascular disease and increased LDL/HDL ratio.
31. Cholesterol regulates the stiffness of cell membranes and is the biosynthetic precursor for vitamin D and steroid hormones.
33. Saturated fats increase LDL, trans fats increase LDL and lower HDL. Vigorous exercise and moderate alcohol increase HDL.

35. The B vitamins and vitamin C are water-soluble. A, D, E, and K are fat-soluble, and can be stored.
37. Fill in the following table for minerals:

Mineral	**Function in the body**
Iron	Incorporated into hemoglobin for oxygen transport
Calcium	Production/maintenance of skeletal system
Sodium and potassium	Fluid balance, nerve and muscle function, energy production
Chloride	Making stomach acid, fluid balance
Copper	Hemoglobin and other protein synthesis
Iodine	Thyroid function for metabolism

39. They do not provide energy, but they do serve various vital roles in metabolism and other bodily functions.

Sources and Further Reading

Fats

Benson, M.K., Kshama, D., and Fattepur, S.R. (2008). Studies on SFA/TFA (saturated/trans fatty acid) rich dietary fats on lipid profile and antioxidant enzymes in normal and stressed rats. *Pharmacognosy Magazine*, 4(16), 320–28.

Berbert, A.A., Kondo, C.R., Almendra, C.L., *et al.* (2005). Supplementation of fish oil and olive oil in patients with rheumatoid arthritis. *Nutrition*, 21(2), 131–36.

Crupkin, M., Zambelli, A. (2008). Detrimental impact of trans fats on human health: stearic acid-rich fats as possible substitutes. *Comprehensive Reviews in Food Science and Food Safety*, 7(3), 271–79.

Dry, J., and Vincent, D. (1991). Effect of a fish oil diet on asthma: results of a 1-year double blind study. *International Archives of Allergy and Applied Immunology*, 95(2–3), 156–57.

Erkkila, A.T., Lichtenstein, A.H., Mozaffarian, D., *et al.* (2004). Fish oil is associated with a reduced progression of coronary artery atherosclerosis in postmenopausal women with coronary artery disease. *American Journal of Clinical Nutrition*, 80(3), 626–32.

Lecerf, J.M. (2009). Fatty acids and cardiovascular disease. *Nutrition Reviews*, 67(5), 273–83.

Zevenbergen, H., deBree, A., Zeelenberg, M., *et al.* (2009). Foods with a high fat quality are essential for healthy diets. *Annals of Nutrition and Metabolism*, 54, 15–24.

http://www.nlm.nih.gov/medlineplus/druginfo/natural/patient-fishoil.html

Cholesterol

www.americanheart.org

http://www.nlm.nih.gov/medlineplus/cholesterol.html

Vitamins

http://www.nlm.nih.gov/medlineplus/vitamins.html

http://www.hsph.harvard.edu/nutritionsource/what-should-you-eat/vitamins/

Vitamin Enrichment

Farris, P. (2005). Topical vitamin C: a useful agent for treating photoaging and other dermatologic conditions. *Dermatological Surgery*, 31, 814–18.

Lin, J.Y., Selim, M.A., Shea, C.R., *et. al.* (2003). UV photoprotection by combination topical antioxidants vitamin C and vitamin E. *Journal of the American Academy of Dermatology*, 48(6), 866–74.

9 The Body as a Beaker—Nutrition, Health, and Wellness

While it's difficult to generalize across the board, the trend in many industrialized Western countries in recent decades has been toward increasing obesity. Simultaneously, citizens of industrialized nations are exposed more than ever before to airbrushed and computer-generated images portraying impossible (or nearly impossible) to attain standards of musculature, athleticism, slenderness, and shape. At the juxtaposition of these two observations lies what we will call the **Western Paradox**—the fact that Western societies are simultaneously consuming a generally atrocious diet and putting a premium on physical perfection. This has spurred the rapid growth of the weight-reduction industry, made room for questionable diet gimmicks touting rapid results, and all the while failed to slow the pace at which obesity has been overtaking smoking as the number one cause of preventable death. People all over the world share the need to provide their bodies with appropriate reactants (in the form of food) so as to synthesize necessary cellular products, generate energy, and maintain life. Along the way, however, we are faced with challenges—some individuals are unable to secure enough food to meet their energy needs, while for others, tempting choices are overabundant. The lifestyles of some expose them, by choice or by necessity, to regular vigorous exercise, while others have more sedentary existences. As a result, the world is full of all manner of different body habitus—some lean and others plump, some muscular and others sinewy. This chapter focuses on the ways in which the molecules of nutrition affect body composition, health, and composition-dependent illness. How do our diets affect how we look and feel? What is the rationale behind the U.S. government-recommended nutritional requirements? For individuals striving to lose weight, what is the chemistry behind weight loss strategies? Most importantly, how does a health-conscious individual construct a balanced, healthy lifestyle amid desk jobs, socially prescribed notions of appropriate body composition, and a plethora of fast food restaurants?

9.1 BODY FAT—TOO MUCH, TOO LITTLE, AND JUST RIGHT

Before we begin a discussion of the disease processes associated with being overly fat, let's first examine the reasons the human body needs fat at all. From

Chapter 8, we know fat is necessary in the diet: It provides a valuable source of energy, allows for absorption of fat-soluble vitamins and maintains immune system function, among other things. In a culture in which periods of plenty of food are followed by times of scarcity, body fat serves as a crucial store of many weeks' worth of energy. In a society in which food is never farther away than a walk to the refrigerator or a trip to the neighborhood grocer, however, is body fat even necessary? Could a very strict diet and exercise regimen strip all the fat off a human body? And if it did, what would be the ramifications? It turns out that fat on the body is as critical to health as fat in the diet: we simply can't live without it.

First, while it might seem that fat energy reserves are of little utility in a society with ready access to food, we nevertheless draw upon these reserves frequently. Sleep, a long day at work without a lunch break, and extended moderate-intensity exercise are all examples of times during which we rely on mobilization of the energy from our fat stores—without them, we would awaken frequently to eat, become frazzled and fatigued at work, and be unable to complete a workout. In addition, fat is more than simply the vehicle whereby fat-soluble vitamins are absorbed from the digestive tract. It's also a repository for these vitamins in the body. Fat-soluble vitamins build up in fatty tissue, meaning that they need not be consumed daily, since reserves are available. Water-soluble vitamins, on the other hand, must be consumed with greater frequency in order to prevent deficiency. Steroid hormones, as we saw in Chapter 8, are cholesterol-derived. Like cholesterol, they're nonpolar, and like fat-soluble vitamins, they concentrate in the fatty tissues. This is important to normal body function, as concentrations of one hormone often influence the production of others. Finally (and eminently practically), fat provides physical padding. The soles of the feet are cushioned by fat pads, as are the palms of the hands. Fat in the lower back protects the kidneys, which are neither completely enclosed within the rib cage nor the pelvic girdle. As you read this book, you're likely seated—a prospect which would be significantly less comfortable without a bit of fat padding the sitting bones! Too little body fat is not compatible with life.

On the other hand, excess body fat (referred to as either **overweight** or **obesity**, depending upon the degree to which the condition exists) is quite dangerous and associated with a number of disease processes. High blood pressure and high cholesterol, both of which are risk factors for cardiovascular disease and stroke, occur more frequently among the overweight and obese. **Sleep apnea**, a temporary inability to breathe during sleep, can result from fat in the neck collapsing the back of the throat as muscles relax. In order to take a breath, an individual with apnea must partially awaken (often hundreds of times a night) to reopen the airway. Sleep apnea is associated with fatigue and general poor health due to chronic exhaustion. **Arthritis**, a degeneration of the joints, occurs among the overweight and obese due to the physical stress of carrying weight in excess of the body's innate biomechanical capacity. Some forms of cancer are more common in this demographic as well. One of the most prevalent disease processes in the overweight and obese population, however, is **type 2 diabetes**. Unlike type 1 diabetes, which results from an inability to produce the blood-sugar-regulating hormone insulin (and which is treated with injected insulin), type 2 diabetes is the result of a cellular insensitivity to the hormone, which is often (but not always) a product of excess body fat. As a result, blood sugar builds up after meals, and can often become dangerously high. This leads to symptoms of chronic hyperglycemia. Further, since cells are unable to take up glucose in the absence of insulin, symptoms of cellular starvation are also evident. Type 2 diabetes is severe, and left unchecked, can be fatal. Since insulin production occurs in type 2 diabetics to the same extent that it occurs in the nondiabetic population, injected insulin is not a helpful treatment regimen. Rather, disease sufferers are counseled to control their diet, reduce intake of refined sugars and carbohydrates, exercise regularly, and most importantly, lose weight.

The overweight and obesity epidemic in the United States is of impressive proportion, so to speak. From 2001–2004 National Institutes of Health (NIH) data, two-thirds of U.S. adults are overweight, with half of these (one-third of total U.S. adults) falling into the obese category. Less than one-third of U.S. adults are of a healthy weight. Minorities, in particular, are afflicted by the obesity epidemic, as are those of lower socioeconomic standing, most likely due to the availability and low cost of high-fat, highly processed foods, the consumption of which is associated with high body fat.

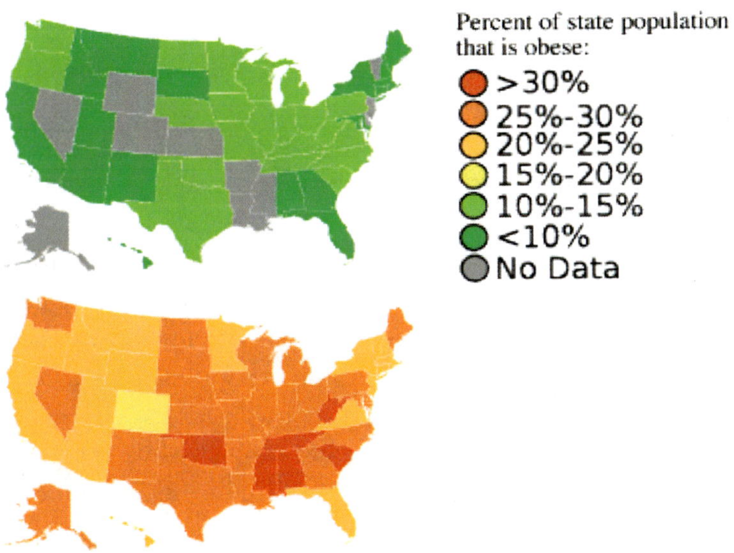

Image 9.1. Between 1989 (top) and 2008 (bottom), the incidence of obesity in the population has increased astronomically, according to the Centers for Disease Control and Prevention records.

From a social perspective, obesity is problematic on the individual, community, and global levels. Of course, individuals are directly affected by their own health. With regard to the community, however, both local and global, the cost of obesity is worth considering. Direct health care costs include preventive, diagnostic, and medical treatment services. These are footed by the government and recouped through taxes in the case of subsidized care, and are footed by insurance companies and recouped via increased premiums across the board in the case of privately funded care. Indirectly, lost wages and work time represent significant individual and group costs. Clearly, obesity is not simply a problem faced by the obese: as members of society, we all pay the price.

9.2 BODY COMPOSITION METRICS

Various health-monitoring organizations, including the World Health Organization (WHO) and NIH, set forth guidelines regarding the percentage of body fat (by mass) appropriate to a healthy human. For men, **essential body fat**, meaning the minimum percentage body fat by mass required to sustain normal function, is 2–4 percent of body weight. For women, this value is higher, at 10–12 percent. The reason for a higher essential body fat in women has to do with the female hormone cycle; in order to produce normal menses, estrogen and progesterone must build up in fatty tissues. Women with very low body fat often experience *amenhorrea*, or lack of a menstrual cycle. As a result, these individuals are unable to become pregnant. **Normal body fat** values are considered to be in the range of 6–26 percent of body weight for men, and 14–31 percent for women, with athletes and fit individuals falling near the lower end of this range. Body fat percentages higher than these indicate overweight or obesity, depending upon the value. A major reason to monitor **body composition** (which falls primarily under the auspices of health care professionals) is to provide appropriate medical care and counseling to individuals who might be at risk for diseases associated with being under- or overweight. For this reason, it's quite common for a physician to check patient weight and height during annual visits. From a practical perspective (and despite the prevalence of electronic scales claiming to do the job), it's quite difficult to measure body fat percentage accurately, and nearly impossible to do so in a physician's office. Some metrics provide for a very rough approximation of percent body fat from **wrist, neck, and waist circumference**. However, particularly extensive (or absent) musculature significantly decreases the accuracy of these measurements. Alternately, three- or seven-site **skin fold tests** may be used to estimate body fat, the theory being that *subcutaneous* (located just under the skin) fat deposits in proportion to total body fat. Calipers, which isolate a pinch of skin and the fat lying just under it, are used to take thickness measurements at various points on the body. Unfortunately, slight variations in caliper placement or technique, hydration level, and other factors can significantly affect the accuracy of the measurement.

The "gold standard" of body fat percentage determination is the **immersion test**, which requires that an individual be weighed while completely submerged in water, having blown all the air out of their lungs.* In addition to the discomfort and cost associated with the procedure, it's rather impractical. For this reason, physicians and other health care providers generally rely on a simple metric based upon weight and height to approximate body composition. Called **body mass index (BMI)**, this metric is calculated as follows and correlated to body composition as per Table 9.1:

$$BMI = \frac{weight\ (kg)}{[height\ (m)]^2}$$

As techniques for approximating weight status go, BMI is both easy to calculate and relatively accurate for most individuals. Where it fails is with the exceedingly muscular (muscle, far denser than fat, can make an individual

***NOTE**

Sound easy? Not so fast! To really get an idea of what this feels like (and why it isn't done often), exhale normally. Now forcibly exhale more (there's more in there, I promise). Keep going—get ALL the air out. Now hold it as long as you can. Generally, people can't hold a forced exhalation long enough for the scale operator to get an accurate weight, so they're given a few minutes to recover, and then the process starts all over again, and again.

Try This:

Calculate the BMI for an individual who is 1.83 m tall, and who weighs 75 kg. Into which weight status category does this individual fall?

Answer: 22.39; normal.

Solution:

$$BMI = \frac{weight}{height^2} = \frac{75\ kg}{(1.83\ m)^2} = 22.39$$

BMI	Weight Status
<18.5	Underweight
18.5–24.9	Normal
25.0–29.9	Overweight
≥30.0	Obese

Table 9.1. BMI scores as they correlate to body weight status.

with normal body fat weigh more than average for their height, resulting in a higher BMI). There's also some suggestion that very inactive, ostensibly skinny individuals with little muscle fall into inappropriately low-weight status categories using BMI. The reason for this is that, as fat is less dense than muscle, a very slight person (these individuals are often female) may actually be carrying a high percentage of body fat while appearing slender and weighing very little, provided that there is minimal muscle density to add to the weight. Despite their appearance and healthy range BMI, these "skinny fats" are nevertheless susceptible to diseases of obesity. As such, BMI values must be regarded with a certain amount of skepticism, or at least verified through a secondary measurement technique, for individuals of abnormal musculature.

Concept Check:

Why is weight alone not a good measure of body composition?

Answer: Weight will vary with height, gender, and musculature.

9.3 NUTRITIONAL RECOMMENDATIONS

Armed with the knowledge that we need protein, carbohydrate, fat, vitamins and minerals to stay healthy, coupled with the information that we should eat enough to stay fit without overeating and gaining excess body fat, how do we go about making informed decisions about what to eat? The U.S. Department of Agriculture (USDA) and Food and Drug Administration (FDA) work jointly to provide the public with nutritional guidelines from which they can extract information for decision-making. These guidelines and standards reflect current research and scientific thinking regarding human nutritional needs.* From the "Basic Four," unveiled in the 1950s and recommending that people consume from each of four food groups (meat, breads and grains, dairy, and fruit and vegetables) each day, through the classic USDA pyramid that comes readily to the minds of many (Image 9.2), and on to the new (2005) USDA pyramid, some guidelines have changed, while others have remained the same.

*NOTE

They also sometimes reflect the efforts of special-interest lobbyists—for instance, the current standards are based upon a 2000 Cal/day diet, but some information regarding a 2500 Cal/day diet is also listed on nutritional labels. This is the result of a compromise made between the FDA (who wanted the 2000 Cal/day label) and the USDA who, looking out for the interests of meat and dairy farmers, sought the higher value.

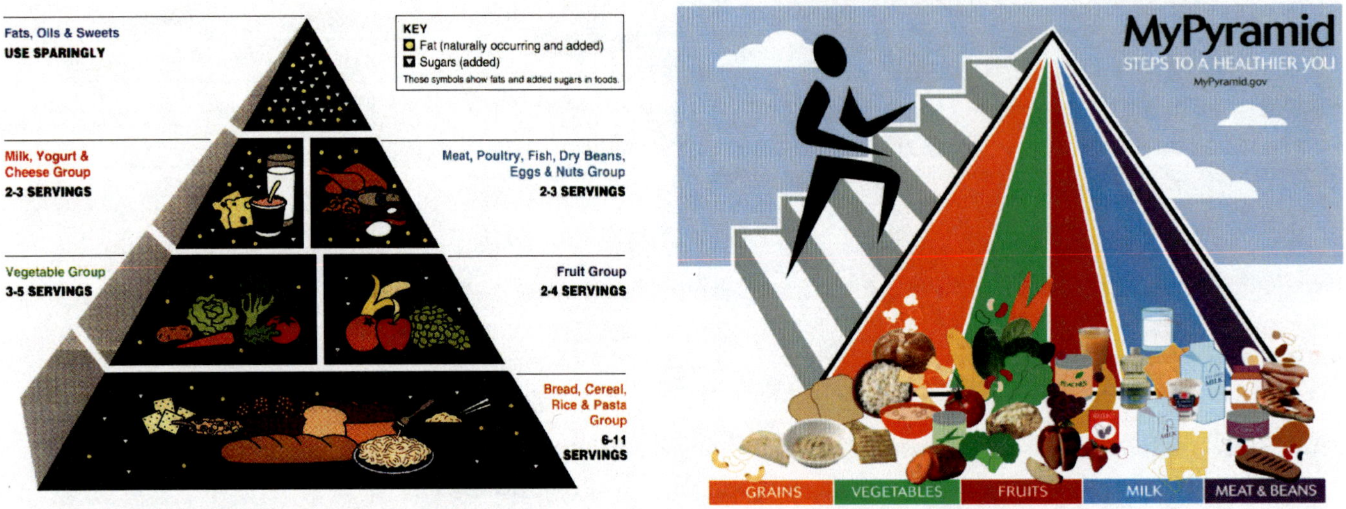

Image 9.2. An older USDA food pyramid put bread and grains at the base, fats and sweets up top. The new pyramid tries to move away from the idea that one particular food group is more important (the "base" of the pyramid, so to speak), and instead structures the groups in bands from base to top. It also incorporates a representation of exercise.

While nutritional recommendations have traditionally been uniform across gender, age, and activity level, a hallmark of the latest set is that they include a certain amount of personalization. This is a nod to research revealing that caloric needs generally decrease with age (among adults), are higher for men than for women, and are higher among more active individuals. The latest USDA pyramid has also been praised for avoiding nebulous references to "servings"—the earlier pyramid advised, for instance, the consumption of 6 to 11 servings per day from the bread, cereal, rice, and pasta group, but no indication was made as to what constituted a "serving." New guidelines provide recommended amounts in weight and volume measurements, reducing ambiguity.

In addition to providing daily consumption guidelines, the USDA and FDA also require that most commercially prepared and processed foods be printed with ingredient lists (in order of prevalence by weight) and nutritional facts (Image 9.3).* These nutrition labels are based upon a 2000 Calorie/day diet, and provide both the mass of each nutrient and the percentage that mass represents of the recommended daily quantity, called the **Percent Daily Value**. Of course, for individuals who generally consume significantly more or less than 2000 Calories each day, these percentages will be inaccurate. Regardless, a consumer can obtain at least a general idea of what exactly is in their food by using the information provided.

A Nutrition Facts label may contain more information than that shown, but by law, it may not contain less. It's important to understand that the label does not necessarily tabulate all the nutrients in a given package: If the package is meant by the manufacturer to be subdivided into servings, the serving size and number of servings per container must be specified. Some macronutrient subdivision is required—not only total fat, for instance, but also quantities of saturated and trans fat must appear so that consumers seeking to limit these

***NOTE**

The exceptions include spices, water, and alcoholic beverages; the latter was the result of intense lobbying by the industry, presumably because they feared falling sales if the caloric content and nutritional vacuity of their products were made clear to consumers.

are able to avoid them. Note that cholesterol and sodium, both of which are prevalent and commonly over-consumed in the Western diet, are listed directly under fat. This conveniently puts all the nutrients that should be limited in the same place for ease of reading. Carbohydrate is always subdivided into fiber (indigestible but important) and sugar. Note that sugar is not subdivided into naturally occurring and added—in other words, whole fruit (if it had a nutrition facts label) would contain sugar. Vitamins and minerals appear at the bottom of the label. Information regarding vitamin A, vitamin C, iron, and calcium is required by law. Other vitamins and minerals may be listed or omitted at the discretion of the manufacturer. Finally, the label contains a brief synopsis of the USDA and FDA recommendations for maximum quantities of fat, cholesterol, and sodium to be consumed each day, as well as minimum recommended carbohydrate and fiber.

The inclusion of trans fat quantitieson nutrition labels is relatively new; the legislation requiring this information took effect in 2008. As discussed in Chapter 8, research indicates that even more so than saturated fat, trans fat is detrimental to cardiovascular health. For this reason, new guidelines recommend limiting consumption to fewer than 2.5 grams daily. Unfortunately, the government allows the food industry a few loopholes in trans fat reporting. For instance, a value of zero may be reported for all quantities of trans fat less than 0.45 g per serving. A bit of simple math reveals that if one were to consume 6 servings a day of foods containing 0.45 g per serving, one would quickly exceed the daily recommended maximum—all without realizing any trans fat had been consumed at all! Astute label readers watch for the words *partially hydrogenated* in a food's list of ingredients; if those words appear, trans fat is present, regardless of what the nutrition label says.

A second technique for producing trans-like fat has recently become more common in the food industry. This technique avoids partial hydrogenation (and therefore the use of the words *partially hydrogenated* in food labels), but fats produced in this way can be identified by the word *interesterified*. Preliminary research suggests that interesterified fats not only act like trans fat in food, they act like trans fat in the body—increasing LDL, decreasing HDL, and negatively affecting cardiovascular health. One further trans fat pitfall lies in the chemistry of the double bonds themselves. While naturally occurring unsaturated (and therefore healthy) oils contain *cis* double bonds (hence their low melting points), these are less stable than *trans* double bonds. As a result, exposure to intense heat over long periods of time can cause the *cis* bonds to break and reform as *trans*. As a result, oils that have been kept at high heat for long periods (such as in commercial deep-fat fryers) often contain heat-produced trans fats. Because the oil that went into the fryer was unsaturated (and trans fat-free), the government allows the food industry to claim that the product is trans fat-free—despite the fact that it may contain many grams of trans fat per serving. While home fryers are unlikely to reach sufficient temperatures (over sufficient time periods) to result in the production of trans fats in any significant amount, restaurant and commercially fried food may be a source.

Nutrition Facts

Serving Size 1 cup (228g)
Servings Per Container 2

Amount Per Serving	
Calories 250	Calories from Fat 110
	% Daily Value*
Total Fat 12g	**18%**
Saturated Fat 3g	**15%**
Trans Fat 3g	
Cholestrol 30mg	**10%**
Sodium 470mg	**20%**
Total Carbohydrate 31g	**10%**
Dietary Fiber 0g	**0%**
Sugars 5g	
Protein 5g	
Vitamin A	4%
Vitamin C	2%
Calcium	20%
Iron	4%

* Percent Daily Values are based on a 2,00 calorie diet. Your Daily Values may be higher or lower depending on your calorie needs.

	Calories	2,000	2,500
Total Fat	Less than	65g	80g
Sat Fat	Less than	20g	25g
Cholesterol	Less than	300mg	300mg
Sodium	Less than	2,400mg	2,400mg
Total Carbohydrate		300g	375g
Dietary Fiber		25g	30g

Image 9.3. A sample nutrition label, color-coded for emphasis. Items highlighted in yellow are generally nutrients to limit, while those highlighted in blue are generally nutrients of which sufficient quantities are necessary.

Concept Check:

Which nutrients should most Westerners avoid over-consuming?

Answer: Fat (especially saturated and trans), cholesterol, and sodium. (And sugar, as we will soon see.)

> **CONCEPT CHECK:**
>
> Why should trans fat be avoided?
>
> Answer: It lowers HDL and raises LDL; it negatively impacts cardiovascular health.

9.4 WHAT WE'RE EATING—THE TYPICAL AMERICAN DIET

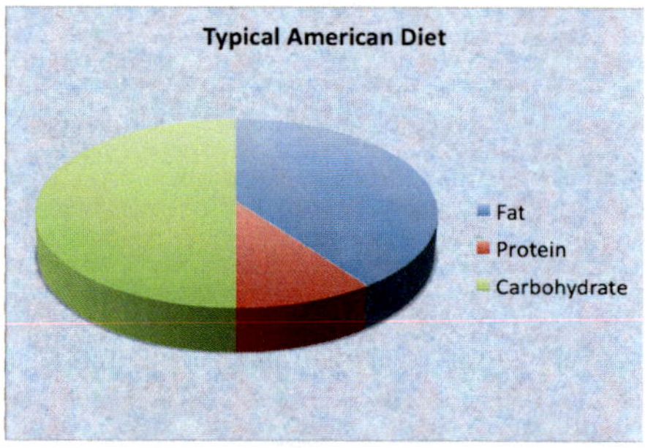

Image 9.4. A breakdown by macronutrient of the typical American consumption pattern.

The average American gets approximately half of his or her calories from carbohydrate, about 10 percent from protein, and the remaining 40 percent from fat (Image 9.4). They eat out about five times per week, three meals of which are composed of hamburgers and French fries. While some sources cite the typical American intake nearing 4000 Cal/day, this fails to account for spoilage, discarded food, and loss through other mechanisms. In reality, estimates closer to 2750 Cal/day are more likely (averaged across men and women)—a value still significantly higher than the recommended 1800 Cal/day or 2400 Cal/day for relatively sedentary females and males respectively.

Purely on the basis of its macronutrient breakdown, the American diet doesn't look too bad—it's higher in fat than most nutritional experts recommend, and lower in protein, but not particularly deviant with regard to carbohydrate recommendations. Why, then, is our national consumption pattern so detrimental to health? The answer lies more in the source of the carbohydrate, protein, and fat (and in the amounts!) than the percentages.

Fast food and processed food have become staples of the American way of life. Unfortunately, these foods are high in saturated and trans fat, refined carbohydrate, and salt, while being low in fiber, protein, and micronutrients. Some processed foods (particularly those that advertise themselves as "healthy" options, like energy bars) boost their vitamin and mineral content through additives, but the efficacy of these added micronutrients is questionable (see The Last Word—Vitamin Supplements). Further, even those fast food options that purport to be healthy often present nutritional pitfalls lurking beneath a wholesome veneer. The FDA regulations allow for considerable leeway with regard to advertising statements made by food companies. For instance, a label identifying "whole wheat bread" means only that the bread contains whole wheat flour among the ingredients—even if it's just a small percentage of the total flour.* A fast food company's "healthy" sandwich, consisting of grilled chicken on a whole wheat bun, could easily be composed of inexpensive, fattier chicken cuts on a bun consisting mostly of refined flour. The claim that a fried food is free of trans fats holds no assurance whatsoever of legitimacy, as we've already explored. Eating as many meals away from home as we do, we have little knowledge of (or control over) what goes into our food. Most of us get less than half the fiber necessary to maintain health, and three times the recommended amount of salt. Where guidelines call for four to five cups of fruits and vegetables a day, most of us get less than one and a half. Even those individuals who would like to control consumption and attempt to

> ***NOTE**
>
> You can get an idea of how much of each ingredient is in a food by noting the order in which they appear on the label—by law, ingredients must be listed from greatest quantity (by mass) to least.

keep track of how much they're ingesting have difficulty with meals prepared outside the home (for which nutritional information is not available): Studies show that people routinely underestimate the caloric content of a meal by 35–50 percent. This effect is more noticeable in larger meals, as those served in restaurants tend to be.

Caloric beverages, too, present a problem. According to USDA data, the average American gets more than half their daily calories from soda and other sweetened beverages. These represent a huge source of **empty calories**—meaning that there is little nutritional benefit apart from energy—and also present the problem of being non-satiating. It seems that our ability to sense the caloric content of food, and the intrinsic mechanisms that promote feelings of hunger in response to a caloric deficit and feelings of satiety in response to caloric balance, are derailed by calorie-containing beverages. Sugared beverages also raise blood sugar quickly, resulting in a strong insulin signal and everything that follows. This combines to make them a nefarious source of calories—they provide no nutritional benefit, not even lasting energy, since the sugar is quickly taken up and stored as fat. The quantity of sugar in a soda is staggering—a 20-oz. bottle contains 17 teaspoons of added sugar, or 69 grams. Lately, fancy coffee drinks have become commonplace. While coffee itself is calorie free, the calories in fancy drinks containing added sugars, full-fat dairy, and whipped cream add up quickly. A 16-oz. blended coffee and cream drink topped with whipped cream can contain a whopping 500 calories—and that's for the smallest size available! In simple terms, that's one beverage providing more than a quarter of a woman's (or about a fifth of a man's) daily recommended calories—without any lasting sense of satisfaction or reduction in hunger. All of this adds up to an approximately 2-lb.-per-year weight gain among average American adults, making it no small wonder that an estimated 80 million people in the United States go on a weight-loss diet each year.

9.5 TRYING TO CHANGE WHAT WE'RE EATING—WEIGHT-LOSS DIETS

In an interesting twist of language, the word *diet* has taken on an additional (and in many cases quite disparate) meaning than that listed as the primary definitionin most dictionaries. **Diet**, from its etymological origins, refers to the food and drink that is taken regularly or habitually—in other words, one's diet is the sum of what one routinely eats and drinks to support life and health. In common parlance, however, the expression *on a diet* connotes a reduction in or departure from one's regular consumption pattern, often for the purpose of losing weight. Further, the "diet" is often designed to produce weight loss quickly—to starve the body of nutrients in order to force it to consume itself. Because self-consumption is not sustainable over time (one can't survive habitual nutritional deprivation), being "on a diet" is incompatible with the concepts of regularity and habit.

CONCEPT CHECK:

Why are calorie-containing drinks problematic in terms of diet?

Answer: They generally contain calories without other nutrients, and they don't provide lasting fullness or satisfaction.

Semantics aside, however, diets (and in deference to common idiom, this section of the text will use the word in the popular manner) and diet books, diet advice and diet methods, abound. While the concept behind weight loss is mathematically simple—when energy expended surpasses energy consumed, the body is forced to make up the difference by burning stored fat—it's chemically quite complex. The reason for the complexity (and the reason weight loss is so difficult for many individuals to achieve) is that the body is designed to maintain homeostasis, or balance. Just as some mechanisms help maintain body temperature and pH, others regulate habitus. To a certain extent, when we consume more than we expend energetically, cells become "sloppy" with their energy use, and calories are wasted. For this reason, occasional overeating doesn't lead to significant weight gain. Likewise, when we expend more than we consume, cells respond by increasing their fuel economy—energetic needs decrease somewhat—to avoid having to burn fat. It is only in the case of regular energy overabundance or deficiency that significant weight change occurs. Further complicating this equation is the fact that the body, evolutionarily engineered to survive periods of famine by storing fat during periods of plenty, much more quickly and easily gains fat than loses it. These factors combine to produce a human body that gains weight relatively easily in a society in which food is plentiful, and loses it with difficulty. Because the most realistic and effective weight loss technique, which involves a minor reduction in calories consumed coupled with an increase in calories expended, and which must take place slowly and gradually, does not appeal to many—diet fads promising quick and dramatic weight loss abound. Some of these rely on "miracle pills" promising results with no exercise or dietary restriction, while others encourage altered consumption patterns. The chemistry of these diets is quite interesting, and worth examining.

9.6 THE LOW-FAT DIET

Based upon the theory that fat is the most calorically dense of the nutrients, low-fat diets often work toward overall caloric reduction through limitation of fat consumption to less than 15 percent of total calories.

Unfortunately, there are a number of short-term and long-term flaws in any attempt to significantly reduce dietary fat. In the short-term, fat is an important **satiety signal**, meaning that its presence in the digestive tract helps to communicate a sense of fullness and satisfaction to the brain. This is accomplished through the action of a hormone called *cholecystokinin (CCK)*, which is only released in response to ingested fat. While there are other satiety signals, including the sensation of fullness that comes from the physical stretching of the stomach, these are generally delayed* and weaker than the CCK signal. Fat also slows digestion and absorption, meaning it helps to regulate blood sugar—in this regard, it acts just like fiber. Furthermore, our evolutionary drive to seek out foods with high energy content leads us to derive sensory pleasure from fat, in much the same way that we derive pleasure from sugar. In fact, studies designed to quantify gustatory pleasure in subjects eating ice

***NOTE**

If you've ever felt fine while finishing a meal, only to feel uncomfortably full 10–15 minutes later, you're familiar with the delayed nature of the stomach stretch signal.

cream or whipped cream suggest that as the fat content of a food decreases, its sugar content must increase in order for its consumption to produce a similarly pleasurable experience. Manufacturers must therefore add far more sugar to low-fat versions of foods in order for the foods to taste good to consumers. This, of course, increases the caloric content of these foods. The importance of fat as both a satiety signal and as a provider of gustatory satisfaction means that dieters employing a low-fat strategy often take in more calories—and far more sugar—than they would if they consumed a normal nutritional distribution while reducing total caloric intake. Sugar, of course, can be converted to fat, such that a low-fat/high-sugar diet still leads to fat (and weight) gain if over-consumption occurs. An amusing and poignant maxim on that note is *low-fat in the box doesn't mean low-fat in the body!*

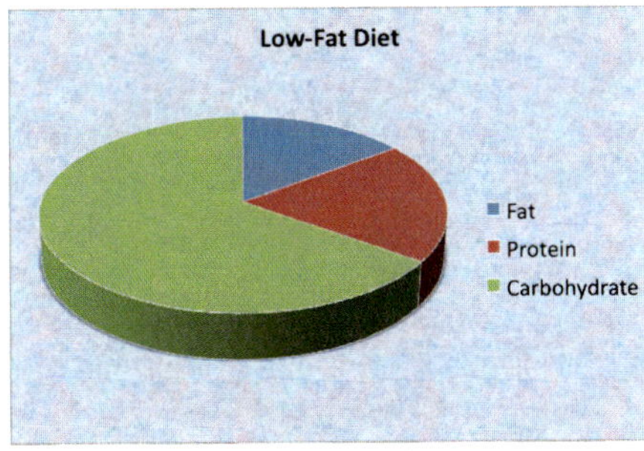

Image 9.5. Low-fat diets are composed largely of carbohydrate. Most nutritionists recommend more protein, a bit more fat, and much less carbohydrate than the typical low-fat diet provides.

CONCEPT CHECK:

Which would be more satisfying, a 500-calorie meal containing some fat or a 500-calorie meal containing no fat? Why?

Answer: The meal containing fat, because it is an important satiety signal.

9.7 STIMULANT DIET PILLS

While FDA-approved pharmaceutical weight loss aids certainly exist (and are becoming more common), here we will focus on over-the-counter diet pills that are advertised to produce rapid results without lifestyle modification. Exceedingly popular for their promise to produce weight loss (and increased energy) without exercise or dietary restrictions, diet pills can be found everywhere from specialty nutrition stores to the local drugstore. They're prescribed by alternative health-care practitioners and advertised on late-night television. A stunning variety of ingredients are featured in these pills, and a complete discussion of the array of chemical strategies they employ (and chemical results they produce) is beyond the scope of this text. However, it's worth focusing on one compound that was nearly ubiquitous in commercial weight loss agents until 2004, when it was banned from dietary supplement use in the United States. **Ephedrine**, a chemical compound isolated from herbs of the *Ephedra* family, is a **sympathomimetic** alkaloid.

Sympathomimetic agents are ones that mimic adrenaline and stimulate the *sympathetic nervous system*, or the branch of the nervous system designed to produce the "fight or flight" response to life-threatening situations. Adrenaline is a major hormone of the sympathetic nervous system, and the symptoms of its release in the body, including pounding heart, heightened response time, and generalized agitation, are also produced by sympathomimetic agents. While ephedrine-containing herbs have been used in Chinese medicine for thousands of years to treat symptoms of asthma and bronchitis,* their adrenaline-like action can also be used to produce weight loss, since activation

***NOTE**

Ephedrine, like adrenaline, stimulates responses that ready the body for action, including dilating the airways. Asthma and bronchitis produce wheezing through irritation and constriction of airways, so the dilating effect of ephedrine helps to alleviate symptoms.

Image 9.6. An Ephedra plant, the source of ephedrine.

of the sympathetic nervous system requires the expenditure of large amounts of energy. As a result, overall energy expenditure increases dramatically—even without increased activity level or exercise—and if caloric consumption remains unchanged while caloric expenditure increases, weight loss occurs.

There is no question that, in sufficient doses, ephedrine produces weight loss. The reason for ephedrine's mimicry of adrenaline is revealed through a cursory glance at their structures; the shared pharmacophore is quite evident (Image 9.7). Disturbingly, however, the increasingly popular street drug methamphetamine also shares this pharmacophore—in fact, like ephedrine, methamphetamine produces rapid and dramatic weight loss. The similarity among these compounds is also evident in the undesirable side effects of ephedrine, which can produce rapid and irregular heartbeat, nausea, confusion and hallucinations, pulmonary edema, and metabolic disturbances, all of which are side effects of methamphetamine use as well.

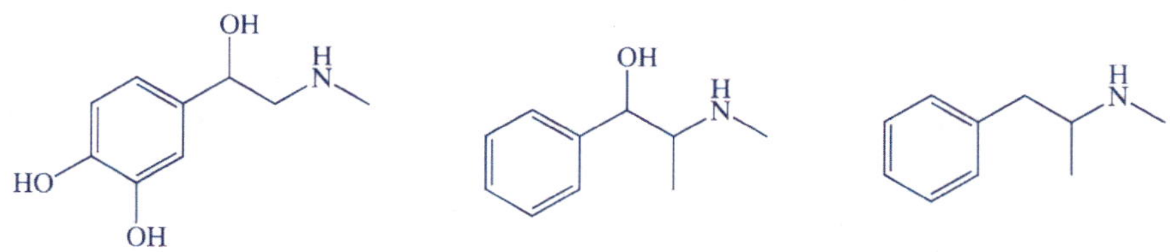

Image 9.7. Adrenaline, also called epinephrine (left), shares much in common structurally with ephedrine (center) and methamphetamine (right).

The side effects of ephedrine can be life threatening at worst, and even at best, adversely affect quality of life. Pill users commonly report feeling nervous, faint, and unable to concentrate. Additionally, like many street drugs, ephedrine use results in *habituation* over time, meaning that it is necessary to take higher and higher doses of the drug in order to produce the same response. Both because of the habituation and because sympathomimetics stimulate release of dopamine, a neurotransmitter that produces feelings of pleasure, addiction is common. Withdrawal from ephedrine can lead to lethargy, depression, and (most disturbing to many dieters) rebound weight gain.

Due to the FDA's determination in the last decade that ephedrine is not a safe dietary supplement, its inclusion in over-the-counter pills has been banned. Still, however, supplement manufacturers utilize the principles of pharmacology to produce new sympathomimetic agents containing the adrenaline pharmacophore. These substances are often added to the banned-compound list shortly after they become available for purchase, but regardless, supplements promising rapid weight loss and increased energy without the need for caloric reduction or exercise generally contain ephedrine-like compounds—and produce ephedrine-like adverse effects.

Concept Check:

Why do the similar shapes of adrenaline, epinephrine, and methamphetamine suggest that they'll have similar effects on the body?

Answer: The shared pharmacophores mean they'll bind to the same receptors. (See Chapter 7 to review this concept.)

9.8 LOW-CARBOHYDRATE DIET

While lower-carbohydrate diets show a great deal of promise in effecting weight loss (particularly if the consumed carbohydrates fall into specific categories—see Sustainability below), extremely low-carbohydrate diets (<20 g carbohydrate per day during early phases, Image 9.8) can have some odd biochemical effects upon the body. Occasionally called **ketogenic diets** in research and literature, these diets call for a drastic reduction in carbohydrate consumption for the duration of the weight loss attempt.

The theory behind ketogenic diets is that strong insulin signals, which result from consumption of sugars and refined carbohydrates, increase body fat storage, as discussed in Chapter 8. Avoiding carbohydrate, then, eliminates the possibility that a strong insulin signal will be produced by a meal. Proponents of these diets suggest that the body can be "trained" to burn fat, and that reducing carbohydrate intake (which reduces glycogen stores) forces cells to rely more heavily upon fat metabolism to meet energy needs. This, in theory, trains the body to utilize fat to a greater extent even upon cessation of the diet. Unfortunately, while both these theories are correct to a point, they fail to fully address the biochemistry of extreme carbohydrate reduction.

Upon beginning a ketogenic diet, individuals are generally pleased with the rapid and dramatic weight loss they experience. A closer look at the chemistry of carbohydrate deprivation, however, reveals that very little of the lost weight is actually fat. Most, in fact, is water. The reason is that carbohydrate is an important energy source for almost all cells in the body. In particular, glucose is important to proper brain function; while most other cells can utilize fatty acids in the absence of carbohydrate, the brain cannot. As carbohydrate deprivation ensues, blood sugar begins to drop. A functioning brain requires upwards of 100 g of glucose daily; with carbohydrate consumption in the region of 20 g per day, glycogen must make up the difference. The hormone glucagon causes the breakdown of glycogen into glucose, which is released into the bloodstream and travels to the brain. Without a constant source of glucose from food, muscles are also forced to break down glycogen stores (they have their own) for energy. Both liver and muscle glycogen become depleted rapidly, and because glycogen is highly polar and co-stores with water in both liver and muscle, the stored water (many pounds of it) is excreted as glycogen breaks down.

Early symptoms of glycogen depletion are common among dieters and include fatigue and an inability to focus their attention. Muscles tend to burn and feel weak with even slight exertion. Exercise is simply not possible in most

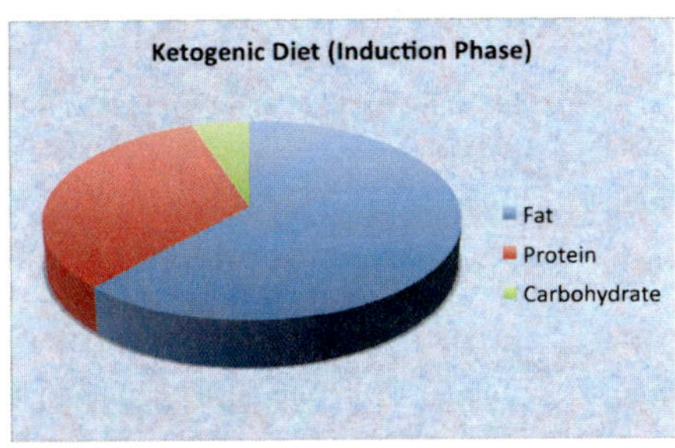

Image 9.8. The caloric distribution in the induction phase of a ketogenic diet is vastly different than that recommended by most nutritionists.

cases. As glycogen stores disappear entirely, the body shifts its chemistry in order to provide an alternate source of nutrition to the starving brain.

CONCEPT CHECK:

Why does glycogen co-store with water, but fat does not?

Answer: Glycogen is polar, but triglycerides (storage fat molecules) are not. See Chapter 8 to review this concept.

In the absence of glucose, there is one other energy source that the brain can utilize. **Ketone bodies**, a group of chemicals including acetone (the compound that gives fingernail polish its familiar smell), are produced through metabolism of fatty acids. Generally, they are mere intermediates in fatty acid breakdown, and do not enter systemic circulation. In the case of extreme glucose deprivation, however, they can be circulated and taken up by brain cells.* While diet-induced **ketosis** has not been shown to be significant enough to be harmful (people generally find this experience so uncomfortable that they moderate their carbohydrate intake before any real damage is done), it nevertheless takes a toll on the body. From a purely aesthetic perspective, ketone bodies in the blood stream dissolve in exhaled air, giving dieters an odd, chemical smell. Additionally, some research suggests that ketosis is interpreted as a starvation signal, causing cells to down-regulate their energy usage and making weight gain upon cessation of the diet much more likely. There is also evidence that depression can occur with a higher prevalence among these dieters—low blood glucose results in decreased serotonin production, which results in depressed mood and decreased feelings of well-being. Regardless, most ketogenic diets begin allowing increased amounts of carbohydrate shortly after ketosis is achieved, meaning that this extremely odd chemistry is short-lived.

***NOTE**

The other circumstances in which circulation of ketone bodies would take place include starvation and uncontrolled diabetes. It's important to note that in general, ketones only circulate when something is terribly wrong.

Research studies on ketogenic diets show mixed results—some suggest these are among the most effective weight-loss strategies available, others find that they are no more effective than simple caloric restriction. Most studies reveal that the strict limits on consumption of carbohydrate-containing foods result in severe nutritional deficiencies. In particular, the avoidance of fruits and whole grains leads to low fiber intake, which can result in severe constipation. Further, while many studies suggest that initial weight loss is greater with ketogenic diets than low-fat or calorie-reduction diets, most diets show similar maintained weight loss over time (six months or longer).

9.9 SUGAR SUBSTITUTES

Nearly all weight-focused diets that involve a modification of consumption agree on one thing—sugar provides empty calories, and should be limited. Unfortunately, the avoidance of sugar is at odds with human instinct, and is contrary to our tendency to seek out pleasurable tastes and eating experiences. For this reason, the **non-nutritive** (meaning they do not provide us with calories) **sweeteners** are hot chemical commodities. The first of these to be developed was *saccharin*, which was discovered serendipitously in 1879 when a chemistry student who'd been in the lab all morning licked his fingers during lunch and noticed they tasted sweet.* While the saccharin molecule is shaped differently from sucrose, it nevertheless has the key atoms of the pharmacophore in the appropriate positions, giving it the ability to bind to sweetness receptors on the tongue.

> ***NOTE**
>
> Lab safety protocol in those days obviously wasn't what it is now, as evidenced by the fact that he traced down the source of the sweetness by returning to the lab and tasting each of the chemicals he'd been working with!

In fact, saccharin binds very tightly to sweetness receptors, making it intensely sweet—about 500 times as sweet as sucrose—such that only a tiny amount need be consumed to provide a similar flavor sensation to that of table sugar. It's often mixed with glucose (which is a bit less sweet than sucrose) to dilute its sweetness so that consumers can measure an appropriate amount into their food or drink. Because the molecule is not one we were designed to consume, we have no enzymes or metabolic pathways capable of digesting it. As such, it has no caloric value. Some researchers have expressed concern that saccharin may be carcinogenic (commercially marketed packages carry a warning to this effect), as the molecule has been shown to cause cancer in lab rats. The rats in question, however, were fed tremendous amounts of saccharin in a short period of time, and no evidence has yet suggested that it shows any sort of carcinogenic activity in people. Despite this, saccharin has lost popularity in recent years as other non-nutritive sweeteners have become available. The newer sweeteners are more stable, don't carry cancer-warning labels (which, despite the lack of evidence, understandably concern consumers), and don't share saccharin's unpleasant, metallic aftertaste.

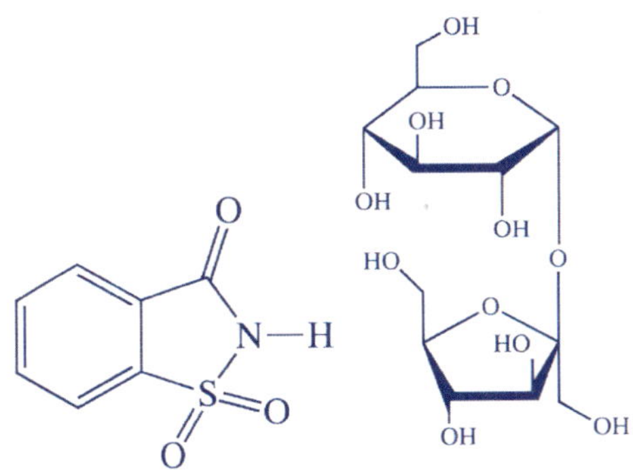
Saccharin Sucrose

Image 9.9. The saccharin molecule compared to sucrose (right).

One such sweetener, *aspartame*, was discovered (again serendipitously) by Jim Schlatter in 1965. Schlatter was working toward the development of a tetrapeptide (four-amino acid) ulcer drug, which he was synthesizing piece by piece. The story goes that he discovered that a modified two-amino acid piece of the drug tasted sweet when he licked his fingers to turn the pages of his lab notebook (again, a case in which poor lab safety led to a lucrative discovery). Like saccharin, aspartame is ostensibly a very different molecule from sucrose. However, again, a few key atoms in a few key locations give aspartame the ability to bind to the sweetness receptor quite tightly—it's 200 times sweeter than sucrose. Researchers have investigated a possible link between aspartame consumption and cancer, and while these studies have indicated a possible

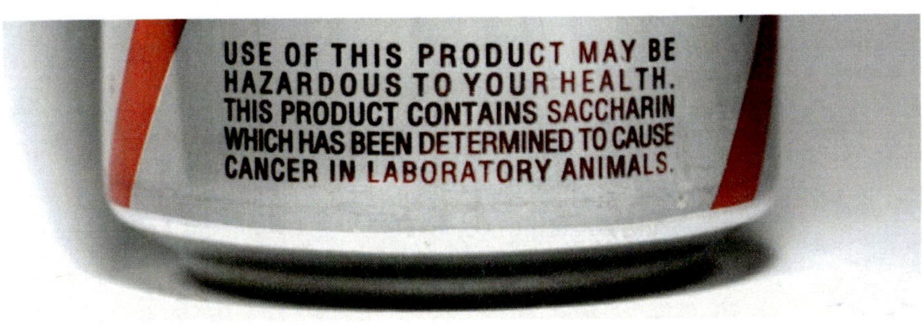

Image 9.10. A can of saccharin-containing soda bearing a warning label.

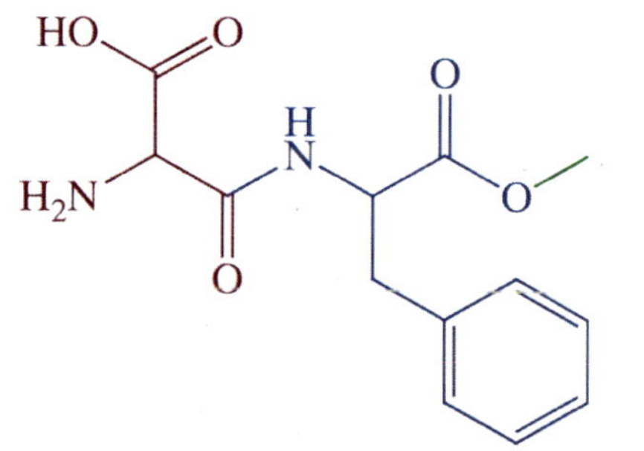

Aspartyl-phenylalanine methyl ester

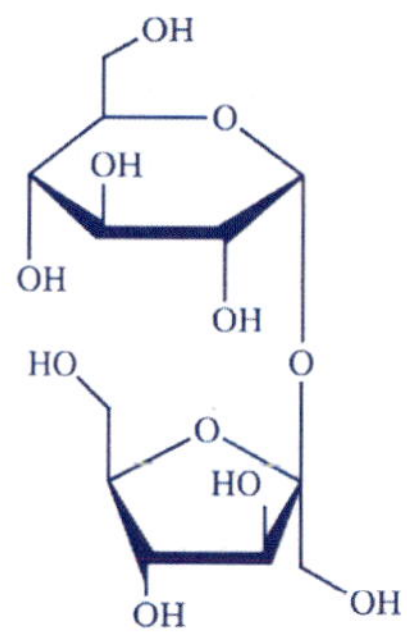

Sucrose

Image 9.11. Aspartame consists of two amino acids (aspartic acid and phenylalanine), with a methoxy (-OCH_3) group forming an ester on phenylalanine instead of the usual OH, for a carboxylic acid. For shape comparison, sucrose is shown on the right.

connection in rats, no data has suggested a causal relationship in humans. Part of the concern comes from the fact that aspartame is not stable over long periods of time, and is particularly unstable if exposed to heat, which is why aspartame-containing diet soft drinks lose their sweetness over time. For this reason, not only the parent molecule, but its breakdown products as well, have been (and continue to be) scrutinized for potential toxicity.

While aspartame is not required to carry a cancer warning, it does have its own warning label, instructing individuals with the genetic disease *phenylketonuria (PKU)* to avoid it. Those with PKU are unable to metabolize phenylalanine, as they lack the liver enzyme responsible for its breakdown. Un-metabolized phenylalanine accumulates in the blood and disrupts brain development, leading to impaired function and mental retardation. To avoid effects of the disease, individuals with PKU must not consume foods rich in phenylalanine, particularly during childhood. Meat, fish, cheese, legumes, and nuts are among the foods that must be avoided—as, of course, are aspartame-containing foods and sweets.

Among the newest of the synthesized sweeteners is *sucralose*, a chlorinated disaccharide. This molecule is much more similar structurally to sucrose than are the other artificial sweeteners (Image 9.12). In fact, the only differences (although they are crucial, as they render the molecule indigestible and therefore non-nutritive) are that the glucose unit is replaced with galactose, and three of the OH groups are replaced with chlorine atoms.

The differences between sucralose and sucrose also render the former 600 times sweeter than table sugar. Unlike aspartame, sucralose has the benefit of being heat and pH stable, meaning that it can be used in baked goods and has a longer shelf life. Sucralose has been evaluated for toxicity and carcinogenic potential by several studies, none of which has found evidence of a risk to human health in reasonable doses. As new as this sweetener is, however, it may take some time before long-term effects of routine consumption (should there be any) begin to reveal themselves.

Image 9.12. Sucrose and sucralose, with arrows to indicate differences between the molecules.

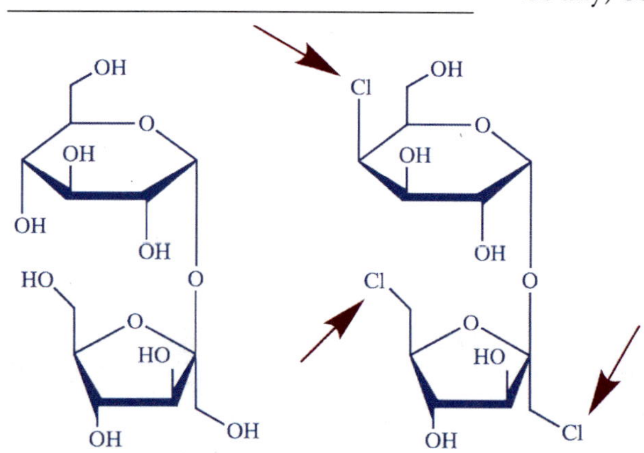

Sucrose Sucralose

There has been some research evidence that artificial sweeteners may cause a paradoxical weight gain by increasing the desire for sweet foods, thus increasing the consumption of sugar-sweetened items. Whether this is true, only time will tell, though some studies suggest that increased sugar consumption increases desire to consume sugar, which suggests that there may be some legitimacy to the claim regarding artificial sweeteners. Regardless of the general absence of research demonstrating detrimental effects of non-nutritive sweeteners, nutritional experts are split in their opinions on this topic. Some believe that a desire to consume sweet food is inevitable, and that replacing caloric with noncaloric sweeteners is beneficial. Others believe that reduced sugar intake reduces desire to consume sweet food, and suggest that sweetened foods should be largely avoided.

From a chemical perspective, it is reasonable to posit that man-made molecules to which we are not evolutionarily adapted can have strange and unanticipated effects on the body; trans fats provide evidence of this. A cost-to-benefit ratio, therefore, might be of use to an individual in determining whether or not to use these chemicals in food.

CONCEPT CHECK:

Why do some of the non-nutritive sweeteners taste sweeter than regular sugar?

Answer: They bind more tightly to sweetness receptors on the tongue.

9.10 SUSTAINABILITY

While only a few weight-loss diets and strategies are addressed in this chapter, there are hundreds of them being marketed. Most, however, share a fatal flaw—they focus upon weight rather than health. Another way of saying that is that they attempt (by design) to produce rapid weight loss through unsustainable eating and/or lifestyle practices. Scores of research studies show that, diet strategy notwithstanding, the single greatest predictor of weight-loss success is duration of a diet: The longer it can be maintained, the more successful it will be. Since most diet strategies are too drastic and uncomfortable to be maintained for long, they tend to be abandoned as soon as the desired weight loss (or some of it, anyway) is achieved. Most plans do not provide any instruction whatsoever regarding appropriate eating practices for weight maintenance, and for this reason, weight rebound after cessation of a diet plan is a common phenomenon. Of course, there are times when weight becomes an emergency, and a physician might recommend a drastic (temporary) eating plan or interventional procedure to achieve a rapid reduction. For most individuals, however, the best road to a healthy weight (and to maintenance of that weight) is a sustainable eating pattern—a balanced *diet*, in the true sense of the word.

Most research suggests (and many nutritionists agree) that a well-balanced diet consists of about 45–65 percent carbohydrate, 10–35 percent protein, and 20–35 percent fat, with plenty of fruits and vegetables. Macronutrient percentages in these ranges provide enough carbohydrate for glycogen production and brain function, sufficient protein to provide for metabolic needs while avoiding unnecessary hardship to the kidneys (the organs responsible for protein waste management),* and enough fat to meet needs and serve as a satiety signal without overindulgence. There is, of course, quite a bit of variability in those guidelines—different people have slightly different energy needs and personal biochemistry, and what works well for one may not be identical to what works well for another. Regardless, a profile of a recommended diet broken down by macronutrient (Image 9.13) looks quite different from any of the weight-loss diets discussed earlier.

***NOTE**

Metabolic waste (urea) from protein degradation is filtered into the urine by the kidneys. Effective elimination of urea is extremely water dependent, meaning that high protein consumption can easily result in dehydration. Since kidneys function suboptimally in low-water conditions, overconsumption of protein can stress these delicate organs.

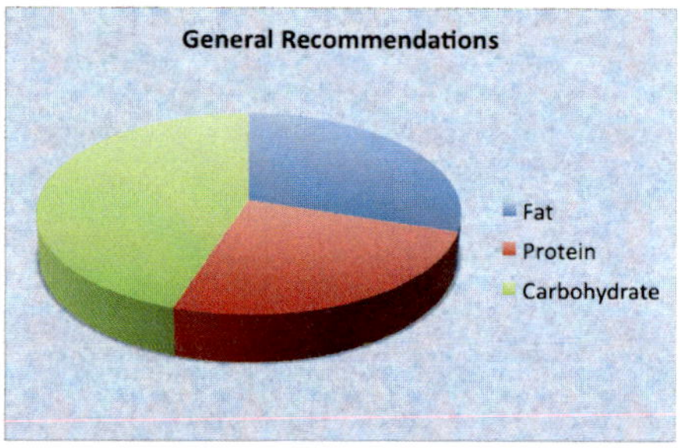

Image 9.13. Most nutritionists agree that a healthy, balanced diet breaks down to around 45–65 percent carbohydrate, 10–35 percent protein, and 20–35 percent fat, depending somewhat upon the individual.

Another common thread among nutritionists is the agreement that not all fats and not all carbohydrates are equally healthy. While 20–35 percent of calories consumed should be from fat, the vast majority of this should be from unsaturated fat. Some even go so far as to suggest that certain sources of unsaturated fat (olive oil, for instance, and cold-water fish) are superior to others. The theory here is that some oils work toward shifting the omega-3/omega-6 ratio into a healthier direction—toward a bit more omega-3 and a little less omega-6 than most Western diets provide. Other nutritionists, however, feel that simply avoiding saturated and trans fats improves a diet so significantly over the traditional American fare that one need not worry about the source of unsaturated fat, as long as it comprises the majority of fat in the diet. In many ways, it comes down to the individual—some people have a deep dedication to crafting the healthiest diet they can and are happy to keep track of sources and types of oils in order to achieve this goal. Others find it challenging (and time consuming) enough just to avoid fast food, and simply aren't willing or able to microanalyze their nutrients.

With regard to carbohydrates, again, most nutritional experts agree that some are superior to others. In Chapter 8, we discussed the important role of fiber in regulating the digestive system, slowing digestion, and decreasing the rate at which blood sugar increases in response to a meal. The latter, critical to optimizing fuel utilization and preventing unnecessary fat storage, is actually quantified through what is called a food's **glycemic index (GI)**. GI measures the rate at which a carbohydrate food enters the blood stream as sugar and compares this with the rate for glucose, which (as the standard) is considered to have a GI of 100. To determine a food's GI, human subjects consume a standard-size sample of food (all foods being tested must contain the same net mass of carbohydrate). Over a period of two hours, their blood sugar is tested repeatedly; at the end of that time, the change in blood sugar is compared to the change that would have resulted from an equal-mass sample of glucose. If the food increases blood sugar at half the rate observed for glucose, the food is assigned a GI of 50. A GI of 25 would correspond to an increase in blood sugar at one-quarter the rate observed for glucose, and so forth. Foods with GI values greater than 100 increase blood sugar at a rate greater than that of pure glucose. Some common foods and their GI values are listed in Table 9.2. Nutritionists suggest that lower GI foods are higher in quality than higher GI foods, although as we'll soon see, there are caveats to this.

Note that while the list of foods in the table is not exhaustive, some trends are nevertheless apparent. Refined versions of foods (white rice, white bagel) have significantly higher GI values than their unrefined counterparts (brown rice, multi-grain bread). This is due to the role of fiber in slowing the rate at which sugar enters the bloodstream. In addition, fat or protein in food slows absorption of sugar—cheese pizza, for instance, has a lower GI than plain crust (white bread) would have. Of course, one must weigh the positive effect of the lower GI value against the negative effect of the saturated fat in cheese when determining whether the pizza is a healthy food choice. Some so-called "health" foods, like tofu-based ice cream substitute, have surprisingly high GI

Low GI (up to 55)		Moderate GI (56–69)		High GI (70 and higher)	
Plain yogurt	14	Orange juice	57	White bagel	72
Cherries	22	White flour pasta	58	Watermelon	72
Kidney beans	27	Cheese pizza	60	Jelly beans	80
Bran cereal	30	Oatmeal	61	White rice	80
Multigrain bread	31	Canned apricots	64	Corn flake cereal	92
Orange	33	Raisins	64	French baguette	95
Ice cream (full fat)	36	Whole-grain cracker	67	Baked potato	111
Brown rice	55	Croissant	67	Tofu frozen dessert	115

Table 9.2. GI values for some common foods.

scores. While traditional ice cream contains saturated fat, it also has a much lower glycemic index than the tofu version.

All of this information (which sometimes seems contradictory) can get confusing: How does a health-conscious consumer make appropriate food choices? In the end, it helps to remember that as complex as human bodies (and their nutritional requirements) seem, they're really nothing more than large beakers in which a number of different reactions are running. Consumption of food is a mechanism for supplying the reagents for these reactions. Given that, we can look at nutrition as a chemical equation that needs to be balanced. The reagents are high-quality carbohydrate, protein and high-quality fat. While a low GI may indicate that a food contains high-quality carbohydrate, if the GI is low because the food contains lots of saturated fat, it's probably not a good reagent! If a food has a low GI and contains unsaturated fat, it's probably a great reagent. Of course, psychology and the desire to enjoy the food we eat also plays a role in all of this—the healthiest diet in the world becomes problematic if it consists of foods that an individual doesn't find satisfying. However, as omnivores, we have much more variety at our disposal than an animal without the ability to choose from a variety of foods, cultivate foods, and seek out those items it prefers most. With a little ingenuity, label reading, and consideration of the chemistry of nutrition, it becomes quite feasible to put together a healthy and tasty diet. It's noteworthy, too, that nutritional experts agree on one more thing—even the best and healthiest diet still has room for the occasional treat, caloric splurge, or nutritionally devoid (but very tasty) dessert!

Concept Check:

If a food has a low GI, does it necessarily contain high-quality carbohydrate?

Answer: No—it may contain low-quality carbohydrate combined with fat or protein.

9.11 THE OTHER HALF OF THE EQUATION—CALORIC EXPENDITURE

While a complete discussion of the biochemical pathways that require caloric input is beyond the scope of this text, it's worthwhile to spend a little time on the other half of the caloric balance equation: Where do the calories go? The vast majority of our caloric expenditure goes toward maintaining cellular and organ function.* The number of calories an individual burns each day on maintenance function alone is called **resting metabolic rate (RMR)**, and varies with age, gender, body composition, and many other factors. While there are several equations that can be used to approximate RMR, most do not take body composition into account, making them less accurate for individuals with higher- or lower-than-normal body fat percentages. Because muscle burns more calories than fat even at rest, two individuals of the same weight and gender will have different RMRs if one is leaner than the other. Assuming that body habitus is relatively normal, however, the Harris-Benedict equation (which has been used for some time and shows very reasonable accuracy) for calculation of RMR is as follows:

***NOTE**

That is, for most people under normal circumstances. A high-caliber athlete or an individual with an extremely physical job (like a fireman or logger) burns far more calories through activity than required to maintain basic functions.

$$\begin{aligned} &Women: \ RMR = 655 + (4.35 \times weight \ \ in \ pounds) \\ &\qquad + (4.7 \times height \ \ in \ inches) - (4.7 \times age \ \ in \ years) \\ &Men: \ RMR = 66 + (6.23 \times weight \ \ in \ pounds) \\ &\qquad + (12.7 \times height \ \ in \ inches) - (6.8 \times age \ \ in \ years) \end{aligned}$$

TRY THIS:

Calculate the RMR for a 35-year-old, 165-lb. man, 6′ 3″ tall.

Answer: 1808.45 Cal/day.

Solution:

RMR = 66 + (6.23 x weight in pounds) + (12.7 x height in inches) – (6.8 x age in years)

RMR = 66 + (6.23 x 165 pounds) + (12.7 x 75 inches) – (6.8 x 35 years)

RMR = 1808.45

Note that RMR does not take into account any sort of physical activity (including exercise and normal daily activity). In individuals of normal activity level, RMR accounts for about 70 percent of calories burned each day.

A second (albeit very minor) caloric expenditure that is independent of activity is the digestion of food. Called the **thermic effect of food**, this phenomenon arises from the fact that it takes a certain amount of energy to digest and process the nutrients we consume. On average, this accounts for a loss of about 5–10 percent of total consumed calories, with fat having a very low thermic effect and protein having the highest effect of the three macronutrients.

Physical activity energy expenditure (PAEE) is the term for calories burned above and beyond RMR due to activity. This includes both routine daily activity (cooking, working, walking to and from the car) and exercise. In normally active individuals, PAEE will account for about 20 percent of total calories burned. Of course, highly active individuals or habitual exercisers can burn significantly more than this.

9.12 EXERCISE

Despite minor squabbles in the arenas of optimal diet, vitamin supplementation, and so forth, government organizations, physicians, and health experts unite on the subject of exercise—Americans (and members of many industrialized Western societies) simply don't get enough! Surveys from the late 1990s and early 2000s show that only about a quarter of U.S. adults engage in vigorous physical activity (not associated with work) lasting 10 minutes or more at least three times a week. About 60 percent of U.S. adults do no physical activity at all, apart from whatever is required by their jobs. Compared to the NIH recommendation of at least 30 minutes of exercise or vigorous physical activity, five days per week, it's clear that most adults aren't getting even close to the amount they need.

Exercise is important to maintaining health for several reasons, not the least of which being that it burns calories, helping to maintain energy balance (or in the case of those needing to lose weight, to shift the balance toward a caloric deficit). Many of us are familiar with the concept that calories are burned during exercise—the more vigorous the activity, the greater the caloric burn. However, there are actually many mechanisms through which a caloric expenditure is accomplished in association with exercise; the so-called **direct effect**, meaning calories burned during the activity itself, is only one of these. That actually comes as good news to most people, since the number of calories burned during activity can appear shockingly paltry. For instance, a 170-lb. man running 11.5 minute miles (5.2 mph) burns approximately 675 Cal/hr, meaning that to burn off the calories in a single meal consisting of a fast-food quarter-pound burger, large fries, and large soda, he'd have to run for two solid hours!* For this same man to lose a pound of body fat, which requires creating a deficit of 3500 Cal, he'd have to run for five hours. Values for the direct effects of other exercise activities are listed in Table 9.3. Thankfully, the direct effect is only a small part of the overall contribution of exercise toward health and negative caloric balance.

A second, very significant effect of exercise is that it temporarily increases resting metabolic rate. This **elevated post-exercise metabolic rate** can last for quite some time—research suggests that RMR can be up to 10 percent higher than usual the morning following exercise. If, for instance, a man with an 1800 Cal/day RMR lifts weights for half an hour, he burns 180 Cal. If his RMR increases by 10 percent for a period of 24 hours, he burns another 180 Cal—without any further exercise. It's worth noting that while this effect is approximately equivalent among different exercise activities several

***NOTE**

Actually, most fitness machines claiming to tally caloric expenditure are quite inaccurate. Some allow the user to input weight, but for those that don't, caloric estimates are inaccurate purely on the basis that caloric expenditure is directly proportional to weight. Furthermore, most machines don't allow the user to indicate gender, and a man burns more calories during exercise than a woman of the same weight. This can result in a 30 percent (or more!) systematic error in the fitness machine's direct effect calculations; generally, the error is on the side of overestimation, rather than underestimation.

Activity	Cal/hr (M)	Cal/hr (F)	Activity	Cal/hr (M)	Cal/hr (F)
Aerobics	630	480	Skiing (Nordic)	636	480
Cycling (10 mph)	444	336	Swimming (fast)	720	522
Racquetball	642	456	Swimming (slow)	288	210
Running (7.5 mph)	900	678	Weights (moderate)	360	264
Running (5.2 mph)	600	456	Walking (3.5 mph)	354	270

Table 9.3. Approximate Cal/hr burned due to direct effect of various activities for an average (170-lb.) male and average (123-lb.) female. Values will be higher for individuals of greater weight and lower for individuals who weigh less or are highly conditioned.

hours post-exercise, it's particularly emphasized in the first few hours following weight lifting (presumably because of the significant stress placed upon muscles by strength training). For this reason, many fitness experts encourage individuals attempting to lose weight to incorporate not only aerobic exercise (owing to its cardiovascular benefits—see below), but also weight training into their regimen. A final (and the most significant) effect of exercise with regard to caloric balance is that it increases muscle mass. Since muscle is highly metabolically active, increased muscle increases RMR—not just post-exercise, but all the time. This is yet another reason that fitness experts encourage weight training, since it tends to increase muscle mass very efficiently.*

***NOTE**

Women sometimes resist the idea of weight training on the grounds that it will make them look big or bulky—in fact, the opposite is true. Because they don't produce significant levels of testosterone, women simply aren't capable of developing the bulky, bulging muscles that male body builders cultivate. Instead, women who strength-train tend to look lean and toned, all while enjoying the benefits of increased muscle mass and a faster metabolism.

CONCEPT CHECK:

What does "elevated post-exercise metabolic rate" mean, and about how long does it last?

Answer: The term means that RMR is higher (for up to a day) following exercise than it would have been otherwise.

Apart from its contribution to negative caloric balance, exercise positively impacts cardiovascular and general health. Like any muscle, the heart needs to be exercised. **Aerobic exercise** has been found to be the most effective means of maintaining good cardiovascular health, for a host of reasons. Aerobic exercise is defined from a metabolic perspective. The lower end of the **aerobic zone** is the heart rate at which the body begins working hard enough that metabolic waste products start to build up in the blood. It's quite difficult to determine this heart rate without metabolic testing, but for most people, it occurs at about 65 percent of **maximum heart rate (HR_{max})**, which is the highest heart rate an individual can sustain and is generally related to age:

$$HR_{max} = 220 - age$$

The upper end of the aerobic zone is again defined metabolically—it's the point at which the body starts working so hard that waste products in the blood increase dramatically—but again, it's difficult to determine precisely without testing. For most people, this occurs at 85 percent of HR_{max}, meaning that 75 percent of HR_{max} is right in the middle of the aerobic zone for most individuals.

TRY THIS:

Calculate HR_{max} and the limits of the aerobic zone for a 25-year-old male.

Answer: HR_{max} = 195 BPM (beats per minute), aerobic zone = 127–166 BPM.

The benefits of aerobic exercise are many. First, forcing the muscles to work harder increases their demand for energy and oxygen. This requires faster circulation, which in turn causes the heart to beat faster and harder, exercising it. Cardiovascular exercise helps reduce risk of stroke, heart attack, atherosclerosis, and other cardiovascular diseases. Also of benefit to the heart, exercise increases HDL production, which favorably affects blood cholesterol. Further, to saturate the blood with oxygen, breathing becomes deeper and faster. *Capillaries* (tiny blood vessels) in the lungs open up to allow blood to flow in more efficiently. Vessels in the muscles also dilate, allowing more efficient blood flow. Enhanced blood flow to the muscles means waste products are flushed away to a greater extent than usual.* Endorphins, which are natural painkillers, are released into the bloodstream. This reduces any sensation of pain, and in conditioned individuals, can result in a feeling of well-being (the so-called "runner's high"). The immune system is activated, accounting for the fact that regular exercisers get sick less frequently than sedentary individuals. There's even evidence that exercise is good for maintaining a sharp mind, a finding of importance to aging adults. A final, and significant, benefit of aerobic exercise lies in the fuel source utilized. It turns out that either of the two major stored fuels—fat and glycogen (which is broken into glucose)—can be burned for energy in the presence of oxygen as per the equations below (the fat equation is not balanced since different fatty acids require oxygen and produce products on different ratios):

***NOTE**

You've noticed this effect if you've ever observed that a light workout the next day can help ease muscle soreness from exercise. The soreness comes in part from built-up metabolic waste products, and these are flushed from the muscles during exercise.

$$fat + x\,O_2 \rightarrow y\,CO_2 + z\,H_2O + energy$$
$$C_6H_{12}O_6 + 6\,O_2 \rightarrow 6\,CO_2 + 6\,H_2O + energy$$

Because the body stores so much more energy in the form of fat than it does in the form of carbohydrate, it preferentially relies on fat metabolism to fuel energy expenditure. The word *aerobic* means "with air," and a second metabolic definition of the aerobic zone is that it is of an intensity high enough to require the body to work hard, but low enough to be able to provide for the muscles' oxygen needs. Of course, when we engage in aerobic exercise, we breathe harder than normal (because the muscles are working harder and need more oxygen to burn fuel), but we're not truly short of breath—it's possible to get enough air. During aerobic exercise, then, the majority of calories burned are fat calories—the body burns exactly the source of fuel most people would like to have a little less of. While many people who are new to exercise feel out of breath and uncomfortable even during moderate efforts—mostly because the feeling of exertion is new to them—a relatively fit individual can tell when they're in their aerobic zone even without checking their heart rate because their breathing will be faster and deeper than usual, but not labored. In fact, a great "quick and dirty" aerobic zone check is being able to talk, but not sing.*

***NOTE**

Singing requires prolonged exhalations, which aren't possible at aerobic intensities—if you can sing, you're below your aerobic zone and aren't reaping the benefits. Being unable to talk, or able to talk only in short bursts, means that the body isn't getting enough oxygen. By definition, this means you're outside your aerobic zone.

As we have all experienced from time to time, there is a level of exertion above the aerobic zone. This is referred to as the **anaerobic zone**, and occurs when muscles are forced to work so hard that the body can't supply sufficient oxygen (*anaerobic* means "without air"). Despite our best efforts to breathe fast and the heart's efforts to pump blood fast and hard, it simply isn't possible to get enough oxygen to the muscles to allow for aerobic fat and carbohydrate burn during very intense efforts. As a result, the muscle cells must switch over to an anaerobic metabolic strategy—they must burn fuel without oxygen. Only glucose can be burned in this way, meaning fat is not utilized as an energy source for very intense efforts. Furthermore, anaerobic carbohydrate burn is not nearly as efficient as aerobic metabolism—the energy yield is approximately 15 times less under anaerobic conditions. This means lots of glucose is burned very quickly to provide energy, and anaerobic efforts rapidly result in depletion of fuel:

$$C_6H_{12}O_6 \rightarrow 2 \text{ lactic acid} + \text{small amount of energy}$$

Note also the production of **lactic acid** during anaerobic metabolism. This is a metabolic waste product that can build up in muscles and in the blood. It is partially responsible for the burning sensation in muscles during very hard efforts, as well as some of the soreness that follows. Generally, the anaerobic zone is identified as any effort about 85 percent of HR_{max}, but an individual accustomed to exercise can tell by feel when they enter this zone, as breathing becomes much more labored and talking becomes difficult (or impossible).

TRY THIS:

If a 20-year-old woman is exercising with a heart rate of 175 BPM, is she likely in an aerobic or anaerobic zone? Can she likely talk? Sing?

Answer: She is anaerobic. She likely can't talk or sing.

Solution:

HR_{max} = 220 – age = 200 BPM

$$\frac{175}{200} \times 100 = 87.5\%; > 85\% = \textit{anaerobic}$$

While anaerobic exercise is incredibly effective at burning large numbers of calories in small periods of time, it's not often emphasized by fitness experts as much as aerobic exercise. There are several reasons for this. First, it is incredibly uncomfortable. While it provides many of the same fitness benefits as aerobic exercise, the discomfort of anaerobic exercise can make it unattractive. Second, it is much harder on the heart and body than aerobic exercise, making it inappropriate for many novice or out-of-shape individuals. While some "experts" claim that aerobic exercise does a better job of training the body to burn fat (since anaerobic metabolism relies purely on glycogen), research shows that the tremendous caloric deficit created by anaerobic work leads to high levels of fat metabolism post-exercise, so this is not a legitimate deterrent. Regardless, the best exercise program is the one a person is able to stick with, and for that reason alone, aerobic exercise is far more popular: after a brief acclimation phase, most individuals find it to be not only tolerable, but actually enjoyable.

CONCEPT CHECK:

Which fuels can be burned to provide energy for aerobic exercise?

Answer: Both glucose and fat—fat is burned preferentially.

9.13 MUSCLE TYPES

As part of our discussion of exercise and muscles, let's take a moment to ponder why conditioned athletes in some sports tend to look very sinewy, while in other sports they appear much bulkier. Even within the sport of running, differences can be striking. Marathon runners tend to have very slender, long muscles, while sprinters almost look like bodybuilders. The reason for this lies in the different muscle types that these athletes have cultivated.

Muscle can be divided into different categories based on the type of metabolism upon which it primarily relies. **Type I muscle fibers**, also called **slow twitch fibers**, are largely aerobic in their metabolic strategy. As a result, they are found in muscle groups that are responsible for endurance efforts. All of us, for instance, have lots of type I fibers in the long muscles that run down either side of the spine and maintain our upright posture. Because we sit and stand for long periods of time every day, these muscles are constantly working against gravity. They often don't have to work very hard, but they do have to maintain a moderate effort for long periods of time. It's a bit difficult to tell (especially since trying to do so risks injury), but they're also not as powerful as other similarly sized muscles—their capacity for strength efforts is much less than their capacity for endurance. On the other hand, **type II muscle fibers**, also called **fast twitch fibers**, rely primarily upon anaerobic metabolism. They are capable of extremely fast (and extremely powerful) movements, but are unable to sustain those efforts for long. We have type II muscle fibers in our hands and eyelids: they allow us to make very quick and precise motions. We also find them elsewhere in the body, where they function in explosive movements (like jumping) and power efforts (like lifting a very heavy box). At this point, there is no conclusive research to indicate whether muscle fiber types are purely genetic or can be trained. Accomplished marathon runners have a very high proportion of type I muscle, though whether this is because their muscle fiber distribution gave them a predisposition to endurance sport greatness, or

Image 9.14. Despite their similarly low body fat percentages, a marathon runner (left) looks very thinly built compared to the bulk of a sprinter (right).

because they trained in endurance efforts resulting in a proliferation of type I fibers, is not known.

Regardless of when or how they developed, the fact remains that muscle analysis of accomplished endurance athletes shows enrichment in type I fibers over the average individual, and analysis of accomplished power athletes shows enrichment in type II fibers. It is the difference in the predominant fiber type that leads to the observable difference in *morphology* (shape) between these athletes. Because type II fibers are primarily involved in anaerobic efforts,* they store large amounts of glycogen. Since glycogen is highly polar and co-stores with water, the stored fuel takes up quite a bit of space. This plumps up muscle groups containing high proportions of type II fiber, and gives them the bulky look typical of a bodybuilder or sprinter. Type I fibers, however, are primarily aerobic and, therefore, are designed to allow for maximum availability of their preferred fuel (fat), and the oxygen needed to burn that fat. All muscle cells store small amounts of fat for use during aerobic efforts, but type I fibers store a particularly large amount since they specialize in fat metabolism. Since fat is hydrophobic and doesn't co-store with water, the stored fuel takes up very little space in the muscle. This results in muscle groups composed of predominantly type I fibers having a long, lean look to them, as observed in endurance athletes. Additionally, type I fibers contain large quantities of the chemical **myoglobin** (Image 9.15).

***NOTE**

It's worthwhile to note that just because a muscle specializes in anaerobic respiration doesn't mean it doesn't need any oxygen—all cells use oxygen to maintain normal function. Instead, the working muscle doesn't require supplemental oxygen for fuel combustion above and beyond its ordinary requirements (which are provided for by oxygen in the bloodstream).

Like its close chemical relative hemoglobin, myoglobin has a reddish color and binds oxygen. Unlike hemoglobin, however, it doesn't circulate through the body in red blood cells. Instead, it is localized to muscles—primarily type I fibers. The job of myoglobin is to take oxygen from circulating hemoglobin and hold that oxygen until the muscle needs it. Essentially, myoglobin provides a muscle with its own personal oxygen supply, allowing aerobic metabolism to proceed even during efforts hard enough that breathing alone can't supply all the oxygen a muscle requires.

CONCEPT CHECK:

Why don't type II fibers contain myoglobin?

Answer: Type II fibers specialize in anaerobic efforts, which burn fuel in the absence of oxygen.

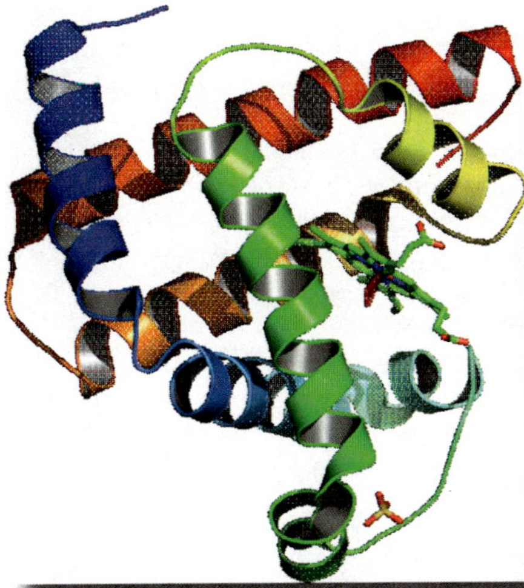

Image 9.15. Like hemoglobin, myoglobin is a protein (shown as a curled ribbon). Also like hemoglobin, it contains a heme group (the molecule in green and red in the center of the protein). The heme is responsible for binding to oxygen.

In addition to producing the distinct physical appearance of athletes who are enriched in one type of muscle fiber, the differences between fiber types also affects the appearance and nutritional value of muscle that we consume for food. For example, many of us know that chicken breast is lower in fat and calories than a similar quantity of chicken drumstick or thigh. The difference lies in the fat content of the muscle—a given mass of chicken breast will be higher in protein and lower in fat than the same mass of chicken thigh (Table 9.4).

	Chicken Breast	Chicken Thigh
Protein (g)	43	36
Fat (g)	5	15
Total Calories	231	293

Table 9.4. Nutritional information for 140 g serving of chicken meat.

Based upon an understanding of muscle fibers, we might conjecture that chicken breast contains a higher proportion of type II fibers, while chicken thigh contains more type I fibers. In fact, that is exactly the case. Further, the dark color of chicken leg meat ("dark meat") comes from the reddish-brown myoglobin in the type I muscle fibers. With no need for myoglobin, the type II fibers that predominate the breast appear pale ("white meat"). All of this makes even more sense when we stop to think about the daily routine of a chicken. As ground birds, chickens don't do much flying. Instead, they spend most of their time walking around, pecking for food. Their legs don't need to produce extraordinary bursts of speed (chickens rarely run), but they do need to support the chicken as it goes about its business. The thighs and drumsticks, therefore, are specialized for endurance efforts, and are composed primarily of type I fibers. When a chicken is startled, it flies a short distance. These flights are intense efforts of limited duration, and so the flight muscles—the pectorals, or what we call chicken "breast"—are specialized for power, and consist mostly of type II fibers.

Ducks, on the other hand, are migratory birds capable of flying thousands of miles. They would not do well with a chicken's type II fiber pectorals. Instead, ducks have primarily type I fibers comprising their muscles of flight. From a nutritional perspective, one might conjecture that duck breast is dark meat—and probably fatty—which, in fact, it is.

CONCEPT CHECK:

Why is "dark meat" higher in fat than "white meat"?

Answer: It is made up of primarily type II fibers, which store fat for energy.

Image 9.16. Chickens and ducks have different habits, thus the composition of their muscles (particularly their flight muscles) is very different. This affects not only their abilities, but also the nutritional profile of their meat.

9.14 THE LAST WORD—VITAMIN AND MINERAL SUPPLEMENTS

Nutrition is a heavily researched field. Not only is it of interest to researchers across many areas and specializations of science, its study is also funded by countless special interest groups and food and supplement manufacturers for product promotion and marketing purposes. While mainstream experts may express minor differences in recommendations regarding quantity and quality of the macronutrients, nowhere in nutrition are opinions as varied and disparate as they are with regard to supplementation. On one hand, it's no secret (nor is there any debate) that the typical Western diet is low in vitamins and many minerals; for this reason, some experts suggest that a daily multivitamin is simply good insurance for health maintenance. On the other hand, however, some experts feel that supplementation encourages a "band-aid" treatment of a poor diet, providing the veneer of good health without encouraging individuals to change their eating habits in order to provide for all their macro- and micronutrient needs. Complicating all of this, there is little conclusive evidence to support long-term supplementation. For each study that shows a positive effect, there's another study that shows no effect whatsoever. A cursory scan of the recent supplement literature turns up the following batch of contradictory results:

- Higher folic acid consumption decreases risk of breast cancer (Rohan *et al.*, 2000).
- A supplement including folic acid, vitamin B_6, and vitamin B_{12} does not affect breast cancer risk (Zhang *et al.*, 2008).
- Vitamin E supplementation shows a 30–40 percent reduction in risk of heart disease in women (Stampfer *et al.*, 1993).
- Vitamin E supplementation shows no reduction in risk of heart disease in women (Lee *et al.*, 2005).
- Vitamin C supplementation shows a reduction in heart disease risk (Knekt *et al.*, 1994).
- Vitamin C supplementation shows no reduction in risk of heart disease (Sesso *et al.*, 2008).

- Selenium supplementation helps prevent prostate cancer in men (Duffield-Lillico *et al.*, 2003).
- Selenium supplementation has no effect upon prostate cancer incidence in men (Lippman *et al.*, 2009).

Further muddying the waters are recent studies demonstrating that supplementation with synthetic vitamins may actually increase mortality. The mechanism for this is as yet unclear, but some researchers suspect that the proportion of vitamins relative to one another may be more important than the quantity, such that artificial supplementation (which often delivers vitamins in proportions not normally encountered in nature) could produce unintended chemical signals, and therefore unanticipated chemical consequences. While this theory has yet to be proven in the case of supplemental micronutrients, it certainly holds true in other areas of physiological chemistry—trans fats, for instance, while similar to saturated fats in their physical properties, have odd and unanticipated chemical properties in the body that probably stem from the fact that we didn't evolve the ability to process them normally. Where does this leave us with regard to supplementation? From a chemical perspective, empirical evidence suggests that we don't fully understand all the reactions taking place in the body: some have layers of complexity that we haven't yet uncovered. As a result, trying to adjust the extent to which reactions happen by supplementing (or withholding) a reagent can have unanticipated, and potentially severe, consequences. Ideally, of course, we'd all get the nutrients we need from the sources we evolved with. Ideally, too, we'd still live the physically active lives our evolutionary ancestors did—though hopefully with the improvements of medicine and technology afforded by our modern existence. Obviously, such an idyllic marriage of evolutionary roots and the modern world is often simply not possible. Our jobs force us into sedentary lifestyles, we have little time to cook wholesome meals from scratch, and a combination of psychology and food science has provided food manufacturers with all the information they need to tempt us into purchasing their products. The vitamin quandary—to supplement or not to supplement—is just one in a long line of questions for which there are no real answers, at least not yet. Perhaps the best compromise in the meantime can be achieved through an understanding of personal chemistry—we're really just big beakers. Where we lack reagents, we can't produce products. If there is a product we need, or a reaction we need to run, we must supply the reagents (and if there is a product we want to avoid, we must avoid supplying them). We bear in mind that the best sources of reagents are those that occur in nature, and from there, we do the best we can.

SUMMARY OF MAIN POINTS

- The Western Paradox is the idea that Western societies are increasingly consuming an atrocious diet while putting a premium on physical perfection.
- An appropriate amount of body fat (neither too much nor too little) is necessary for normal function.
- Excess body fat leads to disease—overweight and obese are two sub-categories of excess body fat, and are associated with apnea, arthritis, cardiovascular disease, and type 2 diabetes.
- Essential body fat is the minimum required to sustain normal function.
- Body fat percentage is monitored by health care professionals to ensure an appropriate amount of body fat—circumference techniques, skin-fold techniques, and BMI are commonly utilized metrics.
- The USDA and FDA are responsible for providing nutritional guidelines for healthy eating.
- These organizations require most foods to carry Nutrition Facts labels to help consumers make informed choices.
- Trans fat in quantities less than 0.45 g/serving may be listed as 0 g on a nutrition facts label. Trans fat in a food is identified through the words *partially hydrogenated* or *interesterified* in the ingredients list.
- Unsaturated fats can chemically rearrange to become trans fats under intense heat conditions, as in a commercial deep-fat fryer.
- Americans (and many Westerners) are eating too much fat, refined carbohydrate, sugar, and salt, as well as too little fiber, lean protein, and micronutrients.
- Eating meals away from home makes it difficult to keep track of what (and how much) we're eating.
- Sugared beverages provide lots of empty calories, and don't create lasting feelings of fullness; they are associated with obesity.
- Many weight-loss diets attempt to produce rapid loss through unsustainable practices that can be detrimental to body chemistry.
- Low-fat diets are generally too high in sugar and—because fat is an important satiety signal—in calories.
- Ephedrine-containing stimulant diet pills can have adverse, methamphetamine-like effects and are addictive.
- Low-carbohydrate diets produce ketosis, which affects mood, energy level, and ability to focus.
- No conclusive evidence links non-nutritive sugar substitutes to cancer in humans, but concern over their long-term effects abounds.
- The best health-promoting diet is one that includes appropriate portions of all macronutrients, as well as plenty of fruits and vegetables.
- Many nutritional experts suggest that balance diets should consist of 45–65 percent carbohydrate, 10–35 percent protein, and 20–35 percent fat. The fat should be mostly unsaturated, and the carbohydrate mostly low to moderate GI (glycemic index).
- A food's GI is a measure of the rate at which it increases blood sugar—higher-quality carbohydrates have lower GI values.
- RMR (resting metabolic rate) is the daily caloric expenditure required to maintain cellular and organ function. It's higher for taller, heavier, younger people, and higher in males than in females.
- The thermic effect of food is the daily caloric expenditure required to digest food.
- PAEE (physical activity energy expenditure) is the daily caloric expenditure from routine activity and exercise.
- Exercise burns calories through a direct effect as well as elevated post-exercise metabolic rate and increased muscle mass.
- Aerobic exercise occurs at moderate intensity and burns fat primarily. It is beneficial to health in many regards.
- Anaerobic exercise occurs at high intensity and burns sugar primarily.

- Type I muscle fibers specialize in endurance efforts, contain high amounts of fat, and are rich in myoglobin.
- Type II muscle fibers specialize in power efforts and contain high amounts of glycogen.

Questions and Topics for Discussion

1. What observations can you make to support (or challenge) the idea of the Western Paradox?
2. What do you think are the primary reasons the United States (and other Western cultures) is becoming fatter? Why do you think this isn't happening in other parts of the world?
3. Why is being overweight/obese a society-wide problem rather than simply an individual problem?
4. Why is essential body fat lower in men than in women?
5. Young female athletes like Olympic gymnasts have a tendency to look much younger than they are. Using an understanding of body fat, comment on why this might be.
6. Why does BMI fail to accurately estimate weight status in certain portions of the population? For what sorts of individuals does BMI fail?
7. Go to www.mypyramid.gov (click "MyPyramid Plan" under Interactive Tools) and enter information requested. Do you think the suggested diet plan would be easy to conform to? Why or why not?
8. What are some of the benefits of the new USDA food pyramid (www.mypyramid.gov) over the older pyramid?
9. How do you feel about the influence lobbying and special interest groups can have upon government organizations and their nutritional suggestions? Do you feel you can trust the recommendations of the USDA?
10. How do you feel about the fact that alcoholic beverage companies don't have to put Nutrition Facts labels on their products? Would you want access to that information?
11. Is it possible to consume the maximum recommended quantity of trans fat in a given day even if all foods consumed say 0 g trans fat on the Nutrition Facts label? How?
12. How is the typical American (Western) diet different from what is commonly recommended?
13. Why are sweetened beverages considered such strong contributors to obesity?
14. What are some ways in which food manufacturers can trick a consumer into thinking a food is healthier than it really is? What are ways to avoid these pitfalls?
15. Discuss low-fat, ephedrine, and ketogenic diets. What are the unique problems associated with each? What is a common problem shared by all three?
16. How would you counsel a friend who came to you asking your opinion on a good weight-loss diet?
17. What might your advice be to a friend who asked your opinion on diet soda?
18. A friend of yours puts lots of butter on his white bread toast. When you ask him about it, he says he is trying to reduce the GI of the toast. What is your opinion on his choice? What might you recommend?
19. While studying late one night, you decide you really need a sweet treat. Your local convenience store has three options—full-fat ice cream, fat-free fruit sorbet, and a frozen tofu dessert. Assuming your objective is to make the best nutritional choice, which do you pick and why?
20. What are the health benefits of aerobic exercise?
21. A friend who doesn't exercise asks you for advice on exercise frequency and intensity. What do you tell her?
22. A friend who exercises regularly and would like to improve his caloric balance asks you for advice on exercise frequency and intensity. What do you tell him?
23. Why is weight training considered such a positive addition to many exercise programs?
24. Why do marathon runners and sprinters look so different?

25. In your opinion (based upon your understanding of chemistry), what is the best way to construct a healthy, balanced, sustainable lifestyle in terms of diet and exercise?

Problems

1. Why is body fat important? What role(s) does it serve?
2. What are the major health risks of being overweight/obese?
3. What is the difference between type 1 and type 2 diabetes?
4. What percentage of U.S. adults (as of 2008) are overweight? Obese?
5. What are the three mechanisms addressed in the text to measure body fat percentage? Which is the most (and least) accurate?
6. What is BMI? How is it calculated?
7. A woman weighs 41 kg and is 1.65 m tall. What is her BMI? What is her weight status?
8. A man weighs 114 kg and is 1.72 m tall. What is his BMI? What is his weight status? Do you trust this weight status on the basis of the available information?
9. For what daily caloric consumption is a standard Nutrition Facts label written?
10. Go to www.mypyramid.gov (click "MyPyramid Plan" under Interactive Tools) and enter information requested. How many calories does the plan suggest you consume each day? Based upon this, is a standard Nutrition Facts label going to over- or underestimate the percentages of each nutrient in what you eat, compared to recommendations?
11. What information is required by law to be on a Nutrition Facts label? Of these, which are nutrients most people get too much of? Which are nutrients most people need more of?
12. Why can deep-fat fryers produce trans fats?
13. What words does a savvy consumer looking to avoid trans fat look for on food labels?
14. What makes fat an important satiety signal?
15. Based upon the structures of ephedrine, adrenaline, and methamphetamine, why is it no surprise that they produce similar effects in the body?
16. Define: Satiety signal; sympathomimetic; ketogenic; and ketone bodies.
17. Imagine that a friend starts a low-carbohydrate diet and loses 10 pounds in two weeks. Are you impressed? What might you tell her about the weight loss?
18. The friend from problem 17 notices that she is feeling lethargic and her muscles burn when she walks up stairs. She asks you to explain what is happening. What would you say?
19. The friend from problem 17 manages to stay on her diet for three weeks, and you start to notice an odd smell, a little like nail polish, when she talks. What is happening?
20. Even though saccharin and aspartame are shaped very differently than sucrose, they fit into the sweetness receptor. Explain this phenomenon.
21. What specific behavior, with regard to the sweetness receptor, causes a molecule to taste sweeter than table sugar?
22. Why must individuals with PKU avoid aspartame?
23. Sucralose is very similar, structurally, to sucrose. Why is it non-nutritive (hint: think enzymes)? Why is it sweeter than sucrose?
24. What is the single greatest predictor of weight-loss or weight-maintenance success?
25. What do most nutritional experts agree comprises a balanced diet?
26. What is meant by high-quality fat? What kinds of fat are high-quality?
27. What is meant by high-quality carbohydrate? What kinds of carbohydrate are high-quality?
28. What is a glycemic index? How is it measured?

29. A food increases blood sugar twice as rapidly as glucose—what is its GI?
30. A food increases blood sugar at one-third the rate of glucose—what it its GI?
31. In general, what sorts of foods have low GI values? High GI values?
32. Why does brown rice have a lower GI than white rice?
33. A high-GI food can be "turned" into a low-GI food through the addition of fat or protein. Why is this so? What are the nutritional concerns associated with this?
34. Define: RMR; thermic effect of food; and PAEE.
35. Calculate the expected RMR for a 19-year-old, 5' 2", 110-lb. female.
36. Calculate the expected RMR for a 45-year-old, 5' 8", 300-lb. male.
37. Under what circumstances would you expect RMR calculations to be inaccurate?
38. What percentage of total caloric expenditure does RMR normally account for? Under what circumstances would this be different?
39. What percentage of total caloric expenditure does the thermic effect of food normally account for?
40. What percentage of total caloric expenditure does PAEE normally account for? Under what circumstances would this be different?
41. Define: Direct effect and elevated post-exercise metabolic rate.
42. Why is increased muscle mass an important effect of exercise?
43. What is aerobic exercise? What is the mathematical way to determine the upper and lower limits of the aerobic zone? What's a "quick and dirty" way to determine if exercise is aerobic?
44. What is anaerobic exercise? What is the mathematical way to determine the lower limit of the anaerobic zone? What's a "quick and dirty" way to determine if exercise is anaerobic?
45. What fuels are used in aerobic exercise? In anaerobic exercise?
46. Calculate HR_{max} and the limits of the aerobic zone for an 18-year-old female.
47. Calculate HR_{max} and the limits of the aerobic zone for a 75-year-old male.
48. A 25-year-old female is exercising with a heart rate of 120 BPM. In what type of exercise is she engaging?
49. An 85-year-old female is exercising with a heart rate of 120 BPM. In what type of exercise is she engaging?
50. What are the differences between type I and type II muscle fibers? Where in the body might the average individual have each?
51. What kind of muscle fiber would you expect a power lifter to be enriched in? A long-distance cyclist?
52. What is myoglobin? In which type of muscle fiber is it found?
53. What does "dark meat" mean in terms of nutritional content of the meat? In terms of type of muscle fiber? What about "white meat"?

Answers to Problems

1. Body fat cushions the organs and stores fat-soluble vitamins and hormones (and, of course, acts as a source of stored energy).
3. Type 1 diabetes (juvenile onset) occurs because the body can't produce insulin. It is treated with insulin injections. Type 2 diabetes occurs because the cells become insensitive to insulin. It is treated through diet and weight management.
5. Wrist, neck, and waist circumference is least accurate. Skin fold is more accurate, and immersion is the most accurate.
7. BMI = 15.1; she is underweight. It's hard to know whether she's truly underfat or not without more information. She could have very little muscle and be low weight, but overfat.
9. 2000 Cal/day (with some information provided for 2500 Cal/day).
11. Nutrition labels must have serving size, servings per container, total Calories, Calories from fat, grams of fat, grams of saturated fat, grams of trans fat, cholesterol, sodium, total carbohydrate, grams of fiber, grams of sugar, grams of protein, vitamin A, vitamin C, calcium, and iron. Most people get too many calories and too much fat, cholesterol, and sodium. Most people get too little fiber, protein, and micronutrients.
13. The words *partially hydrogenated* and *interesterified* indicate the presence of trans fat.
15. Ephedrine, adrenaline, and methamphetamine have similar shapes—they share the same pharmacophore.
17. This weight loss is almost entirely due to water: As glycogen becomes depleted, the water that co-stores with glycogen is lost.
19. The odd smell is from ketone bodies, including acetone, which are circulating in the bloodstream to supply her brain with fuel.
21. Molecules that bind to the sweetness receptor more tightly than sucrose taste sweeter than sucrose.
23. Sucralose is different enough that it can't be digested by sucrose enzymes and sucrose pathways, so it has no caloric content. It is sweeter than sucrose because the changes in structure make it bind to sweetness receptors more tightly than sucrose.
25. Most nutritional experts recommend about 45–65 percent carbohydrate, 10–35 percent protein, and 20–35 percent fat, with plenty of fruits and vegetables.
27. High-quality carbohydrate generally has a low GI—it's high in fiber and low in refined carbohydrate and sugar.
29. GI = 200.
31. In general, fruits and whole grains have low GI. Refined carbohydrates and sugary foods have high GI.
33. Protein and fat turn high-GI foods into low-GI foods by interfering with (and slowing the rate of) carbohydrate hydrolysis and absorption. As long as the added fat is healthy (unsaturated) and the protein is lean, this is a reasonable strategy. However, it would be questionable to add saturated fat, for instance, to white bread. Not only does this introduce extra saturated fat into the diet, it also encourages eating a less healthy choice (white bread), where a better option might be whole grain bread.
35. 1335.6 Cal/day.
37. RMR calculations are likely inaccurate for individuals with extremely low or high amounts of lean tissue as compared to the average individual.
39. The thermic effect of food accounts for about 10 percent of calories expended.
41. Direct effect is the calories burned during an exercise; elevated post-exercise metabolic rate is the higher-than-normal RMR that occurs for some time after exercise.
43. Aerobic exercise is enough exertion to cause an increase in metabolic waste products, but is at moderate enough intensity that adaptations (breathing, heart rate) can supply sufficient oxygen to muscles. It occurs at 65–85 percent of HR_{max}. Exercise is aerobic if the individual can talk but not sing.

45. Aerobic exercise uses fat and glycogen—fat is preferred. Anaerobic uses glycogen only.
47. HR_{max} = 145 BPM, aerobic zone = 94–123 BPM.
49. She's at 89 percent of HR_{max}, so she's anaerobic.
51. A power lifter will have more type II than average, while a long-distance cyclist will have more type I than average.
53. Dark meat is primarily type I; it is higher in fat and calories than white meat (and lower in protein for a given mass). White meat is primarily type II; it is lower in fat and calories and higher in protein than the same mass of dark meat.

Sources and Further Reading

Body Fat

http://www.nhlbisupport.com/bmi/

http://www.nlm.nih.gov/medlineplus/obesity.html

http://win.niddk.nih.gov/statistics/index.htm

National Institutes of Health. (1998). Clinical guidelines on the identification, evaluation, and treatment of overweight and obesity in adults. *The Evidence Report.* No. 98–4083.

Nutritional Recommendations

Grandgirard, A., Sebedio, J.L., and Fleury, J. (1984). Geometrical isomerization of linoleic acid during heat treatment of vegetable oils. *Journal of the American Oil Chemists' Society*, 61(10), 1563–68.

Nestle, M. *Food Politics: How the Food Industry Influences Nutrition and Health.* (2002). Berkeley: University of California Press.

Sims, L.S. (1998). *The Politics of Fat: Food and Nutrition Policy in America.* Armonk, NY: M.E. Sharp, Inc.

www.mypyramid.gov

Typical American Diet

Lansky, D., and Brownell, K.D. (1982). Estimates of food quantity and calories: errors in self-report among obese patients. *American Journal of Clinical Nutrition*, 35(4), 727–32.

Pollan, M. (2006). *The Omnivore's Dilemma.* New York: Penguin Group.

Schlosser, E. (2001). *Fast Food Nation.* New York: HarperCollins.

Schulze, M.B., Manson, J.E., Ludwig, D.S., *et al.* Sugar-sweetened beverages, weight gain, and incidence of type 2 diabetes in young and middle-aged women. *JAMA*, 292(8), 927–34.

Wansink, B., and Chandon, P. (2006). Meal size, not body size, explains errors in estimating the calorie content of meals. *Annals of Internal Medicine*, 145(5), 326–32.

Diets

Bravata, D.M., Sanders, L., Huang, J., *et al.* (2003). Efficacy and safety of low-carbohydrate diets: a systematic review. *JAMA*289(14), 1837–50.

Foster, G.D., Wyatt, H.R., Hill, J.O., *et al.* (2003). A randomized trial of a low-carbohydrate diet for obesity. *New England Journal of Medicine*, 348(21), 2082–90.

Freedman, M.R., King, J., and Kennedy, E. (2001). Popular diets: a scientific review. *Obesity Research*,9, 1S–40S.

Kirby, R.C. (2005). Atkins diet—discussion paper. *Discussion paper developed for the AAFP Commission on Public Health.*

Staszkiewicz, J., Horswell, R., and Argyroupolos, G. (2007). Chronic consumption of low-fat diet leads to increased hypothalamic agouti-related protein and reduced leptin. *Nutrition*, 23(9), 665–71.

Wansink, B., and Chandon, P. (2006). Can "low-fat" nutrition labels lead to obesity? *Journal of Marketing Research*, 43(4), 605–17.

Westman, E.C., Mavropoulos, J., Yancy, W.S., *et al.* (2003). A review of low-carbohydrate ketogenic diets. *Current Atherosclerosis Reports*, 5, 476–83.

Sugar Substitutes

Blundell, J.E., and Hill, A.J. (1986). Paradoxical effects of an intense sweetener (aspartame) on appetite. *Lancet*, 1(8489), 1092–93.

Cohen, S.M., Aral, M., Jacobs, J.B., *et al.* (1979). Promoting effect of saccharin and DL-tryptophan in urinary bladder carcinogenesis. *Cancer Research*, 39(4), 1207–17.

Gurney, J.G., Pogoda, J.M., Holly, E.A., *et al.* (1997). Aspartame consumption in relation to childhood brain tumor risk: results from a case-controlled study. *Journal of the National Cancer Institute*, 89(14), 1072–74.

Olney, J.W., Farber, N.B., Spitznagel, E., *et al.* (1996). Increasing brain cancer rates: is there a link to aspartame? *Journal of Neuropathology and Experimental Neurology*, 55(11), 1115–23.

Prodolliet, J., and Bruelhart, M. (1993). Determination of aspartame and its major decomposition products in foods. *Journal of AOAC International*, 76(2), 275–82.

Rodero, A.B., Rodero, L.D., and Azoubel, R. (2009). Toxicity of sucralose in humans: a review. *International Journal of Morphology*, 27(1), 239–44.

Soffritti, M., Belpoggi, F., Tibaldi, E., *et al.* (2007). Life-span exposure to low doses of aspartame beginning during prenatal life increases cancer effects in rats. *Environmental Health Perspectives*, 115(9), 1293–97.

Tsuda, H., Fukushima, S., Imaida, K., *et al.* (1983). Organ-specific promoting effect of phenobarbitol and saccharin in induction of thyroid, liver, and urinary bladder tumors in rats after initiation with N-nitrosomethylurea. *Cancer Research*, 43(7), 3292–96.

Sustainability

http://www.glycemicindex.com/

Fosterpowell, K., and Miller, J. B. (1995). International tables of glycemic index. *American Journal of Clinical Nutrition*, 62(4), S871–90.

Caloric Expenditure

D'Alessio, D.A., Kavle, E.C., Mozzoli, M.A., *et al.* (1988). Thermic effect of food in lean and obese men. *Journal of Clinical Investigation*, 81(6), 1781–89.

Rosa, A.M., and Shizgal, H.M. (1984). The Harris Benedict equation reevaluated: resting energy requirements and the body cell mass. *American Journal of Clinical Nutrition*, 40(1), 168–82.

Exercise

Bielinski, R., Schutz, Y., and Jequier, E. (1985). Energy-metabolism during the post-exercise recovery in man. *American Journal of Clinical Nutrition*, 42(1), 69–82.

Burleson, M.A., O'Bryant, H.S., Stone, M.H., *et al.* (1998). Effect of weight training exercise and treadmill exercise on post-exercise oxygen consumption. *Medicine and Science in Sports and Exercise*, 30(4), 518–22.

Lethbridge-Çejku M., and Vickerie, J. (2005). Summary health statistics for U.S. adults: National Health Interview Survey, 2003. National Center for Health Statistics. *Vital Health Statistics*, 10(225).

Melby, C., Scholl, C., Edwards, G., *et al.* (1993). Effect of acute resistance exercise on post-exercise energy expenditure and resting metabolic rate. *Journal of Applied Physiology*, 75(4), 1847–53.

Reebok Instructor News. (1991).4(2).

Vitamin Supplements

Bjelakovic, G., Nikolova, D., Gluud, L.L., *et al.* (2007). Mortality in Randomized Trials of Antioxidant Supplements for Primary and Secondary Prevention; Systematic Review and Meta-analysis. *JAMA*, 297(8), 842–57.

Duffield-Lillico, A.J., Dalkin, B.L., Reid, M.E., *et al.* (2003). Selenium supplementation, baseline plasma selenium status and incidence of prostate cancer: an analysis of the complete treatment period of the Nutritional Prevention of Cancer Trial. *BJU International*, 91(7), 608–12.

Knekt, P., Reunanen, A., Jarvinen, R., *et al.* (1994). Antioxidant vitamin intake and coronary mortality in a longitudinal population study. *American Journal of Epidemiology*, 139(12), 1180–89.

Lee, I.M., Cook, N.R., Gaziano, J.M., *et al.* (2005). Vitamin E in the primary prevention of cardiovascular disease and cancer: the Women's Health Study: a randomized controlled trial. *JAMA*, 294(1), 56–65.

Lippman, S.M., Klein, E.A., Goodman, P.J., *et al.* (2009). Effect of selenium and vitamin E on risk of prostate cancer and other cancers: the selenium and vitamin E cancer prevention trial. *JAMA*, 301(1), 39–51.

Rohan, T.E., Jain, M.G., Howe, G.R., *et al.* (2000). Dietary folate consumption and breast cancer risk. *Journal of the National Cancer Institute*, 92(15), 266–69.

Sesso, H.D., Buring, J.E., Christen, W.G., *et al.* (2008). Vitamins E and C in the prevention of cardiovascular disease in men: the Physicians' Health Study II randomized controlled trial. *JAMA*, 300(18), 2123–33.

Stampfer, M.J., Hennekens, C.H., Manson, J.E., *et al.* (1993). Vitamin E consumption and the risk of coronary disease in women. *New England Journal of Medicine*, 328(20), 1444–49.

Zhang, S.M., Cook, N.R., Albert, C.M., *et al.* (2008). Effect of combined folic acid, vitamin B6, and vitamin B12 in women: a randomized trial. *JAMA*, 300(17), 2012–21.

Unit 4: Global Chemistry

Changing the World with Chemistry 10

We've spent the last three units of this text discussing the myriad ways in which chemistry affects our communities and our bodies. There's no doubt, at this point, that the local water and air quality or the type of food we eat affect our lives and those of the people around us. But can chemistry affect the global community as a whole? Are there discoveries and processes so momentous that their effects are felt around the planet? In this chapter, we'll explore some of the major ways in which chemistry has changed the world—for better or for worse—and the discoveries and issues that affect us all, no matter where we live. We'll also begin to discuss the challenges associated with these issues, particularly where it comes to making global decisions or changes. How, for instance, do we portion out global responsibility for effecting change to developed and developing nations, given their very disparate economic welfares? How do we know whether lab-derived evidence applies in the real world, and how do we strike a balance between too much caution and not enough when it comes to implementing (or withdrawing) a technology worldwide? Most importantly, where do we draw the line between concern for the individual and utilitarianism—the responsibility to do the most good for the most people?

10.1 REFRIGERATION

Hopefully we're all comfortable at this point with the idea that science isn't simply what takes place in the lab: The relationship between the lab and the real world is reciprocal and inextricable. Scientific exploration in the lab models the real world, which allows researchers to make predictions. Animal studies, for instance, are used to determine the safety of a new pharmaceutical before it's ever tested on humans. Unfortunately, the lab has its limitations. It's not the "real world," but is instead a microcosm—an incomplete facsimile that allows us only a glimpse of the full ramifications of a technology. When a chemical, an invention, a discovery, or a technology graduates from the lab and debuts in the world, we are often greeted with unanticipated effects.

Until the twentieth century, refrigeration as a household commodity was virtually unknown. The theory behind it was firmly established, however, and was based upon some well-understood laws regarding the behavior of liquids and gases. The common experience of sweating gives us insight into some of these laws. When we exert ourselves or are exposed to high ambient temperatures, we secrete liquid sweat onto the surface of the skin. The water component of this liquid (which also contains dissolved salts) evaporates into a gas. As it does so, it absorbs a great deal of heat, lowering the temperature of the body. The heat is carried away with the evaporating water, leaving us feeling much cooler. The reason an evaporating liquid absorbs heat is that the process of **vaporization** is endothermic—it requires energy. We discussed in Chapter 8 that for every process, a reverse process is possible. Furthermore, if the forward process requires energy, the reverse will release it. We can conjecture, then, that if water vapor were to condense into liquid on the skin, it would release heat. In fact, we are subjected to this phenomenon during the unpleasant experience of obtaining a steam burn. Boiling water, if it splashes onto the skin, burns. Steam from the boiling water, however, burns much more severely. The steam released from boiling water is the same temperature as the water—once it escapes the pot, it doesn't continue to heat up. However, in addition to transferring the heat energy of the water vapor to the skin, energy from the **condensation** process is also transferred as vapor turns to liquid. For any substance, the process of vaporization absorbs energy (heat), while the process of condensation releases energy (heat).

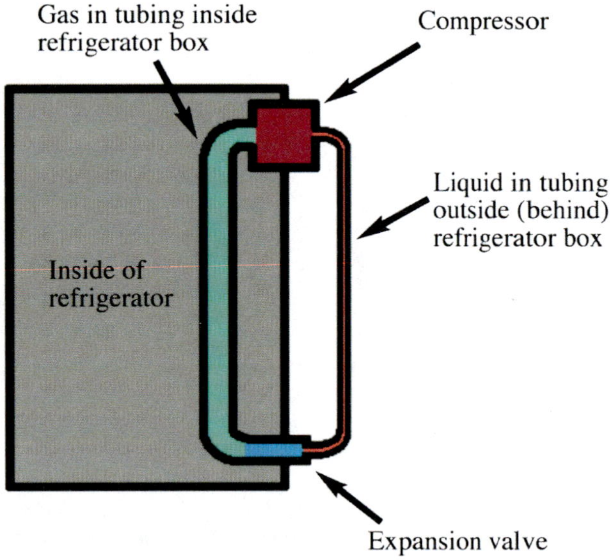

Image 10.1. A refrigerator schematic. The tubing is coiled rather than straight in an actual fridge to allow more surface area. This increases the extent to which the cold gas absorbs heat from the inside of the box, and the extent to which the warm liquid dissipates its heat outside of the box.

This theory is put to use in the design of a refrigeration device. A liquid is confined to a closed tubing system, part of which is inside and part of which is outside a sealed refrigeration box. The liquid circulates through the tubing, and as it enters the refrigeration box, it runs through an expansion valve. The expansion causes it to vaporize, a process that absorbs heat and decreases the temperature inside the refrigeration box. Why does expansion cause vaporization? While a full discussion of gas behavior is beyond the scope of this text, in short, matter expands to fill available space. This phenomenon is echoed in the maxim "Nature abhors a vacuum." If the volume of a tubing system is suddenly greater than the volume of liquid in the tubing, the molecules comprising the liquid will separate to fill the space. As the molecules separate from one another, liquid turns to gas. The chemical (now a vapor) continues circulating through the tubing, eventually exiting the refrigeration box. Upon exiting, it is condensed back into a liquid by a compressor, which forces the molecules closer together. During this process, heat is released. However, since the condensation takes place outside the refrigeration box, the heat is dissipated to the air—you can actually feel the heat dissipating from the coils on the back of the refrigerator. The liquid then circulates back to the expansion valve, and the cycle starts again.

The exact temperature that can be achieved through a vapor-compression cycle depends upon the boiling point of the chemical in the tubing. For a refrigeration system to be able to freeze water, the boiling point of the refrigerant would have to be lower than the melting (freezing) point of water (0°C).

CONCEPT CHECK:

How does a vaporization-compression cycle produce refrigeration?

Answer: As a liquid expands and vaporizes, it absorbs heat.

Several early refrigerants with appropriate boiling points were common in the late 1800s and early 1900s—in particular, SO_2 (-10°C), NH_3 (-33°C), and methyl chloride (CH_3Cl, -24°C) were used. Unfortunately, these were toxic, flammable, or both, resulting in many accidents. In 1928 Frigidaire (a division of General Motors) began supporting development of a new refrigerant. Thomas Midgley* was given the task of identifying or developing a refrigerant that would be stable, nontoxic, nonflammable, noncorrosive, cheap, and of course have a boiling point lower than 0°C. Midgley ended up preparing substituted methane and ethane molecules, in which the usual hydrogen atoms were replaced with fluorine and chlorine. He named his compounds *Freon*, differentiating them from one another by use of a cryptic numbering system. In fact, all the Freons belong to a larger class of molecules called **chlorofluorocarbons (CFCs)**.

***NOTE**

Remember him? He brought us tetraethyl lead in 1921.

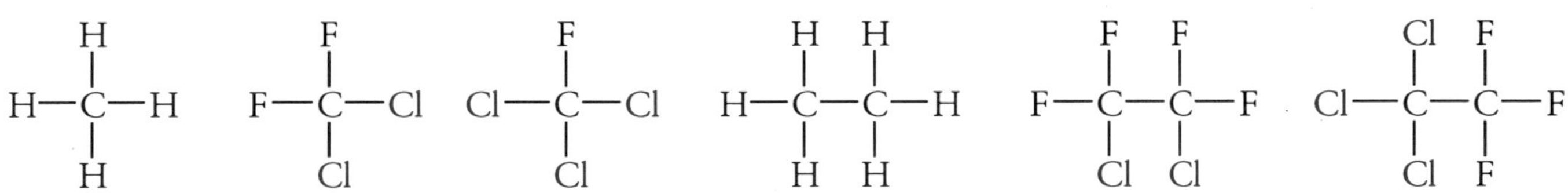

Methane | Freon 12 | Freon 11 | Ethane | Freon 114 | Freon 113

Image 10.2. Methane, ethane, and some of Midgley's Freon molecules.

The new refrigerants seemed to fit the bill nicely—they had appropriate boiling points and were evidently quite safe. Midgley even went so far as to demonstrate the safety of his compounds at a meeting of the American Chemical Society, in a stunt reminiscent of his tetraethyl lead hand-washing display—he inhaled a lungful of Freon (to demonstrate that it was nontoxic), and then exhaled it upon a candle flame, which promptly went out, demonstrating that the Freon was not flammable. Inexpensive to make, CFCs were responsible for making home refrigerators and air conditioning (both of which relied upon the same technology) standard amenities. The Sun Belt, a strip of the southern United States characterized by mild temperatures in the winter and uncomfortably hot summers, experienced a population explosion as air conditioning rendered the region at least tolerable. Owing to their inert chemical nature, CFCs found applications outside of the refrigeration industry. They were used in fire extinguishers to displace oxygen from a flame, they made excellent solvents for cleaning purposes, and they were condensed and used as propellants in spray paint, hairspray, and even asthma inhalers.

The heyday of CFCs continued until the early 1970s, when several researchers made some disturbing discoveries about the compounds. Evidence began to amass that while CFCs are, in fact, nontoxic and nonreactive in the lower atmosphere, they are quite active in the stratosphere, because it is here that they encounter one of the few things with which they do react—UV light.

10.2 UV LIGHT AND EMR

It turns out that our Sun emits many different kinds of radiation—some are responsible for lighting the planet and allowing us to see color, others provide us with warmth, still others can burn us. Understanding the different types of radiation and why some are damaging while others are not necessitates examining the nature of the Sun's rays.

What we call "light" is actually a form of **electromagnetic radiation (EMR)**. Formally defined as a traveling electrical and magnetic disturbance, EMR is more simply defined as radiant energy. The energy has some wave-like properties and some particle-like properties—this dual nature is one of the fundamental enigmas of physics. Suffice it to say, EMR behaves in some regards like an ocean wave, with measurable wavelike properties. For instance, just as it would be possible to measure the size of an ocean wave, it's possible to measure an EMR wave. In either case, the distance from one wave crest to the next is called the **wavelength** (λ), which is measured in meters (Image 10.3). EMR wavelength can vary tremendously depending upon the kind of radiation; it can be as short as a few femtometers (fm, 10^{-15}m), or billions of meters long.

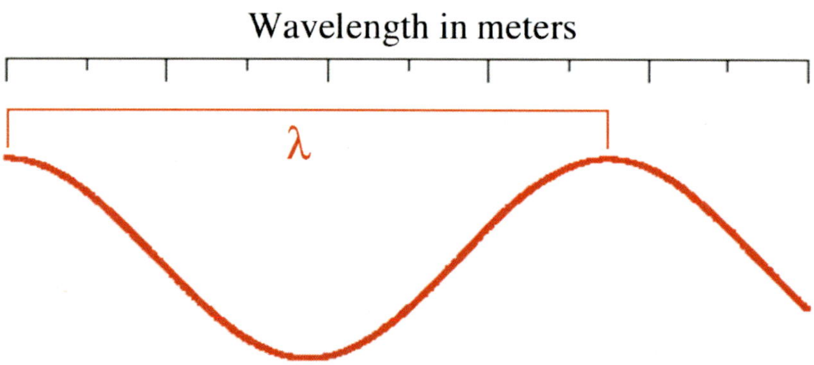

Image 10.3. The wavelength of an EMR wave, like that of an ocean wave, is the distance from one crest to the next.

A second measurable property of EMR is its **frequency** (ν). This is the number of crests that pass a given point in a given period of time: If we continue the comparison to ocean waves, the frequency would be the number of waves per second that crash on the shore. Frequency is measured in units of Hertz (Hz, 1/s, s^{-1}). For instance, if three crests passed a given point in one second, the frequency of the wave would be 3 Hz. As with wavelength, the frequency of EMR depends upon the kind of radiation, and varies greatly—anywhere from less than 1 Hz to more than 10^{24} Hz.

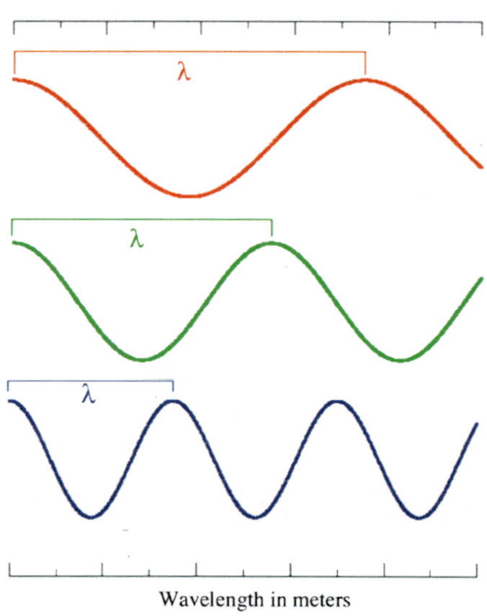

Image 10.4. Wavelength and frequency are inversely related for waves traveling at the same speed.

An interesting property of EMR waves that makes them very distinct from ocean waves is that regardless of their wavelength or frequency, they all travel at exactly the same speed. In fact, EMR travels at the speed of light, which is given the symbol c in equations ($c = 3 \times 10^8$ m/s). A ramification of this is that for all EMR, wavelength and frequency are inversely related—EMR of longer wavelength has lower frequency, and vice versa. To get an idea of why this is so, take a look at the three waves in Image 10.4, and imagine them all moving at equal speed toward the right. If you were to count the number of crests of the red wave (which has the longest wavelength) that passed a given point in one second, you'd find it to be less than the number of crests of the green wave and much less than the number of crests of the blue wave (with the shortest wavelength) passing the same point in one second. If the three waves must travel at the same speed, the one with the longest wavelength has the lowest frequency, and the one with the shortest wavelength has the highest frequency.

This relationship is captured in the wave equation:

$$c = \lambda\nu$$

As we see in the equation above, wavelength (λ) and frequency (ν) are inversely related, and must always multiply to give the same number—c, the speed of light. For any EMR wave, it is possible to calculate either frequency or wavelength, as long as the other value is known.

<u>Example</u>: What is the wavelength of an EMR wave with a frequency of 10^6 Hz?

$c = \lambda\nu$

3×10^8 m/s $= \lambda\ (10^6$ Hz$)$

$\lambda = 300$ m

TRY THIS:

What is the frequency of an EMR wave with a wavelength of 5×10^{-10} m?

Answer: = 6×10^{17} Hz.

Solution:

$c = \lambda\nu$

3×10^8 m/s $= (5 \times 10^{-10}$ m$)\nu$

$\nu = 6 \times 10^{17}$ Hz

TRY THIS:

Which has higher frequency, an EMR wave with a wavelength of 3×10^8 m or one with a wavelength of 7×10^4 m?

Answer: Because 7×10^4 m represents a shorter wavelength, this wave will have a higher frequency.

A mathematical discussion of EMR leaves us with the (somewhat deceptive) sense that all types of radiation are very similar, differing only in wavelength and frequency. While it is true that the fundamental nature of a high-frequency EMR wave is identical to that of a low-frequency EMR wave, these waves affect their surroundings in very different ways. This is because frequency is also related to energy,* a relationship described by another wave equation:

$$E = h\nu$$

In this equation, E is the wave's energy (measured in Joules), and h is called **Planck's constant** ($h = 6.626 \times 10^{-34}$ J·s). This constant is used in several different equations in physics and chemistry, but we'll see it here only in the context of relating frequency and energy to one another. The relationship between

***NOTE**

When we calculate the energy of a "wave" of EMR, we're actually calculating the energy of an individual particle of radiation with a given frequency and wavelength. How can a particle have a wavelength (or frequency, for that matter)? Remember that EMR has a dual nature—it is both wave-like and particle-like. A particle of light is called a *photon*.

frequency and energy is directly proportional—higher frequency waves have higher energy and vice versa, and we can calculate either value from the other.

Example: What is the energy of a wave with a frequency of 3.00×10^{16} Hz?

$$E = h\nu$$
$$E = (6.626 \times 10^{-34}\ J{\cdot}s)(3.00 \times 10^{16}\ Hz)$$
$$E = 1.99 \times 10^{-17}\ J$$

Try This:

What is the frequency of a wave with energy equal to 5.4×10^{-12} J?

Answer: = 8.1×10^{21} Hz

Solution:

$E = h\nu$

$5.4 \times 10^{-12}\ J = (6.626 \times 10^{-34}\ J{\cdot}s)(\nu)$

$\nu = 8.1 \times 10^{21}$ Hz

Try This:

What is the energy of a wave with $\lambda = 2.3 \times 10^{-10}$ m?

Answer: $E = 8.6 \times 10^{-16}$ J.

Solution:

$c = \lambda\nu$

$3 \times 10^{8}\ m/s = (2.3 \times 10^{-10}\ m)\nu$

$\nu = 1.3 \times 10^{18}$ Hz

$E = h\nu$

$E = (6.626 \times 10^{-34}\ J{\cdot}s)(1.3 \times 10^{18}\ Hz)$

$E = 8.6 \times 10^{-16}$ J

Since we know that wavelengths and frequencies of EMR span many orders of magnitude, we can conjecture that the energies of EMR waves must also encompass a large range. This is quite true, and furthermore, the physical effect of EMR upon the world (and the way in which we experience the radiation) depends entirely upon its energy. Image 10.5 shows the spectrum of EMR, from very high energy radiation (high frequency, short wavelength) like gamma rays, to very low energy radiation (low frequency, long wavelength) like radio waves.

Gamma rays, which we'll see again in Chapter 12, are a form of extremely high-energy radiation that can be released during nuclear reactions. They are very damaging to living tissue, and can cause radiation sickness, mutations, and death. Slightly lower in energy are **X-rays**. They are not quite as penetrating as

Concept Check:

Why are the highest energy EMR waves shortest in wavelength?

Answer: High energy waves have high frequency, and, therefore, short wavelength.

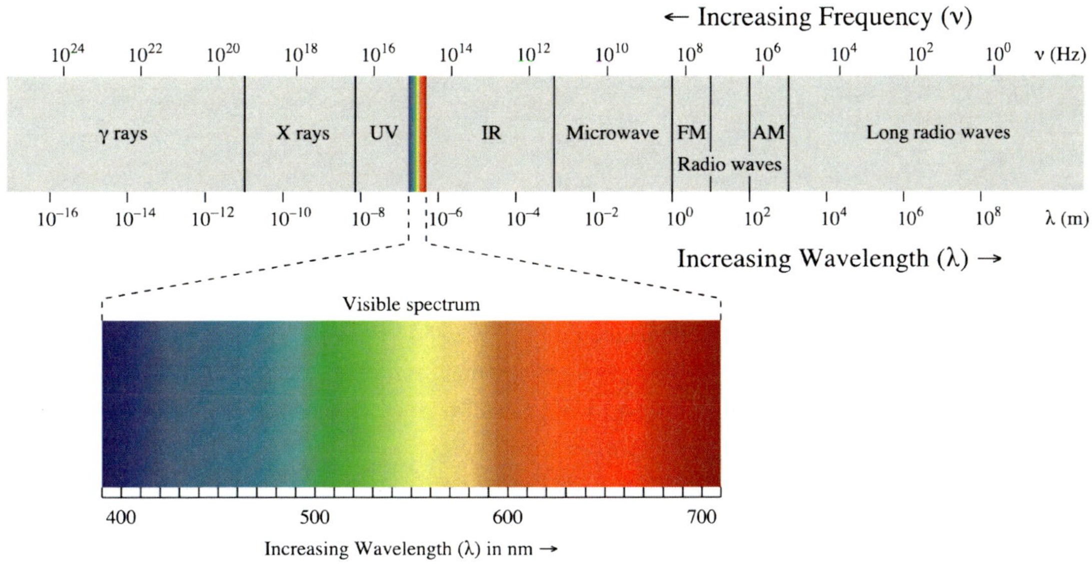

Image 10.5. Types of EMR. Higher energy radiation is on the left end of the spectrum.

gamma rays,* but exposure in sufficient quantity can still lead to tissue damage, including mutation. **Ultraviolet radiation (UV)**, lies between X-rays and visible light on the EMR spectrum. The penetrating power of UV light is less than that of X-ray radiation—generally, UV can penetrate the skin, but can't go deeper into the body. Still, it can damage and cause mutations in living tissue. Why do these three types of radiation possess the ability to damage tissue, and why do mutations result? The answer lies in the mechanism by which the radiation interacts with molecules in cells. UV, X-rays, and gamma rays are collectively called **ionizing radiation**. When such radiation strikes a molecule, it has the ability to weaken and break bonds (Image 10.6).♦

It's useful to model bonds as springs when talking about the effect of radiation. In a very real way, bonds are actually much like springs—they have a natural resonance with which they stretch and compress, bringing atoms closer together and allowing them to drift further apart while retaining their attachment to each other. Bonds can bend and flex a bit without breaking—which to some extent they do naturally, but outside forces can exacerbate this. It's not far from the truth to imagine that UV and other ionizing radiation

***NOTE**

In fact, X-rays penetrate soft tissue, but are reflected by dense tissue like bone. This has led to their incorporation as a diagnostic tool in medicine.

♦NOTE

The name *ionizing radiation* comes from the fact that these wavelengths can also cause an electron to be knocked out of a molecule, leaving the molecule positively charged (ionic). Since electrons are normally found in pairs, the loss of one electron leaves a lone electron—a free radical—behind. As we've seen, free radicals are highly reactive species.

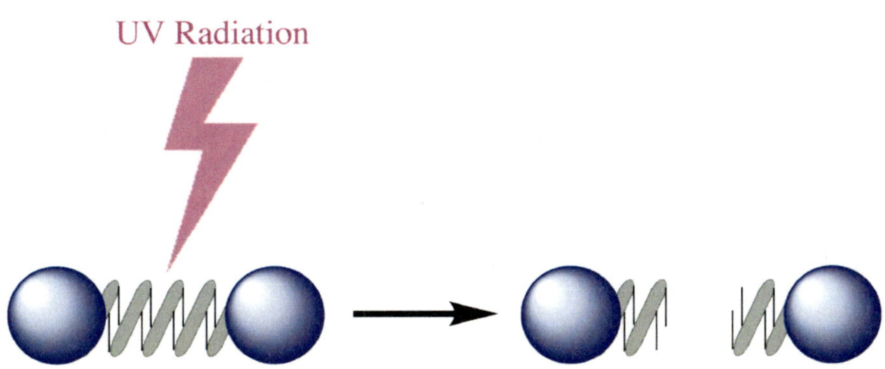

Image 10.6. The effect of UV (and other ionizing radiation) on bonds.

*NOTE

Visible light and other EMR with frequency lower than UV does not have sufficient energy to break bonds. These types of radiation can affect molecules, but not by ionizing them. We'll see the effects of these different types of radiation later in this chapter and throughout the rest of the text.

stresses a bond with such force that it causes it to snap. In fact, as discussed in Chapter 8, breaking bonds always requires energy—ionizing radiation has sufficient energy to effect this process.* Of course, we must keep in mind that breaking the bond has freed two atoms, which are now free to form new bonds and rearrange into different molecules.

The biological ramifications of exposure to ionizing radiation stem directly from its ability to break bonds in molecules—if the broken bonds occur in **DNA** (our genetic material), mutations and cancer can result. The type of cancer that occurs is related to the type of ionizing radiation. Very penetrating radiation, like gamma rays, can cause deep tissue mutations, including leukemia and other blood cancers. The far less penetrating UV radiation can't cause mutations in deep tissues, but it can certainly lead to skin cancer. DNA isn't the only molecule susceptible to ionization—proteins and many other biomolecules can be affected as well. For instance, the protein **collagen** is responsible for maintaining the skin's elasticity. Over time and with age, this protein begins to degrade, leaving the skin with less resilience (and more wrinkles). This natural process is dramatically accelerated by repeated exposure to UV radiation: Sunbathing can lead to prematurely wrinkled skin.

As anyone who's ever had a bad sunburn can attest, cellular damage from UV exposure can be acute; severe sunburn results in the death of several layers of skin cells. This provides us some insight into the importance of protection from UV—if we were exposed to the full strength and spectrum of the Sun's rays (almost 10 percent of which are UV; Image 10.7), all living organisms would be forced to retreat underground or underwater to avoid catastrophic radiation exposure. Thankfully, the ozone layer protects us from most UV light that is incident upon the Earth.

CONCEPT CHECK:

Is visible light ionizing radiation?

Answer: No; it's lower in energy than UV.

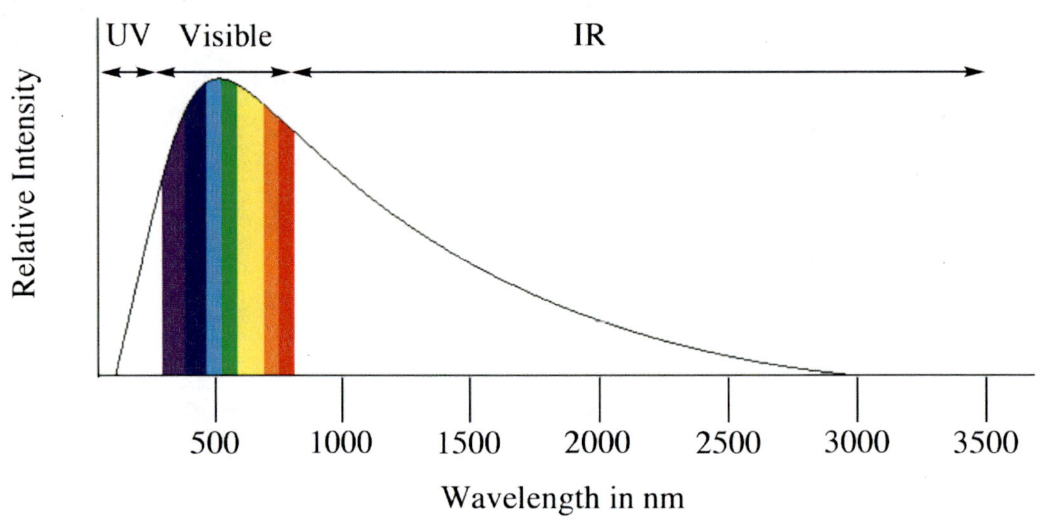

Image 10.7. The Solar spectrum.

10.3 THE OZONE LAYER

We first encountered the ozone molecule in Chapter 6, as a secondary product of combustion. Tropospheric ozone is not naturally occurring; it's a pollutant. In the stratosphere, however, ozone is a natural component of the atmosphere, and serves the vital function of protecting us from the Sun's ultraviolet radiation. As we learned in Chapter 6, ozone is an allotrope of oxygen, but its odd and asymmetrical arrangement of atoms (Image 10.3) makes it much less stable and much more reactive than O_2.

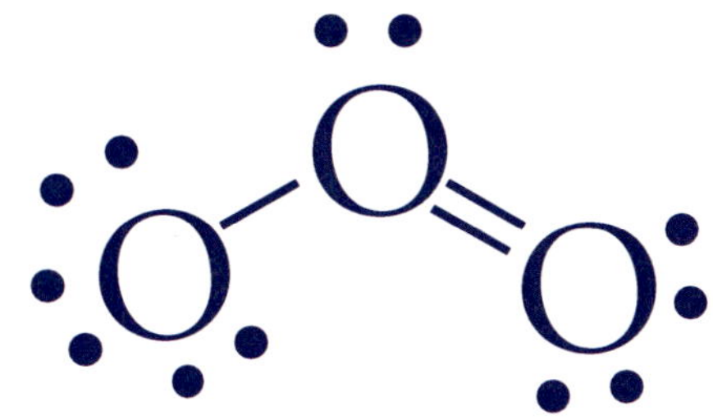

Image 10.8. The ozone molecule.

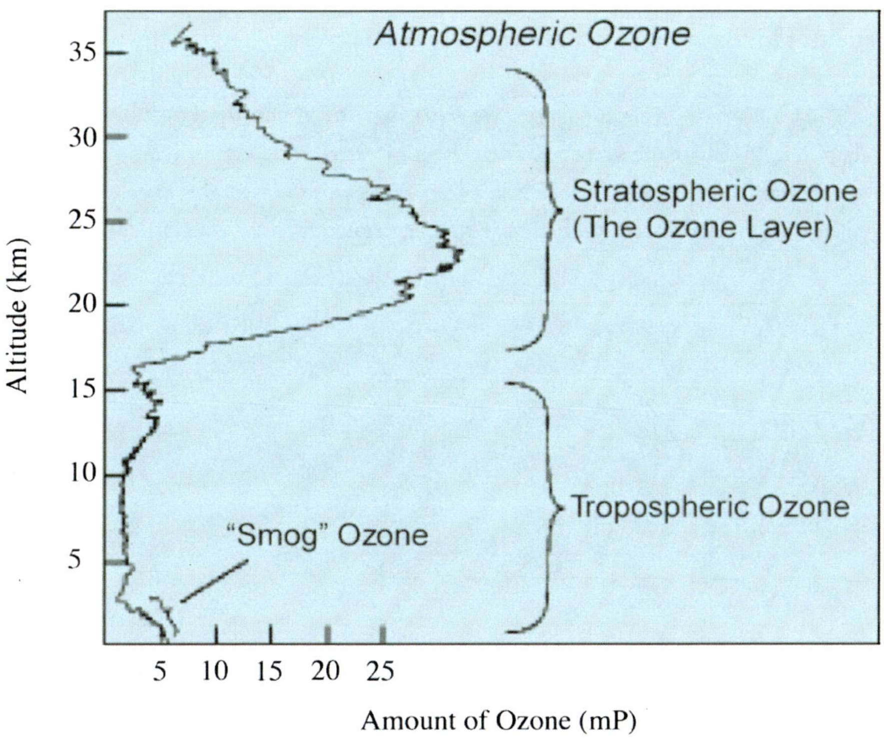

Image 10.9. The ozone layer is a band of higher ozone concentration occurring between 15 and 35 km above the Earth's surface. Tropospheric ozone is a pollutant; it does not occur naturally.

A layer of ozone 20 km thick spans the stratosphere (Image 10.9). Since atmospheric pressure is much lower at those elevations, however, gases are far less dense and molecules much farther apart than they are near Earth's surface. In fact, if the entire ozone layer were brought down to sea level, it would be a mere 3 millimeters in thickness! Regardless, the ozone layer is an important shield—one without which life would be very different.

Sydney Chapman, a British chemist, determined the mechanism by which the ozone layer shields us from UV light, which is now known as the **Chapman Cycle** (Image 10.10). In fact, the chemistry of the cycle is somewhat complex, involving two separate processes, each of which takes place in two steps:

> ***NOTE**
>
> Recall that in chemical equations, hv is taken to mean "light of the appropriate energy." Now we know why: The wave equation E = hv reminds us that hv is, literally, energy.

*Destruction of ozone:**

$O_3 + h\nu \rightarrow O + O_2$

$O + O_3 \rightarrow 2\ O_2$

Production of ozone:

$O_2 + h\nu \rightarrow 2\ O$

$O + O_2 \rightarrow O_3$

In one process, ozone absorbs UV radiation and a bond is broken, producing an oxygen atom and O_2. The oxygen atom is highly reactive, and can combine with more ozone to form two additional molecules of O_2. A chemical species that is formed in one step of a process and consumed in the next step is called an **intermediate**. While we wouldn't expect to find significant quantities of atomic oxygen in the stratosphere (since it's used as a reactant in the second step of the process), it would certainly be present in at least small concentrations. The other Chapman Cycle process requires the absorption of UV by O_2, which splits the molecule into two atoms of oxygen; here again, atomic oxygen is an intermediate. Each of these can react with more O_2, resulting in the production of ozone. In each case, UV light is providing the energy necessary to break the bonds and allow for molecular rearrangements.

The Chapman Cycle is a **dynamic equilibrium**, meaning that while individual molecules of ozone are constantly being destroyed, new molecules of ozone are constantly being formed. These two processes occur at the exact same rate in a healthy atmosphere, such that ozone concentration is constant over time.

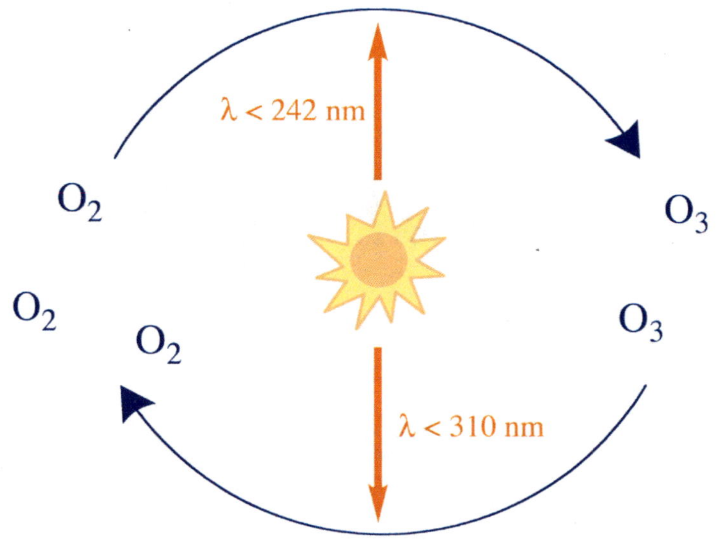

Image 10.10. The Chapman Cycle, for which a simplified chemical reaction can be written: $3\ O_2 \rightarrow 2\ O_3$, $2\ O_3 \rightarrow 3\ O_2$.

While both processes of the Chapman Cycle require the absorption of UV light, destruction of ozone depends upon a different wavelength (and energy) of radiation than does its production. The reason for this is that ozone is much less stable than O_2, meaning that it takes more energy (shorter wavelength radiation) to break the bond in O_2 than to break the bond in O_3. Lower energy UV—radiation with wavelengths from about 310–242 nm—is absorbed by ozone, resulting in the production of O_2. Higher energy UV, with wavelengths less than 242 nm, is absorbed by O_2, initiating the reactions that culminate in the production of ozone. Since both O_2 and ozone are capable of absorbing UV, and since both are present in the stratosphere, much of the UV radiation incident upon the Earth never reaches its surface; most is absorbed in the upper atmosphere through the reactions of the Chapman Cycle.

UV light can actually be divided into three categories: **UV-A** has the longest wavelength (400–315 nm) and is the least damaging. **UV-B** has an intermediate wavelength (310–280 nm), while **UV-C** has the shortest wavelength (280–100 nm) and would be most damaging to living organisms. Thankfully, UV-C is absorbed in both the destruction and production of ozone: As a result, no UV-C penetrates the upper atmosphere under normal circumstances. UV-B is partially absorbed in the stratosphere—it has a wavelength appropriate for effecting the destruction of ozone. While the ozone layer shields us from most UV-B, a small percentage still reaches the Earth. Because this form of radiation is so penetrating, even small amounts are capable of causing tremendous damage; UV-B is responsible for most sunburns and cancers. UV-A is not significantly absorbed in the upper atmosphere—it's too low in energy to break bonds in either ozone or O_2—and the majority of it reaches the Earth's surface. While not as damaging as the other forms of UV radiation, it still has the ability to produce changes in DNA over time.

10.4 CFCS AND UV LIGHT

As we've seen, UV light has the ability to break bonds in lots of different molecules: O_2, O_3, DNA, proteins, and a wide variety of other compounds. Generally, it's absorbed by and reacts with the first appropriate molecule it encounters (with the proviso that high-energy UV is capable of breaking stronger bonds than low-energy UV). Thanks to the ozone layer, not much UV reaches the troposphere, but it is intense in the upper stratosphere. To return to our discussion of CFCs, their nonreactive nature in the troposphere means that they have a long lifetime—they enter the global atmospheric cycle, and eventually drift upward into the stratosphere. Here they encounter intense UV, with which they quickly react to form free atoms of chlorine (Image 10.11).

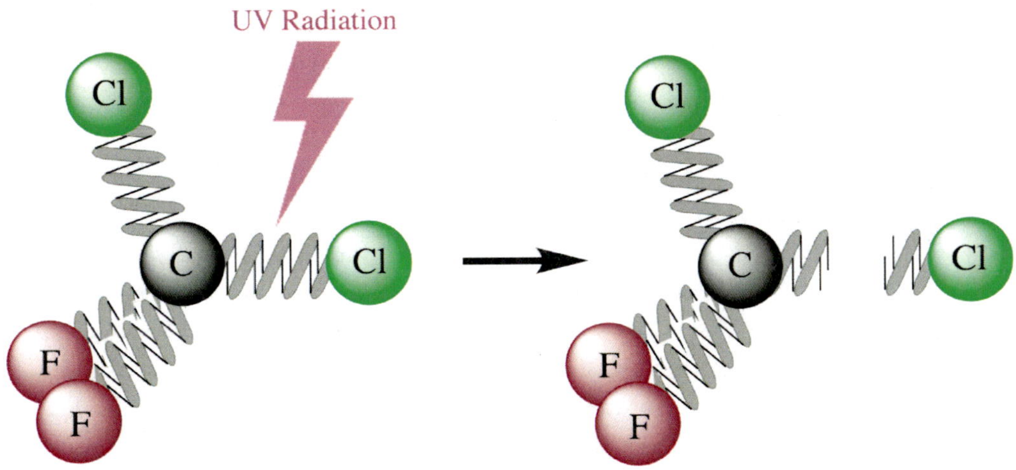

Image 10.11. A CFC reacts with UV light. The result is the liberation of anatom of Cl, a highly reactive species.

The chlorine atom split from its parent CFC is very high in energy, and goes on to react with the first unstable molecule it encounters, generally O_3:

$$Cl + O_3 \rightarrow ClO + O_2$$

The products of this reaction are O_2 and chlorine monoxide, ClO, itself a reactive species. The ClO is capable of reacting with an atom of oxygen (recall that oxygen atoms are intermediates in both parts of the Chapman Cycle, so they are present in the upper atmosphere), producing more O_2 and releasing the original chlorine atom.

$$ClO + O \rightarrow O_2 + Cl$$

The ramifications of these reactions are not immediately disturbing—it appears on the surface that chlorine has simply reacted with ozone, producing two molecules of O_2 as a result. The problem, however, is that the chlorine atom responsible for initiating the ozone destruction was reformed in the process, meaning that it can go on to react with another molecule of O_3 (producing more O_2), and then another, and so forth. A reactant that is regenerated in this fashion is called a **catalyst**, which means a chemical species that accelerates a reaction without being consumed itself. Small amounts of the chlorine catalyst are capable of reacting with large numbers of ozone molecules very quickly. In fact, a single CFC in the upper atmosphere can destroy ozone molecules numbering in the tens and even hundreds of thousands. The result is a massive **perturbation**, or disturbance, of the natural ozone-oxygen cycle that is held in dynamic equilibrium by the Sun's radiation. CFCs accelerate ozone destruction without affecting the rate of ozone formation.

The theory that CFCs possessed the ability to react with ozone arose in the early 1970s. Paul Crutzen, an atmospheric chemist, was studying the reaction of NOx with O_3. He also reported that oxides of other elements, including chlorine, should be able to do the same. But the relevance of this report went unrecognized, as there was no known source of ClO or other chlorine oxides in the upper atmosphere. A few years later, NASA (National Aeronautics and Space Administration) asked researchers to evaluate the potential for and ramifications of chlorine released into the upper atmosphere by space shuttle exhaust. It was determined that there was, indeed, a mechanism by which ClO could react with ozone, and that the reaction was catalytic, dramatic, and rapid. An alarming breakthrough followed shortly thereafter, when Mario Molina and Sherwood Rowland, a pair of researchers working in California, proposed that CFCs were a plentiful source of stratospheric chlorine. Further, they found that there were no natural processes by which CFCs were removed from the environment other than that mediated by UV light and resulting in ozone destruction.* In response to these findings, the *New York Times* ran a front-page story about the connection between CFCs—ubiquitous in refrigeration devices and aerosols—and ozone depletion. In response, Americans voluntarily cut their purchase and use of CFC-propelled aerosols by 67 percent.

CONCEPT CHECK:

If UV light also destroys ozone to produce O_2, why are CFCs a problem?

Answer: UV light simultaneously destroys and produces ozone, keeping the cycle in balance. CFCs only destroy ozone; they don't produce it.

***NOTE**

For this accomplishment they, together with Crutzen, received the 1995 Nobel Prize in chemistry.

For several years after thereafter, worldwide CFC usage fell. The U.S. Congress banned CFCs in aerosol cans in 1978, but continued to allow their use as refrigerants. DuPont initiated a program to begin searching for CFC alternatives, but over the next several years as concern for the ozone layer retreated from the front page of the newspaper (and out of the public eye), production of the compounds began to increase once more, and worldwide use soared. The next breakthrough came in the analysis of some unusual data gathered by a British research team in Antarctica. They were measuring the thickness of the ozone layer over the continent and were shocked to discover a depletion cycle that showed seasonal fluctuation—strangely, the hole seemed to get worse through the summer months and recover partially during winter months. The depletion was so significant and unexpected that they collected data for three years before finally publishing their discovery. Some of the confusion owed to the fact that NASA satellites had also been monitoring the thickness of the ozone layer, and had detected no depletion over the same period. The discrepancy was eventually traced to the satellite data having been "fitted" to remove outlying data points—particularly low measurements were systematically thrown out, as they were deemed inaccurate. When satellite data was reexamined with all data points intact, the measurements correlated with those taken by the British team. There was no longer any doubt: a large hole had opened up in the ozone layer. Image 10.12 is 28 years of ozone layer thickness over Halley Bay, Antarctica. Notice that while year-to-year fluctuations in thickness occur, the trend has been toward a dramatically thinner layer.

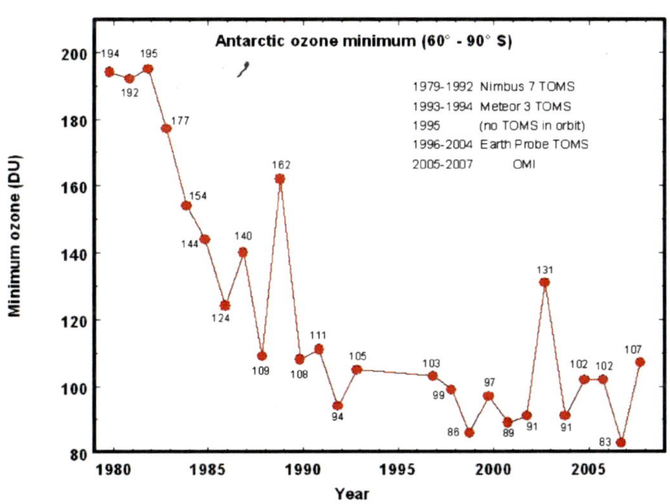

Image 10.12. The thickness of the ozone hole (yearly minimum values) over Halley Bay, Antarctica. TOMS is the Total Ozone Mapping Spectrometer employed by NASA to monitor the layer. Ozone thickness is measured in Dobson units (DU), where the normal thickness is 300–500DU.

While monitoring revealed the hole to be present year-round, it was particularly large in the austral summer months (winter in the Northern Hemisphere). It was eventually determined that both the seasonal variation

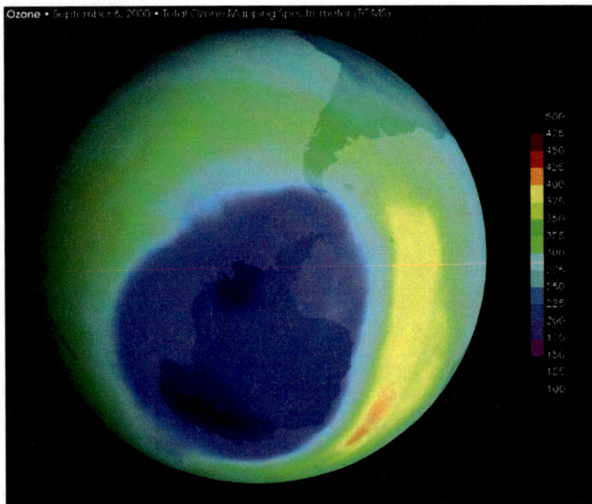

Image 10.13. The ozone hole in 2000.

Image 10.14. Polar stratospheric clouds are responsible for the location and seasonal fluctuation of the ozone hole.

and the polar location of the hole had to do with temperature. The Antarctic winters have the lowest average winter temperatures of any place on the planet. These low temperatures promote the formation of **polar stratospheric clouds**.

Due to the low water content of the stratosphere, cloud formation at that elevation is rare. Only in the case of exceedingly low temperatures, such as in the midst of an Antarctic winter, can clouds form, and when they do, they capture and provide a reaction surface for stratospheric ClO. Near the beginning of the austral summer (December), the increasingly intense sunlight causes rapid warming of the upper atmosphere, increasing air circulation. Ozone-depleted air moves away from the pole, and ozone-rich air from surrounding areas migrates southward. The result is a rapid catalytic destruction of ozone and an expanded area of depletion.

10.5 THE MONTREAL PROTOCOL

In September of 1987, the **Montreal Protocol on Substances That Deplete the Ozone Layer** became available for international ratification. The protocol called for a phase-out of the production and use of CFCs and other **ozone-depleting substances (ODS)**. There have been many revisions to the original protocol that have generally served to increase the rate at which the phase-out will occur; the most recent of these was signed in 1999 in Beijing.

As of September 2009, the protocol became the first United Nations treaty to receive universal ratification. That the Montreal Protocol has been able to engender global cooperation on an environmental problem of this magnitude—the solution to which has both developmental and economic ramifications—is beyond remarkable. In formulating a global treaty of this sort, there are multiple obstacles that must be overcome. The first of these is the disparity between developed and developing nations. For the Montreal Protocol to be effective, it required a commitment to ODS phase-out the world over. Developing nations, however, initially expressed feeling that they were being put at an economic and developmental disadvantage in order to fix a problem they hadn't been a party to causing—the United States, Canada, the European nations and so forth had benefited from ODS, and had enjoyed economic growth as a result of their implementation. These countries had also been directly responsible for the initial ozone depletion. Developing nations, while concerned about the environmental ramifications of continued depletion, nevertheless expressed an understandable desire to pursue their own economic growth, including the requisite utilization of refrigerants in homes and in industrial applications. The United Nations addressed these concerns by writing the Montreal Protocol to allow a grace period for developing nations—where developed countries were to phase CFCs out entirely by the end of 1995, the developing world was given an additional 15 years. Other ODS, including carbon tetrachloride and methyl bromide,* were also scheduled for phase-out by the mid-1990s in developed nations, with an additional 15–20 years allowed for developing nations.

A second concern in authoring the protocol—and another whose solution can be credited with contributing to the global acceptance of the treaty—had to do with tabulating usage of the substances and sanctioning trade. Of concern, any country that did not sign the protocol would be in the powerful position of holding a monopoly on ODS production. In order to encourage universal compliance while fostering trade and economic development during the period of time in which developing nations were still allowed to utilize ODS, the protocol defined use as consumption plus production, minus export. This allowed developing countries continued access to ODS during their grace period—at least until such a time as replacement substances became available—without encouraging or inadvertently rewarding noncompliance. Additionally, a World Bank fund was established in 1990 to aid developing nations in their reduction of ODS use.

Taken together, the measures instated by the Montreal Protocol have had a significant effect. Atmospheric data is beginning to show a positive impact upon the environment: effective atmospheric chlorine levels are falling (Image 10.15). Projections suggest that by 2050, mid-latitude ozone will have largely

***NOTE**

While CFCs represent a huge class of ODS, there are others. Carbon tetrachloride, for instance, was commonly used as an industrial solvent until its phase-out, while methyl bromide found application as a pesticide. The Montreal Protocol sought to address all possible ODS, regardless of application or chemical class.

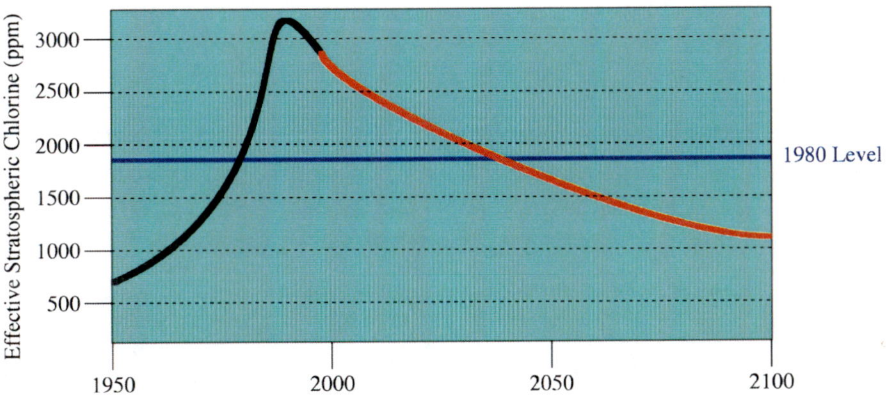

Image 10.15. Effective stratospheric chlorine levels show the effects—and projected effects—of the Montreal Protocol (data from NOAA [National Oceanic and Atmospheric Administration]).

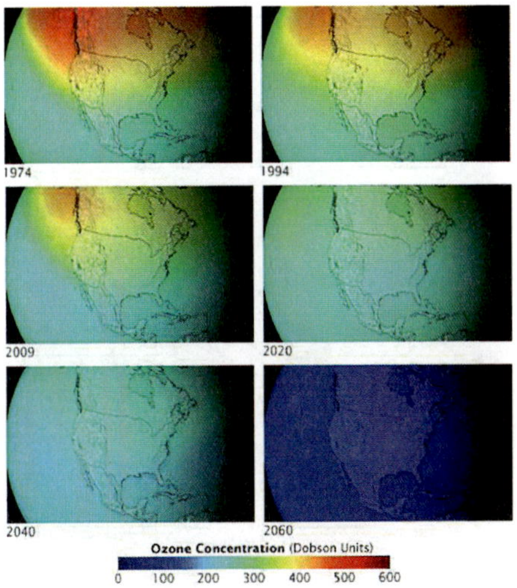

Image 10.16. NASA data projecting the ozone depletion had there been no Montreal Protocol.

recovered, though it is expected that polar ozone will require an additional fifteen years to regain its normal thickness.

Worldwide cooperation and emerging evidence of the treaty's effectiveness make the Montreal Protocol one of the most effective global environmental treaties ever developed—one need only glance at the NASA projection for the ozone destruction timetable had the treaty never been implemented (Image 10.16) to appreciate the magnitude of the accomplishment.

10.6 REFRIGERANTS—MOVING FORWARD

Still, while it is tempting to write off the Montreal Protocol as an unmitigated success that has heralded a new age of global cooperation and environmental security, the story doesn't end there. Once a technology has been implemented, it's nearly impossible to take it away again. As such, not one of the signatories expected that CFCs and other ODS would be phased out without replacements. DuPont and other companies quickly provided the first-generation substitute refrigerants—**hydrochlorofluorocarbons (HCFCs)** (Image 10.17). These compounds are more reactive than CFCs and have shorter lifetimes in the troposphere, meaning their ozone-depleting potential is somewhat less than that of the CFCs. The HCFCs were accepted as temporary substitutes and scheduled for phase-out by 2020 in developed nations (2040 in developing nations). Unfortunately, they are exceedingly potent greenhouse gases, and contribute to climate change (see Chapter 13).

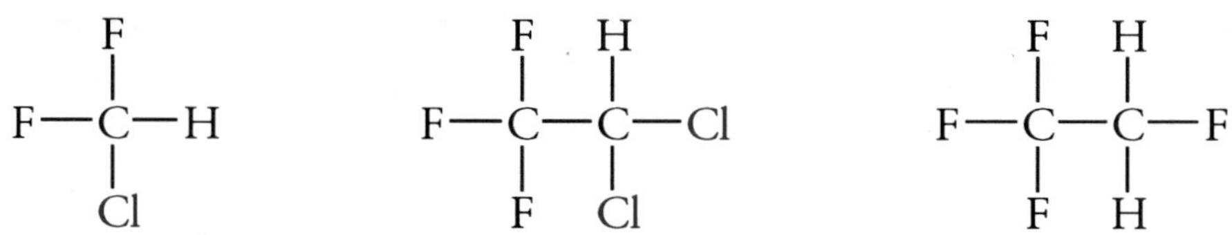

Freon 22 Freon 123 Freon 134a

Image 10.17. Two HCFCs (left and center) and an HFC (right).

The second-generation replacement substances are **hydrofluorocarbons (HFCs)** (Image 10.17), whose use is expected to rise sharply in the coming decades as developing nations reach the end of their CFC and HCFC grace periods. These compounds lack any chlorine, and have no ozone-depleting potential. Regardless, their astronomical potential to contribute to climate change (they are powerful greenhouse gases) has begun to cause concern in the scientific community. The United States and several other countries have recently proposed adding HFCs to the list of banned compounds under the Montreal Protocol—not for their effect upon the ozone layer, but for their environmental impact elsewhere.

One promising refrigerant, called **greenfreeze**, is a mixture of propane and isobutane (Image 10.18). Not only do these compounds have zero ozone-depleting potential and minimal global warming potential, their energy efficiency in refrigerators and air conditioners is as good or greater than that of HFCs.

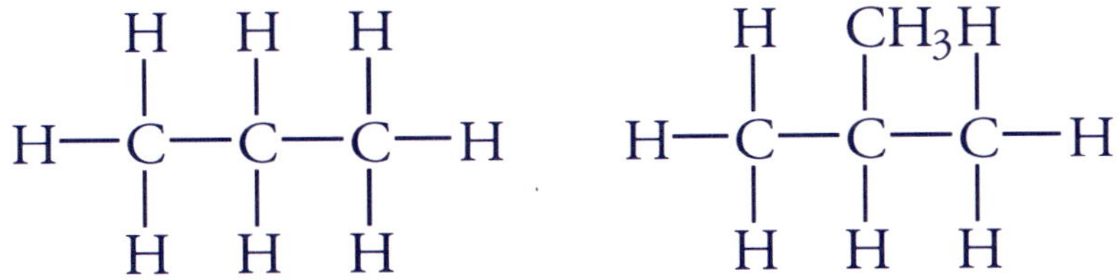

Propane Isobutane

Image 10.18. The components of greenfreeze.

The politics of greenfreeze have been interesting—and revealing. Originally developed in Germany, the technology captured the attention of the public: a consortium of fridge manufacturers immediately responded by denouncing greenfreeze on the grounds that it was not a feasible refrigerant. Regardless, one company proceeded with the manufacture of a greenfreeze-reliant refrigerator, which ended up winning an award from the German Ministry for the Environment. In 1993 the fridge manufacturers who had initially denounced the technology responded to public pressure and unveiled plans to produce fridges utilizing greenfreeze. Since that time, most European nations and many other countries around the world have switched to the greenfreeze coolant. The technology has largely been ignored in the United States, ostensibly because of the flammability of hydrocarbons (despite the empirical observation that widespread implementation of the technology elsewhere has proceeded without incident). It cannot be discounted that there may be pressure from chemical companies in the United States to avoid switching to greenfreeze-reliant refrigerators and air conditioners, on the grounds that the coolant compounds are not patentable.

10.7 THE NITROGEN CYCLE

The ozone-oxygen cycle is not the only cycle that has been perturbed during the last century. As we'll see in Chapter 13, the carbon cycle has also been badly disrupted. Perhaps no elemental cycle has been so altered, however, as that of nitrogen. The natural nitrogen cycle is maintained by a delicate balance of life and death, of assimilation and release. Abundant in the atmosphere, N_2 is nevertheless nonreactive—most living organisms have no ability to incorporate it chemically. An essential atomic ingredient for many of the molecules of life, including proteins and DNA, nitrogen must first be *fixed*, or made into a reactive form, before it can be taken up and utilized by plants. While a limited amount of nitrogen is fixed through lightning activity, most enters the cycle thanks to the efforts of **nitrogen-fixing bacteria**, which are found both in soil and in the root nodules of leguminous plants (including alfalfa, soy, and beans). The nitrogen-fixing bacteria produce NH_3, a basic compound that quickly reacts with water to form **ammonium, NH_4^+** (Image 10.19). This compound is converted to nitrite (NO_2^-) and then to **nitrate (NO_3^-)** by soil bacteria. Both ammonium and nitrate can be assimilated by plants, which use the compounds to produce the building blocks of DNA, amino acids, and other biomolecules.

CONCEPT CHECK:

Why can't plants and animals take up N_2 directly from the atmosphere?

Answer: N_2 is very unreactive and must be chemically transformed into a reactive species before it can be used by plants and animals.

Image 10.19. The nitrogen cycle.

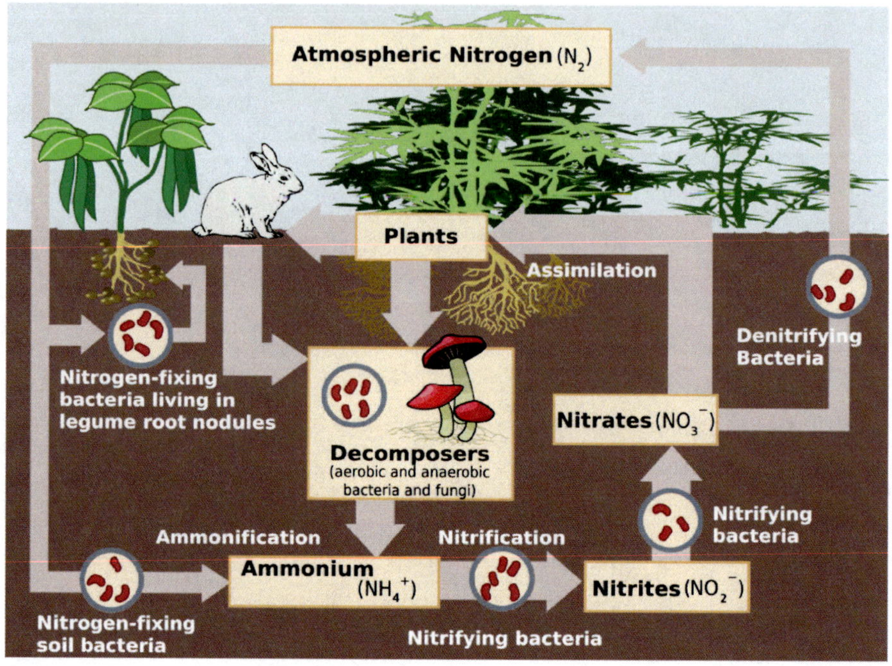

The nitrogen cycle is actually somewhat more entangled and complex than presented above. Reactive nitrogen not assimilated by plants can be utilized and released into the atmosphere as N_2 by still other varieties of soil bacteria. Plants (and the herbivores that feed on them and the carnivores that feed on the herbivores) die and decay. When they do, decomposers including bacteria and fungi return the nitrogen to the soil as NH_4^+. Similarly, nitrogenous waste excreted by animals returns reactive nitrogen to the soil.

Regardless of the nitrogen cycle's complexity, two fundamental truths remain. First, in a naturally operating nitrogen cycle, all nitrogen incorporated into any molecule in any multicellular organism can be traced back to nitrogen-fixing bacteria. Second, the cycle represents a dynamic equilibrium, just like that observed between ozone and oxygen; while reactions occur constantly, they are in balance with one another, and total concentrations of each nitrogen species do not change with time.

Our ability to affect the nitrogen cycle was minimal until the twentieth century; we simply lacked the technology to reproduce the nitrogen fixation carried out by soil bacteria. Nevertheless, reactive nitrogen was an important commodity. Farmers, for instance, needed it to fertilize their crops, a feat generally accomplished by spreading manure on fields. It was also common to grow crops in cycles, rotating plants that depleted the soil of nitrogen, like corn, with leguminous plants that enriched the soil, like soy. While this limited the growing capacity of the land—a plant grows within the confines of the raw materials available to it—the system seemed to work well enough.

10.8 THE HABER-BOSCH PROCESS

The real leap forward in nitrogen-fixing technology was inspired not on the farm, but on the battlefield. In addition to providing for the needs of plants,

Image 10.20. A saltpeter mine in the Chilean desert.

reactive nitrogen is also a crucial ingredient in the production of explosives. Specifically, nitric acid (HNO_3), which can be made either from ammonia or nitrate, is essential to the manufacture of TNT, nitroglycerine, and similar compounds. Natural, concentrated sources of nitrates are uncommon—in fact, one of the few known reservoirs of nitrate in the early twentieth century was the arid desert of Chile, with its huge saltpeter ($NaNO_3$) deposits.

During the first World War, Germany found the British Navy firmly ensconced between them and Chile, leaving the Germans in desperate need of an alternate source of reactive nitrogen. Fritz Haber, a German chemist, solved the problem by using high pressure and high temperature to accomplish what hitherto only soil bacteria had been able to do—he fixed nitrogen:

$$N_{2(g)} + 3\ H_{2(g)} \rightarrow 2\ NH_{3(g)}$$

This reaction, initially called the Haber process but later renamed the **Haber-Bosch process** in honor of the chemist who industrialized it, had far-reaching ramifications. Throughout World War II, the United States was among the many countries manufacturing huge quantities of ammonium nitrate (NH_4NO_3) for explosives manufacture.* At the end of the war, the United States found itself sitting on a large pile of reactive nitrogen, with a greatly reduced need for explosives. More importantly, a generation of young men coming home from the war needed jobs during peacetime. The problem of what to do with the ammonium nitrate—and how to create new jobs for the soldiers—was eventually solved by maintaining and expanding the ammonium nitrate factories, putting the young men to work in manufacturing, and

***NOTE**

The fixation process complete, transformation of ammonia to ammonium (and to nitrate) is a synthetically trivial process, easily accomplished in the lab.

CONCEPT CHECK:

What does the Haber-Bosch process do?

Answer: It chemically fixes nitrogen to a reactive species.

distributing the product to farmers as fertilizer. This particular policy forever changed not only how food is grown, but from what it's grown.

10.9 RISKS AND BENEFITS OF THE HABER-BOSCH PROCESS

Prior to the advent of chemical fertilization, plants were very nearly sources of "free" energy—they took up CO_2, water, and sunlight (a process that will be examined in greater detail in Chapter 11), and they grew. While the farmer did need to spread manure from his cows or other stock on the field from time to time, and while he certainly expended personal energy in the pursuit of agriculture, no fuel was required to effect plant growth—all the necessary power was provided by the Sun. Chemical fertilization changed the energetics of agriculture drastically. The Haber-Bosch process, for all its utility, is a tremendous consumer of energy. The requisite pressures and temperatures to react elemental nitrogen with hydrogen are achieved only through the combustion of large quantities of fuel. In the same way that fuel is burned to run an automobile engine, fuel—generally petroleum—is now burned to fix nitrogen. The production and utilization of chemical fertilizer shifted the energy balance of food production, for the first time, into a deficit: It takes more than a calorie of petroleum to produce a calorie of food.

Industrial nitrogen fixation has had other effects as well. Cheap, abundant fertilizer encourages over-fertilization of fields, leading to huge quantities of reactive nitrogen entering the soil. Some of this is converted to NO_2, augmenting the formation of acid rain. Reactive nitrogen is also washed out of the soil through the process of field irrigation and can enter local municipal water supplies. Nitrite, an anion produced from nitrate in aqueous solution, binds to hemoglobin and can lead to inadequate oxygenation of tissues in individuals drinking the water. Also, reactive nitrogen washed off fields eventually works its way into lakes, rivers, and the ocean, where it acidifies aqueous environments and kills aquatic organisms. Finally, recall that many tropospheric pollutants (and greenhouse gases, as we'll learn in Chapter 13) are reactive nitrogen species themselves—NO, NO_2, and N_2O all occur in higher concentration where nitrogen fertilizer has been applied.

The end result of the Haber-Bosch process is not to increase the amount of nitrogen in the cycle—to do so is simply not possible, as atoms of nitrogen can't be created through chemical processes. Rather, the process accelerates the fixation of nitrogen, increasing the concentration of reactive species without accelerating the denitrification reactions that restore N_2. There is no concern that we will run out of N_2—it's tremendously abundant and serves merely as a "holding tank" for nitrogen atoms. Rather, the acceleration of nitrogen fixation increases the concentrations of reactive nitrogen species in the world, with myriad effects. The Haber-Bosch process has significantly accelerated global nitrogen fixation, and has become the major mechanism for formation of reactive nitrogen in the developed world. As a means of illustration, recall that a century ago, every atom of nitrogen in an individual's body would have been fixed by soil or root-nodule bacteria. Today, however, only about 2 percent* nitrogen in a person's body at any one time was ever fixed by bacteria—the remaining 98 percent was fixed in an industrial lab!

Unfortunately, where science and society intersect on a global scale, complications abound. It would be short sighted to characterize the Haber-Bosch process as having had a negative effect upon humanity. With the global population increasing rapidly, starvation and the availability of food are pressing issues. There is no doubt that chemical fertilization increases the growing capacity of the land (crop yields per acre of land are now as much as 15 times what they were 100 years ago), which increases the quantity and decreases the price of food. Whether the population would have stayed smaller or, perhaps more likely, starvation and disease become more rampant had the Haber-Bosch process never been conceived is anyone's guess.

Speculation aside, there are important lessons to be taken from this discussion. First, it is difficult (if not impossible) to anticipate all the ramifications of a new technology. Second, global cycles are interconnected, and a disturbance in one arena tends to show its effects elsewhere; nitrogen fixation is a thread connecting pollution and ozone, energy, and climate change. Finally, many of our global dilemmas were conceived as solutions to problems, and have since become the problems for which we are seeking solutions. CFCs allowed for population expansion and industrial growth, at the expense of the ozone layer. HFCs appeared to be a promising replacement refrigerant, until attention turned to climate change. The same nitrogen-fixing technology that has fed the world since the 1940s is now revealing itself to be responsible for many burgeoning crises. As much as we owe our global dilemmas—the ozone hole and climate change, acid rain and pollution, destruction of aquatic life and deforestation—to technology, we must remember that the innovations were implemented for a reason.

***NOTE**

Assuming the individual in question is a resident of a developed country and eats a "typical" industrialized diet.

10.10 THE PRECAUTIONARY PRINCIPLE REVISITED

In our discussion of tetraethyl lead (Chapter 6), we saw an early attempt to use the precautionary principle as a means of forestalling an innovation for which there was little scientific evidence with regard to risk. A concerned physician suggested that the burden of proving tetraethyl lead's safety lay upon those wishing to manufacture and distribute it—and that in the absence of proof, caution should prevail. In the end, the substance was approved on the grounds that there was no evidence of risk, and history bears witness to the result.

Environmental advocates often point to this example to demonstrate the power of the precautionary principle: had it been heeded, lead pollution and lead-related deaths would have been avoided. This is a compelling argument. Looking back, few would suggest that tetraethyl lead should ever have made it out of the development phase and into the hands of the public. Of course, in hindsight, such judgments are easily made. It's much harder to accurately assess risk when a technology is new.

Image 10.21. A U.S. soldier being sprayed prophylactically with DDT.

The precautionary principle's strength in hindsight is its greatest weakness when looking forward. As opposed to risk-benefit analysis, which takes into account the contribution an innovation will make, the precautionary principle focuses solely on risk. Further, it demands a logical impossibility—the proof that no risk exists. In reality, no activity is without risk, and had the precautionary principle been applied rigorously throughout the course of human civilization, we could not have progressed beyond building fires in caves (if we had ever allowed ourselves to build fires at all!). Risk is unavoidable, inherent in day-to-day life, and certainly inextricable from technological advancement. A risk-benefit analysis makes allowances for the inevitability of risk—in performing such an analysis, we compare an innovation's contribution to its (anticipated) potential for harm, and then we acknowledge and accept reasonable risk. As discussed earlier in this chapter, a risk-benefit analysis is most powerful when it is most flexible: we do well to adapt our analysis as new evidence presents itself.

Different paths appear sensible depending upon the type of analysis—precautionary principle or risk-benefit—used to determine a course of action. For instance, a 2001 United Nations convention (Stockholm) addressed the use of the pesticide **dichlorodiphenyltrichloroethane (DDT)**. Originally applied during the second half of World War II, DDT was highly effective at preventing epidemics of the diseases typhus and malaria, both of which are spread by mosquito vectors and which threatened U.S. troops in the South Pacific.

Later, DDT was used as an agricultural pesticide, with remarkable effectiveness. Regulation was nonexistent, resulting in farmers applying up to 25 times the necessary quantity to crops. A *persistent organic pollutant*, DDT concentrates in the soil and degrades slowly. The compound is stored in fatty tissue and bioaccumulates, meaning that it concentrates up the food chain. Insects exposed to nonlethal doses of DDT store the chemical. The insects are eaten by small birds, which end up storing the combined quantity of DDT from the insects. Large predator birds, then, concentrate the compound further as they feed on small birds, and so forth. The result of this bioaccumulation is that animals nearest the top of the food chain are exposed to the highest doses of the chemical. In 1962 the American biologist Rachel Carson published *Silent Spring*, an exposé of DDT's environmental impact. A U.S. ban on the insecticide followed shortly thereafter, and in the time since, the environment has largely recovered. The United Kingdom continued using DDT for many years after the U.S. ban—albeit in much lower doses—and while wildlife testing revealed some of the compound in large predators, no wildlife deaths or ill effects were ever recorded.

Unfortunately, many replacements for DDT are more toxic, more expensive, or less effective. This has not presented a problem in developed countries (which relied on the compound mainly as an agricultural pesticide and which have the economic means to utilize more expensive substitutes), but developing nations have struggled to curtail DDT use. The UN restricts widespread agricultural spraying, which uses large amounts of the compound, has the potential to adversely affect the environment, and can lead to DDT resistance

CONCEPT CHECK:

How is DDT similar to mercury in its environmental effects?

Answer: Both bioaccumulate and are found in highest concentration in top predators (see Chapter 6 to review this concept).

in mosquitoes. However, developing nations continue to find the insecticide highly effective in the prevention of malaria. It was this use of DDT—in which small amounts of the chemical are sprayed on interior house walls—that was addressed in the 2001 Stockholm Convention.

Using the precautionary principle as a guide, one would be left with no choice but to categorically ban DDT for all uses: the risks are clear and unequivocal, the compound is pollutant and toxic, and the dangers to wildlife are known. While human toxicity effects of DDT have never been demonstrated, it is reasonable to extrapolate that risk from observed animal toxicity.

A risk-benefit analysis, however, leads us to a somewhat different conclusion. While the benefits of agricultural use of DDT are outweighed by the known risks, the risk of interior wall spraying is much lower, since far less chemical is used. Further, the benefits are tremendous. With nearly 300 million cases of malaria per year, a million of which lead to death (mostly in Africa, and mostly among children), there is no doubt that malaria is a major public health concern. Several countries including South Africa and Ecuador have reduced antimalarial DDT use in response to global pressure, only to watch the disease skyrocket. No other pesticide has the combination of effectiveness, relative safety, and low cost afforded by DDT. After careful consideration, the Stockholm Convention—to which there are currently 169 parties—determined to allow the use of DDT for disease vector control, while stipulating strict safety measures and routine reevaluation of the risk-to-benefit ratio. It is nearly impossible to estimate the number of lives that have been saved through this decision.

In facing global challenges, there is much we can learn from the Montreal Protocol, which serves not simply as an important and successful treaty, but as an example of the need for flexibility and adaptation. Important as technology is to the advancement of our society, we must remember that risk is ubiquitous—no activity, technology, or chemical is without it—and that advancement comes only through conceding to acceptable risk. We must constantly revise our risk-benefit analysis as risks, which can never be fully anticipated, present themselves, without forgetting that once a technology is introduced, our way of life is transformed, and it can be difficult or impossible to backtrack. We must be willing to add or subtract technologies from our "phase-out" list as new information becomes available, and we must be sensitive to the burden this places on the developing world.

10.11 THE LAST WORD—SUNLIGHT IN THE TROPOSPHERE

For all the importance of the ozone layer in protecting us from too much of the Sun's ultraviolet rays, life on Earth would not be possible without our star and the full spectrum of its radiation. While UV rays can destroy tissue, causing sunburn and worse, they also allow for the synthesis of vitamin D from cholesterol in the skin. UV's ability to break bonds leads to chemical reactions—some to our detriment, and some that we couldn't survive without.

UV is not the only radiation from the Sun, however. Also emitted is infrared (IR) radiation, which plays an important role in maintaining the planet's livable temperature, as we'll see in Chapter 13. Perhaps most familiar to our everyday experience, though, is the visible light from the Sun. This provides a mechanism for sensing our environment, is the fuel that powers plant growth, and translates into the rich palette of colors that paints our world. What we observe as "colorless" light is, in fact, a mix of the entire visible spectrum. Our eyes respond to wavelengths from approximately 380 nm (violet, highest in energy) to 750 nm (red, lowest in energy). Some molecules do not absorb visible light, and hence appear colorless. For this reason, the air around us (mostly N_2 and O_2, neither of which absorbs in the visible portion of the spectrum) has no color. Other molecules, however, absorb some or all of the visible spectrum and appear colored as a result.

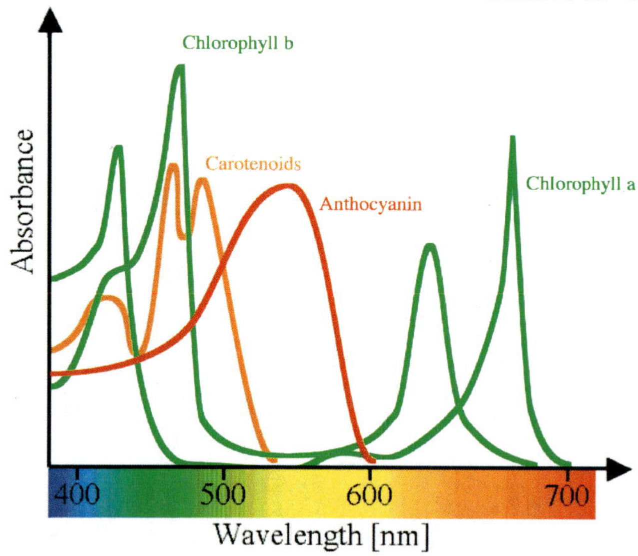

Image 10.22. The light absorption of various leaf pigments.

Tree leaves, for instance, are a familiar green during spring and summer months. A major light-absorbing molecule in leaves is *chlorophyll*, which plants use to collect sunlight for energy. Chlorophyll appears green not because it absorbs green light, but because it absorbs other colors of visible light, leaving green (and yellow) behind (Image 10.22). The unabsorbed light is reflected off the molecule and into the eye of whomever or whatever happens to be looking at the leaf. Instead of sensing a full spectrum of light, the eye senses only yellow and green, leaving the observer with an impression of the leaf's color.

Image 10.23. Fall color brings out brilliant yellows in aspen leaves and fiery reds in maple. (Aspens © Scott Lefler, 2007. Maple leaves © Scott Lefler, 2007.)

Chlorophyll, it turns out, is not the only pigment present in a leaf. A second class of pigments is the *carotenoids*, so named for their presence in carrots, and with a predictable color given the source. Carotenoids assist chlorophyll in collecting sunlight (this will be discussed more in Chapter 11), but their absorption spectrum is somewhat different than that of chlorophyll. Rather than absorbing orange and red light, as chlorophyll does, carotenoids absorb only violet, blue, and green. Generally, we don't see the color of carotenoids in leaves—the orange and red light they reflect is absorbed by chlorophyll, and neither molecule absorbs much yellow. Where carotenoids make their appearance is in the fall.

Dropping temperatures and decreasing light cause trees to slow their growth in fall and winter months. Chlorophyll, normally produced in large amounts and regularly recycled by the plant, is no longer synthesized. As chlorophyll concentrations in the leaves decrease, the orange and red light normally absorbed by the molecule is instead reflected. Carotenoids persist longer than chlorophyll in the leaf, meaning that for a brief and dramatic period at the end of the growing season and before the leaves are lost for the winter, the trees sport a fiery palette of yellow, orange, and red.

Augmenting the fall color is a third class of compounds, the anthocyanins. These are not generally produced through the growing season, nor are they manufactured by all species. Instead, synthesized from residual sugar in the leaf at the beginning of fall as a sort of arboreal sunscreen, the anthocyanins absorb the highest-energy visible light—violet through yellow—leaving behind the more benign orange and red wavelengths. The purpose is to prevent excess light (which is no longer needed to power growth) from being absorbed by and damaging inactive light-harvesting machinery. It is an eminently practical, and breathtakingly beautiful, phenomenon.

Our relationship with the Sun is a bit like our relationship with technology: it is at once essential and detrimental, exposing us to countless risks and securing incalculable benefits. It is creative and destructive, providing for life and taking it away again. We must be conscientious and take the necessary precautions, but the fact is unavoidable that the Sun, like technology, is absolutely essential to our way of—and our experience of—life on our blue-green (violet and yellow, and red and orange) planet.

SUMMARY OF MAIN POINTS

- Refrigeration depends upon a cycle of vaporization (absorbs heat) and condensation (releases heat) of a refrigerant liquid.
- Good refrigerants must not be toxic, reactive or explosive, and must have boiling points less than the melting point of water.
- The Freons (CFCs) were developed by Thomas Midgley in the late 1920s as a class of refrigerants; they are very chemically inert.
- The ozone layer is located in the stratosphere and shields us from UV light.
- UV is a form of EMR: All EMR has the same speed (written *c* in equations) and an inversely related wavelength and frequency.
- The wave equation $\mathbf{c} = \lambda\nu$ relates measurable quantities for EMR.
- The energy of EMR is directly related to frequency by the equation $E = h\nu$, where $h = 6.626 \times 10^{-34}$ J·s.
- High-energy EMR (gamma rays, X-rays, and UV) are ionizing radiation and can break bonds in molecules (including DNA, leading to cancer).
- Low-energy EMR (visible light, IR, microwaves, and radio waves) is not ionizing radiation and can't break bonds.
- The Sun emits UV, visible light, and IR; the UV is far too intense to be compatible with life in the absence of the ozone layer.
- The Chapman Cycle describes the dynamic equilibrium whereby ozone is converted to O_2, and O_2 back to ozone—both portions of the cycle require the absorption of UV light.
- Because of the Chapman Cycle, O_2 in the upper atmosphere absorbs much of the UV-C and UV-B incident upon the Earth, while O_3 absorbs UV-B.
- No UV-C, very little UV-B, and a fair amount of UV-A (which is the least damaging of the three types of UV light) reach the surface of the Earth.
- CFCs react with UV light, producing chlorine atoms which catalytically convert ozone to O_2.
- The idea that CFCs led to ozone destruction was theoretical at first, then verified by the discovery of the hole in the ozone layer.
- The ozone hole formed over Antarctica because it is the coldest place on Earth. The hole shows seasonal variation because polar stratospheric clouds form in winter, trapping chlorine and allowing for catalytic ozone destruction in the austral spring and summer.
- The Montreal Protocol and its amendments have put restrictions on CFCs and other ODS. Restrictions are more stringent for developed rather than developing nations.
- HCFCs were accepted as temporary substitute refrigerants, but also deplete the ozone layer.
- HFCs do not deplete the ozone layer, but are powerful greenhouse gases.

- Greenfreeze, a hydrocarbon refrigerant, has no potential to deplete the ozone layer or to affect the global climate, and has been widely accepted in many countries (but not in the United States).
- The Montreal Protocol has begun to show a positive effect upon the ozone layer, with full recovery by 2065 (assuming full compliance with the Montreal Protocol).
- Nitrogen is required by all living organisms, but must be fixed to a reactive form by soil bacteria before it can be assimilated.
- The Haber-Bosch process is an energy-dependent industrial process for fixing nitrogen that has greatly perturbed the global nitrogen cycle.
- Food production dependent upon chemical fertilizer represents a negative energy balance.
- The Haber-Bosch process has been responsible for greatly increasing global food production, which has had a positive effect upon starvation and disease worldwide.
- The precautionary principle is an examination of risk without consideration of benefit, and fails to allow for the inherent risk associated with any technological advancement.
- Risk-benefit analysis can be used to weigh risks, which are unavoidable, against benefit in the attempt to produce the best possible solution to a global dilemma.

Questions and Topics for Discussion

1. How is the relationship between chemistry and society different on the global level than on the individual or community level? How is it the same?
2. What does it mean to say "When a chemical, an invention, a discovery, or a technology emerges from the lab and debuts in the world, we are often greeted with unanticipated effects"? Is it possible to remove the element of surprise?
3. On a molecular or atomic level, why do you think vaporization is endothermic? What is happening? (Hint: think about intermolecular interactions).
4. Why is the air behind a refrigerator warm?
5. Why is it important that a refrigerant gas not be explosive or toxic?
6. How is EMR similar to an ocean wave? How is it different?
7. Based only upon our discussion of EMR so far, do you think exposure to microwaves can cause cancer? Why or why not?
8. What is a dynamic equilibrium, and why is it important to our discussion of ozone?
9. Why does more UV-A reach Earth's surface than UV-C?
10. Why do you think that the initial public response to the *New York Times* article about ozone description didn't have a long-term effect?
11. Why did NASA satellite data miss spotting the ozone hole?
12. What elements of the Montreal Protocol made it successful? What lessons can we learn from this?
13. What are the special challenges associated with international environmental treaties? How can they be overcome?
14. Has the Montreal Protocol been successful? Explain your reasoning.
15. Why are HCFCs less ozone-depleting than CFCs? Why are HFCs not ozone-depleting? Why are HFCs considered a "dangerous" substance?
16. What does the response to greenfreeze in Germany and in the United States suggest about its potential as a refrigerant and about the politics of environmental science?
17. What is the nitrogen cycle, and why is it important?
18. Should the Haber-Bosch process be banned? Defend your position.

19. Apply the precautionary principle to an existing technology (microwave ovens, airplanes, vaccines). According to the principle, should the technology have been implemented?
20. Apply risk-benefit analysis to an existing technology (microwave ovens, airplanes, vaccines). According to the analysis, should the technology have been implemented?
21. Do you think that the precautionary principle has utility in global decision-making? Why or why not?

Problems

1. Define: condensation, vaporization, CFC.
2. Name several ways in which CFCs changed the world (for the better).
3. Why do we call ozone protective in the stratosphere, but a pollutant in the troposphere?
4. Define: EMR; wavelength; and frequency.
5. What is the ramification of all EMR traveling at the same speed?
6. Place the following waves in order from highest frequency to lowest frequency: Wave 1, ν = 1.1×10^{22} Hz; Wave 2, ν = 5.6×10^{15} Hz; Wave 3, ν = 7.6×10^{3} Hz.
7. Place the following waves in order from highest frequency to lowest frequency: Wave 1, λ = 5.5×10^{2} m; Wave 2, λ = 3.1×10^{-6} m; Wave 3, λ = 7.5×10^{0} m.
8. Place the following waves in order from longest wavelength to shortest wavelength: Wave 1, λ = 1.2×10^{-15} m; Wave 2, λ = 9.1×10^{-4} m; Wave 3, λ = 8.2×10^{-11} m.
9. Place the following waves in order from longest wavelength to shortest wavelength: Wave 1, ν = 8.4×10^{0} Hz; Wave 2, ν = 5.1×10^{20} Hz; Wave 3, ν = 2.2×10^{12} Hz.
10. What is the frequency of an EMR wave with λ = 5.4×10^{-16} m?
11. What is the wavelength of an EMR wave with ν = 4.1×10^{16} Hz?
12. Place the following waves in order from highest energy to lowest energy: Wave 1, ν = 3.6×10^{12} Hz; Wave 2, ν = 1.1×10^{8} Hz; Wave 3, ν = 2.4×10^{20} Hz.
13. Place the following waves in order from highest energy to lowest energy: Wave 1, λ = 7.0×10^{-10} m; Wave 2, λ = 6.3×10^{-20} m; Wave 3, λ = 7.1×10^{8} m.
14. What is the energy of an EMR wave with ν = 8.4×10^{12} Hz?
15. What is the energy of an EMR wave with λ = 7.6×10^{-14} m?
16. What is the frequency of an EMR wave with E = 4.1×10^{-15} J?
17. What is the wavelength of an EMR wave with E = 8.6×10^{-20} J?
18. Describe the effect of UV light on bonds in molecules.
19. Why does UV exposure lead to premature aging and cancer?
20. Why is the ozone layer important? Is ozone the only important component of the ozone layer?
21. How is the ozone layer maintained naturally? What is/are the overall reaction(s)?
22. What are the three types of UV, and which is most damaging? Which is highest in energy? Which has the longest wavelength? The lowest frequency?
23. Describe the reaction that a CFC undergoes in the stratosphere.
24. Why do CFCs cause catalytic destruction of ozone?
25. Why is the hole in the ozone layer located over Antarctica?
26. Why does the hole in the ozone layer get larger in the austral summer?
27. What alternatives were or are available to CFCs? What are the problems associated with each? Which is the most promising?
28. Why is it going to take so long for the ozone hole to repair itself?
29. What is the difference between reactive and nonreactive nitrogen?
30. What is the Haber-Bosch process, and what does it do?

31. What are some of the negative effects of chemical fertilization?
32. What does it mean to say that the energy balance of food production is negative?
33. What is the difference between the precautionary principle and risk-benefit analysis?

Answers to Problems

1. Condensation is gas turning to liquid; vaporization is liquid turning to gas; CFCs are chlorofluorocarbons—they were introduced as refrigerants in the 1920s.
3. In the stratosphere, ozone protects us from UV. The small amount of ozone in the troposphere isn't able to shield us from much UV, but it is a respiratory irritant and is a component of smog.
5. If all EMR travels at the same speed, then frequency and wavelength must be inversely related.
7. Wave 2, $\lambda = 3.1 \times 10^{-6}$ m; Wave 3, $\lambda = 7.5 \times 10^{0}$ m; Wave 1, $\lambda = 5.5 \times 10^{2}$ m.
9. Wave 1, $\nu = 8.4 \times 10^{0}$ Hz; Wave 3, $\nu = 2.2 \times 10^{12}$ Hz; Wave 2, $\nu = 5.1 \times 10^{20}$ Hz.
11. $\lambda = 7.3 \times 10^{-9}$ m.
13. Wave 2, $\lambda = 6.3 \times 10^{-20}$ m; Wave 1, $\lambda = 7.0 \times 10^{-10}$ m; Wave 3, $\lambda = 7.1 \times 10^{8}$ m.
15. $E = 2.6 \times 10^{-12}$ J.
17. $\lambda = 2.3 \times 10^{-6}$ m.
19. When bonds break in DNA (genetic material), cancer can result. When bonds break in structural proteins like collagen, wrinkles and premature aging of the skin can result.
21. $3\ O_2 \rightarrow 2\ O_3$ (high energy UV), $2\ O_3 \rightarrow 3\ O_2$ (lower energy UV).
23. CFC reacts with UV to release Cl. Cl reacts with ozone to produce ClO and O_2. ClO reacts with O to produce O_2 and release Cl.
25. Antarctica is the coldest place on Earth. In the winter, polar stratospheric clouds form above the continent and trap chlorine.
27. HCFCs were the first-generation alternatives—they were ozone-depleting and were greenhouse gases. HFCs are a second-generation alternative—they are not ozone-depleting, but are powerful greenhouse gases. Greenfreeze (propane and isobutane) is the newest alternative, and has no ozone-depleting or greenhouse potential. It is flammable, which has stalled its implementation by the United States.
29. Reactive nitrogen (any species except N_2) can be taken up by plants and assimilated in biomolecules. Nonreactive nitrogen must be fixed by bacteria before it can be used by plants.
31. Chemical fertilization leads to pollution of drinking water, surface water, and ocean water. It can increase acidity of soil, acid rain, and the concentration of pollutant and greenhouse gases.
33. The precautionary principle looks only at risk and requires the proof that a technology be risk-free prior to implementation. Risk-benefit analysis looks at both risk and benefit, assumes no activity is risk-free, and accepts a reasonable risk in exchange for a desired benefit.

Sources and Further Reading

Ozone Layer

http://ozonewatch.gsfc.nasa.gov/

Cicerone, R.J., Stolarski, R.S., and Walters, S. (1974). Stratospheric ozone destruction by man-made chlorofluoromethanes. *Science*, 185(4157), 1165–67.

Crutzen, P. (1974). Review of upper atmospheric photochemistry. *Canadian Journal of Chemistry*, 52(8), 1569–81.

Farman, J.C., Gardiner, B.J., Shanklin, J.D. (1985). Large losses of total ozone in Antarctica reveal seasonal ClOx/NOx interaction. *Nature*, 315(6016), 207–10.

Grabiel, D.F. (2007, September). Crucial crossroads. *Our Planet: the Magazine of the United Nations Environment Programme.*

James, R.W., and Missenden, J.F. (1992). The use of propane in domestic refrigerators. *International Journal of Refrigeration*, 15(2), 95–100.

Kowalok, M.E., (1993). Common threads—research lessons from acid rain, ozone depletion, and global warming. *Environment*, 35(6), 12–38.

Molina, M.J., and Rowland, F.S. (1974). Stratospheric sink for chlorofluoromethanes—chlorine atomic-catalyzed destruction of ozone. *Nature*, 249(5460), 810–12.

Stolarski, R.S., and Cicerone, R.J. (1974). Stratospheric chlorine—possible sink for ozone. *Canadian Journal of Chemistry*, 52(8), 1610–15.

Sullivan, W. (1974, September 26). Tests show aerosol gases may pose threat to Earth. *New York Times*, p. A1.

Nitrogen Fixation

Le Couteur, P., and Burreson, J. (2003). *Napoleon's Buttons: 17 Molecules That Changed History*. New York: Penguin Group.

Pollan, M. (2006). *The Omnivore's Dilemma*. New York: Penguin Group.

Smil, V. (2001). *Enriching the Earth: Fritz Haber, Carl Bosch, and the Transformation of World Food Production*. Massachusetts Institute of Technology.

DDT and Malaria

www.who.int/topics/malaria/en/

Carson, R. (1962). *Silent Spring*. New York: Houghton Mifflin.

Timbrell, J. (2005). *The Poison Paradox*. Oxford: Oxford University Press.

United Nations Environment Programme. (2001). *Stockholm Convention on Persistent Organic Pollutants (POPs)*. UNEP/Chemicals/2001/3. 50 p.

"Current" Events—Electrochemistry 11

Electricity is so commonplace that we rarely stop to think about what it is and how it's generated. Upon reflection, we know it's a form of energy, since it can be used to power machines. We can even surmise that electricity must therefore be transformed from another form of energy, because the first law of thermodynamics reminds us that energy can never be created or destroyed. Still, our everyday experiences with electricity tend to leave us a bit mystified with regard to its true nature—lightning strikes seem to come out of thin air and vanish without a trace. A mere flip of a switch turns a flashlight on, and just as quickly turns it back off again. What's really going on in the clouds or inside the device to produce these events? Why do a few AA batteries make a flashlight work—and why can we recharge some batteries, but not others? Finally, how can sunlight be used to produce electricity, and is there a future for solar power as an alternative energy source? In order to answer these questions, we must first look at the nature of electrical energy.

Image 11.1. Lightning, power lines, and batteries are just a few of the phenomena and objects associated with electrical energy.

11.1 REDOX

We saw in Chapter 8 that a box held above the ground has gravitational potential energy: the box is at a higher energy state when separated from the Earth, and the force of gravity will act upon the box to move it toward a lower energy state. Like any high-energy to low-energy transition, this happens spontaneously, with a release of energy. In much the same way, the electrostatic force between charged particles can result in particle motion. If separated from a negative charge, a positive charge has **electrical potential energy** due to the electrostatic attraction between particles. The charges will spontaneously rush toward each other, and as they start to move, their potential energy is transformed first into energy of motion (kinetic energy), and then into other forms of energy, depending upon the specific situation. We generally refer to moving charge simply as a **current**, which has **electrical energy**. The only difference between this scenario and that of the falling box is that electrostatic forces are responsible for both attraction and repulsion, depending upon the charges involved. For instance, two negatively charged particles near each other in space also have electrical potential energy—their proximity represents a high-energy state, and they will spontaneously move away from each other. The same is true for two positively charged particles. In the end, though, the result is the same: Potential energy can be transformed into other forms of energy, and used to do work. Of course, electrical energy has to come from somewhere (all energy does!), and one common source is chemical energy.

Not all chemical reactions have the potential to create electrical energy. In fact, while all reactions involve electrons (in the sense that bonds are broken and formed), only some involve the actual transfer of electrons from one atom to another, and it is this transfer of electrons that can be used to generate electrical energy. Reactions involving electron transfer have two components. One atom or molecule must lose electrons—this process is called **oxidation**. Another atom or molecule must gain electrons through a process called **reduction**.* Let's revisit an old reaction that we saw for the first time in Chapter 4—the formation of ions. We looked at the ionization of Na and F, which took place through an electron transfer:

***NOTE**

There's a mnemonic to help you remember this—*LEO the lion says GER. LEO* stands for "Loss of Electrons—Oxidation," and *GER* stands for "Gain Electrons—Reduction." Apparently, aside from helping us remember which reaction is which, we also learn from this mnemonic that LEO the lion can't spell!

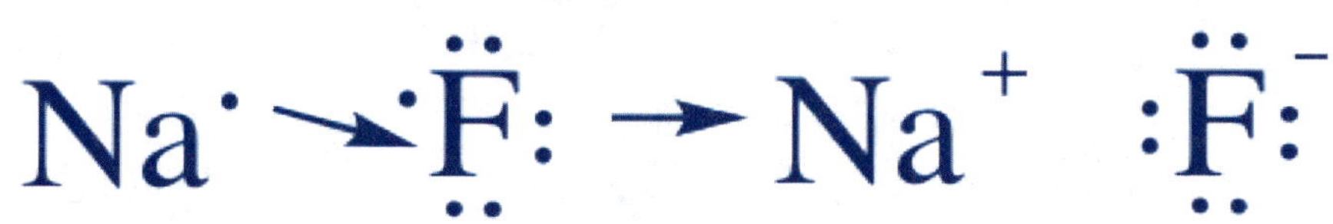

It wouldn't be possible for Na to form the Na^+ ion without F (or some other atom) present to accept the electron—electrons can't just disappear! Further, F couldn't ionize to F^- without the Na (or some other atom) present to donate an electron, since electrons can't come out of thin air. In fact, the reaction between Na and F is an oxidation/reduction reaction in which Na is oxidized and F is reduced. This same logic holds true for all electron transfer reactions: oxidations and reductions must always occur in pairs. For this reason, electron transfers are often called **redox** reactions, which is shorthand for reduction/oxidation.

We could write out the redox reaction between Na and F as a complete chemical reaction:

$$Na + F \rightarrow Na^{+} + F^{-}$$

Alternately, we could keep explicit track of the electron being transferred in the reaction by separating it into two parts:

$$Na \rightarrow Na^{+} + e^{-}$$
$$F + e^{-} \rightarrow F^{-}$$

CONCEPT CHECK:

How do we know that sodium is losing one electron when it reacts with fluorine?

Answer: It has one electron in its valence; it forms a cation with a charge of +1.

The above are called **half-reactions**, and they give us insight into the electron transfer taking place. We know sodium is losing an electron, because we understand the principles of ionization. However, even if we didn't understand the principles of valence that drive sodium's loss of its outermost electron, we could still surmise (using conservation of charge) that for Na to become Na^{+}, a single electron must be lost. Similarly, even without our understanding of ionization, conservation of charge tells us that F can only become F^{-} through gaining an electron. Half-reactions can be written from redox reactions simply by analyzing the change in charge. It's important to note that correctly written half-reactions can be added up to give the original redox equation.* Further, the oxidation half-reaction (the one in which electrons are a product) must always contain the same number of electrons as the reduction half-reaction (the one in which electrons are a reactant) in order to sum to a correctly balanced chemical equation. Let's revisit another familiar ionization reaction, this time finding the overall balanced equation by finding and summing the individual half-reactions. This can be determined by following a series of steps:

***NOTE**

Though doing so may involve some multiplication. In the example involving Mg and F, the reduction half-reaction may be written either as $F + e^{-} \rightarrow F^{-}$, or as $2\ F + 2\ e^{-} \rightarrow 2\ F^{-}$.

Note that since the second reaction is the first multiplied through by a factor of two, it's really the same reaction written a different way.

1) Determine the individual half-reactions involved:

We can use principles of ionization to determine the relevant half-reactions.

Oxidation: $Mg \rightarrow Mg^{2+} + 2\ e^{-}$
Reduction: $F + e^{-} \rightarrow F^{-}$

Note that these half-reactions do not contain equal numbers of electrons. When we looked at the magnesium and fluorine reaction in Chapter 4, we used Lewis structures to help us determine how many of each atom would be involved in a balanced reaction. Now, however, we can use the half-reactions—as long as the number of electrons lost in the oxidation half-reaction is equal to the number required in the reduction half-reaction, our half-reactions will add up to a correctly balanced overall equation.

2) Multiply one or both half-reactions by a whole number value to obtain equal numbers of electrons in the two half-reactions:

In this case, we multiply the fluorine reaction by 2; now both half-reactions involve two electrons.

$2\ F + 2\ e^- \rightarrow 2\ F^-$

3) Add the half-reactions to determine the overall reaction:

$$
\begin{array}{l}
\ \ Mg \rightarrow Mg^{2+} + 2\ e^- \\
\underline{+\ 2\ F + 2\ e^- \rightarrow 2\ F^-} \\
\ \ Mg + 2\ F + 2\ e^- \rightarrow Mg^{2+} + 2\ e^- + 2\ F^-
\end{array}
$$

4) Just as in a regular addition problem, eliminate quantities appearing on both sides of the equation:

$Mg + 2\ F + \cancel{2\ e^-} \rightarrow Mg^{2+} + \cancel{2\ e^-} + 2\ F^-$

Note that this does not mean the electrons weren't involved, it just means they were transferred from Mg to F and aren't net reactants or products.

5) Rewrite the equation in its final form:

$Mg + 2\ F \rightarrow Mg^{2+} + 2\ F^-$

TRY THIS:

Write a balanced equation for the reaction of Al with O, using the half-reaction method.

Answer: $2\ Al + 3\ O \rightarrow 2\ Al^{3+} + 3\ O^{2-}$

Solution:

Oxidation: $Al \rightarrow Al^{3+} + 3\ e^-$

Reduction: $O + 2e^- \rightarrow O^{2-}$

Oxidation, multipled through: $2\ Al \rightarrow 2\ Al^{3+} + 6\ e^-$

Reduction, multiplied through: $3\ O + 6e^- \rightarrow 3\ O^{2-}$; electrons in half-reactions are now equal.

Combined: $2\ Al + 3\ O + 6\ e^- \rightarrow 2\ Al^{3+} + 3\ O^{2-} + 6\ e^-$

Final: $2\ Al + 3\ O \rightarrow 2\ Al^{3+} + 3\ O^{2-}$

We can use these same principles to determine what overall reaction takes place between reactants in more complicated cases, as long as we are given the half-reactions. For instance, in nickel-cadmium batteries, electrons are transferred from cadmium to nickel:

Oxidation: $Cd \rightarrow Cd^{2+} + 2\ e^-$
Reduction: $Ni^{3+} + e^- \rightarrow Ni^{2+}$

These half-reactions are slightly different looking from what we've seen so far; for one thing, both species are metallic. For another thing, the nickel reaction involves changing an ion into another ion, not into a neutral species. These curiosities, however, are incidental, and in no way affect our ability to build a balanced chemical reaction out of the two halves. Using our electron-balancing technique, we'd need to multiply the nickel reaction by two for the number of electrons to be equal. The full reaction would read:

$$Cd + 2\ Ni^{3+} \rightarrow Cd^{2+} + 2\ Ni^{2+}$$

More Active →	$Li \rightarrow Li^{+} + e^{-}$
	$Na \rightarrow Na^{+} + e^{-}$
	$Mg \rightarrow Mg^{2+} + 2\ e^{-}$
	$Zn \rightarrow Zn^{2+} + 2\ e^{-}$
	$Cd \rightarrow Cd^{2+} + 2\ e^{-}$
	$Ni \rightarrow Ni^{2+} + 2\ e^{-}$
	$Pb \rightarrow Pb^{2+} + 2\ e^{-}$
	$Fe \rightarrow Fe^{3+} + 3\ e^{-}$
← Less Active	$Cu \rightarrow Cu^{2+} + 2\ e^{-}$
	$Fe^{2+} \rightarrow Fe^{3+} + e^{-}$
	$Ni^{2+} \rightarrow Ni^{3+} + e^{-}$
	$Ag \rightarrow Ag^{+} + e^{-}$
	$2\ H_2O \rightarrow O_2 + 4\ H^{+} + 4\ e^{-}$
	$Au \rightarrow Au^{3+} + 3\ e^{-}$
	$F^{-} \rightarrow F + e^{-}$

Table 11.1. Activities of common chemical species.

Since these are both metallic elements, we wouldn't necessarily expect them to react with one another. In fact, we know from principles of chemical bonding that they certainly can't form an ionic compound like the elements in the redox reactions we've discussed so far. However, it turns out that some metals are easier to oxidize than others—this has to do with a combination of many chemical properties—and can transfer electrons to less active species, where **activity** is just a measure of the ease with which a chemical is oxidized (Table 11.1).

Reactant species near the top of the table are easier to oxidize than reactant species near the bottom. Conversely, product species near the bottom of the table are easier to reduce than product species near the top (where a reduction would simply involve reversing the reaction shown in the table). For this reason, we can use the table to predict which reactions can be paired up (one as an oxidation, the other as a reduction); as long as the species being oxidized is more active than the species that will be formed through reduction, the energetics of redox are favorable, and a spontaneous reaction will occur. With regard to our nickel-cadmium battery, this theory bears out: since Cd is much more active than Ni^{2+} (which is what will be produced when Ni^{3+} is reduced), the reaction will proceed spontaneously. Why and how this generates electrical energy, we'll see shortly.

TRY THIS:

Predict whether a spontaneous reaction will occur between Pb and Ag^{+}, and write out the balanced overall reaction.

Answer: Reaction is spontaneous: $Pb + 2\ Ag^{+} \rightarrow Pb^{2+} + 2\ Ag$.

Solution:

Since Pb is more active than Ag, reaction will occur spontaneously. Half-reactions must be balanced with regard to electrons before determining overall reaction.

Oxidation: $Pb \rightarrow Pb^{2+} + 2\ e^{-}$

Reduction: $2\ Ag^{+} + 2\ e^{-} \rightarrow 2\ Ag$

It's worth noting that not all redox reactions are **spontaneous**, meaning that they don't all happen on their own. If we try to react a less active species

Image 11.2. Copper is more active than water (the species produced when oxygen is reduced). For this reason, green copper oxides form on the surface of exposed metal; this is sometimes used to aesthetic advantage in architecture (left). On the other hand, the oxidation product of iron—rust—is not desirable. Regardless, it's hard to avoid, given that iron is more active than water.

with one whose product is more active, no spontaneous reaction will occur. However, we can write equations for **non-spontaneous** reactions—meaning reactions that don't happen on their own—as we can discuss a case in which a box is lifted into the air, which itself is non-spontaneous. How does a non-spontaneous chemical reaction take place? The same way that a box ends up in the air—through an input of energy! Spontaneous processes release energy and non-spontaneous processes require it. Since spontaneous processes, including falling boxes and spontaneous chemical reactions, release energy, it's actually possible to harness that energy and transform it into a useful form. This is true whether we drop a box on a plank to launch a rock in the air, as we discussed in Chapter 8, or burn molecules in oxygen to release energy. In

TRY THIS:

Predict whether a spontaneous reaction will occur between Au and Ag^+, and write out the balanced overall reaction.

Answer: Reaction is not spontaneous: $Au + 3\ Ag^+ \rightarrow Au^{3+} + 3\ Ag$.

Solution:

Since Au is less active than Ag, reaction will not occur spontaneously, but we can still write the reaction. Half-reactions must be balanced with regard to electrons before determining overall reaction.

Oxidation: $Au \rightarrow Au^{3+} + 3\ e^-$

Reduction: $3\ Ag^+ + 3\ e^- \rightarrow 3\ Ag$

Image 11.3. Alessandro Volta (left) and his contemporary Luigi Galvani (center). Both worked on the physics of electricity, and Galvani discovered the similarity between electrical and neural impulses through his frog experiments (right).

the specific case of spontaneous redox reactions, we can transform chemical energy into electrical energy, as we'll see shortly."

11.2 GALVANIC CELLS

From a historical perspective, much of early electrochemistry was discovered and explored by two Italians in the late eighteenth century, Alessandro Volta and Luigi Galvani. It was Volta who built the first electrochemical cell (the precursor to the modern battery), and Galvani who noted the similarity between inorganic electricity and neural conduction. The story goes that one day while Galvani was working at his lab bench with a freshly dissected frog's leg, the sciatic nerve (a large nerve innervating the leg) touched a piece of metal on Galvani's desk and received a small shock of static electricity. The leg kicked as though the frog were still alive, leading Galvani to begin an exploration of the nature of bioelectricity. In further experiments, he set up a lightning rod and metal wire in his yard, and reportedly took to hanging frog legs from the wire during lightning storms. The resulting chorus line of kicking disembodied limbs must have been quite a sight—and perhaps made Galvani a bit of an unpopular neighbor!

These two physicists are remembered today by lending their names to common electrical units and techniques—a *volt* is a unit of electrical current, while *galvanized* metal has been electrically coated with another metal to prevent corrosion. Perhaps ironically, given their famously adversarial relationship, they are both immortalized in the common name for the energy-generating electrochemical cell, which is referred to either as a **Galvanic cell** or a **voltaic cell**. The principle behind the cell is the conversion of chemical energy into electrical energy through the harnessing of energy released by spontaneous redox reactions. Take, for instance, the reaction of solid zinc ($Zn_{(s)}$) and Cu^{2+}. From Table 11.1, we can determine that these two species should react spontaneously as:

Oxidation: $Zn_{(s)} \rightarrow Zn^{2+}_{(aq)} + 2\ e^-$
Reduction: $Cu^{2+}_{(aq)} + 2\ e^- \rightarrow Cu_{(s)}$

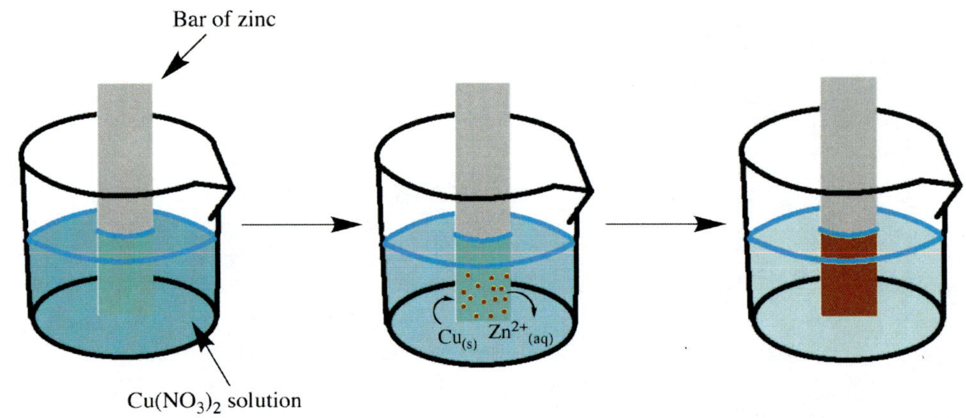

Image 11.4. A bar of zinc in $Cu(NO_3)_2$ solution (copper nitrate has a blue color) will react with the Cu^{2+}. Over time, solid copper starts to collect on the zinc bar, and the solution becomes enriched in Zn^{2+}.

Balanced Reaction: $Cu^{2+}_{(aq)} + Zn_{(s)} \rightarrow Cu_{(s)} + Zn^{2+}_{(aq)}$

If we were to put a bar of solid zinc in a beaker of $Cu(NO_3)_2$, we'd expect a spontaneous reaction. Here, nitrate is a spectator ion and is left out of the chemical reaction. In fact, the only reason it's there at all is that it's not possible to find a sample of Cu^{2+} on its own: as we learned in Chapter 4, ionic compounds are always found in pairs. In fact, there are many compounds of Cu^{2+}, but the nitrate salt is commonly used because it's very water-soluble.

The beakers in Image 11.4 represent what our zinc bar and copper solution would look like with the passage of time. $Cu^{2+}_{(aq)}$ that came in contact with the zinc bar would be reduced by the zinc, and solid copper would begin to appear on the surface of the bar. The copper solution, originally blue, would slowly lose its color as the concentration of $Cu^{2+}_{(aq)}$ decreased. The zinc bar would shrink over time, because the oxidized zinc (Zn^{2+}) would dissolve in the solution. Since the overall zinc-copper reaction is spontaneous, all of this would happen on its own, and would proceed until either the zinc bar was completely covered with copper or the solution became depleted in Cu^{2+}.

Concept Check:

Would a copper bar react in a solution of $Zn(NO_3)_2$?

Answer: No; that would be non-spontaneous. A reaction would only occur with an input of energy.

Now, while this is an interesting phenomenon to observe, it's not necessarily useful unless we particularly want a zinc bar covered with copper. Even though the reaction is releasing energy (since it's spontaneous), we have no way to collect that energy—it's all being dissipated slowly as heat. However, through a few small adjustments to the system, we can produce something much more useful.

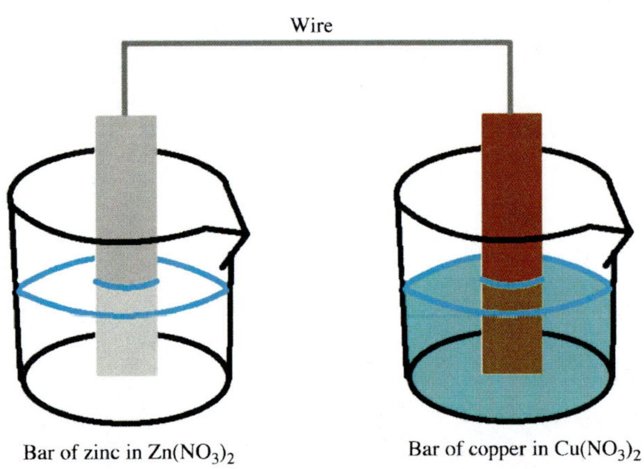

Image 11.5. While the zinc bar and copper solution are now physically separated from one another, a wire allows the free movement of electrons between beakers.

In Image 11.5, the zinc bar has been separated from the solution of $Cu(NO_3)_2$. At first, this separation of reactants seems counterproductive—it's hard to imagine a reaction taking place without the reactants touching! HCl, for instance, could never react with NaOH if we put them in separate beakers; by necessity, the reactants must physically touch in order to rearrange:

$$HCl + NaOH \rightarrow H_2O + NaCl$$

It's worth taking a moment to notice that this reaction is not a redox. It's more like an "ion swap," in which each ion maintains its charge, and therefore neither gains nor loses electrons. In order to see this, it can be helpful to keep explicit track of the ions, remembering that an acid acts ionic in solution:

$$H^+ + Cl^- + Na^+ + OH^- \rightarrow H^+ + OH^- + Na^+ + Cl^-$$

(Where H^+ and OH-come together as a pair to form H_2O, and Na^+ and Cl^- come together as a pair to form NaCl.)

Since the reaction requires a recombination of ions, it would be impossible to achieve without a physical juxtaposition of the species. Even a wire connecting the beakers wouldn't allow the reaction to proceed, since ions can't travel through wires. Electrons, however, can, which is what differentiates our zinc-copper reaction from the neutralization above. In the case of the zinc-copper reaction, there are only two requirements for the reaction to proceed: zinc must be in contact with something that will carry its electrons away, and the copper solution must be in contact with a source of electrons. This is accomplished by putting a zinc bar in a beaker of copper solution—the zinc is directly in contact with the copper ions, and they take electrons from the zinc. However, the same conditions are fulfilled if the reactants are in separate beakers, connected by a wire. The zinc metal bar (called an **electrode**) can lose electrons, which travel through the wire toward the copper metal bar. The copper bar itself doesn't react with them, since copper metal can't be reduced to an anionic species, so the electrons travel through the bar (also an electrode) to its surface, where they are collected by copper ions in solution. Those copper ions then plate out on the copper bar, just as they did in the single-beaker system. Since the only thing that must happen for the reaction to proceed is an electron transfer, it's perfectly plausible to run a redox reaction this way.

The two necessary parts of this electrochemical cell are the zinc bar and the copper solution. The solution in the zinc beaker doesn't have to contain zinc ion—it could contain any unreactive aqueous solution. The bar in the copper beaker doesn't have to be made of copper—it can be anything that's conductive and doesn't react. However, by convention, the bars and solutions in each beaker are often (but not always) matched up by element. The one problem with our system as shown in Image 11.5, though, is that a charge imbalance will quickly develop. As Zn^{2+} builds up in the zinc beaker, it becomes harder and harder for more zinc to oxidize, because forcing more positive charge into an already positively charged solution is quite difficult on account of the electrostatic repulsion between the charges. Similarly, as Cu^{2+} leaves the copper solution, an excess of negative charge from the nitrate begins to accumulate. This makes it difficult for more copper ion to be reduced. The reaction begins to become unfavorable fairly quickly under these conditions. To remedy this, it's possible to add a **salt bridge** between the cells, as in Image 11.6. The salt bridge is made up of spectator ions (often KNO_3 or $NaNO_3$) that can flow freely into either beaker to balance charge. In the case

Image 11.6. A copper-zinc cell.

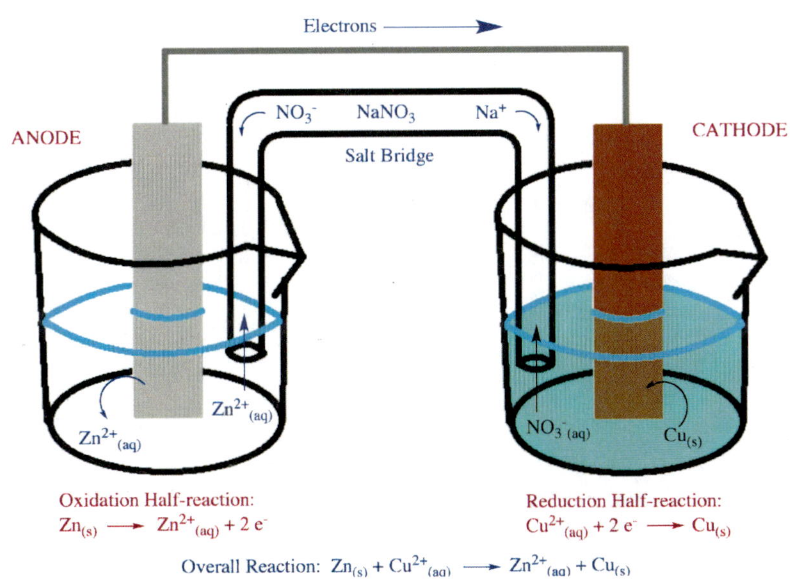

***NOTE**

A common mnemonic is *An Ox and a Red Cat; An* (for anode) and *Ox* (for oxidation) go together, while *Red* (for reduction) and *Cat* (for cathode) go together.

of the zinc-copper cell in the Image, NO_3^- would flow into the zinc beaker, because that is the beaker in which positive charge is being added to solution. Similarly, Na^+ would flow into the copper beaker, since the redox reaction is removing positive charge from that solution. Further, Zn^{2+} ions from the zinc beaker could flow into the salt bridge, which would help remove excess positive charge from that solution, and nitrate ions from the copper beaker (left over from the $Cu(NO_3)_2$) could flow into the salt bridge.

The two electrodes in the cell above are referred to as the **anode** and the **cathode**. For any electrochemical cell, the anode is where oxidation takes place, while the cathode is the site of reduction.*

So, certainly separating the zinc and copper from one another in beakers is interesting, but why is it chemically useful? In so doing, we've forced the electrons to travel through a wire to get from the anode to the cathode, and in forcing electrons to move, we've generated an electrical current. We can use that current to do work: for instance, we could attach a light bulb

TRY THIS:

Using Table 11.1, determine how a Galvanic cell could be made from lead and silver. Write out the half-reactions and overall reaction, indicate which way electrons would flow, and identify the cathode and anode.

Answer:

$Pb + 2\ Ag^+ \rightarrow Pb^{2+} + 2\ Ag$.

Oxidation: $Pb \rightarrow Pb^{2+} + 2\ e^-$ (this would be the anode)

Reduction: $2\ Ag^+ + 2\ e^- \rightarrow 2\ Ag$ (this would be the cathode)

Electrons would flow from the lead beaker to the silver beaker.

to the wire, and our zinc-copper **battery** (another word for a spontaneous electrochemical cell) would light up the bulb. Why does an electrical current do work? A light bulb provides a fairly simple example. As electrons flow through a wire, they produce friction, which generates heat. As the light bulb filament (an extension of the wire) heats up, it starts to glow, producing light! This is precisely what happens when you put batteries in a flashlight, for instance.

11.3 ELECTROLYTIC CELLS

Batteries are spontaneous electrochemical cells, but there are non-spontaneous cells as well. At first glance, the utility of this sort of cell is questionable—why would we be interested in setting up a redox reaction that requires an input of energy? There are actually several useful applications of non-spontaneous cells, also called **electrolytic cells**. First, they can be used to decompose compounds into their constituent elements. Like NaF, NaCl forms from elementals through a highly spontaneous redox reaction. The reverse reaction, therefore, must have two characteristics: it must be a redox reaction (if it took an electron transfer to go in the forward direction, it will take a transfer of electrons to reverse the reaction), and it must require an input of energy. By using an external source of electrical energy to force electrons out of chloride and into sodium, it's possible to reverse the reaction:

$2\ NaCl \rightarrow 2\ Na + Cl_2$
Oxidation: $2\ Na^+ + 2\ e^- \rightarrow 2\ Na$
Reduction: $2\ Cl^- \rightarrow 2\ e^- + Cl_2$

Because chlorine gas is so reactive, it (like sodium metal) is rarely found in nature. However, it has a wide variety of industrial uses, including the production of some plastics, so the ability to produce it through **electrolysis** ("splitting with electricity") is quite important. Another common electrolytic reaction involves the decomposition of water into H_2 and O_2, also a non-spontaneous process. The hydrogen derived from this reaction can be used to produce power (more on this in Chapter 13) or used for chemical synthesis reactions.

Most relevant to our discussion of batteries, however, is the applicability of electrolysis to recharging a spent cell. We all know from experience that batteries have a limited lifetime—in light of their chemistry, this makes sense. At some point, either the anode will run out of material to be oxidized or the cathode will run out of material to be reduced and the reaction will cease, which will cease the flow of electricity. In theory, however, it should be possible to recharge a battery by applying electricity to it, forcing the reaction to proceed in the reverse direction, and regenerating the starting materials. For instance, with our zinc-copper cell, we'd use an outside source of electrical energy to force the electrons to flow from the copper side toward the zinc side, achieving the following non-spontaneous reaction:

$Zn^{2+}_{(aq)} + Cu_{(s)} \rightarrow Zn_{(s)} + Cu^{2+}_{(aq)}$
Oxidation: $Cu_{(s)} \rightarrow Cu^{2+}_{(aq)} + 2\ e^-$
Reduction: $Zn^{2+}_{(aq)} + 2\ e^- \rightarrow Zn_{(s)}$

This would regenerate the zinc bar and increase the concentration of Cu^{2+} in the copper solution. In theory, once we removed the outside source of electricity, the reaction would resume in the original (spontaneous) direction, and would once again produce a current.

While the theory is sound, in reality, not all batteries are rechargeable. Typical alkaline batteries, for instance, are powered by a fairly complex redox reaction involving zinc powder and manganese dioxide. Completely recharging these would require returning each atom to its original location in order to maximize surface area for the chemistry to take place. This simply isn't possible with existing technology. While rechargers for "non-rechargeable" batteries do exist, they confer limited additional lifetime upon the cells. Further (and more important), batteries not intended for recharge can heat up during an attempt to reverse the reaction, which leads to the potential for explosion. True rechargeable batteries are assembled in such a way that when external current is applied, the reaction is reversed quickly and efficiently. Car batteries, which use a lead-acid reaction, were among the first rechargeable cells. If a car's engine is off, the battery supplies power to the accessories, like the lights:

$Pb_{(s)} + PbO_{2(s)} + 2\ H_2SO_{4(aq)} \rightarrow 2\ PbSO_{4(s)} + 2\ H_2O_{(l)}$
Oxidation: $Pb_{(s)} + H_2SO_{4(aq)} \rightarrow PbSO_{4(s)} + 2\ H^+_{(aq)} + 2\ e^-$
Reduction: $PbO_{2(s)} + H_2SO_{4(aq)} + 2\ H^+_{(aq)} + 2\ e^- \rightarrow PbSO_{4(s)} + 2\ H_2O_{(l)}$

The half-reactions above are much more complicated that what we've seen so far in the chapter, and while we can see that electrons are in fact being transferred, it's not immediately obvious where they're coming from and where they're going. To fully dissect the half-reactions, we must look more closely at the individual species involved. In the oxidation half-reaction, neutral lead becomes Pb^{2+}—a particular that we can determine only by noting that the lead cation is paired with an SO_4^{2-} on the product side. We can therefore represent the heart of the oxidation half-reaction as:

$$Pb_{(s)} \rightarrow Pb^{2+} + 2\ e^-$$

The reduction half-reaction reactants contain Pb^{4+}, which we determine by noting the pairing with 2 oxide anions (O^{2-}). In the products we find $PbSO_4$, which contains Pb^{2+}. The reduction half-reaction can thus be simplified to:

$$Pb^{4+} + 2\ e^- \rightarrow Pb^{2+}$$

The other species involved in the overall reaction (and complete half-reactions) do not change their charges. It's fairly easy to see this with regard to SO_4^{2-}, but the dispositions of H^+ and O^{2-} are a bit more difficult to track without an understanding of *oxidation numbers*,* a discussion of which is

***NOTE**

An oxidation number is a theoretical charge given to a neutral atom in a molecule. For instance, O is not anionic in the covalent molecule H_2O, but it has an oxidation number of -2 which serves as an indicator that even though it's not anionic anymore, it hasn't actually lost electrons if water is formed from the oxide anion. A full discussion of the rules for determining oxidation numbers can be found in any general chemistry textbook.

CONCEPT CHECK:

How do we know in which direction the lead-acid battery reactions are running?

Answer: If the engine is off, the reactions are running in the spontaneous direction. If the engine is on, the reactions are being forced in the reverse direction.

beyond the scope of this text. Suffice it to say, these species participate in the chemistry of a lead-acid battery, but are not involved in electron transfer.

Like any battery, a car battery will cease to provide power if it depletes its oxidation (or reduction) reactant. If you've ever parked your car with the lights on and returned to a dead battery, you're familiar with this phenomenon. Generally, however, a car is not left to run on battery power for long stretches of time, and the battery is designed so that it is recharged while the motor is running. Since recharging a battery requires an input of energy, the car's engine is called upon to power the electrolytic reaction. While the motor is running, the battery is recharged by a complete reversal of the reactions observed during discharge:

$2\ PbSO_{4(s)} + 2\ H_2O_{(l)} \rightarrow Pb_{(s)} + PbO_{2(s)} + 2\ H_2SO_{4(aq)}$
Oxidation: $PbSO_{4(s)} + 2\ H_2O_{(l)} \rightarrow PbO_{2(s)} + H_2SO_{4(aq)} + 2\ H^+_{(aq)} + 2\ e^-$
Simplified oxidation: $Pb^{2+} \rightarrow Pb^{4+} + 2\ e^-$
Reduction: $PbSO_{4(s)} + 2\ H^+_{(aq)} + 2\ e^- \rightarrow Pb_{(s)} + H_2SO_{4(aq)}$
Simplified reduction: $Pb^{2+} + 2\ e^- \rightarrow Pb_{(s)}$

The relationship between car battery and engine is reciprocal: the battery powers the starter motor and ignition, which allow the engine to start. The engine, in turn, recharges the battery. In the case of a dead battery, the only way to start the motor is with an external source of electrical energy, as in a jump-start, which utilizes the power from another vehicle's battery. If a dead battery is jump-started, the car must be left running for a period of time to allow the engine to recharge the battery, so that it can start the motor the next time the vehicle is used.

Image 11.7. A large array of solar panels.

11.4 PHOTOVOLTAIC CELLS

Like any voltaic (Galvanic) cell, a **photovoltaic** cell produces electricity from another form of energy. Unlike typical voltaic cells, however, photovoltaic cells do not rely on chemical energy in order to accomplish this: instead, they harvest and transform radiant energy from the Sun.

The major reason solar power is so attractive as an energy source is that, aside from being completely renewable and nonpolluting, it is tremendously

***NOTE**

Global incident solar energy is almost 11 times higher than accessible solar energy, but current technology precludes the harvesting of radiation from oceans and high mountain regions.

•NOTE

If gasoline is used to fuel an automobile, about 20 percent of the energy from the gas creates forward motion. If electricity is used to power the vehicle by way of a battery, up to 86 percent of the electrical energy creates forward motion.

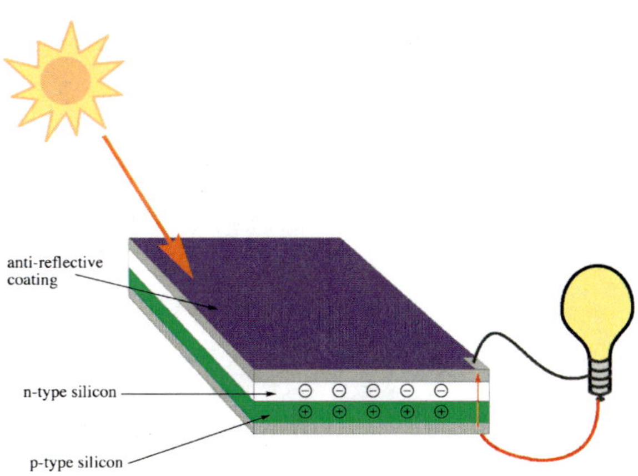

Image 11.8. A schematic of a photovoltaic cell. Note that the electrons will move from n-type to p-type silicon, but that by convention, the direction of current is said to be opposite the direction of electron flow (red arrow).

abundant. Currently, global energy consumption is estimated at about 12.5 terawatts (10^{12} watts; a watt is 1 J/s). That number is expected to approach 17 terawatts in the next two decades. Incident solar radiation upon accessible terrain* is approximated at nearly 600 terawatts—more than 30 times the projected demand. Further, proponents of solar power note that the global energy requirement estimates are based upon fossil fuel reliance. Coal and other combustion fuels are fairly inefficient sources of energy, meaning that the replacement of combustion-generated power with solar-generated electrical power would actually decrease global energy needs.• The same is true, incidentally, for other sources of sustainable electrical energy, including wind power.

Photovoltaic electrical generation relies upon the use of silicon semiconductors, materials with somewhat less ability to conduct electricity than would be observed in a metal. While there are many different kinds of photovoltaic cells currently being utilized (with frequent new developments in technology), one type in particular is called the **silicon p-n junction solar cell**. In a p-n junction solar cell, silicon is separated into two distinct semiconducting regions, each of which is **doped** with small quantities of another element. **N-type silicon**, for instance, is purposely contaminated with group 5 elements

CONCEPT CHECK:

Why do electrons flow from n-type silicon to p-type silicon?

Answer: Compared to p-type silicon, n-type is electron-rich.

like arsenic or phosphorus. Compared to silicon's four valence electrons, arsenic and other group 5 elements have a valence of five—this incorporates "extra" electrons into the silicon matrix. **P-type silicon**, on the other hand, incorporates small amounts of group 3 elements like gallium or boron. The "missing" valence electrons in the silicon matrix owing to the presence of the group 3 elements create **electron holes**. When sunlight strikes the solar panel, electrons absorb the radiant energy and become free to move through the silicon matrix—this is the nature of a semiconductor. Because the n-type silicon has a higher electron density than the p-type silicon, electron flow

***NOTE**

Why molten salt rather than water? Among the reasons are that it's possible to heat salt to a higher temperature than water; salt holds the heat much longer than water, allowing for continued power generation well into the night.

Image 11.9. A parabolic trough solar-thermal power apparatus at Sandia National Laboratories in New Mexico.

will occur in only one direction: from n to p across the junction, shown in Image 11.8. One-way flow of electrons is an electrical current and a source of electrical power.

A major challenge associated with expansion of photovoltaic technology is that it can't be used to provide power at night. Recent advancements in another form of harvesting solar power, however, may offer a solution. **Parabolic trough solar-thermal power plants** concentrate the Sun's heat onto a pipe of fluid using a long parabolic mirror. The hot fluid (often a molten salt)* can be used to generate renewable heat from non-combustion sources, or can be used to produce steam, which drives steam-powered electrical generators.

There are several hurdles to overcome in the large-scale implementation of solar power. Some materials required for solar cells, for instance, are not abundant on the planet. However, the variety of different cell types (and the disparate selection of elements from which they are produced) offer opportunities to optimize production and efficiency of those cells for which materials are more available, and phase out or redesign cells for which materials are scarce. Also, cell reuse and recycling could play a valuable role in maximizing the lifetime of supplies. Proponents of solar energy expansion suggest that, even using only existing technology, it would be economically and logistically feasible to end reliance upon fossil fuels in the next two decades. Some solar advocates suggest that a combined approach, relying on wind- and water-generated power in addition to solar, would be most effective, while others argue that the sheer abundance of solar energy makes it the single most appropriate source of global energy.

Those who oppose further development of solar power sometimes cite the inefficiency of the technology with regard to energy transformation. Most photovoltaic panels convert radiant energy with an approximately 12–20 percent efficiency, with some newer panels approaching 30–40 percent. Regardless, even using the conservative estimate of 12 percent, harvested energy from solar power could be in the realm of 72 terawatts—more than four times the predicted global needs of the future. When one takes into account the reduced global energy requirements using an electricity-dependent (rather than fossil-fuel-dependent) system, the existing availability of cells with far more than 12 percent conversion efficiency and the likelihood of future technologies with increased efficiency, it becomes apparent that low efficiency is simply not a tenable argument against solar power. Not only is the "wasted" energy not necessary to meet global requirements, solar energy is also fully renewable and nonpolluting. In other words, a 40 percent efficient coal-fired power plant is distressing, since the 60 percent of energy lost as heat still results in the release of pollution and carbon emissions from combustion. A loss of the majority of sunlight incident upon a solar panel, however, is neither polluting nor of any environmental consequence—the sunlight would hit Earth's surface whether we collected it or not.

Image 11.10. The environmental impact of Hoover Dam (to say nothing of the 250-sq.-mi. artificial lake it creates) is tremendous compared to the small footprint and emission-free operation of a solar plant.

How much land would be required (and how many solar plants would need to be built) in order to provide for global energy needs based upon current technologies? Several estimates suggest that a square array approximately 100 miles on a side would be sufficient to power the United States—that's an area slightly smaller than the state of Massachusetts. It bears reminding, however, that continued technological advancement will reduce the necessary footprint. For comparison purposes, Hoover Dam generates electricity as water falls through the dam and spins generator turbines; Lake Mead, which was produced through building the dam, covers 250 square miles. A concentrated solar plant generating the same amount of annual electricity as the dam would require only about 10 square miles. Further, when one considers the footprint of a coal-fired power plant (which must include the land required to mine the coal), solar plants occupy comparatively little land.

It's important to note that many of the arguments against development of solar technology come from fossil fuel unions, companies invested in petroleum, and special interest lobbyists. On one hand, it's understandable that a coal miner or oil field worker would want to protect his or her interests, and one feels sympathy for the plight of individuals faced with losing jobs in an industry that global sentiment has begun to turn against. On the other hand, it's also important to remember that the global dependence upon fossil fuels has tremendous environmental costs, and that (despite the utilitarian sound of such a statement) in the end, the global community must make a decision as to whether the livelihood of a select group of workers and businessmen warrants continued environmental strain.

11.5 PLANTS—NATURAL SOLAR POWER GENERATORS

The ability to harness solar power and convert it to electrical energy is neither new nor is it a human invention. In fact, plants and a few other organisms have been doing it for billions of years. Unlike man-made photovoltaic cells, plants don't use doped silicon in their light-harvesting panels. Still, similarities exist: photosynthetic organisms utilize a system designed to allow unidirectional electron movement in the presence of sunlight.* The term **photosynthesis** comes from the ability of these organisms to collect light energy and store it as chemical energy.

***NOTE**

Those who feel man-made photovoltaic cells are inefficient would be horrified by plants—typical crops convert solar energy with about 6 percent overall efficiency!

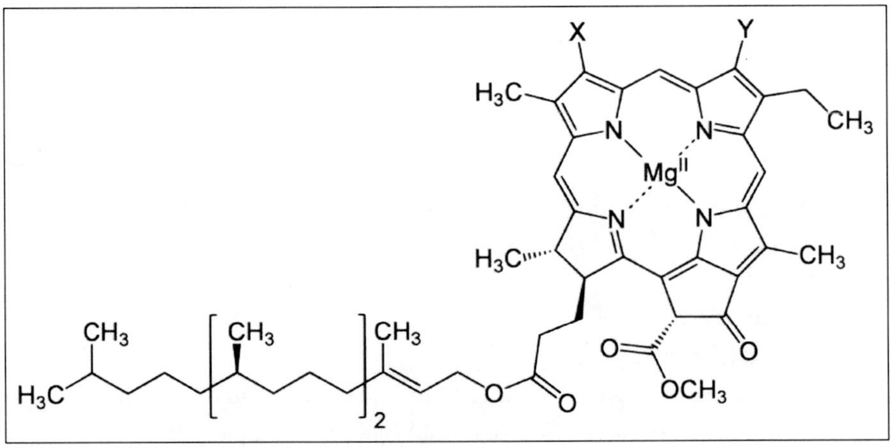

Image 11.11. A general structure of chlorophyll. Differences at X and Y differentiate between forms, but all collect light energy.

Let's take a moment to focus on a major difference between plants and animals. Animals eat to fulfill their energy needs, while plants get their energy from the Sun. This, however, fails to address the source of a plant's carbon. The molecules of life are carbon based: In order to grow, animals utilize the carbon building blocks from the nutrients they consume. Plants, on the other hand, obtain carbon in another way. The cellulose (structural material) and amylose (energy storage) that make up a plant can't be obtained through the roots—instead, they have to be synthesized in the plant itself, using CO_2 and H_2O as starting materials. As we saw in Chapter 9, glucose can be burned for energy by the reaction:

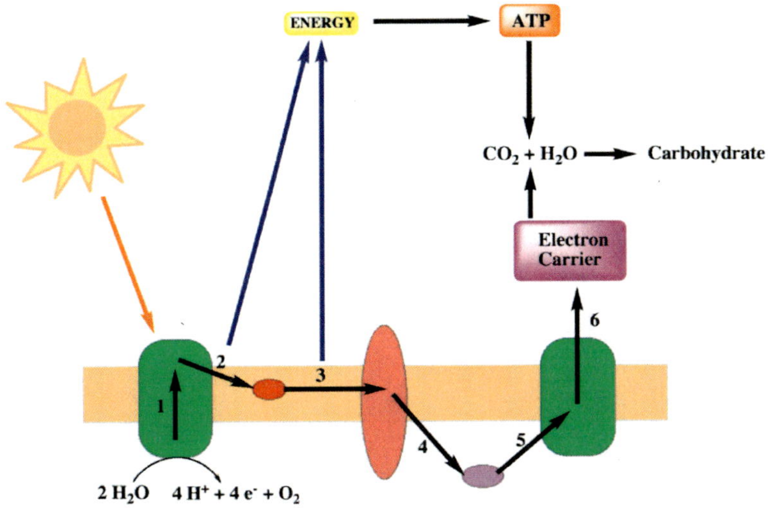

Image 11.12. A photosystem (green, embedded in a tan membrane) is an important part of photosynthesis, shown here in a schematic representation.

$$C_6H_{12}O_6 + 6\,O_2 \rightarrow 6\,CO_2 + 6\,H_2O$$

The reverse of this reaction, therefore, is endothermic. Since both cellulose and amylose are polymers of glucose, and the formation of glucose from starting materials requires energy, we see that some of a plant's energy must go to building carbohydrate; this is where photosynthesis comes in.

The process of photosynthesis starts with the action of a water-splitting enzyme, which oxidizes water by the reaction:

$$2\,H_2O \rightarrow 4\,H^+ + 4\,e^- + O_2$$

The electrons from water are transferred to a molecule of **chlorophyll** (Image 11.11). Chlorophyll is a light-absorbing molecule—it's responsible for the green color of leaves—and while it is the molecule most commonly associated with photosynthesis, it's by no means the only light-harvesting compound. As we saw in Chapter 10, carotenoids and anthocyanins can also absorb light.

A combination of many proteins, enzymes, and light-harvesting pigments make up a **photosystem**, of which the water-splitting enzyme is a part. Together, many photosystems and a number of additional enzymes and proteins comprise the cellular structure responsible for photosynthesis, called a **chloroplast**. The photosystems and most of the auxiliary enzymes are embedded in a membrane inside the chloroplast.

> ***NOTE**
>
> Recall from Chapter 10 that all EMR, including visible light, has energy (which can be calculated as $E = hv$).

The schematic in Image 11.12 above represents the step-by-step processes of photosynthesis. The redox reaction whereby electrons are transferred from water to chlorophyll is non-spontaneous and requires energy. In order to accomplish this reaction, chlorophyll and other light-harvesting pigments in the photosystem absorb light energy from the Sun,* which powers the electron transfer (step 1). The energy also excites the electrons, which loosens them within the molecular structure (just as light energy loosens electrons within the silicon matrix of a photovoltaic cell). The loose electrons flow

Image 11.13. ATP—short for adenosine triphosphate—is a chemical currency molecule.

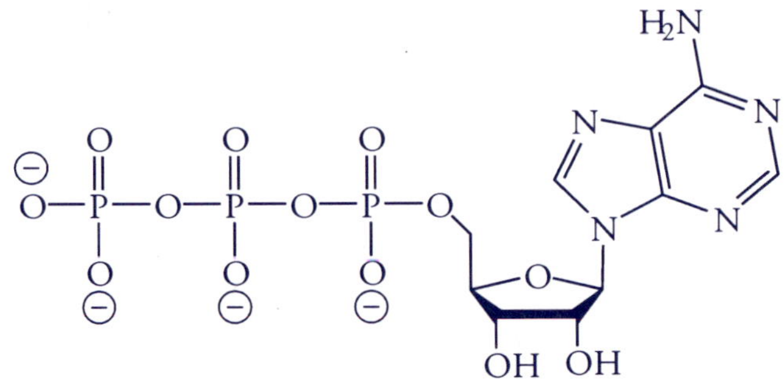

from chlorophyll to another molecule (called pheophytin), and from there to another molecule (called a quinone), and so forth down a long line of molecules (steps. 2–5). In each case, the molecule accepting the electrons forms a product that is less active than the molecule donating the electrons, meaning that each of these transitions is spontaneous and releases energy. Some steps release small amounts of energy, which are essentially lost. Other steps release larger amounts of energy, which are represented in the schematic (steps. 2 and 3). All these electron-accepting molecules combine to make up a unidirectional molecular "wire," which allows the flow of electrons from the photosystem to its eventual destination, via an electron carrier molecule (step 6). The energy released in step. 2 and 3 is then used to synthesize **ATP**, an energy-transfer molecule.

ATP (Image 11.13) is sometimes called "chemical currency" because of its ubiquity in the living world. A relatively small molecule, it nevertheless has very high-energy bonds owing to the tremendous electrostatic repulsion between negatively charged atoms, meaning that it takes quite a bit of energy to make it. Conversely, the breakdown of ATP releases lots of energy, which makes it ideal for short-term storage and transfer of energy from one place to another in a cell. While animals use molecules like glycogen and fat to store large amounts of energy for later harvesting, it's not the glucose or fatty acids themselves (the breakdown products of glycogen and fat) that allow muscles to function and cells to grow. Instead, glucose, fatty acids, and other molecules are metabolized, and the energy is used to make ATP. The ATP is then directly responsible for fueling processes in the cell. Plants use the Sun's energy to make ATP, which is then broken down to provide the energy to synthesize carbohydrate.

CONCEPT CHECK:

What are the sources of energy and electrons that a plant uses to build carbohydrate?

Answer: Energy is from the Sun, electrons are from water.

In the final portion of photosynthesis, carbohydrate is produced from water and CO_2. This requires both energy (from ATP) and electrons. The electrons are required to build the new bonds that will combine many smaller molecules into one larger one. With the completion of this synthesis, the energy transformation and electron transfer are complete—energy from the Sun has been stored as carbohydrate, and electrons from water have been transferred to carbohydrate through a long series of redox reactions. In the process, O_2 has been produced, which is important to animals and other oxygen-dependent life forms. The stored solar energy, having been transformed into chemical energy, is now available for the plant itself (which stores some of the carbohydrate to fulfill its own energy needs) and for organisms that consume the plants.

Further, the very important process of **carbon fixation** has taken place. This is the transformation of atmospheric CO_2 (which animals can't use to build the molecules of life) into macronutrients, which animals are able to utilize.

11.6 HUMAN BATTERIES—RESPIRATION

We know from our discussion of nutrition in Chapters 8 and 9 that carbohydrate consumption yields energy. In fact, we could literally burn glucose—just as we could burn methane in a Bunsen burner—to provide light and heat. In the cells, however, the process whereby glucose is converted to energy is a bit more complex. Remember that simply because a reaction releases energy doesn't mean that energy is useful; burning glucose in a beaker on the kitchen counter would make a little flame that we could use to heat water, or we could simply let the fire burn, and the energy would be "lost," or at least rendered useless. It's very important that our cellular process for "burning" macronutrients takes place in such a way that the energy released is converted to a usable form—namely, to ATP. This is accomplished through a process that shares many similarities with photosynthesis.*

***NOTE**

Frankly, the similarity should come as no surprise. It's common in nature to see one successful process or structure duplicated (in a modified fashion) in other organisms. For instance, despite significant differences in function, the form of the skeletal structures of human, bat, and whale forelimbs are remarkably similar!

♦NOTE

We sometimes use the word *respiration* to refer to our process of breathing, but technically, it's a metabolic term. Oxygen-utilizing bacteria rely on respiration, despite the fact that they don't breathe.

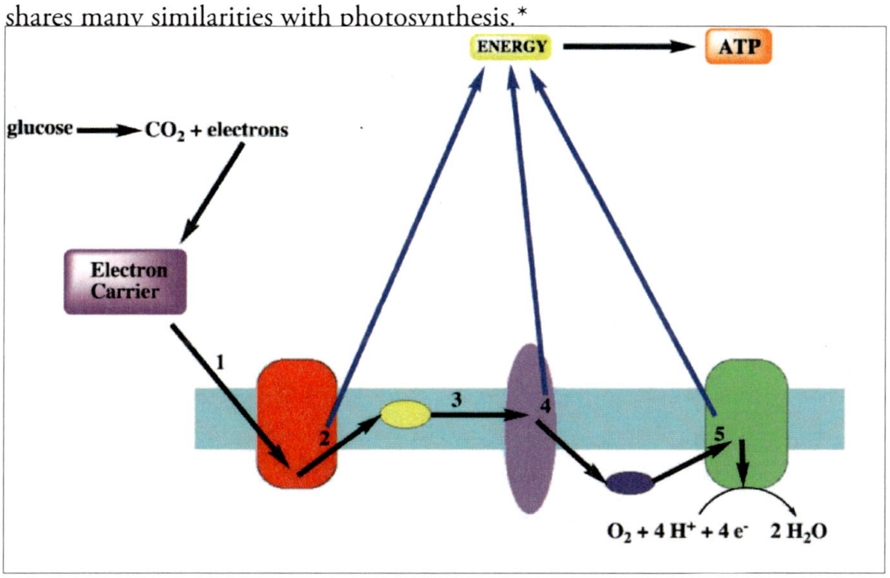

Image 11.14. A schematic representation of respiration. Note the similarities to photosynthesis, despite the fact that one process is powered by carbohydrate and the other produces it.

While a complete discussion of the chemical process by which glucose is broken down into CO_2 is beyond the scope of this text, we know a few things about the process just from our discussion of photosynthesis: it must yield energy, and it must release electrons. The electrons required to produce carbohydrate from starting materials in photosynthesis are released during the reverse reaction, and this begins the process of **cellular respiration**,♦ which is the conversion of chemical energy in macronutrients to chemical energy in the form of ATP. The schematic in Image 11.14 provides a step-by-step representation of respiration, beginning with the transfer of electrons from carbohydrate to an electron carrier, which in turn passes them to a respiratory enzyme (step 1).

CONCEPT CHECK:

Why do we need oxygen?

Answer: It's the last electron acceptor in the series of transfer reactions that are used to collect energy for production of ATP.

There are many similarities between respiration and photosynthesis. First, as in photosynthesis, the enzymes of respiration are clustered together in the membrane of a cellular structure called the **mitochondrion** (plural mitochondria).This membrane is shown in blue in Image 11.14, and the embedded enzymes are shown in various colors. Electrons are passed from one molecule to the next in a unidirectional fashion; this forms a molecular "wire." Also as in photosynthesis, for each electron transfer reaction, the product formed when a molecule accepts the electrons is less active than the molecule that donated the electrons, meaning each redox reaction is spontaneous, and yields energy. Finally, just as with photosynthesis, some of this energy is lost to unusable forms, while some of it is collected and used to produce ATP. So far, there are two major differences between respiration and photosynthesis. First, the initial redox reaction in respiration is spontaneous and requires no energy input. The initial redox reaction in photosynthesis, on the other hand, requires energy from the Sun. Additionally, while both processes share the ultimate goal of producing ATP, photosynthesis also uses the electrons collected from water to synthesize carbohydrate. In respiration, carbohydrate is broken down rather than produced. The only goal is to make ATP, and the electrons are essentially waste. Whether they're needed or not, however, the electrons have to go somewhere—if they weren't removed from the last respiratory enzyme (light green in the schematic), it wouldn't be able to continue to accept electrons, and the entire process would stop. Ceasing the flow of electrons would cease ATP production, which would defeat the purpose of the reaction. As a result, electrons from the last enzyme must be disposed of. This is where oxygen comes into play. In a reverse of the reaction that started photosynthesis, the last respiratory enzyme passes electrons to O_2, reducing it to water. The water, then, is a metabolic waste product (which can be eliminated from the cell or used for other things), and as long as a continuous supply of O_2 is available, the energy production will continue. This is the reason that we, and other oxygen-dependent organisms, need O_2!

Incidentally, the processes whereby amino acids and fatty acids are metabolized for energy are analogous to that shown for carbohydrate: the breakdown of the nutrients yields electrons, which are transferred along the same molecular wire to oxygen, and energy from those transfers is collected to make ATP.

11.7 ANAEROBIC METABOLISM—FERMENTATION

What happens, though, if there is insufficient O_2 to allow the electron transfer reactions of respiration to continue to function?

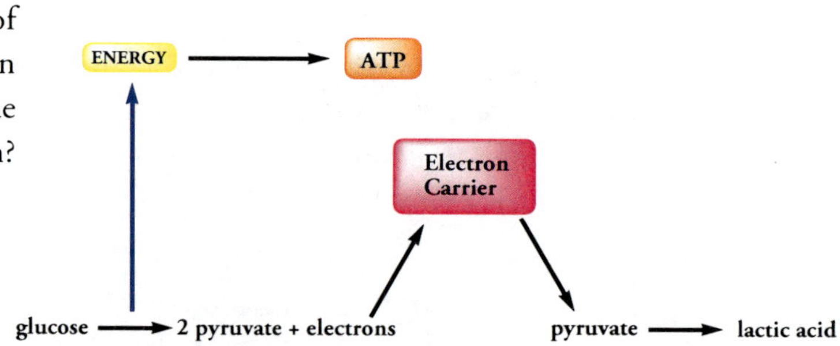

Image 11.15. Lactic acid fermentation.

Conditions such as these occur in the case of intense exercise. As discussed in Chapter 9, power efforts require energy production (which we now know is the production of ATP) in excess of the body's ability to supply oxygen. The result is that cells must utilize an anaerobic strategy for metabolism of glucose. The chemical difference between aerobic and anaerobic metabolism lies purely in the oxygen supply—as long as sufficient oxygen is available, respiration can proceed. This is highly desirable, because respiration is the most efficient mechanism for production of ATP. In the absence of sufficient oxygen, respiration ceases to function. Without a final electron acceptor to oxidize it, the last respiratory enzyme cannot continue to oxidize the enzyme before it, which in turn can no longer accept electrons from its donor enzyme, and so on. As an alternative energy production strategy, a **fermentation** reaction is used, as represented in Image 11.15. Unfortunately, the energy production of fermentation is far below that of respiration: about 15 times less energy is produced, meaning 15 times less ATP is generated per molecule of glucose. It is for this reason that fermentation is generally avoided by multicellular organisms as a metabolic strategy, except in cases of dire need.

***Note**

This is why muscles burn during hard efforts. Anaerobic work requires cells to produce energy by fermentation rather than respiration. Fermentation produces lactic acid, which leads to a burning sensation.

Concept Check:

When does a cell (that normally respires) use fermentation rather than respiration?

Answer: Under low oxygen conditions.

Despite the fact that fermentation doesn't involve a long series of electron transfer reactions, the electron carriers must still participate—for glucose to be oxidized, something must be reduced. The oxidation of glucose yields energy, providing for production of a small amount of ATP. However, the process of fermentation doesn't end there. If electron carriers were continually reduced without ever dropping off their electrons, all available carriers would soon become full. This would be a disastrous state of affairs—without carriers to reduce, glucose oxidation, and therefore energy production, would cease. In order to solve this problem, something must accept electrons from the carriers. **Pyruvate**, the product of glucose oxidation, is able to accept electrons. The product (aside from empty electron carriers that can return to react with more glucose) is lactic acid.* Note that production of lactic acid from pyruvate does not increase the energy yield of fermentation, nor is it a particularly desirable outcome; it's a waste product, and must be disposed of. However, its formation is critical to continued energy production under anaerobic conditions and as such, it's unavoidable.

While larger organisms simply can't produce sufficient energy to sustain function under anaerobic conditions—at least not for any length of time—there are many unicellular organisms capable of doing so, owing to their lower energy needs. Some bacteria and yeast utilize fermentation as a primary (or at least optional) energy-production strategy, a fact we use to our advantage in the cultivation of the interesting fermentation-based food products discussed in Chapter 5. Bacteria like *Lactobacillus acidophilus* ferment sugars to lactic acid, which reacts with milk to produce many of the fermented dairy products we enjoy. Yeast, on the other hand, ferment sugars to produce CO_2 and ethanol (Image 11.16), both of which are utilized in the production of various comestibles. The chemistry of ethanol-producing

Image 11.16. Ethanolic fermentation.

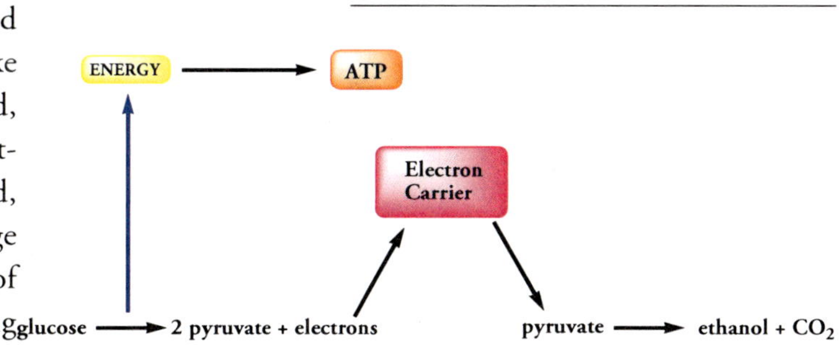

Image 11.17. Lightning results from the separation of charge within thunderhead clouds; this can result in either cloud-to-ground or cloud-to-cloud discharge.

fermentation is quite similar to that of lactic acid fermentation. The difference lies only in the specific enzyme responsible for reduction of pyruvate; the yeast enzyme results in an ethanol product rather than lactic acid.

Regardless of the specifics, either type of fermentation accomplishes two important goals: production of energy, and regeneration of empty electron carriers. In the end, also, it's just frankly intriguing to realize that the same metabolic strategy that leaves us with burning muscles after a sprint or a session in the weight room is also responsible for the bubbles (and the alcohol) in a beer!

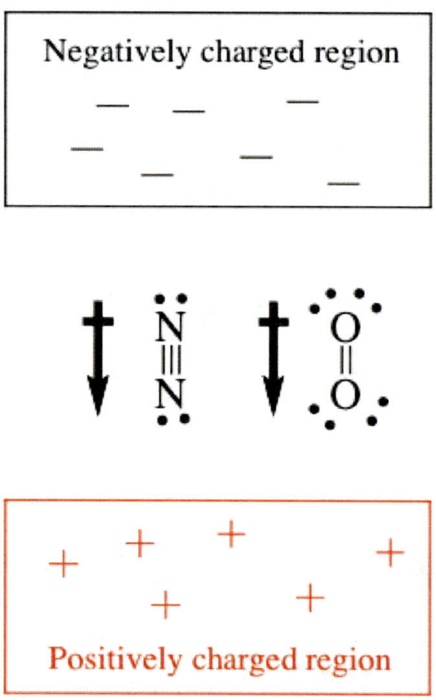

Image11.18. Separation of charge is induced in otherwise neutral molecules

11.8 THE LAST WORD—LIGHTNING

Lightning is one of nature's most interesting phenomena, both beautiful and destructive, admired and feared, intensely studied, and still incompletely understood. While theories abound regarding the specifics of the phenomenon, it has been known for some time that lightning is an electrical discharge—recall Galvani and his Franken-frogs. Within thunderclouds, falling drops of rain and ice become electrically polarized. One theory holds that this is a consequence of the droplets passing through Earth's natural electrical field. Other theories suggest the involvement of forces including wind, humidity, and atmospheric pressure. Also debated is the mechanism whereby these ions are segregated within a cloud. Regardless, theories converge upon the observation that charge does, indeed, separate. Large areas of positive and negative charge form within clouds, and this separation represents a tremendous potential for electrical current.

In the photovoltaic cells discussed earlier in the chapter, electrons were held within a silicon matrix, and energy from sunlight loosened them enough to allow their movement from slightly negative to slightly positive regions of doped silicon. In clouds, however, no such energy input is required; significant separation of charge results in the buildup of massive electrostatic forces. In the same way that plastic can be used to coat and insulate a wire, preventing it from transferring electricity if it's touched, the insulating medium of the air initially prevents the discharge of electrical potential energy from charged clouds. The areas of accumulating charge begin to exert electrostatic forces upon molecules of air between the bottom of the cloud and the ground, with electrons pulled toward positively charged regions and pushed away from negatively charged regions. This creates dipoles in molecules that would otherwise have none, as in Image 11.18.

Eventually, the electrostatic forces upon air molecules produce such significant dipoles that ionization results, and a variety of interesting species not normally found in the atmosphere are formed. These include O^+, N_2^+, O_2^+, and N^+. Just as ions (electrolytes) in water allow the solution to conduct electricity, charged particles in the air allow current to flow. This produces the massive electrical discharge that we see in a lightning strike. Further, the high-energy charged particles quickly combine with neutral molecules, initiating a sequence of reactions that concludes with the production of NO_2. As discussed in Chapter 6, NO_2 can react to form ozone, the odor of which can sometimes be detected in the air shortly after a lightning strike.

With lightning striking the Earth 100 times a second (nearly three trillion times each year), and delivering an awesome electrical discharge with every strike, it's easy to understand why the phenomenon generates so much interest. At a time when many researchers are turning their attention to harnessing the energy of the Earth, it's worthwhile to stop sometimes and simply admire the forms of tremendous power that exist all around us, and defy not only our ability to harness them, but even our ability to fully understand them.

SUMMARY OF MAIN POINTS

- Separation of opposite charge produces electrical potential energy, which can be used to do work.
- Moving charge is electrical energy and is called a current.
- Chemical reactions involving transfer of electrons can be used to generate electricity and are called redox reactions.
- Redox reactions must include both an oxidation (electron loss) and a reduction (electron gain); the two parts of the reaction are called half-reactions.
- Correctly written half-reactions can sum to give the overall chemical reaction.
- The spontaneity of a redox reaction can be determined using the oxidation potential, or activity, of the species involved: as long as the reactant being oxidized is more active than the product of the reactant being reduced, the reaction will be spontaneous.
- Non-spontaneous redox reactions will only occur if energy is added to the system. It's still possible to write out reactions for (and balance half-reactions for) non-spontaneous redox reactions, regardless of the energetics.

- A Galvanic (voltaic) cell converts chemical energy into electrical energy by utilizing a spontaneous redox reaction.
- In a Galvanic cell, the reactant being oxidized is isolated from the reactant being reduced—the only connections between them are a wire (to allow electron flow) and a salt bridge (to balance charge).
- As electrons flow from the material being oxidized (the anode) to the material being reduced (the cathode), they can be utilized to do work.
- Spectator ions are present in a reaction mixture but are not involved in the reaction.
- Galvanic cells have both anodes (where oxidation takes place) and cathodes (where reduction takes place); electrons always flow from anode to cathode.
- An electrolytic cell is a non-spontaneous cell that requires an input of energy to run a redox reaction—these are used to produce elements for industrial purposes.
- Recharging a battery involves turning a Galvanic cell into an electrolytic cell by applying outside energy and running the spontaneous redox reaction backward.
- Batteries designed to be rechargeable are capable of reversing the redox reaction with greater efficiency than non-rechargeable batteries.
- Cars work through a reciprocal relationship between battery and motor. When the car is off, the battery powers the accessories and the starter motor, but while the engine is running, it recharges the battery.
- Photovoltaic cells transform radiant solar energy into electrical energy.
- Silicon p-n junction solar cells use group 5 element-doped silicon (n-type) layered with group 3 element-doped silicon (p-type) to create regions of electron density and electron paucity.
- Solar energy loosens electrons in silicon, allowing unidirectional flow from n-type to p-type silicon—this is an electrical current.
- Parabolic trough solar-thermal power plants concentrate solar energy onto a pipe of fluid that is heated. This can be used to provide heat or produce electricity through steam generation.
- Solar power plant land requirements are low compared to many other technologies.
- Photosynthesis is the conversion of solar energy into chemical energy (production of ATP, and then carbohydrate) through unidirectional electron flow; oxygen is a by-product.
- The electrons transferred during photosynthesis are eventually used to make carbohydrate, which provides structural material for the plant and energy for both the plant and plant consumers.
- Cellular respiration is the conversion of chemical energy from macronutrients into chemical energy in the form of ATP through unidirectional electron flow.
- The electron transport chain of respiration transfers electrons from carbohydrate to oxygen (forming water), and the energy released during electron transfer is used to make ATP.
- Fermentation is anaerobic energy production; energy yield is lower than for respiration.
- Electron carriers used in fermentation must react with the product of glucose oxidation (pyruvate) to be reoxidized to lactic acid or ethanol (both of which are waste products).

Questions and Topics for Discussion

1. What is electricity?
2. Why must electron transfer reactions include both oxidations and reductions?
3. Why is our discussion of ionization from Chapter 4 so relevant to a discussion of redox?
4. Why do electrons show up in half-reactions, but not in overall reactions?
5. What does it mean to say a species is very active with regard to redox?
6. Why can silver become tarnished (react with air), but gold can't?

7. Platinum is prized as a jewelry metal because it doesn't react with air. What can you surmise about its activity?
8. When determining whether a redox reaction is spontaneous, why do we compare the activity of the species being oxidized to the activity of the species produced through the reduction? Why don't we just compare the activities of the reactants?
9. Why do we separate the species being reduced from the species being oxidized in a voltaic cell?
10. Why can we separate reactants in a voltaic cell, but we can't separate reactants in the reaction of HCl and NaOH?
11. What is the purpose of a salt bridge in a voltaic cell?
12. What is the purpose of the wire in a voltaic cell?
13. What is the difference between a voltaic cell and an electrolytic cell?
14. How can we be sure there is no electron transfer involving SO_4^{2-} in the lead-acid battery (car battery) reaction: $Pb_{(s)} + PbO_{2(s)} + 2\ H_2SO_{4(aq)} \rightarrow 2\ PbSO_{4(s)} + 2\ H_2O_{(l)}$?
15. How are photovoltaic cells similar to voltaic cells? How are they different?
16. What are the challenges to solar energy as a global alternative energy strategy?
17. Do you think solar energy is a feasible global alternative energy strategy?
18. What advantages do parabolic trough solar-thermal power plants have over photovoltaic plants?
19. Why do plants save/use the electrons harvested from water during photosynthesis, but we eliminate the electrons harvested from carbohydrate during respiration?
20. If plants use sunlight to make ATP, why do you think they store carbohydrate?
21. Why do plants produce both cellulose and amylose from the glucose they synthesize?
22. Why do you think plants use light-harvesting molecules besides (in addition to) chlorophyll?
23. Why do you think that photosystems and other enzymes and proteins involved in photosynthesis are all located in a chloroplast-contained membrane?
24. What about the structure of ATP suggests that it's a high-energy molecule?
25. How are photosynthesis and respiration similar? How are they different?
26. Why do we breathe?
27. Why do our muscles burn when we engage in an intense physical effort? Use more detail than presented in Chapter 9.
28. Why do we end up with an alcoholic, carbonated beverage when yeast contaminate a solution of starch and water? Use more detail than presented in Chapter 5.

Problems

1. Define each of the following: Electrical potential energy; current; redox; oxidation; reduction; and half-reaction.
2. Label each of the following as an oxidation or reduction reaction:
   $Na \rightarrow Na^+ + e^-$
   $Cl_2 + 2e^- \rightarrow 2\ Cl^-$
   $Fe^{2+} \rightarrow Fe^{3+} + e^-$
3. Label each of the following as an oxidation or reduction reaction:
   $O_2 + 4\ e^- \rightarrow 2\ O^{2-}$
   $Ca \rightarrow Ca^{2+} + 2\ e^-$
   $Ce^{4+} + e^- \rightarrow Ce^{3+}$
4. For each of the following redox reactions, indicate which species is being oxidized and which is being reduced:

$Cl_2 + 2\ Li \rightarrow 2\ Li^+ + 2\ Cl^-$

$Mg + Cu^{2+} \rightarrow Mg^{2+} + Cu$

5. For each of the following redox reactions, indicate which species is being oxidized and which is being reduced:

   $Zn + 2\ H^+ \rightarrow Zn^{2+} + H_2$

   $4\ K + O_2 \rightarrow 4\ K^+ + 2\ O^{2-}$

6. Write out half-reactions for each of the following:

   $Cl_2 + 2\ Li \rightarrow 2\ Li^+ + 2\ Cl^-$

   $Mg + Cu^{2+} \rightarrow Mg^{2+} + Cu$

7. Write out half-reactions for each of the following:

   $Zn + 2\ H^+ \rightarrow Zn^{2+} + H_2$

   $4\ K + O_2 \rightarrow 4\ K^+ + 2\ O^{2-}$

8. Combine the following half-reactions to produce a balanced redox reaction:

   $Sn \rightarrow Sn^{2+} + 2\ e^-$

   $Cu^{2+} + 2e^- \rightarrow Cu$

9. Combine the following half-reactions to produce a balanced redox reaction:

   $Na \rightarrow Na^+ + e^-$

   $I_2 + 2e^- \rightarrow 2\ I^-$

10. Combine the following half-reactions to produce a balanced redox reaction:

    $S + 2e^- \rightarrow S^{2-}$

    $Al \rightarrow Al^{3+} + 3\ e^-$

11. Combine the following half-reactions to produce a balanced redox reaction:

    $Li \rightarrow Li^+ + e^-$

    $2\ H^+ + 2\ e^- \rightarrow H_2$

12. How could the following half-reactions be combined to produce a spontaneous redox reaction? (Hint: use Table 11.1.)

    $Zn \rightarrow Zn^{2+} + 2\ e^-$

    $2\ H_2O \rightarrow O_2 + 4\ H^+ + 4\ e^-$

13. How could the following half-reactions be combined to produce a spontaneous redox reaction? (Hint: use Table 11.1.)

    $Ni^{2+} \rightarrow Ni^{3+} + e^-$

    $Fe^{2+} \rightarrow Fe^{3+} + e^-$

14. How could the following half-reactions be combined to produce a non-spontaneous redox reaction? (Hint: use Table 11.1.)

    $Mg \rightarrow Mg^{2+} + 2\ e^-$

    $Pb \rightarrow Pb^{2+} + 2\ e^-$

15. How could the following half-reactions be combined to produce a non-spontaneous redox reaction? (Hint: use Table 11.1.)

    $2\ H_2O \rightarrow O_2 + 4\ H^+ + 4\ e^-$

    $Na \rightarrow Na^+ + e^-$

16. If an iron bar were immersed in a solution of $NaNO_3$, would sodium metal plate out on the bar? What if the solution were $Au(NO_3)_3$? $Cd(NO_3)_2$?
17. If a magnesium bar were immersed in a solution of $NaNO_3$, would sodium metal plate out on the bar? What if the solution were $Au(NO_3)_3$? $Cd(NO_3)_2$?
18. Define each of the following: voltaic cell; cathode; anode; salt bridge; and spectator ion.

19. For a voltaic cell consisting of magnesium and silver electrodes in beakers of soluble $Mg(NO_3)_2$ and $AgNO_3$ respectively, answer each of the following. (Hint: use Table 11.1.):
    a) What is the overall reaction?
    b) What are the half-reactions?
    c) Which bar is the cathode and which is the anode?
    d) Which electrode dissolves over time?
    e) If the salt bridge contains KNO_3, in which direction does each salt ion flow?
20. Imagine a battery is used to recharge the voltaic cell in question 19 and answer each of the following:
    a) What is the overall reaction?
    b) What are the half-reactions?
    c) Would the cathode and anode change, or stay the same?
    d) Which electrode dissolves over time?
    e) If the salt bridge contains KNO_3, in which direction does each salt ion flow?
21. For a voltaic cell consisting of gold and lead, answer each of the following: (Hint: use Table 11.1.):
    a) What is the overall reaction?
    b) What are the half-reactions?
    c) Which occurs at the cathode, and which at the anode?
    d) Which electrode dissolves over time?
    e) If the salt bridge contains KNO_3, in which direction does each salt ion flow?
22. Imagine a battery is used to recharge the voltaic cell in question 21 and answer each of the following:
    a) What is the overall reaction?
    b) What are the half-reactions?
    c) Would the cathode and anode change, or stay the same?
    d) Which electrode dissolves over time?
    e) If the salt bridge contains KNO_3, in which direction does each salt ion flow?
23. Write out the overall reaction for the electrolysis of water. Which species is oxidized, and which is reduced?
24. Write out the half-reactions and overall reaction for the electrolysis of KI.

For questions 25–28, use the lead-acid reactions given in the chapter.

25. In a lead-acid car battery, what half-reaction would take place at the anode when the car engine was off, but the lights were on (use half-reactions given in the chapter)?
26. In a lead-acid car battery, what reaction would take place at the cathode when the car engine was off, but the lights were on (spontaneous overall reaction is $Pb_{(s)} + PbO_{2(s)} + 2\ H_2SO_{4(aq)} \rightarrow 2\ PbSO_{4(s)} + 2\ H_2O_{(l)}$)?
27. In a lead-acid car battery, what reaction would take place at the anode when the car engine was on (spontaneous overall reaction is $Pb_{(s)} + PbO_{2(s)} + 2\ H_2SO_{4(aq)} \rightarrow 2\ PbSO_{4(s)} + 2\ H_2O_{(l)}$)?
28. In a lead-acid car battery, what reaction would take place at the cathode when the car engine was on (spontaneous overall reaction is $Pb_{(s)} + PbO_{2(s)} + 2\ H_2SO_{4(aq)} \rightarrow 2\ PbSO_{4(s)} + 2\ H_2O_{(l)}$)?
29. Define each of the following: doping; n-type silicon; p-type silicon; and silicon p-n junction.
30. Why does a p-n junction photovoltaic cell utilize alternating sheets of Si doped with group 3 and group 5 elements?
31. In photosynthesis, water is oxidized. What is reduced (initially)? What is the final electron acceptor?
32. Why is it important that each reduced enzyme in photosynthesis is lower in activity than the enzyme that reduced it?
33. What is/are the major goal(s) of photosynthesis?
34. Why does photosynthesis require energy, but respiration does not?

35. Glucose is oxidized during respiration. What is reduced (initially)? What is the final electron acceptor?
36. What is/are the major goal(s) of respiration?
37. Why must electron carriers be involved in fermentation?
38. What are the differences between ethanolic and lactic acid fermentation?
39. What is/are the major goal(s) of fermentation?

Answers to Problems

1. Electrical potential energy is the potential for electron movement/work from the separation of opposite charges or close positioning of like charges; current is the flow of electrons; redox is a chemical reaction involving electron transfer; oxidation is the loss of electrons; reduction is electron gain; half-reactions are the oxidation and reduction parts of a redox reaction.
3. Reduction; oxidation; reduction.
5. Zn is being oxidized, H^+ is being reduced; K is being oxidized, O_2 is being reduced.
7. Half-reactions shown under complete reaction below.

   $Zn + 2\ H^+ \rightarrow Zn^{2+} + H_2$
   Oxidation: $Zn \rightarrow Zn^{2+} + 2\ e^-$
   Reduction: $2\ H^+ + 2\ e^- \rightarrow H_2$

   $4\ K + O_2 \rightarrow 4\ K^+ + 2\ O^{2-}$
   Oxidation: $4\ K \rightarrow 4\ K^+ + 4\ e^-$ (can also be written as 1:1:1)
   Reduction: $O_2 + 4\ e^- \rightarrow 2\ O^{2-}$
9. $2\ Na + I_2 \rightarrow 2\ Na^+ + 2\ I^-$
11. $2\ Li + 2\ H^+ \rightarrow 2\ Li^+ + H_2$
13. $Fe^{2+} + Ni^{3+} \rightarrow Ni^{2+} + Fe^{3+}$
15. $2\ H_2O + 4\ Na^+ \rightarrow 4\ Na + O_2 + 4\ H^+$
17. Sodium would not plate out; gold would; cadmium would.
19. a) $Mg_{(s)} + 2\ Ag^+_{(aq)} \rightarrow Mg^{2+}_{(aq)} + 2\ Ag_{(s)}$; b) the oxidation is $Mg_{(s)} \rightarrow Mg^{2+}_{(aq)} + 2\ e$, the reduction is $2\ Ag^+_{(aq)} + 2\ e^- \rightarrow 2\ Ag_{(s)}$; c) the magnesium bar is the anode, and the silver bar is the cathode; d) the magnesium bar dissolves; e) the K^+ would flow toward the silver side, and the NO_3^- would flow toward the magnesium side.
21. a) $3\ Pb_{(s)} + 2\ Au^{3+}_{(aq)} \rightarrow 3\ Pb^{2+}_{(aq)} + 3\ Ag_{(s)}$; b) the oxidation is $3\ Pb_{(s)} \rightarrow 3\ Pb^{2+}_{(aq)} + 6\ e^-$, the reduction is $2\ Au^{3+}_{(aq)} + 6\ e^- \rightarrow 2\ Ag_{(s)}$; c) the lead bar is the anode, and the gold bar is the cathode; d) the lead bar dissolves; e) the K^+ would flow toward the gold side, and the NO_3^- would flow toward the lead side.
23. $2\ H_2O \rightarrow 2\ H_2 + O_2$; water is both oxidized and reduced, since it's the only reactant.
25. $Pb_{(s)} + H_2SO_{4(aq)} \rightarrow PbSO_{4(s)} + 2\ H^+_{(aq)} + 2\ e^-$
27. $PbSO_{4(s)} + 2\ H_2O_{(l)} \rightarrow PbO_{2(s)} + H_2SO_{4(aq)} + 2\ H^+_{(aq)} + 2\ e^-$
29. Doping is adding a small amount of contaminant to a substance, as in silicon semiconductors; n-type silicon has extra electrons due to group 5 element doping; p-type silicon has electron holes due to group 3 element doping; a silicon p-n junction occurs when n-type silicon and p-type silicon are put in contact to allow electron flow.
31. The first chemical to be reduced by water is chlorophyll. The final electron acceptors at the end of the transfer process are CO_2 and H_2O, to make carbohydrate.
33. To store chemical energy (ATP) and provide electrons for carbohydrate synthesis.

35. The first chemical to be reduced is an electron carrier. The final electron acceptor at the end of the electron transport chain is O_2.
37. The oxidation of glucose requires a reduction (of electron carriers).
39. To produce some energy (ATP) and reoxidize electron carriers.

Sources and Further Reading

Solar Power

Ashley, Steven. (2008, August). Sunny days for silicon. *Scientific American*, 299(2), 32.

Delucchi, M.Z., and Jacobson, M.A. (2009, November). A path to sustainable energy by 2030. *Scientific American*, 301(5), 58.

http://www.sandia.gov/csp/

http://www.eere.energy.gov/

Photosynthesis

Bugbee, B.G., and Salisbury, F. B. (1988). Exploring the limits of crop productivity. *Plant Physiology*, 88(3), 869–78.

Lightning

Ferguson, E.E., and Libby, W. F. (1971). Mechanism for fixation of nitrogen by lightning. *Nature*, 229(5279), 37.

From Alchemists to Oppenheimer and Beyond—Nuclear Chemistry 12

The alchemists, early physicists and chemists (including such notables as Sir Isaac Newton), spent their lives in pursuit of what are now recognized as mythical concepts. The Philosopher's Stone, for instance, was said to produce the Elixir of Life, a substance that would grant immortality. Further, the alchemists hoped to achieve *transmutation*, or the transformation of less valuable metals (like lead) into gold. We learned in Chapter 1 that transmutation is chemically impossible—the Law of Conservation of Matter states that atoms of one substance may not be transformed into atoms of another substance by any chemical means. In fact, it was **John Dalton**, an English chemist, who set forth the principles of atomic theory (including the Law of Conservation of Matter) in the early 1800s. As Dalton's atomic theory gained acceptance, the pursuit of alchemy declined.

Image 12.1. The word "nuclear" conjures many images, including those of a mushroom cloud from a nuclear blast (top) and a nuclear power plant (middle). The symbol for radioactivity from a nuclear reaction (bottom) is internationally recognized.

12.1 RISE OF NUCLEAR CHEMISTRY

The field of chemistry enjoyed tremendous growth through the nineteenth and early twentieth centuries, as Dalton's atomic theory helped explain the empirical properties of substances and their reactions with one another. The idea that a chemical reaction is a rearrangement of atoms, involving the breaking and forming of bonds, stemmed directly from Dalton. Both inorganic and organic chemistry blossomed into major fields of study, giving rise to whole

Image 12.2. An alchemist in pursuit of the Philosopher's Stone (left). John Dalton, father of atomic theory (right).

> ***Note**
>
> While the medical applications of X-rays were recognized immediately, with the first radiology department already established by 1896 (in Scotland, incidentally), the dangers of routine exposure to radiation were not yet known. For some time after their invention, X-ray machines appeared at fairs and in exhibitions, as a sort of sideshow entertainment.

new industries and breathing life into old ones. The dye and pharmaceutical industries, explosives and plastics, were all developed or advanced during this time period. In 1896 a French physicist named Henri Becquerel noticed an interesting phenomenon during his investigation of uranium salts—they appeared to give off penetrating rays (much like X-rays, which had been discovered approximately 20 years prior).* This discovery intrigued a graduate student, **Marie Curie**, who was looking for a dissertation topic. She investigated uranium further, as well as looking for other elements that emitted similar penetrating rays. In the process, she and her husband, Pierre, discovered the elements polonium and radium. The discovery of these strongly radioactive elements set the stage for later research by Ernest Rutherford, who categorized charged radioactivity into two classes (alpha and beta, as we'll see shortly), and later used alpha radiation to determine the structure of the atom and discover the nucleus.

In addition to allowing for the elucidation of atomic structure, nuclear chemistry and the study of radioactivity went on to turn science upsidedown. Newton's laws of physics were found to have limited applicability to sub-atomic particles, empirical evidence demonstrated that Dalton's theories about conservation of mass and matter didn't apply in nuclear reactions, and Einstein determined that mass and energy were more similar (and more interconvertable) than anyone had ever thought. It was truly the dawn of a new age.

Image 12.3. Giants of physics and chemistry at the Solvay Conference, 1927. Among these great scientists, Marie Curie (seated third from left, front row), and Albert Einstein (seated fifth from left, front row).

12.2 RADIOACTIVE DECAY

Unlike chemical reactions, which involve electrons from the electron cloud being shared or transferred between atoms, **nuclear reactions** take place within an atom's nucleus. A very simple decay reaction takes place when a **parent nucleus*** undergoes **decay** (meaning breaks down). A radioactive particle is lost from the parent nucleus, and the remainder of the parent is called a **daughter nucleus**:

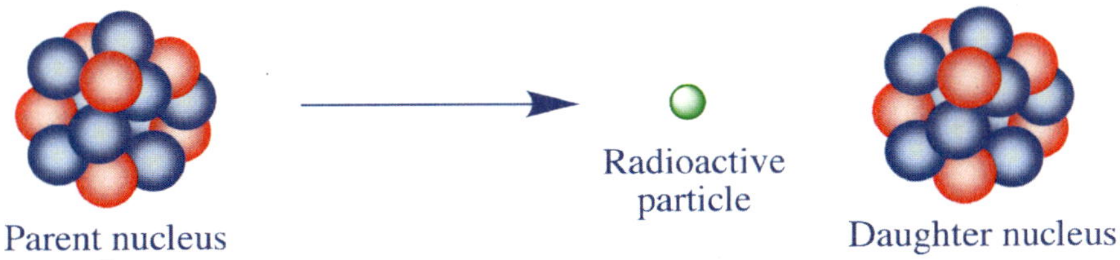

The first decay reactions to be discovered involved the emission of charged particles from a nucleus. Charged particles can be deflected by an electromagnetic field, which is easily produced in the lab, and allows for detection of radiation. The direction in which particle rays are deflected provides information about the charge of the particles. Further, in an electromagnetic field, heavier particles are deflected to a lesser extent, meaning that the mass of the particle can also be established. One such charged particle is called an **alpha particle** (α), and consists of two protons and two neutrons. The alpha particle is essentially a helium nucleus (not a helium atom—it has no electrons!), and is therefore sometimes written as ${}_{2}^{4}He$. It's also quite common to see the particle written as ${}_{2}^{4}\alpha$ in **nuclear equations**. A nuclear equation shares much in common with a chemical equation; just as in a chemical reaction, reactants are written on the left side of an arrow, and products are written on the right. Also as in a chemical reaction, mass is conserved in nuclear reactions. Unlike a chemical reaction, however, matter is not conserved—in other words, if an atom of uranium appears on the reactant side of the equation, it may not appear on the product side. In balancing chemical reactions, we didn't explicitly check to be certain that mass was conserved; because every atom of a given element weighs the same, we could be confident that as long as the type and number of atoms were conserved, mass would be conserved as well. In nuclear reactions, since mass is conserved but matter is not, we must explicitly check for conservation of mass in order to make sure that an equation is balanced. For this reason, it's very important to keep track of not simply the identities of atoms involved in the reaction, but their masses as well. These are sometimes indicated by following up the name of an atom with its mass. Take, for instance, the reaction below, which shows the alpha decay of uranium:

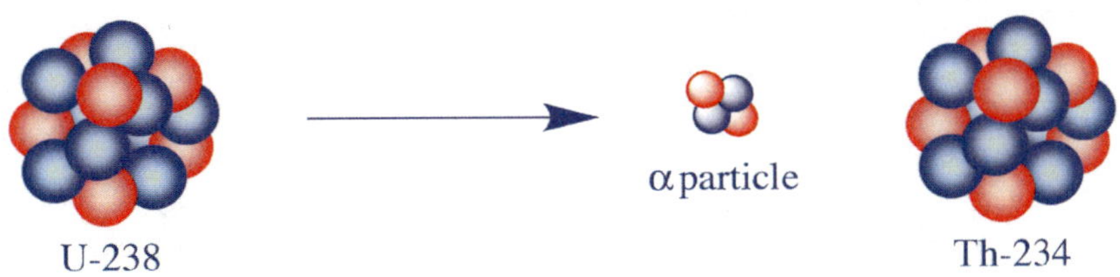

> ***NOTE**
>
> Of course, atoms are not devoid of electrons. The parent nucleus in question has an electron cloud, as all atoms do, but nuclear reactions don't involve the electrons surrounding the nucleus—they involve only the nuclear particles. For this reason, instead of referring to a "parent atom," we generally refer to a "parent nucleus," with the understanding that it is of course a complete atom.

In words, this would read *uranium-238 reacts by alpha decay to form thorium-234*. Like a chemical equation, a nuclear equation is a symbolic representation of a reaction. In a more common representation of a nuclear reaction, we use the full atomic shorthand introduced in Chapter 2, which includes not only the atomic symbol and mass number, but the atomic number as well:

$$^{238}_{92}U \rightarrow {}^{4}_{2}\alpha + {}^{234}_{90}Th$$

Notice that this equation is balanced—the mass of the reactant is 238 amu. An alpha particle, weighing 4 amu, is lost, and the daughter nucleus has a mass of 234 amu. The total mass on either side of the equation is 238 amu. In addition to conservation of mass, nuclear equations (like chemical equations) must demonstrate conservation of charge. Since the nuclear reaction doesn't involve an atom's electrons at all, we ensure that charge is conserved by keeping track of protons. From the atomic number of uranium, we see that there are 92 protons in the parent nucleus. An alpha particle, containing two protons, is lost, and the daughter nucleus has 90 protons. In sum, there are 92 protons on each side of the equation. In all nuclear equations, both mass and charge must be conserved.

Note that on the periodic table, thorium does not have a mass of 234 amu. In fact, thorium's mass number is 232. This does not mean that the daughter nucleus is not a "real" atom of thorium—in fact, it's what is called an **isotope**. Isotopes are different types of atoms of the same element—they all have the same atomic number (number of protons), but their mass numbers may vary, meaning that they will have different numbers of neutrons. Nearly every element on the periodic table exists in several different isotopic forms. An element's mass number on the periodic table (rounded to the nearest integer value) is the mass of the most common or most stable isotope. Most atoms of carbon, for instance, have an atomic mass of 12 amu, meaning that most atoms of carbon consist of 6 protons (the atomic number) and 6 neutrons. Some carbon, however, has a mass of 13.* This isotope of carbon (called carbon-13) has 7 neutrons, which we determine by subtracting the atomic number of carbon—common to ALL atoms of carbon, regardless of mass—from the mass number. Likewise, carbon-11 has 6 protons and 5 neutrons. The isotopes of a given element react identically from a chemical standpoint, but may act very differently in terms of their nuclear reactions.

***NOTE**

How would you know this? It's not really possible to look at the periodic table and anticipate whether and what isotopes of a particular element might exist. Once you know an isotope exists, however, it's possible to use its identity and mass to determine its number of protons and neutrons.

TRY THIS:

Determine the number of protons and neutrons in each of the following: U-235, $^{15}_{8}O$, nitrogen-15, $^{210}_{84}Po$.

Answers: 92 protons, 143 neutrons; 8 protons, 7 neutrons; 7 protons, 8 neutrons; 84 protons, 126 neutrons.

A second type of radioactive decay involves the loss of a **beta particle** (β) from the nucleus. A beta particle is an electron, and the idea that an electron could be coming from the nucleus leaves us with a bit of a quandary—our understanding of atomic structure tells us that the nucleus consists of protons and neutrons, but not electrons. In fact, it may be easier to think of the beta particle as a massless unit of negative charge (which, of course, makes it an electron) that comes from the nucleus. In order to conceptualize where negative charge might be found in the nucleus, let's dissect a proton and a neutron. A proton consists of one unit of mass (1 amu) and 1 unit of positive charge (+1):

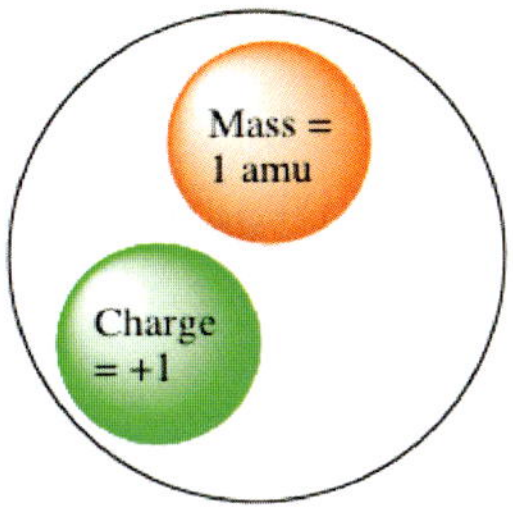

We've thus far been thinking of a neutron as consisting of one unit of mass (1 amu) and 0 units of charge, but instead, let's imagine it slightly differently:

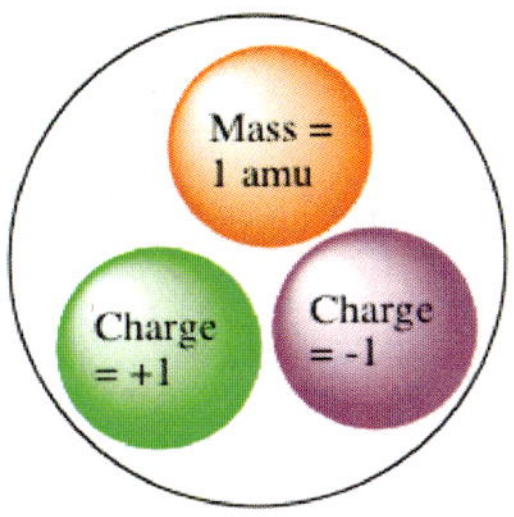

In the picture above, the neutron still has a mass of 1 amu and a charge of 0, but this charge is the net result of a +1 unit and a -1 unit. When we think about a neutron in this way, it's much easier to imagine where in the nucleus a massless particle with a charge of -1 might come from: In beta decay, it's as though a neutron is simply losing its -1 charge unit. When this happens, the former neutron is now a particle with a mass of 1 amu and a charge of +1—it's a proton!

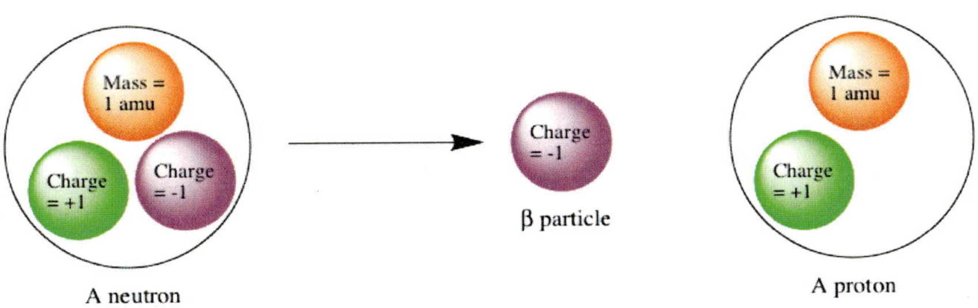

Beta decay does not result in the loss of any mass from the parent nucleus; it's simply the loss of one unit of negative charge or, as we see from the above, the conversion of a neutron to a proton. Lead-214 reacts by beta decay to produce bismuth-214, as seen here:

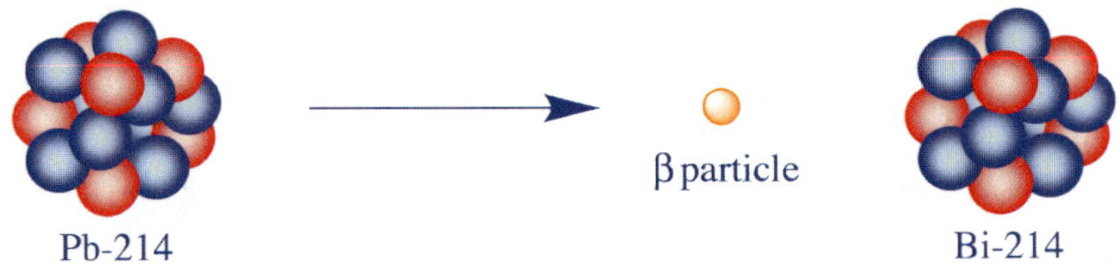

We can write this out as a nuclear equation:

$$^{214}_{82}Pb \rightarrow \, ^{0}_{-1}\beta + \, ^{214}_{83}Bi$$

As an alternative to the symbol for beta decay in the equation above, it's quite common to see the beta particle written in such a way as to indicate that it is, in reality, an electron: . It's important to note the negative sign in front of the "atomic number" for the beta particle. Because this particle does not contain any protons, the number instead represents the particle's charge. As in all chemical and nuclear reactions, both mass and charge must be conserved in beta decay. We must remember that we are now working with both positive and negative charges when balancing the equation. On the reactant side, we have a total mass of 214 amu, and 82 protons. On the product side, we add the masses for a total of 214 amu. We then add the atomic numbers/charges, for a total of 82 protons (83 – 1 = 82). A simple way to think about this is that within nuclear equations, the superscripts must balance, and the subscripts must balance.

Because we always observe conservation of mass and charge in nuclear equations, it's possible to predict the product of a nuclear reaction as long as we know the identity of the parent nucleus (we have to know not only which element, but which isotope) and of the decay particle. For instance, we could determine the daughter nucleus produced by the alpha decay of polonium-218:

1) Write out the equation including the known information, leaving a blank for the unknown information. Make sure to fill in the atomic number for the parent nucleus from the periodic table.

$$^{218}_{84}Po \rightarrow \, ^{4}_{2}\alpha + \_\_\_\_\_$$

2) Making sure to conserve mass and charge, determine the mass number and atomic number for the daughter nucleus.

218 = 4 + (daughter mass number); daughter mass number = 214
84 = 2 + (daughter atomic number); daughter atomic number = 82

$$^{218}_{84}Po \rightarrow \, ^{4}_{2}\alpha + \, ^{214}_{82}\_\_\_\_$$

3) Using the daughter nucleus's atomic number, determine the identity of the daughter. Make sure you use the atomic number, not the mass number—there are isotopes of many elements with a mass of 214 amu, but there's only one element with an atomic number of 82!

$$^{218}_{84}Po \rightarrow \, ^{4}_{2}\alpha + \, ^{214}_{82}Pb$$

Using this same technique, it's possible to determine the type of decay taking place in a nuclear reaction, or the identity of the parent nucleus, as long as the other two reactants or products are known.

TRY THIS:

Determine the daughter nucleus produced when bismuth-214 decays through emission of a beta particle.

Answer: $^{214}_{83}Bi \rightarrow \, ^{0}_{-1}\beta + \, ^{214}_{84}Po$

Solution:

214 = 0 + (daughter mass number); daughter mass number = 214

83 = -1 + (daughter atomic number); daughter atomic number = 84

TRY THIS:

What type of decay produces N-14 from C-14?

Answer: $^{14}_{6}C \rightarrow \, ^{0}_{-1}\beta + \, ^{14}_{7}N$

Solution:

14 = 0 + (daughter mass number); daughter mass number = 14

6 = -1 + (daughter atomic number); daughter atomic number = 7

A third decay particle, called a **gamma particle** (γ), is both massless and chargeless. The gamma particle is written , and is pure electromagnetic energy. Recall from Chapter 10 that gamma rays (beams of gamma particles) are at the highest-energy end of the electromagnetic spectrum. Like all EMR, we can think of them as waves (with a measurable wavelength), or as particles (with a measurable energy). We'll think about them as particles for the time being, because this allows us to imagine their release from a nucleus. These particles are lost from a nucleus as it relaxes energetically, but since they are both massless and chargeless, they do not change either the mass number or atomic number. Often, a high-energy nucleus is indicated with the use of an asterisk, but this is not critical:

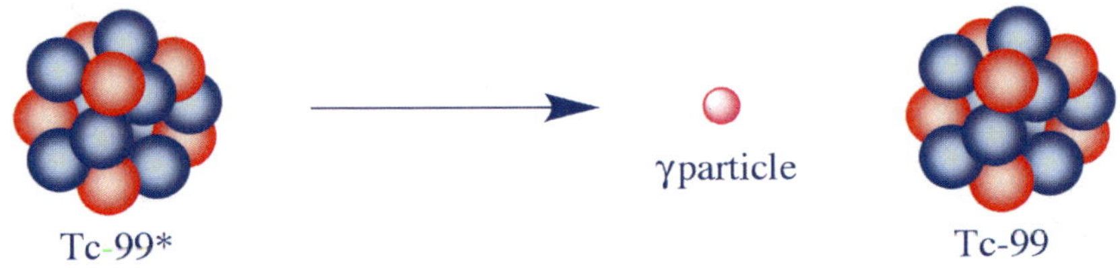

The reaction above could be written as either of the following:

$$^{99}_{43}Tc* \rightarrow {}^{0}_{0}\gamma + {}^{99}_{43}Tc$$

$$^{99}_{43}Tc \rightarrow {}^{0}_{0}\gamma + {}^{99}_{43}Tc$$

If a gamma decay equation is written without an asterisk on the parent nucleus, it's nevertheless taken to mean that the parent was higher in energy, and that energy has been lost in the form of a gamma particle. Because neither mass nor atomic number changes from parent to daughter nucleus, gamma decay equations are easy to recognize.

TRY THIS:

Determine the daughter nucleus produced when Ni-260 decays through emission of a gamma particle.

Answer: $^{60}_{28}Ni* \rightarrow {}^{0}_{0}\gamma + {}^{60}_{28}Ni$ or $^{60}_{28}Ni \rightarrow {}^{0}_{0}\gamma + {}^{60}_{28}Ni$

Solution:

60 = 0 + (daughter mass number); daughter mass number = 60

28 = 0 + (daughter atomic number); daughter atomic number = 28

CONCEPT CHECK:

Which of the three decay particles have mass? Which have charge?

Answer: Alpha and beta have mass and charge.

12.3 RADIOACTIVE DECAY SERIES

Recall from Chapter 2 that nucleons are subject to two different forces within a nucleus. The strong force holds any nucleon to any other nucleon, but is only effective over short distances. Additionally, there is electrostatic repulsion between protons. Therefore, larger nuclei must contain greater numbers of neutrons relative to protons. Despite the extra neutrons, very large nuclei (bismuth and larger) are always radioactive—they're so big, pieces of the nucleus (decay particles) tend to fall off. It seems to make sense, then, that the product of a nuclear decay reaction might be radioactive itself. U-238, for instance, decays by giving off an alpha particle to form Th-234, but thorium is also large, and therefore radioactive. In actuality, it's quite common for decay reactions to occur in a series (Image 12.4), with each daughter nucleus becoming the parent in a subsequent reaction, until a stable nucleus is reached. This is called a **radioactive decay series.**

In general, alpha decay reduces nuclear bulk and is commonly emitted from large nuclei. Beta decay adjusts the ratio of protons to neutrons, and is common both in small nuclei (such as radioactive isotopes of carbon and nitrogen, which are not likely to produce alpha particles) and as an intermediate step in a decay series.

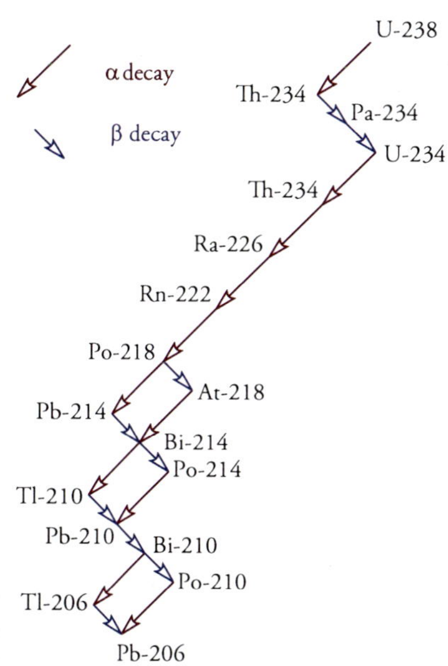

Image 12.4. A radioactive decay series for U-238; Pb-206 is a stable nucleus. Note that there are several branch points from which more than one decay path is possible.

CONCEPT CHECK:

Why do we expect a heavy atom to decay more than once?

Answer: The daughter nuclei are also radioactive (if heavier than Bi).

12.4 NUCLEAR FISSION

An interesting development in nuclear chemistry arose in the early twentieth century, as scientists began attempting to split the atom. The idea came from Rutherford, who had used alpha particles to determine the structure of the atom—by firing alpha particles at gold atoms, he had determined that they often passed through the atom without being deflected (or with minimal deflection), but that they sometimes experienced significant deflection. From this, he surmised that much of the atom was made up of a nearly massless cloud (this, he determined, consisted of electrons). At the very center of the atom, however, was a very massive, positively charged core—the nucleus—that would deflect the positively charged alpha particle. While a monumental discovery in its own right, this led Rutherford to further wonder what would happen if he purposely aimed the alpha particles at the nucleus. Would the nucleus continue to deflect all the radiation or would it be possible, given sufficient acceleration of alpha particles, to hit the nucleus and cause a rearrangement of its own subatomic makeup?* In 1919, he demonstrated that striking a nitrogen-14 nucleus with an alpha particle causes the incorporation of that particle, followed by a fragmentation of the new nuclear group:

$$ {}^{14}_{7}N + {}^{4}_{2}\alpha \rightarrow {}^{18}_{9}F \rightarrow {}^{17}_{8}O + {}^{1}_{1}H $$

While this is a slightly more complex nuclear reaction than the examples we've seen so far, this equation is also balanced with regard to mass and charge: each reactant or product grouping has a mass of 18 amu and a charge of 9. Unlike the decay reactions from earlier in the chapter, the reaction above is induced—nitrogen-14 is nonradioactive under ordinary circumstances. Rutherford wanted to continue to test his theory, and sought smaller,

***NOTE**

Picture this as a game of "atomic billiards": Rutherford was trying to scatter the nucleus using an alpha particle as a "cue ball."

higher-energy particles with which to bombard a nucleus. The particle accelerator was invented shortly thereafter, allowing for the bombardment of stable nuclei with high-energy protons and neutrons, which also induced nuclear reactions. In 1932, Rutherford used the particle accelerator to bombard lithium with protons. The resulting nucleus split into two helium atoms:

$$^{7}_{3}Li + ^{1}_{1}H \rightarrow ^{4}_{2}He + ^{4}_{2}He$$

In addition to the intriguing split of a single larger nucleus into two equal parts (a departure from the normal pattern in which a small particle was released from the nucleus), the lithium reaction also released a tremendous amount of energy, given its atomic scale. Further, and even more bizarre, a very small amount of mass was observed to have been lost in the reaction—about 0.02 amu. This lost quantity is called **mass defect**. While the defect is too insignificant to show up in the balanced nuclear equation—the mass numbers appear to balance perfectly because the lost mass is so small compared to the total—it nevertheless puzzled researchers, as it seemed to violate the law of conservation of mass. Unfortunately, the lithium-splitting reaction was quite difficult to achieve, with nuclear events taking place in only one out of billions of atoms in a sample. For this reason, neither the energy generated nor the mass defect were given much attention.

In 1939 a German group, including the Jewish scientist Lise Meitner, achieved the first true heavy atom **nuclear fission** reaction. Induced to react by bombardment with neutrons, a uranium nucleus did not decay by releasing small particles, but instead broke into two nearly equivalent pieces (as well as several neutrons):

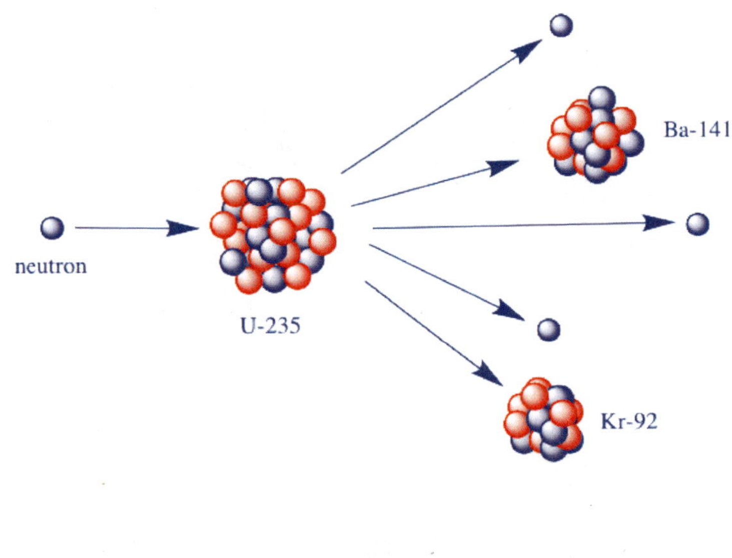

$$^{235}_{92}U + ^{1}_{0}n \rightarrow ^{236}_{92}U \rightarrow ^{141}_{56}Ba + ^{92}_{36}Kr + 3\ ^{1}_{0}n$$

As with the earlier lithium experiments, both a tremendous energy release and mass defect were observed. Meitner refused to dismiss these observations;

despite having to flee Germany and leave her research group for political reasons, she continued to work on the problem of mass defect. The explanation, she eventually reasoned, lay in **Albert Einstein**'s theory of relativity, which had been developed nearly 35 years prior, but which had not yet been proven. Based upon Einstein's observations that a moving body has slightly more mass than an identical body at rest, he theorized that mass and energy were interchangeable quantities, with a relationship determined by the equation:

$$E = mc^2$$

Here, E is energy (J), m is mass (kg), and c is the speed of light (3.0×10^8 m/s). Because the speed of light is such a large quantity to begin with and is amplified by being squared in Einstein's equation, the energy yield from even a small amount of mass would be enormous. Meitner determined that in fact, the energy yielded in the fission experiments correlated exactly as predicted by Einstein's equation to the observed mass defect.

TRY THIS:

What amount of energy (in Joules) would be released if a nuclear reaction resulted in a mass defect of 0.05 kg?

Answer: 4.5×10^{15}J.

Solution:

$E = mc^2$

$E = (0.05 \text{ kg})(3.0 \times 10^8 \text{ m/s})^2 = 4.5 \times 10^{15} \text{ J}$

As experiments in nuclear fission progressed, it was determined that a typical mass defect was in the range of 0.1% of the original fissionable mass: for 1.0 kg of reacted material, this would equate to approximately 1.0×10^{-3} kg (or 1.0 g) of defect. Per Einstein's equation, the energy yield from such a reaction would be 9.0×10^{13}J of energy, or the same amount released by the detonation of 22,000 tons (44,000,000 pounds) of TNT! In the historical context of these discoveries, it's no wonder that the first application of nuclear fission was as an instrument of war.

12.5 THE ATOM BOMB

Developing a weapon from the nuclear fission of uranium posed some challenges. It was immediately apparent to scientists that the reaction had the potential to be self-sustaining, owing to the fact that a neutron was required to start the reaction, and three neutrons were yielded. In the event that each (or at least one) of these neutrons struck another nearby U-235 nucleus, a second reaction would occur, and so on. It was also apparent that the reaction could

Image 12.5. The potential for exponential growth of an induced uranium fission reaction.

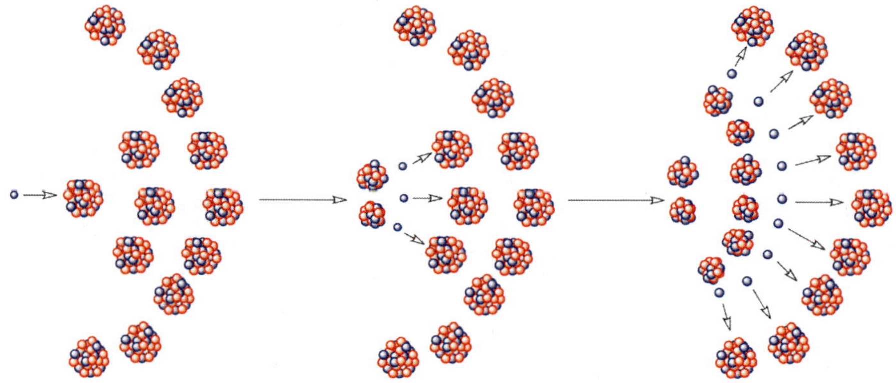

grow exponentially if all three neutrons generated as products struck three nearby U-235 nuclei, each of which produced three new neutrons, causing the reaction of nine new nuclei, and so forth (Image 12.5).

The exponential growth of a nuclear reaction was a desirable outcome with regard to production of a weapon—it would result in the fastest possible release of energy and maximal explosive potential. It was necessary, however, to ensure that the released nuclei would actually contact unreacted nuclei, rather than being lost in one way or another. Therefore, to produce a self-sustaining reaction, called a **nuclear chain reaction**, uranium samples had to contain higher concentrations of the reactive isotope, U-235, than that found in nature. Naturally occurring uranium is a mixture of isotopes, primarily U-235 and U-238. The heavier isotope is capable of absorbing neutrons, but generally does not fission in response, thus rendering these neutrons "lost" with regard to the chain reaction. Furthermore, U-238 represents approximately 99.3 percent of naturally occurring uranium (0.7 percent U-235). Unfortunately for weapons chemists, isotopes of an element are chemically identical to each other, differing only in weight and nuclear properties. This makes the logistics of separating isotopes from one another exceedingly complex. In the end, a method called **gaseous diffusion** was invented, in which gas phase uranium is passed through a filter. The lighter isotope filters faster than the heavier (though the separation is far from pure), isolating a higher percentage of U-235 on one side of the filter (approximately 3.5 percent U-235), and leaving less of the reactive isotope (approximately 0.3 percent U-235) on the other side. The leavings, called **depleted uranium** or **uranium tails** are radioactive waste that must be stored in high-level facilities, and which are difficult and expensive to reprocess into usable uranium fuel.

The hurdle of uranium enrichment overcome, the final step in developing a weapon was to determine what quantity of nuclear fuel—called the **critical mass**—would be required to ensure that released neutrons would strike reactive nuclei a sufficient proportion of the time, and that the reaction would grow in magnitude. It was determined that each nuclear fuel (Pu-239, which was discovered in the process of the uranium weapons studies, was also used in weapons development) displayed a characteristic mass at which a self-sustaining reaction could be observed. For U-235, the critical

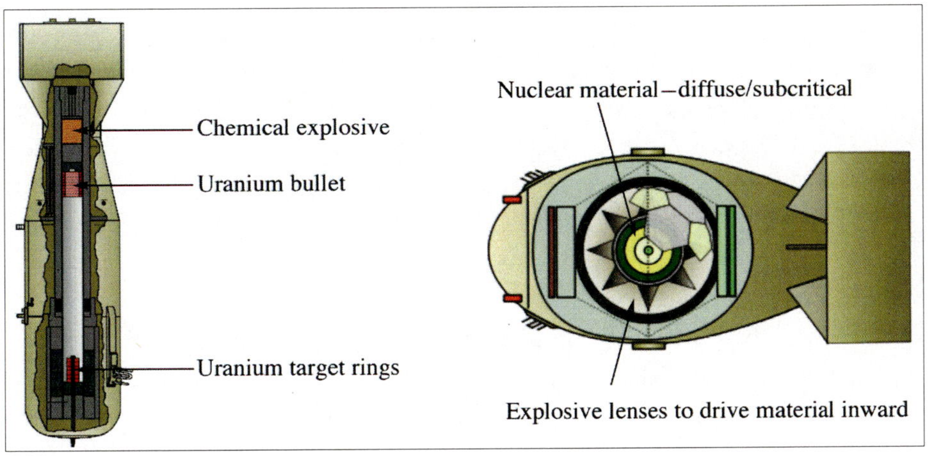

Image 12.6. Little Boy was a "gun type" weapon, using a chemical detonation to drive a subcritical uranium bullet into subcritical uranium rings, creating a critical mass (left). Fat Man was an "implosion type" weapon, using a chemical detonation to compact a diffuse core of plutonium into a critical mass.

mass was approximately 15 kg. One major consideration in the building of a nuclear weapon was that a critical mass of either fuel, once assembled, could be counted upon to react immediately. While the fission reactions of both U-235 and Pu-239 are neutron-induced, it is not necessary to introduce neutrons into a critical mass in order to generate a reaction—stray neutrons from sources such as cosmic rays abound, meaning that weapons would have to be designed so as to ensure that fuel masses were kept subcritical until detonation. Two types of nuclear weapons were developed through the **Manhattan Project**, an undertaking of the U.S. Army, led by physicist Robert Oppenheimer and staffed by a team of chemists and physicists. Both weapons functioned by bringing together subcritical masses of nuclear fuel through chemical detonation, resulting in the assembly of a critical mass that would then undergo nuclear fission.

An "implosion type" bomb was tested at the Trinity Site near Alamogordo, New Mexico, on July 16, 1945. The "gun type" was never tested prior to its use later that summer. The results were impressive: reports of the sound of an explosion came from as much as 200 miles away, with the mushroom cloud rising to a height of over 7 miles.

Image 12.7. An original film shot from nuclear testing at the Trinity Site (left), and the commemorative sign that marks the area along the highway today (right).

CONCEPT CHECK:

Why is the concept of mass defect important to the development of a nuclear weapon?

Answer: The mass defect is the source of a nuclear weapon's explosive energy.

On August 6, 1945, the B-29 bomber *Enola Gay* dropped "Little Boy" on Hiroshima, Japan (which had not been bombed conventionally so as to serve as an ideal experimental site for testing an atomic weapon in war). The explosive force has been estimated at around 15,000 tons of TNT. Three days later, "Fat Man" was dropped on Nagasaki. While reports of the death toll from the two explosions and their aftereffects vary (numbers appear to be well over 300,000), it has been difficult to determine exactly how many lives were taken by the attack. The reason for this is that in addition to the effects of pressure, fire, and radiation resulting in immediate death, lower doses of radiation from the bomb and its fallout (unreacted radioactive nuclear material dispersed into the atmosphere, which later rained down over a large portion of Japan) resulted in increased incidences of various cancers among survivors. The appropriateness of the nuclear attack upon Japan has been debated globally, and never since has a nuclear weapon been used in war.

12.6 NUCLEAR FUSION

In a reaction nearly opposite that of fission, it's possible for smaller nuclei to combine in a process called **nuclear fusion**. The energetic yield of fusion is tremendous—it is this type of reaction that accounts for the heat and energy produced by the Sun and other stars.* The bulk of the Sun's energy stems from the fusion of hydrogen into helium, through a process that can be simplified as follows:

***NOTE**

This process is called *nucleosynthesis*.

$$ {}^{1}_{1}H + {}^{1}_{1}H \rightarrow {}^{2}_{1}H + {}^{0}_{1}e + neutrino $$
$$ {}^{1}_{1}H + {}^{2}_{1}H \rightarrow {}^{3}_{2}He + {}^{0}_{0}\gamma $$
$$ {}^{3}_{2}He + {}^{3}_{2}He \rightarrow {}^{4}_{2}He + 2\,{}^{1}_{1}H $$

Not to be confused with a beta particle, in the fusion series above, ${}^{0}_{1}e$ is the symbol for a **positron**, which is a massless, positively charged particle. Like the similar (but negatively charged) beta particle, positrons must be considered when balancing a nuclear reaction. A *neutrino* is a massless, neutral particle, and as such, doesn't affect the mass or charge of the products. While fusion releases an enormous amount of energy (owing, as in fission, to the conversion of a small amount of mass to a large amount of energy), it also takes an enormous amount of energy to initiate fusion. Because all nuclei are positively charged and repel each other, energy is required to accelerate nuclei into each other at phenomenal speed. In a reactor such as the Sun, fusion becomes self-sustaining—the energy released from fusion reactions supplies the energy required to initiate further fusion reactions. Man-made fusion, however, presents a logistical challenge. With an eye toward utility in power-generation and weapons, physicists and chemists applied themselves to the production of fusion reactions as a side arm of the Manhattan Project. It was determined that, appropriately fueled, deuterium and tritium (both isotopes of hydrogen) could be fused:

$$ {}^{2}_{1}H + {}^{3}_{1}H \rightarrow {}^{4}_{2}He + {}^{1}_{0}n $$

The "appropriate fuel" to produce fusion in desirable yield turned out to be a fission reaction—the phenomenal energy released by fission was sufficient to power fusion. As a result, where the original nuclear weapons ("atom bombs") were fission weapons detonated through chemical reactions, the "hydrogen bombs" initially created and tested in the early 1950s were fusion weapons triggered, essentially, by atom bombs.

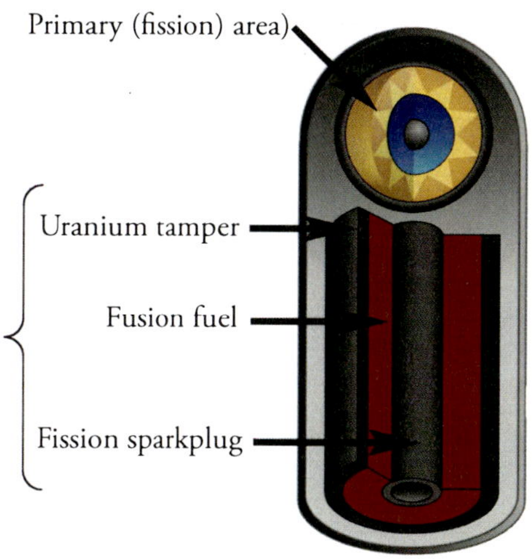

Image 12.8. A likely schematic of the hydrogen bomb.

As shown in Image 12.8, a hydrogen bomb likely consists of two distinct areas (one fission, one fusion), suspended in polystyrene foam. The primary area is made up of a subcritical plutonium core, much like that at the center of "Fat Man," which is compressed into a critical mass through chemical detonation. The fission reaction creates tremendous heat, releases radiation, and compresses the polystyrene foam. These conditions compress the fission sparkplug within the secondary area, initiating a second fission reaction—this time much closer in proximity to the fusion material, which consists of ${}^{6}_{3}Li$ and lithium deuteride, a source of ${}^{2}_{1}H$.

Upon bombardment with neutrons from the secondary fission reaction, lithium-6 fissions, producing tritium, which then fuses with deuterium:

$$ {}^{6}_{3}Li + {}^{1}_{0}n \rightarrow {}^{4}_{2}He + {}^{3}_{1}H $$
$$ {}^{3}_{1}H + {}^{2}_{1}H \rightarrow {}^{4}_{2}He + {}^{1}_{0}n $$

While hydrogen bombs have been tested (generally underwater, so great is their power), none has ever been used in wartime. Given the limited testing, it's difficult to accurately approximate the potential destructive power of such a weapon, but most estimates hold that a hydrogen bomb would have an explosive power of up to 10 megatons (equivalent to 10 million tons of TNT). By comparison, "Little Boy" had a 13,000-ton equivalency. The combined explosive power of every single bomb dropped during the course of World War II equals only 2 megatons.

CONCEPT CHECK:

Fission reactions start without an input in energy; why does fusion require an energy input?

Answer: It takes energy to bring two nuclei (positively charged) very close together.

12.7 NUCLEAR ENERGY

While the atomic bombs required that nuclear fission reactions "go critical" (become exponential chain reactions), it was noted by those individuals working on the Manhattan Project that a similar energetic yield, without the explosion, might be achieved through a controlled fission reaction. In order to produce such an outcome, masses of fuel would need to be

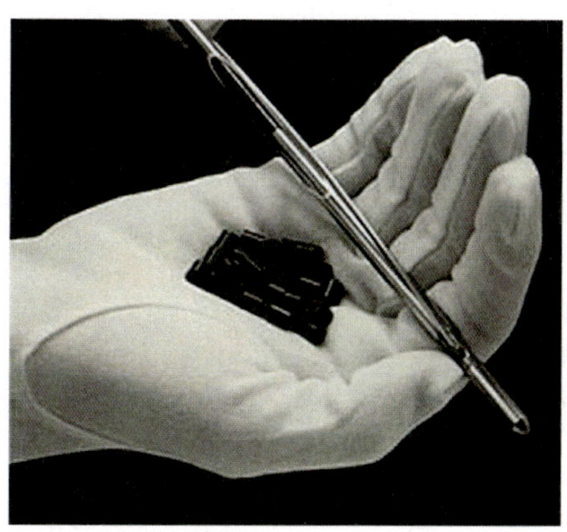

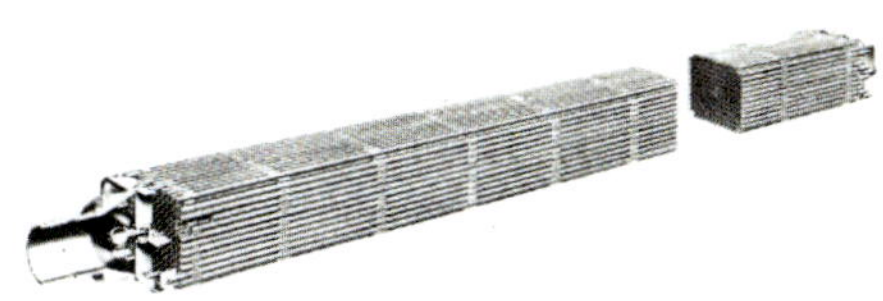

Image 12.9. Fuel pellets (left) are stacked into rods, and then grouped into fuel assemblies (right). Control rods are interspersed to absorb neutrons and control the rate of reaction.

kept subcritical, and neutron-absorbing materials would be necessary to ensure that exponential reaction growth could not result. The development of these safety measures heralded the age of nuclear power—the first nuclear reactor to be used for generation of electricity was built in the USSR in 1954, and 1957 saw the first nuclear reactor built in the United States. Nuclear power generation potential grew rapidly, both owing to expansion of existing reactors and the building of new ones. Worldwide, from the early 1960s to the late 1980s, the amount of nuclear power generated on a yearly basis increased by a factor of 300.

Nuclear reactor fuel is essentially identical to weapons fuel, though the purity may be significantly lower, as over-purified fuel increases the likelihood of a critical reaction. The general design of a nuclear plant involves a controlled reaction taking place within a **nuclear reactor**, which is an isolated reaction vessel**. Fuel assemblies**, made up of pellets composed of oxides of U-235 or Pu-239, are interspersed with **control rods**, made of elements that absorb neutrons without fissioning, such as boron and cobalt.

In addition to fuel assemblies and control rods, the nuclear reactor is filled with **primary coolant**, a liquid (often metal or molten salt) that heats up as it absorbs the energy released by the fission reaction. The hot coolant is then used to generate steam, which rotates a turbine in a steam-powered electrical generator, producing electricity.

It should be noted that neither the water entering a nuclear power plant (which will eventually be vaporized into steam) nor the steam exiting the cooling tower is ever in contact with radioactive material. As Image 12.11 illustrates, primary coolant—which is in direct contact with radioactive material, and which is generally a metal or molten salt—is circulated through a closed loop by means of a pump. As the primary coolant passes through a heat exchanger, its heat is transferred to a **secondary coolant** (also a metal or molten salt). This coolant is also circulated within a completely closed loop. Secondary coolant, contained in its pipes, passes into a steam generator, where it transfers heat to water pumped in from the outside.

The water is vaporized, and the steam exits through the cooling tower via the turbines in an electrical generator. While the primary coolant can take

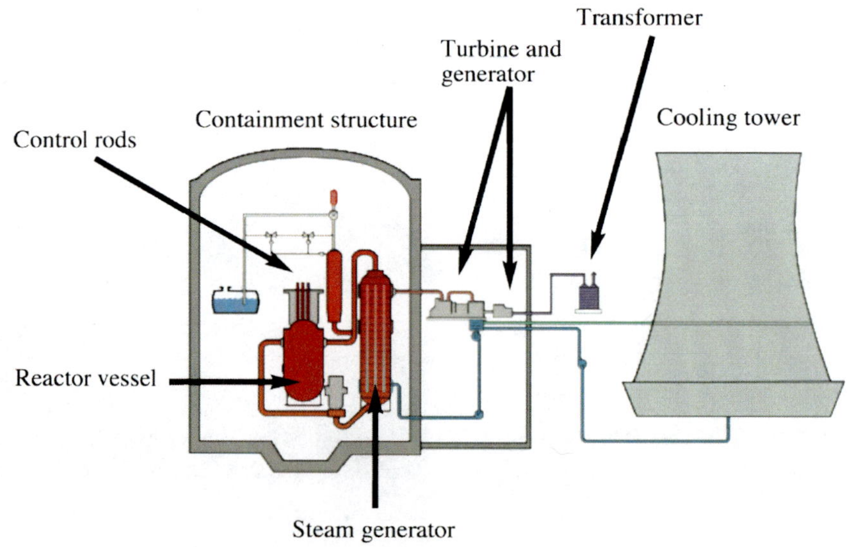

Image 12.10. A schematic of a nuclear power plant. Source of heat aside, a nuclear plant is nearly identical to a coal-fired power plant—heat vaporizes water, which forms steam, which turns a turbine and generates electricity.

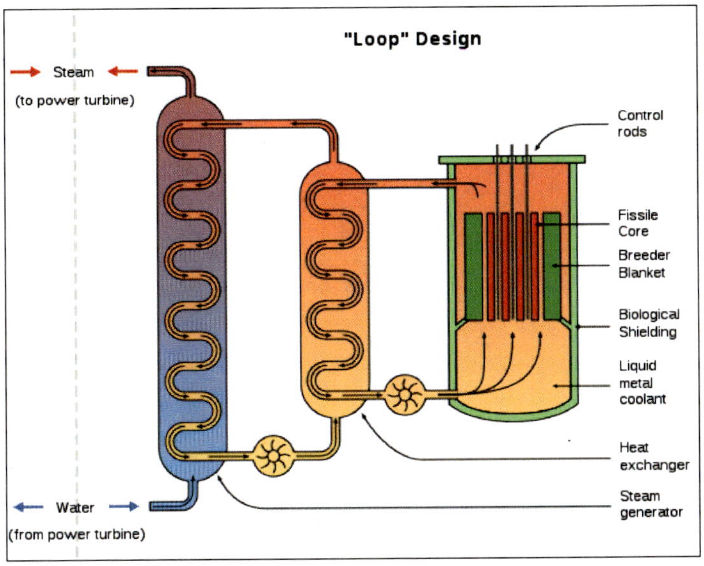

Image 12.11. Water that enters from (or steam that exits to) the environment outside a nuclear power plant is separated from the nuclear reaction by two completely closed loops.

on temporary radioactivity from the fuel, it is not possible to transfer that radioactivity to the secondary coolant,* meaning that the closed loop system prevents water exiting the cooling tower from ever having been exposed to radiation. In fact, individuals living within 50 miles of a nuclear power plant are exposed to an average of three times less radiation than those living within the same proximity of a coal-fired power plant. The naturally occurring radioactive isotopes of carbon and other elements present in coal (which are vaporized and released as particulate matter upon combustion) account for significantly more radiation exposure than that afforded by a nuclear generator.

*Note

Some types of radiation can cause nonradioactive substances to become radioactive—neutrons, for instance, can induce radioactivity (hence the induced radioactivity of the primary coolant). However, exposure to other types of radioactivity, such as gamma radiation, can't induce radioactivity—a gamma ray is nothing more than energy, and it therefore can't change the nucleus of an atom or induce it to fission. For this reason, nuclear pellets (and nuclear weapons) can induce radioactivity in other substances (like primary cool-ant), but primary coolant (which doesn't release neutrons) can't cause secondary coolant or anything else to become radioactive.

Concept Check:

Does the steam leaving a nuclear plant's cooling tower ever come in contact with radiation?

Answer: No; it is separated by several closed loops.

12.8 HALF-LIFE

An important concept in a discussion of radioactivity is that of **half-life**. All **radioisotopes** (meaning radioactive nuclei) decay at predictable rates. Half-life is the amount of time required for half the nuclei in a sample to decay into another nuclear form. Carbon-12 is a stable isotope, and does not have a half-life. Carbon-14, however, is radioactive, and has a half-life of 5730 years. This means that given a sample of 100 grams of carbon-14, half the sample will have reacted to become something else (and 50 grams of C-14 will remain unreacted) after

Image 12.12. Mass of an original 100 g radioactive sample remaining after a given period of time. Note that the decay is asymptotic.

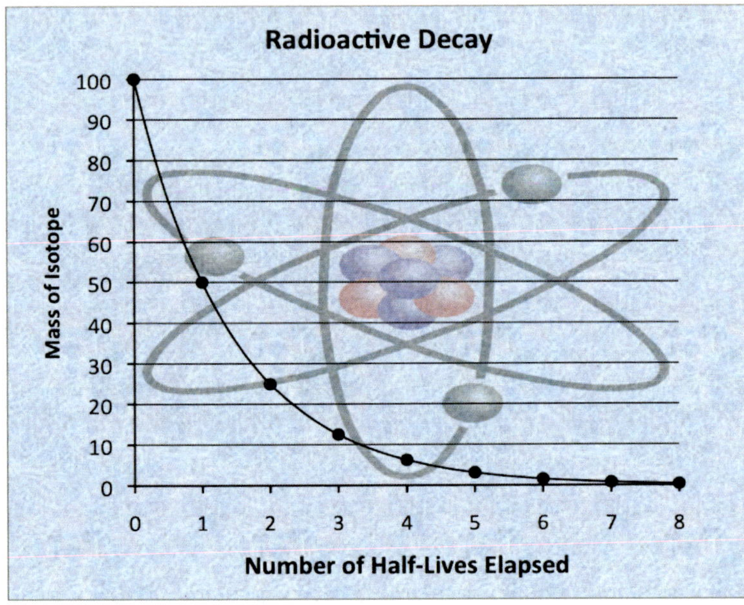

5730 years. Should another 5730 years be allowed to pass, half of the 50 grams remaining of C-14 will react to become something else, leaving just 25 grams of C-14. Note that since half the remaining sample reacts with each subsequent half-life, the decay curve is asymptotic—meaning that mathematically, the mass of isotope approaches but does not actually reach zero—as in Image 12.12

The half-lives of radioisotopes vary tremendously—some decay in a matter of microseconds, while others have half-lives in the billions of years (Table 12.1).

Radioisotope	Half-Life
U-238	4.47×10^9 years
U-235	7.03×10^8 years
Pu-239	24065 years
C-14	5730 years
H-3	12.35 years
Ra-228	5.75 years
Ra-225	14.8 days
Rn-222	3.82 days
C-11	20.4 minutes
N-13	9.97 minutes
O-15	2.04 minutes
N-16	7.13 seconds
Rn-219	3.96 seconds
Po-212	3.05×10^{-7} seconds

Table 12.1 Half-lives of some radioisotopes.

It's noteworthy that while we tend to keep track of the mass remaining of the original radioisotope, the rest of the mass doesn't disappear. Rather, it is present in a different nuclear form. Whether this new form goes on to react is entirely dependent upon what kind of decay is taking place, and which new nuclear form is produced. For instance, C-14 generally decays by beta emission:

$$^{14}_{6}C \rightarrow {}^{0}_{-1}\beta + {}^{14}_{7}N$$

Because N-14 is a stable nucleus, no further decay takes place. One may therefore assume that if 100 grams of C-14 has been allowed to decay for 5730 years (1 half-life), 50 g of C-14 remain, and the remaining 50 g have been converted to N-14. After a second half-life, 25 g of C-14 would remain, and a total of 75 g of N-14 would have been formed. In the case of U-238 (half-life of 4.47×10^9 years), determining the quantity of the original radioisotope remaining is merely a matter of counting half-lives, but figuring out the mass of each decay product is much more complex. U-238 decays to form Th-234, with a half-life of 24 days, which decays to form Pa-234, with a half-life of 1.2 minutes, and so on. Thus, given 100 grams of U-238, we'd be assured that after 4.47 x 10^9 years, only 50 grams of U-238 would remain. The remaining 50 g, however, would be working its way through a chain of nuclear reactions in an asynchronous manner, making it very difficult to determine exactly how much of each radioisotope would be present.

It's very important to realize that half-lives are discrete "check points" in time. That is, we can be assured that half of an initial sample will have decayed after one half-life has passed. The decay takes place continuously throughout the half-life, however, at a constantly decreasing rate. As such, if less than one half-life has passed, there is less than 100 g (but more than 50 g) of an original 100 g sample remaining.

There is a mathematical equation we can use to solve half-life problems:

$$N_t = N_0\left(\frac{1}{2}\right)^{\frac{t}{t_{0.5}}}$$

Here, N_t is the quantity of radioisotope remaining after a given period of time (t), N_0 is the original amount of radioisotope, and $t_{0.5}$ is the length of the half-life. Note that time and half-life may be in any units—seconds, minutes, years, and so forth—as long as they match. For instance, how much of a 55 g sample of Ra-228 remains after 17.25 years? (see Table 12.1 for half-life.)

$$N_t = N_0\left(\frac{1}{2}\right)^{\frac{t}{t_{0.5}}}$$

$$N_t = 55g\left(\frac{1}{2}\right)^{\frac{17.25}{5.75}}$$

$$N_t = 6.875g$$

While the half-life equation can be quite useful, many students will find that an intuitive approach to solving problems is somewhat simpler. Using an intuitive approach, the same problem could be addressed as follows:

1) Determine how many half-lives have elapsed.

$$17.25\ years\ \ total \times \frac{5.75\ years}{1\ half-life} = 3\ half-lives$$

2) Determine how much of the original sample remains after each half-life.

$$55\ g\ initially \times 0.5 = 27.5\ g\ after\ \ half-life\ 1$$
$$27.5\ g \times 0.5 = 13.75\ g\ after\ \ half-life\ 2$$
$$13.75\ g \times 0.5 = 6.875\ g\ after\ \ half-life\ 3$$

TRY THIS:

How much of a 45.6 g sample of N-13 remains after 4 half-lives?

Answer: 2.85 g.

Solution:

$$45.6\ g\ initially \times 0.5 = 22.8\ g\ after\ half-life\ 1$$
$$22.8\ g \times 0.5 = 11.4\ g\ after\ half-life\ 2$$
$$11.4\ g \times 0.5 = 5.7\ g\ after\ half-life\ 3$$
$$5.7\ g \times 0.5 = 2.85\ g\ after\ half-life\ 4$$

TRY THIS:

How much time does it take to reduce 200 g of U-235 to 12.5 g?

Answer: 2.81×10^9 years.

Solution:

$$200\ g\ initially \times 0.5 = 100\ g\ after\ half-life\ 1$$
$$100\ g \times 0.5 = 50\ g\ after\ half-life\ 2$$
$$50\ g \times 0.5 = 25\ g\ after\ half-life\ 3$$
$$25\ g \times 0.5 = 12.5\ g\ after\ half-life\ 4$$
$$4\ half-lives \times \frac{7.03 * 10^8\ years}{1\ half-life} = 2.81 * 10^9\ years$$

12.9 STORING NUCLEAR WASTE

A major logistical problem associated with nuclear power generation is the production of **high-level radioactive waste**, which includes spent fuel and the waste from reprocessed spent fuel. While the rate of fission in this material has slowed to the point that it can no longer be used for energy generation, the spent fuel is still hot (thermally) and highly radioactive: Recall that heavy atoms are all radioactive, and that even if nearly all the U-235 or Pu-239 in a sample has reacted, the daughter nuclei are still reactive.* To date, there is no permanent high-level waste storage facility in the United States, meaning that nuclear plants are responsible for storing their own spent fuel. Additionally, reprocessing of spent fuel is not currently legal in the United States, due to concern that the enriched fuel could be used in weapons. As such, all spent fuel must be treated as waste

***NOTE**

The daughter nuclei are generally not fissionable, making them unusable as fuel.

without any attempt to reprocess. For an average power plant, that comes out to about 20 tons of spent fuel per year—all of which must be stored. Currently, nuclear plants in the United States store their spent fuel in pools initially built to house waste temporarily. Pools ensure that spent rods are under at least 20 feet of water, a sufficient barrier to shield nearby workers from the worst of the radiation. Many once-operational plants have been decommissioned because they've run out of storage. Most operational U.S. nuclear plants are likely to run out of space before 2020. In the meantime, some plants have taken to an alternate, **dry-casking** method of storage in which spent fuel is kept above ground in metal containers within a concrete structure. This practice is of questionable safety.

Yucca Mountain, a site in southern Nevada, has been proposed as a permanent **deep geological repository** for radioactive waste. Targeted for investigation in 1978, the site was originally slated to begin accepting waste in 1998. This date has been delayed many times, with current estimates anticipating that the site will not become operational until at least 2020. The idea of geological repositories has been subject to global debate. Proponents argue that nuclear technology offers an abundance of emission-free power, and that with many facilities out or nearly out of storage space for spent fuel, an alternative must be made available. Opposition to the plan comes from scientists and environmental groups, who cite the potential for radiation exposure. Because the Yucca Mountain site lies on and near several major geological faults, there is concern that existing (and future) fractures in the rock structure could allow radioactive gases to escape as decay processes produce gaseous elements. Further, storage containers will inevitably fail some time in the next several hundred years; the regulations currently in place require only that storage containers be assured 300–1000 years of integrity. When containers do fail, liquid waste will seep easily through the fractured rock structure into the local water supply. While the land in Nevada was originally targeted because it was largely unsettled and *non-arable* (did not support agriculture), the water table under Yucca Mountain serves populated (and agricultural) areas. It is worth noting that the fault lines in the Yucca Mountain area cannot be assumed to be inactive—seismic activity is quite common, and an earthquake would accelerate the dissemination of radioactive material into the air and water. Finally, residents of Nevada (a state with no nuclear power plants of its own), generally object to the idea of a waste repository in their home state.

Image 12.13. Yucca Mountain is located on the western end of the Nevada Test Site.

Even under ideal circumstances, the idea of a geological repository is fraught with debate. Some radioisotopes in the decay chains of U-235 and Pu-239 have short half-lives and would decay quite quickly. Others

Concept Check:

Why is the proposed repository in Nevada?

Answer: The proposed land is non-arable and uninhabited.

have half-lives in the tens of thousands of years. As a result, it is quite difficult to predict to what extent farmers or citizens of nearby populated areas, some time in the future, would be exposed to radiation. While regulation is in place to keep potential exposure levels low for the next several hundred years, it becomes impossible to predict how the factors of time, seismic disturbance, container failure, inevitable leakage, and unforeseen events may combine to produce a hazard.

12.10 NUCLEAR DISASTERS

Despite the surge in nuclear power production from the 1960s through the 1980s, this period was followed by a rapid and dramatic decline. Plans to build new plants were canceled, with about two thirds of those commissioned after the early 1970s never coming to fruition. This was the result of a number of factors. Oil prices fell dramatically during that period of time, which made the promise of vast amounts of nuclear power less attractive in the face of the rising costs of building new reactors. In 1979 the Three Mile Island Nuclear Generating Station in Pennsylvania suffered a partial **meltdown**—meaning containment failure—due to a cooling system malfunction. The U.S. Nuclear Regulatory Commission reports no deaths or injuries from the incident (though it can be difficult to separate the delayed effects of mild to moderate radiation exposure from those due to routine aging and environmental factors). Regardless, the meltdown captured global attention, and significantly dampened national support for expanding the nuclear program. A number of systemic upgrades and changes to prevent similar events were implemented in the wake of the incident.

Image 12.14. The Chernobyl reactor (above). An abandoned house within the exclusion zone around the site of the nuclear disaster (below).

In 1986 the city of Chernobyl, Ukraine (formerly part of the USSR), suffered the most dramatic reactor disaster in history during an experimental procedure designed to test safety feature responses to a potential power outage. A number of factors combined to create the catastrophe—the experiment was timed to take place during a shift change, rather than being seen through by a single group of workers. Inexperienced workers shut off safety features in excess of what was necessary to perform the experiment. An operator inserted control rods too far, bringing the reactor to a near standstill, following which control rods were extracted too much in order to restart the reaction. This initiated a chain of events that culminated in heat and pressure buildup, control system failure, and eventual explosion of the reactor casing. Radiation exposure was intense—many workers received lethal doses within minutes, with death following in the next few weeks. Residents of Chernobyl and the nearby city Pripyat were evacuated, and both remain unpopulated today, as they are within an exclusion zone of continued contamination surrounding the old reactor.

The death toll of the Chernobyl disaster is difficult to quantify accurately. According to the World Health Organization, fewer than 50 people died as a direct result of radiation exposure from the incident. Regardless, a team of scientists working under the auspices of the United Nations has suggested that the number could eventually rise as high as 4000 owing to delayed effects. Some environmental organizations argue that even this estimate fails to fully represent the extent of the disaster.

The Chernobyl event significantly impacted global attitudes regarding the safety of nuclear power generation. Several countries began phasing out nuclear reactor construction and operation in response. It is difficult to gauge the effect of the Chernobyl event in the United States, as attitudes had already been impacted by the Three Mile Island incident seven years prior. In any event, no new power plant has become active in the United States since 1996.

12.11 THE FUTURE OF NUCLEAR POWER

While there are proponents of expanded nuclear power generation in the United States and globally, the logistics of waste disposal, expense of building new reactor sites, and public fear surrounding nuclear technology have prevented any significant growth. Particularly with attention turning to emission-free energy technologies, it is difficult to justify further development of a technology that produces high-level waste that must be stored permanently (for all intents and purposes) and that carries with it the risk of radiation exposure in surrounding areas. Finally, it is problematic to suggest that additional funding should be shunted into fission-based nuclear technology, given our ever-expanding global energy requirements. If funding for energy technology goes toward building nuclear power plants and a waste repository, then that money is not available for development of newer technologies that don't produce radioactive waste. While some might like to see a global nuclear economy (waste containment issues aside), in order to fully address world energy needs with current nuclear technologies, the number of operational power plants would need to increase tremendously. Calculations by one Cal Tech research group suggest that to meet current global energy needs, a power plant would need to be built every other day for the next 50 years (discounting any increase in population or power requirements during that time). Note also that fissionable fuel is nonrenewable and limited in quantity. On a global scale, it appears that current nuclear technologies present a bit of an impasse—it is prohibitively expensive to maintain and expand nuclear power as a partial energy strategy, and problematic on many levels to switch to a fully nuclear global energy economy.

Some researchers are exploring the possibility of a fusion-based nuclear power generator, hoping that the astronomical energy potential can be used to justify years of research and development costs until the technology is fully fledged. The idea under consideration involves producing and trapping *plasma* (an electrically charged gas) within a reactor cell. The hot plasma would act as a source of energy for the fusion reaction. The hydrogen bomb also uses plasma to effect fusion; the plasma in that case is created through extreme heat and pressure from a fission reaction applied to polystyrene foam. Of course, a nuclear reactor could not rely on a fission bomb in order to provide energy for fission; as a result, the production, trapping, and safe maintenance of plasma through other means is under investigation. *Magnetic confinement* involves using electromagnetic fields and radiation to confine hydrogen gas, heat it, and pressurize it sufficiently to induce the plasma state, whereupon fusion would take place. Alternately, *inertial confinement* relies upon lasers to accomplish the same task. In any case, the fusion reaction most frequently targeted for investigation is that of deuterium with tritium, owing to the abundance of deuterium and lithium, which is the nuclear precursor to tritium (see Nuclear Fusion above). This technology is still highly experimental, however, and no plans are currently in place for a fusion-based nuclear power facility.

12.12 HEALTH EFFECTS OF IRRADIATION

As the radiation victims of the atomic bomb and Chernobyl make all too clear, exposure to nuclear radiation is detrimental to health. The types of radiation vary greatly in their abilities to penetrate the human body and the severities of their effects. The three types of decay particles addressed in this chapter (as well as X-rays) are ionizing radiation, capable of breaking chemical bonds and damaging DNA, proteins, and other molecules, as discussed in Chapter 10. In the case of an acute (one-time, massive dose) exposure, extensive cellular damage can result in death. Chronic exposures (more limited doses, repeated over time) can lead to mutations and cancer as well as diseases and symptoms associated with aging. While alpha radiation can cause damage if taken internally or inhaled, the particles themselves are of insufficient energy to penetrate the skin. Radon, an alpha-emitter ubiquitous in Earth's crust, is a major source of alpha particle exposure. Beta radiation is somewhat higher in energy and more penetrating than alpha, but exposure is less common. Carbon-14 is a beta-emitter, but is dispersed in the environment and generally not encountered in concentrations sufficient to cause acute radiation effects. Of the major decay particles, gamma rays are the most energetic and damaging. Acute **radiation sickness**—a set of symptoms resulting from radiation exposure—is almost always due to gamma radiation. Aside from nuclear accidents and use of gamma rays in medical applications (see below), most exposure is through contact with naturally occurring radioisotopes. In fact, we are bombarded with radiation on a routine basis—radon in the ground is a significant natural source, as are isotopes of nitrogen and carbon in the ground and air. Cosmic rays, which are more highly concentrated at higher elevations (including on airplanes), also contribute to **background radiation**, a term for unavoidable radiation from the natural world.

Radiation exposure is measured in units of *rem*, which stands for *roentgen equivalent man*. This unit multiplies a radiation dose, measured in *rads*, or Joules of radiation absorbed per mass of tissue, by a biological effectiveness factor. The idea here is that some types of radiation are more penetrating than others, and specifics of absorption (where on the body, internal versus external, and so forth) partially determine effectiveness of the dose. While it's difficult to define a "safe" dose of radiation (particularly as effects are cumulative), the Environmental Protection Agency (EPA) correlates doses to predictable symptoms (see Table 12.2).

Exposure (rem)	Health Effect
5–10	Changes in blood chemistry
50	Nausea within hours
55	Fatigue
70	Vomiting
75	Hair loss within a few weeks
90	Diarrhea
100	Hemorrhage
400	Possible death within months
1000	Destruction of intestinal lining, bleeding, death within weeks
2000	Damage to central nervous system, unconsciousness within minutes, death within days

Table 12.2. Physical effects of acute radiation exposure. Chronic effects vary somewhat.

Source of Exposure	Annual Dose
Radon (inhaled, from ground—varies with region)	200 mrem
Background (cosmic rays—varies with elevation)	55 mrem
Medical (X-rays, CAT scans, nuclear medicine)	53 mrem
Internal body (food, water, air, incorporated radioisotopes)	39 mrem
Consumer products (TVs, computers, smoke detectors)	10 mrem
Other (nuclear fallout, weapons testing)	3 mrem

Table 12.3. Average radiation absorbed by United States residents from naturally occurring and common man-made sources. Note: 1 rem = 1000 mrem.

While the thought of constantly absorbing radiation—from the ground, the air, the cosmos, and from within our own bodies—is somewhat disconcerting, remember that these are incredibly low dosages. The EPA estimates that the average U.S. resident absorbs on the order of 0.3–0.4 rem/year (Table 12.3), far below the threshold necessary to cause even the mildest radiation symptoms. Further, while radiation effects are cumulative, they also dissipate somewhat over time: provided that exposures don't occur in rapid succession, the whole is less than the sum of the parts. Finally, it may be of interest to those who use consumer products and engage in certain habits that some voluntary activities result in radiation exposure far more significant than that from background sources. For instance, one coast-to-coast flight across the United States results in as much radiation exposure as an entire year's worth of fallout from nuclear energy production, weapons testing, and coal burning—about 3 mrem. Smoking contributes 50–280 mrem of radiation per year, depending upon the type of cigarette and quantity smoked. This means a heavy habit can nearly double annual exposure to radiation! In general, while some people are subject to special circumstances (including cancer treatment, nuclear medicine, and significant nuclear fallout), most individuals will not accumulate enough radiation over a lifetime to experience symptoms of radiation sickness.

CONCEPT CHECK:

Under normal circumstances, what is the source of the majority of radiation to which we are exposed?

Answer: Background (natural) sources such as radon, cosmic rays, etc.

12.13 NUCLEAR MEDICINE

Despite a conflicted global philosophy with regard to nuclear energy production, nuclear technology is nevertheless commonly utilized (and very important) to other applications, including medicine, archaeology, and food science. The X-ray machine was the first of the diagnostic techniques to utilize radiation, and afforded physicians the ability to see inside the human body. **Computed tomography (CT)**, a three-dimensional image created by computer from a series of X-ray images, soon followed. X-rays do not technically fall under the auspices of nuclear medicine, as they are not generated by nuclear phenomena—rather, they are the result of acceleration of electrons within an electromagnetic tube—but they nevertheless result in significant exposure to ionizing radiation, for which reason they are generally avoided in a medical setting without indication. Single X-rays typically result in dosages less than 100 mrem per exposure, while large-area CT scans may provide 1 rem or more. The utilization of tissue-penetrating radiation has since advanced far beyond the X-ray; beta- and gamma-emitting sources now find application in both diagnosis and treatment of certain conditions.

Many cancers are susceptible to radiation treatment. A tumor is composed of cells proliferating at an abnormally high rate, meaning that they copy their DNA more frequently than nearby cells. Since DNA is particularly sensitive to ionizing radiation, rapidly dividing cells are most likely to be affected (and killed) by radiation exposure.* For this reason, beta and gamma rays (and

***NOTE**

Chemotherapeutic pharmaceuticals also specifically target rapidly dividing cells. The common side effects of both chemotherapy and radiation treatment for cancer occur because not all rapidly dividing body cells are cancerous—hair follicles and epithelial cells (in skin and the mucous membranes) also divide rapidly, and can easily be damaged or killed by cancer treatments. For this reason, hair loss and mouth sores are common in cancer patients undergoing treatment even high-intensity X-rays) have been used as cancer treatments. The Gamma Knife, for instance, is used to focus several beams of gamma radiation from cobalt-60 upon a brain tumor. Where the beams meet, the radiation intensifies to the point that a tumor is typically destroyed in a single day. Since beta radiation is not nearly so penetrating, external beams of beta particles can only be used in the irradiation of surface tumors.

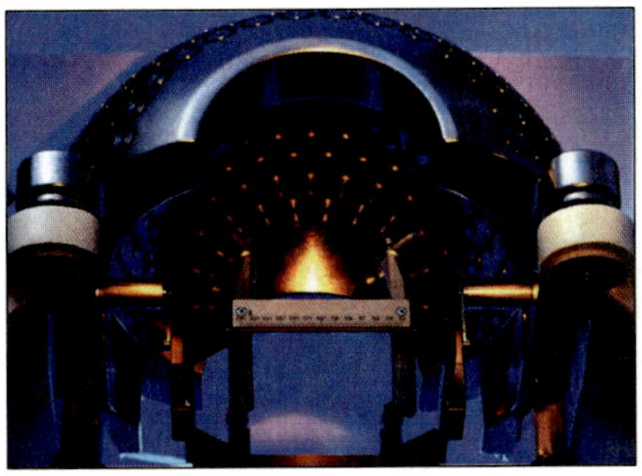

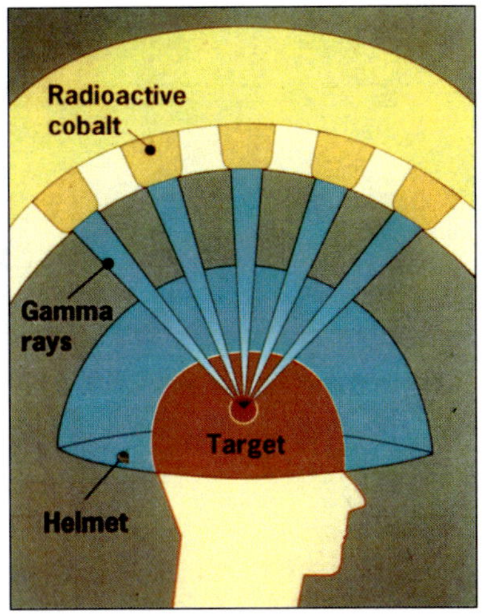

Image 12.15. The Gamma Knife functions by focusing several gamma rays onto a small target.

Frequently, radioisotopes are incorporated directly into the body. **Positron emission tomography (PET)** scans produce many images over a period of time, so that metabolic activity can be gauged. A radioisotope is introduced as part of a larger molecule normally utilized by the body (C-11, for instance, may be incorporated into glucose).* The location of the isotope-containing molecules is then monitored based upon their radioactivity. These scans allow for imaging of tumors, which are highly metabolically active, and tend to take up glucose out of proportion with the cells around them. They can also be used to diagnose some brain diseases, and map heart and brain function.

The thyroid, a gland responsible for production of several hormones and regulation of metabolism, takes up iodine as part of its normal function. In the case of hyperactivity of thyroid cells (such as in Graves' disease) or thyroid cancer, I-131, a beta- and gamma-emitter, can be introduced into the body. This radioisotope has a half-life of eight days, and is concentrated by the target gland. For this reason, systemic effects of radiation are mitigated by the localization of the radioactive substance. I-131 can be used to monitor thyroid activity if given in small doses or, in larger doses, is effective at killing thyroid tumor cells. Other radioisotopes show site-specific activity as well, based upon which body organ tends to concentrate these elements. Strontium-89 and samarium-153, for instance, are both beta-emitters and are particularly effective against bone tumors. Phosphorus-32, also a beta-emitter, is often injected into hollow brain tumors. Bone and brain tumors are far too deep to access using externally administered beta radiation, but internal administration of a radioisotope not only brings the beta-emitter into close proximity with the tumor, it also allows for localization of radiation, preventing unnecessary exposure of healthy tissue.

*Note

Remember that all isotopes of an element are identical in their chemical properties. An atom of C-11 can bond with other atoms of C, H, and O to form. glucose just as well as an atom of C-12 can. The difference, though, is that the C-11 is radioactive, meaning that its location in the body can be tracked by monitoring radioactive decay.

Concept Check:

Why does radiation damage tumor cells more than nearby non-tumor cells?

Answer: Tumor cells are rapidly dividing, and thus are more susceptible to DNA damage.

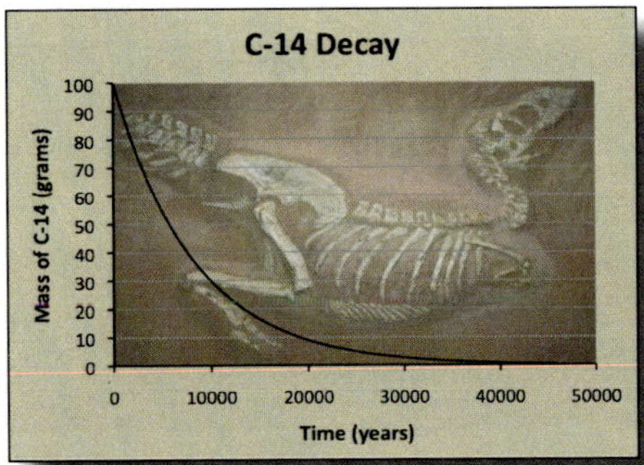

Image 12.16. Carbon-14 decays predictably with a half-life of 5730 years.

12.14 CARBON DATING

Nuclear technology can also be applied to the carbon dating of fossils, textiles, and other objects of historical interest or significance. A naturally occurring isotope of carbon, C-14 is produced in the upper atmosphere when N-14 is bombarded with cosmic radiation:

$$^{14}_{7}N + ^{1}_{0}n \rightarrow ^{14}_{6}C + ^{1}_{1}H$$

The majority of carbon on Earth is C-12 (about 99 percent of total carbon), with C-13, a stable isotope, representing about 1 percent of total carbon. C-14 is present in concentrations on the order of one part per trillion.

There are two important principles to carbon dating. The first is that carbon is taken up by plants during the process of photosynthesis (see Chapter 11). Since the three naturally occurring isotopes of carbon are chemically indistinguishable from one another, plants take up these isotopes in direct proportion to their representation in the atmosphere. During a plant's life, it would be expected to contain approximately 99 percent C-12, 1 percent C-13, and one part per trillion C-14. Because herbivorous animals get their carbon from plants (and carnivorous animals get their carbon from other animals), living animals would also be expected to have carbon isotope ratios identical to those of the atmosphere. The second major principle of carbon dating is that C-14 is radioactive, and therefore decays predictably, with a half-life of 5730 years (Image 12.16). For this reason, it is possible to use the proportion of C-14 remaining in a bit of dead organic matter to determine its year of death.

If, for instance, a mummy was discovered in which C-14 was present at only .5 parts per trillion (half the normal concentration), one would determine that the mummified individual perished one C-14 half-life ago—that is, 5730 years ago. Recent advancements producing highly accurate analytical techniques allow the different isotopes of carbon within a specimen to be sorted and counted, which makes carbon dating quite accurate. A limitation of the procedure, of course, is that it is only relevant to organic (once living) material.

12.15 IRRADIATION OF FOOD

A final application of nuclear technology worth discussing—particularly in light of its generally negative perception by the public—is that of food irradiation. Typically accomplished using a cobalt-60 source of gamma rays, food is irradiated for the purpose of killing viruses, bacteria, and insects that may be present. It is important to recall that gamma radiation consists of high-energy photons. While direct exposure can be harmful or fatal (hence the efficacy of this technique against microbes and insects), gamma radiation cannot induce radioactivity in other nuclei. For this reason, it is simply impossible for irradiated food to become radioactive. Early in the investigation of irradiation in the food industry, there was concern that exposure to gamma rays might decrease the vitamin (or other nutrient) content of foods. A number of studies have examined this issue, and in general, irradiation does not appear to decrease vitamin or nutrient content to any significant degree. Regardless, there are several reasons irradiation of comestibles is globally debated. Proponents argue that reducing microbial and insect loads keeps food fresh longer, meaning less spoilage during transport. This is particularly important in rural and undeveloped countries, where produce may need to be brought in from quite a distance away over a period of days or weeks. Opponents, however, are concerned that irradiation provides a veneer of safety with regard to bacterial proliferation in foods; simply because bacteria have been killed doesn't mean the toxins they produced while alive have been removed from the food. Additionally, the limited shelf life of perishable items has historically prevented them from being transported over particularly long distances (which occurs at a tremendous energy expense, and can be detrimental to the local economy). Irradiation, opponents argue, would simplify the logistics of extremely long-haul food transport, with ramifications upon sustainability in terms of increased fuel combustion and decreased support of local agriculture.

Nuclear science, which lies at the intersection of chemistry and physics, is a rapidly evolving field. Even if the promise of an unimaginable quantity of cheap nuclear power, anticipated since the 1950s, is never realized—whether for logistical reasons or by consensus—many other applications of the technology are and will remain important to our way of life. The lessons of the past remind us that any time a technology holds great potential for effecting positive change, it also has the potential to cause harm, either through intentional misuse or unanticipated consequences. Irrespective of how we as individuals or nations feel about nuclear weapons, energy, and technology, the fact remains that no society can possibly make decisions without affecting others. We share a global responsibility for determining in which direction we will take nuclear science, and in which direction we will allow it take us.

12.16 THE LAST WORD—MICROWAVES

Since gaining household popularity in the 1980s, the microwave has become an integral part of the American kitchen. The technology is completely unrelated to nuclear phenomena: Microwaves are not generated through nuclear reactions, nor are they a form of ionizing radiation. Still, the public perception of microwaves, what they are, and how they cook is frequently confused with nuclear technology, as apparent in common slang when we sometimes refer to microwaving food as "nuking" it.

Microwaves are a form of EMR (Chapter 10). Also in this category are gamma and X-rays (which probably serve to strengthen the misperception that microwaves are a nuclear phenomenon). However, the wavelength of microwave radiation is much longer than that of ionization radiation, meaning that the frequency and energy are much lower. As a result, microwaves, like visible light, are far too low in energy to break molecular bonds. Instead, microwaves cause molecules to rotate,* which produces heat through friction (just as rubbing your hands together leaves them feeling warm).

***NOTE**

Specifically, microwave ovens cause water molecules to rotate, which is why it's not possible to heat completely dry substances in a micro-wave, and why some dishes (specifically ceramics, which can absorb water) are not microwave safe, and get very hot if used in an oven.

This heat is distributed throughout the food, defrosting, warming, or cooking it. Really, a microwave oven cooks in a manner absolutely analogous to your stovetop. The only difference is that on the stovetop, the heat comes from the stove element, while in a microwave, the heat comes from water molecule friction. In no way are the vibrating water molecules chemically altered—there is simply no truth to the common misconception that microwaving renders food radioactive, unfit to eat, or devoid of nutrients (unless, of course, you overcook it, which is always detrimental to vitamin content, even on the stovetop). As for whether microwaves can leak radiation and cause injury, the answer is that it's not likely, since they are designed to stop functioning if the door is opened, and doors are structured to prevent leakage. In any case, the most important consideration is that in the event that microwaves did escape the oven in any significant quantity, the only thing they'd be capable of in a human is what they're capable of in food—production of heat. That is to say, in the same way you'd burn your finger if you touched a hot stove element, you could burn your finger if you found a way to stick it in a functioning microwave. Under no circumstances, though, can microwaves cause cancer—they're simply the wrong type of radiation!

Image 12.17. Microwave radiation causes molecular rotation, which generates heat through friction.

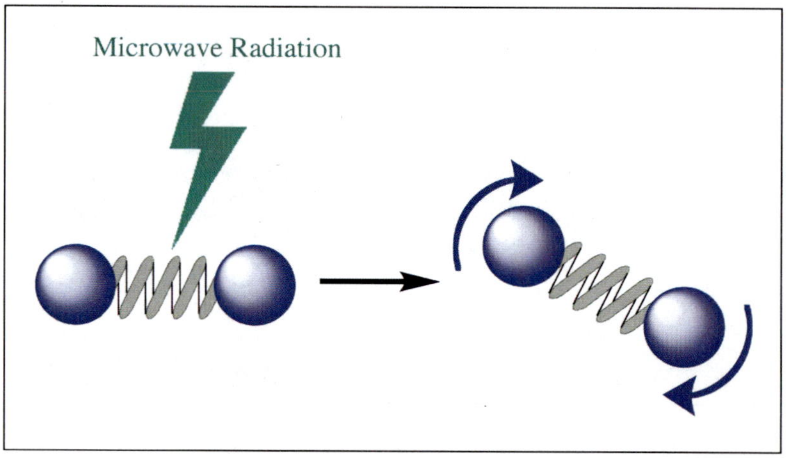

SUMMARY OF MAIN POINTS

- Unlike chemical reactions, nuclear reactions can change the identity of atoms by changing the number of protons in the nucleus.
- Radioactive decay is the loss of a small decay particle from a parent nucleus to form a new daughter nucleus.
- Common decay types include alpha ($^{4}_{2}He$), beta ($^{0}_{-1}e$), and gamma($^{0}_{0}\gamma$), which is pure energy.
- Nuclear equations are balanced with regard to mass (superscript) and number of protons (subscript), but NOT with regard to atoms or atom identities.
- Atoms of the same element with different masses are called isotopes. The mass of an element on the periodic table reflects the mass of the most common isotope.
- All large elements (bismuth and larger) are radioactive: if one decay event does not produce a stable nucleus, a decay series takes place until a stable nucleus results.
- Nuclear fission is the splitting of a larger nucleus into two smaller nuclei.
- Fission takes place with a mass defect—the lost mass is too inconsequential to show up in the balanced nuclear equation, but is converted to energy by the equation $E = mc^2$.
- Nuclei can be induced to fission by bombardment with neutrons and other high-energy particles.
- U-235 and Pu-239 are both fissionable fuels and are used in nuclear weapons and fuels.
- The quantity of fissionable fuel necessary to self-sustain a chain reaction is called the critical mass.
- U-235 must be enriched out of the more common (and non-fissionable) U-238 by gaseous diffusion, a process that leaves radioactive waste in the form of depleted uranium.
- The first nuclear weapons used chemical detonations to bring critical masses of fissionable fuel together; the fuel would then undergo a nuclear reaction.
- Nuclear fusion is the combination of two smaller nuclei to form a larger one—like fission, it results in a mass defect and yields tremendous energy.
- Fusion reactions fuel the Sun and stars, and are responsible for producing heavier elements from hydrogen.
- Fusion reactions require tremendous initial energy input; for this reason, the hydrogen bomb uses a fission reaction to initiate a fusion reaction.
- If a nuclear reaction is controlled, the energy can be harnessed and used to produce electricity. This is accomplished in nuclear power plants.
- Nuclear plants use fissionable fuel (U-235 or Pu-239) in fuel assemblies that are interspersed with neutron-absorbing control rods to keep the reaction from growing exponentially.
- Radioactivity is sequestered from the outside of the plant through two separate closed loops.
- Radioisotopes decay predictably with a given half-life: one half-life is the time for half of a sample to decay.
- Nuclear waste from power plants is radioactive and must be stored. Most plants in the United States are out of or are running out of storage, and are dry-casking waste.
- A geological repository has been proposed in Nevada, but has met with opposition and is decades behind schedule.
- Nuclear disasters like Three-Mile Island and Chernobyl have significantly impacted public support for nuclear power plant proliferation.
- While researchers are exploring the possibility of fusion-generated power, this technology is far from fully developed.
- Radioactive decay produces ionizing radiation—radiation with the ability to break chemical bonds.
- Exposure to ionizing radiation can have severe health ramifications. Some exposure from natural sources is avoidable, while other sources of exposure can be avoided.

- Nuclear technology contributes significantly to medicine through diagnostic and treatment techniques including imaging, targeted destruction of tumors, and radiotherapy.
- The predictable decay of radioisotopes allows for carbon dating, a technique that determines the age of an organic sample based upon its remaining C-14.

Questions and Topics for Discussion

1. How is nuclear chemistry similar to the concept of reactions held by the ancient alchemists? By Dalton? How is it different?
2. How are nuclear reactions similar to chemical reactions? How are they different?
3. Given that nuclei don't contain electrons, where do beta particles come from?
4. Why are all the large nuclei radioactive? Explain in terms of nuclear forces
5. How is it possible that nuclear equations balance with regard to mass, despite the mass defect?
6. Why is the fission of U-235 self-sustaining under the right circumstances?
7. Explain the basics of the atomic bomb. What is the purpose of the chemical detonation?
8. Through what process is U-235 enriched out of U-238? Why are the uranium tails environmentally hazardous?
9. How is nuclear fission similar to nuclear fusion? How are they different?
10. Why does the hydrogen bomb require a primary fission reaction?
11. To what extent is the nuclear material in a power plant separated from the outside environment?
12. How is nuclear fission used to generate power?
13. Why is a discussion of half-life relevant to the issue of high-level radioactive waste?
14. What are the risks associated with building a deep geological repository? What are the benefits?
15. Does risk-benefit analysis support the idea of building a deep geological repository in Nevada? Why or why not? What about the precautionary principle? (See Chapters 6 and 10 to review this concept.)
16. The nuclear disasters presented in the chapter significantly affected public opinion of nuclear power. Do you think this was warranted? Why or why not?
17. Do you think that nuclear power (fission, fusion, or both) should continue to be evaluated as potential energy sources? Why or why not?
18. Based upon what you've learned, would you feel comfortable living near a nuclear power plant? Explain your reasoning.
19. Many people fear nuclear technology and radiation, but willingly submit to X-rays, which are also a source of ionizing radiation. Why do you think this is so?
20. A skeptical scientist tried to disprove the accuracy of carbon dating by trying it out on his titanium wristwatch. He showed that the watch was 60000 years old. Does this convince you that carbon dating is fallible? Why or why not?

Problems

1. What is conserved in a chemical reaction? In a nuclear reaction?
2. What is not conserved in a chemical reaction? In a nuclear reaction?
3. Of the common decay particles, which have mass? Which have charge?
4. What is the net result of alpha decay of a nucleus?
5. What is the net result of beta decay of a nucleus?
6. What is the net result of gamma decay of a nucleus?
7. How many protons and neutrons are in: ^{55}Mn, ^{201}Hg, potassium-39, Ba-141?

8. How many protons and neutrons are in: Bi-214, ^{11}C, Np-239, radium-225?
9. Complete each of the following nuclear equations with the missing particle or nucleus:

$$^{210}_{82}Pb \rightarrow \_\_\_ + ^{210}_{83}Bi$$

$$^{230}_{90}Th \rightarrow \_\_\_ + ^{226}_{88}Ra$$

$$^{210}_{84}Po \rightarrow ^{4}_{2}He + \_\_\_$$

$$\_\_\_ \rightarrow ^{228}_{89}Ac + ^{0}_{-1}e$$

10. Complete each of the following nuclear equations with the missing particle or nucleus:

$$^{226}_{88}Ra \rightarrow \_\_\_ + ^{222}_{86}Rn$$

$$\_\_\_ \rightarrow ^{229}_{90}Th + ^{0}_{0}\gamma$$

$$^{59}_{26}Fe \rightarrow ^{0}_{-1}e + \_\_\_$$

$$\_\_\_ \rightarrow ^{210}_{81}Tl + ^{4}_{2}\alpha$$

11. Write balanced equations for each of the following: Bismuth-214 undergoes beta decay; Pu-242 emits alpha radiation; ^{81}Se emits beta radiation; U-235 undergoes gamma decay.
12. Write balanced equations for each of the following nuclear transformations: I-131 to Xe-131; Ra-226 to Rn-222; ^{201}Hg to Tl-201; rubidium-81 to strontium-81; Th-231 to Ra-227.
13. What is a radioactive decay series, and when does it occur?
14. Define: Nuclear fission; mass defect; and critical mass.
15. A reaction takes place with a mass defect of 0.030 kg. What is the energy yield?
16. The nuclear fission reaction of 2 kg of Pu-239 occurs with a 0.1 percent mass defect. What is the energy yield?
17. What mass defect is required to generate 5.5×10^{16} J of energy?
18. What percent mass defect would be necessary to generate 2.5×10^{14} J of energy from 35 kg of fissionable fuel?
19. The half-life of tritium is 12.3 years. What mass of tritium from a 48.0 g sample remains after 24.6 years? (Use intuitive approach.)
20. Approximately how long does it take for the sample of tritium in Problem 19 to be reduced to 2.0 g of tritium? (Use intuitive approach.)
21. A radioactive sample with an initial mass of 100.0 g has been reduced to 1.56 g of the initial radioisotope in 85 years. How long is the radioisotope's half-life? (Use intuitive approach.)
22. Using Table 12.1, how long does it take 75 percent of a sample of N-16 to have reacted? (Use intuitive approach.)
23. Using Table 12.1, how much U-235 remains from an original sample of 54.3 g after 9.9×10^{9} years? What has happened to the rest of the uranium? (Use equation.)
24. Using Table 12.1, how much C-11 remains of an original 35.0 g sample after 146 minutes? (Use equation.)
25. What are the primary sources of (unavoidable) radiation to which we are routinely exposed?
26. What are some major sources of radiation to which we are exposed that are avoidable?
27. What is ionizing radiation, and what does it do in the body?
28. What are the symptoms of radiation sickness?

29. Why can radiation be used to treat cancer? What are the side effects?
30. Why is carbon dating only useful to date organic materials?

Answers to Problems

1. In a chemical reaction mass, matter (identity of atoms), and charge are conserved. In a nuclear reaction, mass (except for defect) can be said to be conserved, and charge is conserved.
3. Alpha particles have mass, both alpha and beta have charge; gamma particles have neither.
5. Beta decay converts a neutron into a proton with no loss of mass.
7. ^{55}Mn: 25 protons, 30 neutrons; ^{201}Hg: 80 protons, 121 neutrons; potassium-39: 19 protons; 20 neutrons; Ba-141: 56 protons; 85 neutrons.
9. Complete each of the following nuclear equations with the missing particle or nucleus:

$$^{210}_{82}Pb \rightarrow\ ^{0}_{-1}e +\ ^{210}_{83}Bi$$

$$^{230}_{90}Th \rightarrow\ ^{4}_{2}He +\ ^{226}_{88}Ra$$

$$^{210}_{84}Po \rightarrow\ ^{4}_{2}He +\ ^{206}_{82}Pb$$

$$^{228}_{88}Ra \rightarrow\ ^{228}_{89}Ac +\ ^{0}_{-1}e$$

11.

$$^{214}_{83}Bi \rightarrow\ ^{0}_{-1}e +\ ^{214}_{84}Po$$

$$^{242}_{94}Pu \rightarrow\ ^{4}_{2}He +\ ^{238}_{92}U$$

$$^{81}_{34}Se \rightarrow\ ^{0}_{-1}e +\ ^{81}_{35}Br$$

$$^{235}_{92}U \rightarrow\ ^{231}_{90}Th +\ ^{4}_{2}He$$

13. A decay series occurs when the daughter nucleus of a decay reaction is also radioactive, and decays again. These are particularly common with large nuclei.
15. 2.7×10^{15} J.
17. 0.61 kg.
19. 12.0 g.
21. 14.2 years.
23. 0.003 g; the rest of the material has decayed to become daughter nuclei.
25. Radon inhaled and in the ground, cosmic rays, incorporated radioisotopes.
27. Ionizing radiation breaks chemical bonds, damaging DNA, proteins, and other molecules. It's associated with cancer, disease, and aging.
29. Radiation damages cells in their growth phase—cells growing fastest will be affected most. Cancer cells are rapidly dividing, so are very susceptible. Other rapidly dividing cells (like skin and hair cells) are also affected and lead to side effects (loss of hair, ulcers, etc.) of radiation treatment.

Sources and Further Reading

Early Nuclear Science

Heilbron, J. L. (2003). *Ernest Rutherford and the Explosion of Atoms*. New York: Oxford University Press.

Preston, D. (2006). *Before the Fallout: From Marie Curie to Hiroshima*. New York: Penguin.

Willis, K. (1995). The origins of British nuclear culture, 1895–1939. *The Journal of British Studies*, 34(1), 59–89.

Nuclear Bombs

Hughes, J. (2002). *The Manhattan Project: Big Science and the Atomic Bomb*. New York: Columbia University Press.

Nuclear Power and Waste

www.nrc.gov

Lemons, J., and Malone, C. (1989). Siting America's geological repository for high-level nuclear waste: implications for environmental policy. *Environmental Management*, 13(4), 435–41.

Lewis, N. (2007). Powering the planet. *MRS Bulletin*, 32, 808–20.

Mould, R. F. (2000). *Chernobyl Record: The Definitive History of the Chernobyl Disaster*. London, UK: Institute of Physics Publishing.

http://www.ocrwm.doe.gov/

U.S. Nuclear Regulatory Commission. (2002). *Radioactive Waste: Production, Storage, Disposal* (NUREG/BR-0216, Rev.2) [Brochure]. Washington, DC: Office of Public Affairs.

Health Effects

www.epa.gov

http://www.ocrwm.doe.gov

Applications

Duodu, K. G., Minnaar, A., Taylor, J. R. N. (1999). Effect of cooking and irradiation on the labile vitamins and antinutrient content of a traditional African sorghum porridge and spinach relish. *Food Chemistry*, 66(1), 21–27.

Early, P. J., Landa, E. R. (1995). Use of therapeutic radionuclides in medicine. *Health Physics*, 69, 677–94.

Graham, W. D., Stevenson, M. H. (1999). Effect of irradiation on vitamin C content of strawberries and potatoes in combination with storage and with further cooking in potatoes. *Journal of the Science of Food and Agriculture*, 17(3), 371–77.

Graham, W. D., Stevenson, M. H., and Stewart, E. M. (1998). Effect of irradiation dose and irradiation temperature on the thiamin content of raw and cooked chicken breast meat. *Journal of the Science of Food and Agriculture*, 78(4), 559–64.

Orton, C. G., (1995). Uses of therapeutic X-rays in medicine. *Health Physics*, 69, 662–76.

www.cancer.org

13 Energy, Chemistry, and Society

As a global community, the last several decades have seen us forced to confront many challenges. Our expanding population and increasing reliance on technology will require us to face many more such challenges in the coming years, some of which we foresee and some of which will surely catch us by surprise. At present, however, none looms in such pressing and daunting fashion as the global energy crisis. We must power the world, but our current sources of energy are also sources of pollution that perturb global cycles. In previous chapters, we've discussed how combustion reactions produce tropospheric pollutants and nuclear reactions produce high-level waste. We've examined our use of chemical fertilizer—which we rely upon to provide for the nutritional energy needs of a hungry world population—and its effect upon the nitrogen cycle. Now, however, we turn our attention to the big picture, and look at how each of these perturbations (and many others) affect the atmosphere on a much larger scale than previously discussed. In this chapter, we will examine the relationship between human energy requirements and the planet's carbon cycle. We'll use a scientific and skeptical analysis of available data and evidence to determine whether it suggests that human activity is perturbing the cycle, and whether that perturbation is resulting in climate change. Finally, we'll discuss how global, local, and individual actions affect energy balance.

13.1 AN IMPORTANT INTRODUCTORY NOTE FROM THE AUTHOR

The topics we've discussed up to this point in the text are not matters of major debate, either in the scientific world or within the whole of society. There is a global, scientific, and popular consensus on the environmental effects of pollution. The reality of ozone depletion and the steps necessary to reverse it are not questioned. The central topic of this chapter, however—global warming—is hotly debated in political and popular arenas. As we begin our discussion, I ask you to be a skeptic, but I also ask you to remember what that means. A skeptic believes nothing he or she is told, but instead asks to see the evidence, and personally evaluates it on its own merits. A skeptic neither believes in global warming nor categorically dismisses it as a hoax, until the evidence has been reviewed. A skeptic is not swayed by politicians or journalists, friends or Internet bloggers, popular media, or even isolated scientific sources. Rather, a skeptic insists on seeing the data firsthand, and insists that it be scientifically gathered, reproducible, and corroborated. To accept global warming without evidence is foolhardy, and similarly, to dismiss it without examination of the evidence is not skepticism, but simply denial. I encourage you to read this chapter as you would any other source—with a skeptical and discerning eye—and I further encourage you to seek out additional information that corroborates or refutes what is presented here.

13.2 OUR GLOBAL ENERGY DEPENDENCE AND FOSSIL FUELS

As a global community, we consume a tremendous (and ever increasing) amount of energy. In 2008, estimated global energy consumption was nearly 5 x 10^{20} J. While this is a staggeringly large quantity, it is much less than the amount of energy the Earth receives from the Sun each year, which is estimated at 2 x 10^{23} J—nearly 500 times the annual energy consumption. Given that energy input into the planet so greatly outweighs energy from the planet that we consume, why do we so often hear reference to a global energy "crisis"? The reason is that most of our energy requirements are not fulfilled by sunlight—at least, not by the sunlight we're receiving now. Instead, the majority of fuels we burn—called **fossil fuels**—were created millions of years ago when prehistoric organisms used photosynthesis to capture and store energy from the Sun. The fossil fuels, which include coal, petroleum, and natural gas, collectively provide approximately 85 percent of energy currently consumed worldwide.

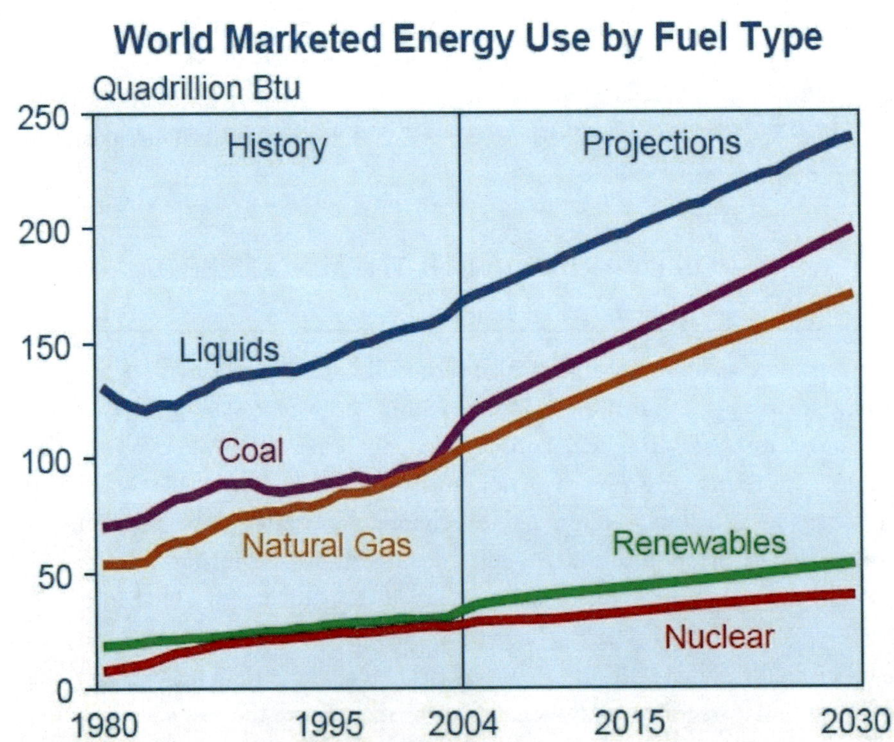

Image 13.1. Global energy use history and projections from the U.S. Energy Information Administration (EIA). A BTU is 1.06 kJ, or 252 calories. The units on the y-axis are as multiples of 10^{15} BTU. "Liquids" refers to petroleum.

In order to understand fossil fuels and our reliance upon them, let's take a few minutes to discuss where they come from and why they are such an important source of energy. From some of our discussions in earlier chapters, we know that energy can neither be created nor destroyed; instead, it must be transformed from one form into another. Incident sunlight, for instance, can be transformed into chemical energy (glucose) by plants. If the sunlight falls upon a man-made photovoltaic panel, it can be transformed into electrical energy. The simplest energy transformations, however, rely purely upon the capture and transfer of heat energy—a coal-fired power plant, for instance, uses the heat released when coal is burned to boil water and produce steam, which can drive a piston (as in the early steam engines) or turn a generator turbine. The same principle is applied in the cases of a parabolic solar power trough, which uses sunlight to heat a liquid, and nuclear power, in which the energy released by fission is used to heat liquid. In both of these cases, the hot liquid is then used to generate steam.

In early times, wood was the most accessible combustion fuel. Unfortunately, wood fires don't burn particularly hot, which limits their ability to provide power. Further, wood is not a very dense source of energy: it takes a large volume of wood to provide a relatively small amount of heat. This means that extracting energy from wood is very labor intensive, given the amount that must be gathered, and also, that it is an impractical source of energy for transportation, given the weight and quantity of fuel that must be carried. Hence, the ability to mine coal was a necessary prerequisite to the Industrial Revolution, as discussed in Chapter 6. Compared to wood, **coal** is a very dense source of energy—it contains more than twice as much energy per unit mass. Like wood, coal represents the transformation of sunlight into chemical energy; coal is the fossilized remains of ancient plant life that died, partially decayed, and was then buried beneath layers of earth. The compression of the organic material stalled its decomposition, meaning that while coal contains much less water than wood (hence its increased energy yield per gram burned, since water represents noncombustible mass), the energy-containing carbon-based molecules are still partially intact. Coal is not a pure substance, but is instead a mixture of many organic molecules and is composed largely of carbon and hydrogen, with some nitrogen and oxygen also figuring into the mix. Trace sulfur—the exact percentage varies significantly from sample to sample—is also present. As discussed in Chapter 6, the presence of nitrogen and sulfur make coal combustion a very polluting process. The major products of coal burning, however, are CO_2 and H_2O, the compounds produced when any hydrocarbon or hydrocarbon derivative burns in oxygen. We'll soon see that CO_2, while not a pollutant that directly impacts human health, has tremendous ramifications as an environmental pollutant.

Like coal, **petroleum** and **natural gas** contain stored solar energy from millions of years ago. Generally found together, these two fossil fuels are the partially decomposed remains of sea organisms. As tiny oceangoing plants and animals died, they fell to the sea floor where they mixed with and were buried in sediment. Over time, the shifting layers of sediment created covered, pressurized pockets of organic material that, like coal, had many of the original energy-containing molecules still partially intact. When these pockets of fossilized organic material are accessed today, they are found to contain a mixture of gas and liquid. The liquid, made up of larger hydrocarbons and other organic material, is petroleum. Due to its low sulfur content, petroleum combustion produces fewer SOx emissions than coal. The lighter hydrocarbons are not found as liquid components of petroleum, but instead make up natural gas. With minimal contamination by non-hydrocarbon organics (and composed primarily of methane, which is too small to produce combustion particulates), natural gas is the cleanest burning of the fossil fuels.

% by Mass	Coal	Petroleum
Carbon	75–90%	83–87%
Hydrogen*	4.5–5.5%	10–14%
Oxygen	5–20%	0.5–1%
Nitrogen	1–1.5%	0.1–0.5%
Sulfur	1–7%	0.5%

Table 13.1. Comparison of the percent composition by mass of coal and petroleum. *Recall that hydrogen is much lighter than the other organic elements, meaning that even significant quantities will represent relatively low percentages by mass.

The challenges associated with mining petroleum and natural gas made coal the first popular fossil fuel. Petroleum followed, gaining popularity in the early- to mid-1900s. Natural gas, however, initially presented a logistical dilemma. It couldn't be carried like coal, or piped like petroleum. It was difficult to collect, store, and transport. As a result, for many years it was common practice to simply light the gas on fire as it escaped from a newly drilled oil well. With the problematic gas out of the way, removal of petroleum could commence. Since the mid-1900s, however, natural gas retrieval and processing technologies have progressed to the point that it is now used in proportions very similar to those of coal. In the United States, petroleum is still the major source of energy, however.

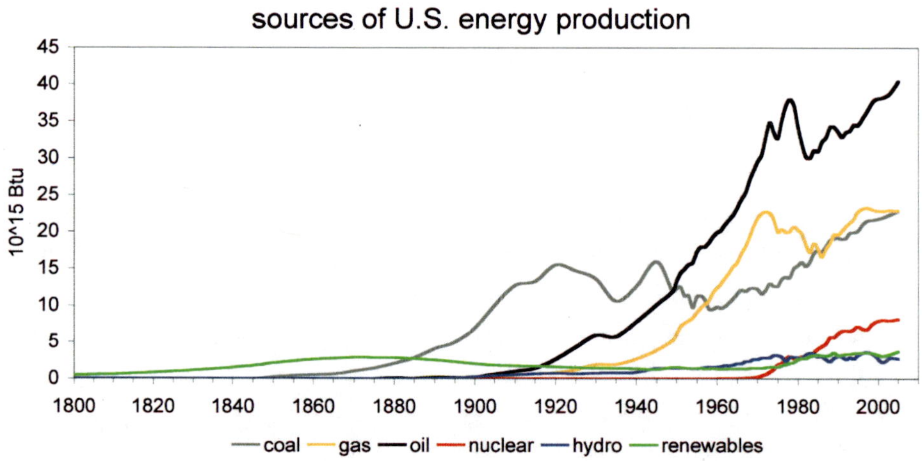

Image 13.2. Sources of energy production in the United States per the Energy Information Administration (EIA). On this graph, "renewables" refers to wood through the 1940s, and thereafter refers to wood, biofuel, wind, and solar. Oil is petroleum.

One problem associated with an energy economy that depends largely upon fossil fuels is that they are **non-renewable**: The partial decomposition and pressurization process that transformed living plants and animals into dense sources of energy took millions of years. Since fossil fuels can't be produced on demand, there's no way to increase the supply in response to increasing need. Our tremendous global energy demand threatens to outstrip the availability of these fuels. This is one reason that the search for alternative energy sources, including solar, has been gaining momentum. While estimates for when oil will be depleted vary tremendously,* there is no doubt that the supply of petroleum will be exhausted in the not-too-distant future. In the likely event that oil availability dwindles dramatically in the next few decades (and in the absence of alternative energy technologies becoming more predominant than they are), coal is oil's most likely successor—there is about 40 times as much retrievable coal as there is petroleum. Even so, the nonrenewable nature of fossil fuels and their ever-diminishing quantity is not the biggest problem they present. Rather, their greatest threat to our welfare comes from one of their combustion products—the hitherto innocuous-seeming CO_2.

***NOTE**

It's difficult to predict exactly what our future energy dependence and available energy sources will be, particularly given that there are new petroleum deposits discovered from time to time. At the same time, existing but currently inaccessible deposits may or may not become accessible in the future, depending upon regulations and advancements in technology.

CONCEPT CHECK:

Which are the three most widely used fuels in the United States? In the world?

Answer: Petroleum, natural gas, and coal.

13.3 THE GREENHOUSE EFFECT

The words *greenhouse effect* are often used synonymously with *climate change*. In fact, nothing could be further from the truth. The **greenhouse effect** is essential to the habitability of our planet—without it, Earth would be an average of about 33°C (60°F) cooler than it is and incapable of supporting life. Much in the same way that a literal greenhouse (a glass building meant to trap the Sun's energy) maintains plants through cool nights or seasons by keeping them warm, some of the molecules comprising our atmosphere naturally trap the Sun's heat and warm the Earth. This process is outlined in Image 13.3. When sunlight hits the planet, some is reflected by the Earth's atmosphere (1), and some by the Earth itself (2). The remaining radiation is absorbed either by the Earth (3) or by atmospheric gases (4). The radiation absorbed by the Earth is released as heat energy, some of which escapes into space (5), and some of which is trapped by atmospheric gases (6).

Image 13.3. The greenhouse effect.

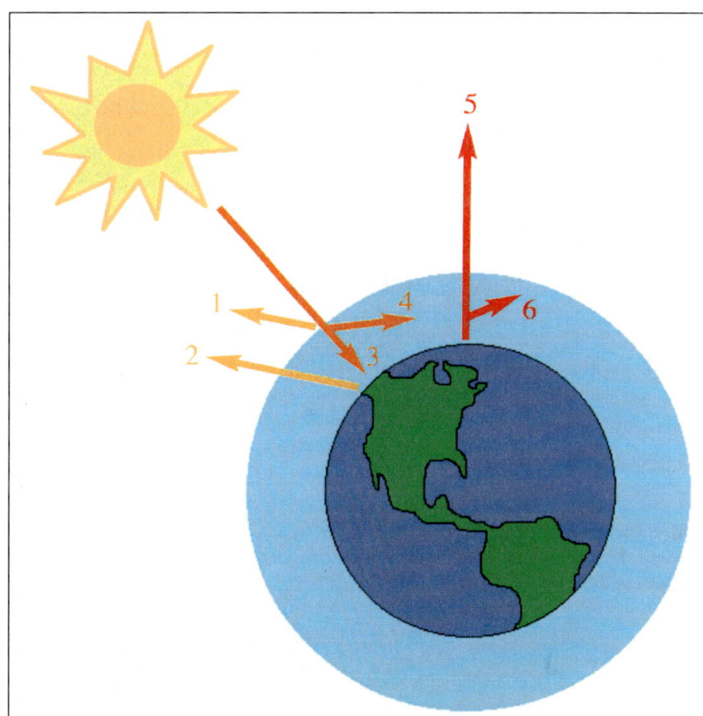

The collective action of atmospheric gas traps about 81 percent of the energy radiated by the Earth, which would otherwise be lost to space. The gases then re-radiate that heat upon the Earth, increasing the temperature. Understanding the mechanisms whereby gases are able to absorb and re-radiate heat requires that we spend a little time discussing yet another type of EMR, because what we feel as heat is actually **infrared radiation (IR)**.

IR has a longer wavelength (and is therefore lower in frequency and energy) than the UV light we discussed in Chapter 10. It is not ionizing radiation and is incapable of breaking bonds. However, here again, our bond-as-spring model serves us well, because IR is capable of compressing and stretching bonds, causing a molecule to vibrate.

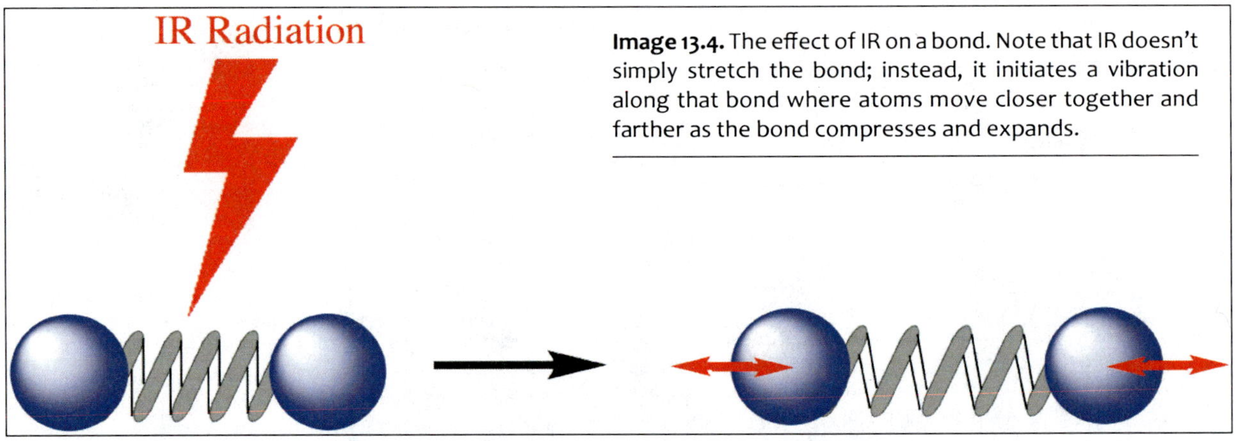

Image 13.4. The effect of IR on a bond. Note that IR doesn't simply stretch the bond; instead, it initiates a vibration along that bond where atoms move closer together and farther as the bond compresses and expands.

Absorption of IR is not permanent—shortly after a bond begins to vibrate, the molecule relaxes. This not only stops the vibration, but also results in the release of IR energy. This absorption and subsequent release of IR is the key to understanding the greenhouse effect, because when a molecule traps IR, it's collecting heat, which is subsequently released and re-radiated upon the Earth.

Actually, not all molecular vibrations (and not all molecules) absorb IR. Table 13.2 shows many of the vibrations of CO_2, some of which absorb IR and others of which do not. It turns out that only vibrations producing a change in net dipole are capable of absorbing IR. As such, molecules that have at least one IR-absorbing vibration can trap heat, and are therefore **greenhouse gases**.

The first vibration listed in the table shows CO_2 bending symmetrically around the central carbon. As this vibration takes place, the oxygen atoms will move together (up and down), while the carbon atom will move in a direction opposite that of the oxygens. During the vibration, the molecule's shape is constantly moving between the extremes of an inverted "V" (as shown in the table) and a "V" (the shape created when carbon is at the bottom of its trajectory, and the oxygen atoms are at the top of theirs) (Image 13.5). Since the C-O bonds are polar, the molecule's net dipole will also vary throughout this vibration process. When the molecule is bent as shown in the table, the net dipole points downward. At the other extreme of the vibration, the dipole points upward (having moved through the neutral position in which CO_2 has

Vibration	Effect on Net Dipole	IR Absorbing?
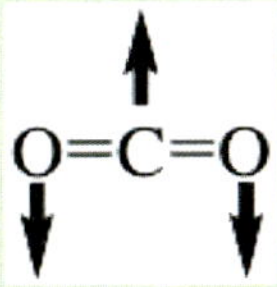 (This can also take place in and out of the plane of the page, to the same effect)	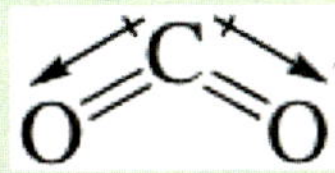 Change in net dipole	Yes
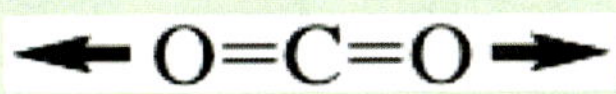	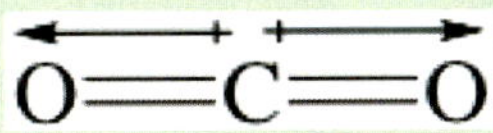 None	No
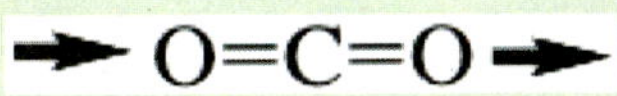	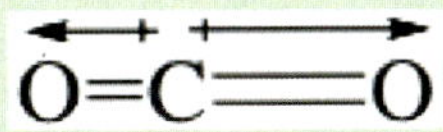 Change in net dipole	Yes

Table 13.2. Vibrations of CO_2 and IR absorption.

no net dipole). Vibrations that change a molecule's dipole are high-energy vibrations—IR supplies the energy to produce this movement, and when the IR is released, the movement ceases.

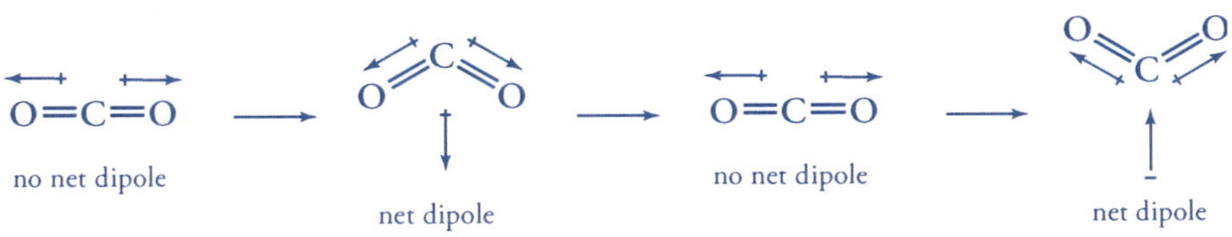

Image 13.5. The variation of a carbon dioxide molecule's net dipole as it vibrate sin one possible mode.

The second vibration listed in the table, in which carbon stays in place and the oxygens move together (away from carbon, then closer to carbon), does change the length of bonds, but does not change the net dipole (Image 13.6). Since the C=O bonds always stretch or compress at the same time (and since there's no change in geometry around the central atom), the dipoles are always equal and opposite. As a result, the net dipole is always zero. This particular vibration, then, does not absorb IR. Regardless, CO_2 has other IR-absorbing vibrations, and is therefore a greenhouse gas.

Image 13.6. A vibration of carbon dioxide in which the net dipole does not change, on account of symmetry. Regardless, carbon dioxide does have other vibrations that affect net dipole, and is therefore capable of IR absorption, though not by this particular vibrational mode.

There are greenhouse gases besides CO_2. SO_2, for instance, was mentioned in Chapter 6 as being not only a tropospheric pollutant, but a powerful greenhouse gas. Making this determination requires that we analyze SO_2's potential for IR-absorption, which requires us to identify at least one dipole-changing vibration. A good technique for going about this process is as follows:

1) Draw a Lewis structure of the molecule in question, including bond and net dipoles.

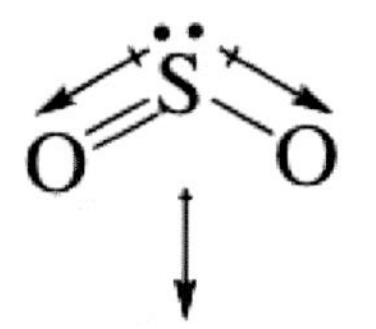

Note that lone pairs on outer atoms do not affect the geometry of the molecule, so they've been omitted from this structure in the interest of simplicity. SO_2 has a bent structure with a net dipole.

2) Try to determine a way in which the molecule could bend or bonds could stretch that would change the net dipole. It may be helpful to sketch this, even though it won't look like a normal or correct Lewis structure.

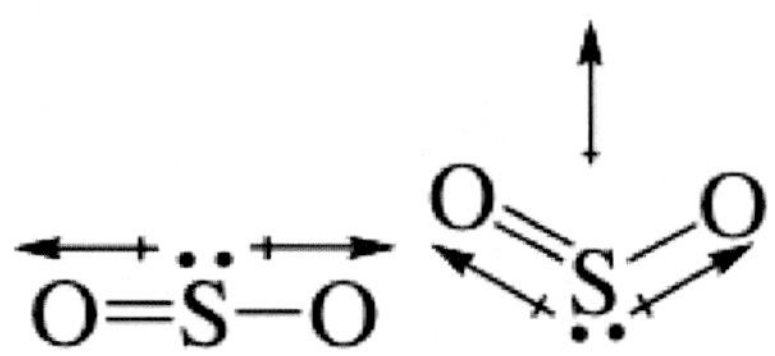

If SO_2 vibrated with the oxygens moving together (up and down) while the sulfur moved in the opposite direction, the molecule would pass through a linear geometry (no net dipole) on its way to a "V" shape (net dipole in the opposite direction of the original). This represents an IR-absorbing vibration. A given molecule will often have many IR-absorbing vibrations—determining that it's a greenhouse gas requires identifying only one such vibration. SO_2 is therefore a greenhouse gas.

TRY THIS:

Using the technique outlined above, determine whether HCl is a greenhouse gas.

Answer: It is a greenhouse gas.

Solution:

Stretching HCl along the bond lengthens (changes) the dipole. This is IR-absorbing.

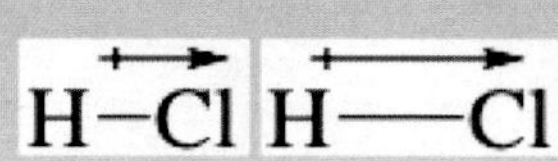

In the end, whether a relaxed molecule has or does not have a net dipole is not the determining factor in whether it's a greenhouse gas—SO_2 has a net dipole, while CO_2 has none. Both, however, are greenhouse gases. Even without sketching dipole-altering vibrations, it turns out that an easy way to identify a greenhouse gas is that it will always have at least one bond dipole. Any molecule with at least one bond dipole can be bent or stretched to produce a change in the net dipole, and that bending or stretching represents an IR-absorbing vibration. By extension, any molecule with no bond dipoles is not a greenhouse gas. N_2, the most prevalent of the atmospheric gases, is not a greenhouse gas. With no bond dipoles, there are no bending or stretching motions capable of IR-absorption:

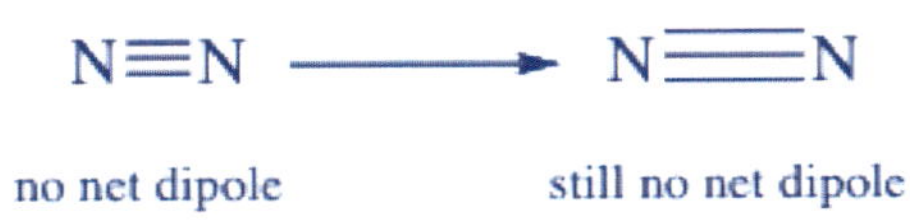

Try This:

Using bond dipoles as your guide, determine whether each of the following is a greenhouse gas: O_2; CH_4; and CH_2Cl_2.

Answer: O_2 is not a greenhouse gas (no bond dipoles), CH_4 is a greenhouse gas*, and CH_2Cl_2 is a greenhouse gas.

*Note

Note that while we considered the C-H bond *essentially* nonpolar for purposes of determining solubility in the earlier chapters of this text, there is a very small dipole since C is slightly more polar than H. The dipole, in this application, is very important—CH_4 is a potent greenhouse gas!

13.4 THE CARBON CYCLE

Atoms of carbon, like atoms of nitrogen, cycle through many different forms on Earth. Inorganic carbon, as found in carbonate rocks or CO_2, can be taken up by living organisms and transformed into organic carbon. Oceans, rocks, organisms, and the atmosphere itself are all important and **sinks** (meaning natural repositories of) and sources of carbon in the world.

Image 13.7. Some important components of the carbon cycle.

The carbon cycle (Image 13.7) includes many transitions and transformations through which an atom of carbon can pass. Atmospheric carbon (CO_2) can dissolve in the ocean (1), where it may be used by photosynthetic organisms, incorporated into carbonate rocks, or simply remain in solution. CO_2 is also released from the surface of the ocean (2). Plants participate in the carbon cycle in several ways. They take up CO_2 and store it in the form of organic molecules during photosynthesis (3). They also release CO_2 when they break down stored glucose for energy through the process of plant respiration (4). When plants are eaten, the consumer assimilates the plant's organic carbon. Plants and animals die and decay, a process that releases carbon back into the atmosphere (5). Under the right circumstances, carbon from decaying organisms can be fossilized, which adds it to the carbon sink of the fossil fuels; what's more, carbon can be taken up to form terrestrial carbonate minerals (6), from which it can later be re-released to the atmosphere (7). Anthropogenic inputs into the carbon cycle include the combustion of biomass and fossil fuels for direct energy (8) and carbon-intensive land use (9), including the production of chemical fertilizing agents. Note that the complexity of the cycle means many actions affect several different processes—deforestation, for instance, reduces the tree population, which reduces the amount of carbon taken up through photosynthesis. The decaying tree releases its carbon to the atmosphere at a relatively slow rate, but if the removed tree is burned, its carbon is released much more quickly. Even subtle perturbations, such as temperature changes, can significantly affect the cycle. Warmer oceans, for instance, don't dissolve as much CO_2 as cooler ones. This disrupts the CO_2 exchange at the ocean surface, and results in an increase of CO_2 in the atmosphere.

Looking at some numbers, our perturbation to the carbon cycle seems insignificant—human activities add about 8 gigatons (Gt, 10^9 tons) per year of carbon to the atmosphere, compared with an ocean-atmosphere exchange of about 115 Gt/year. Plants take up around 120 Gt/year through photosynthesis, and release about 50 Gt/year through the process of respiration. Regardless of our comparatively small contribution to the atmospheric carbon, however, the imbalance compounds over time. As with the perturbation of any other chemical cycle, our addition of carbon to the atmosphere amounts to speeding up one set of reactions (carbon entering atmosphere) without affecting the rate of the others (carbon exiting atmosphere). It's worth noting that while some factions doubt the connection between increasing atmospheric carbon and climate

Image 13.8. Global carbon emissions from various sources. Prior to the mid-1800s, the primary source of anthropogenic carbon emissions was combustion of wood. The emissions were insignificant in quantity.

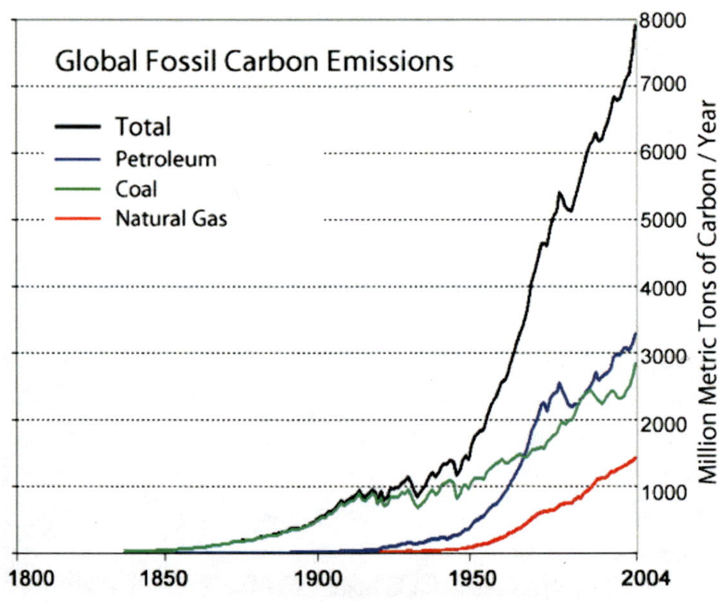

change, the statement that increasing carbon emissions are being released into the atmosphere is a scientific absolute and not a topic of debate. As Image 13.8 shows, our global carbon emissions have been steadily increasing ever since the Industrial Revolution.

Armed with an understanding of the carbon cycle and the science behind the greenhouse effect, we can now begin to address two very important issues: whether the planet is getting warmer, and, if it is, whether it's our fault or part of a natural cycle. Historically, the first indication that certain gases had the ability to affect the global temperature came about in the mid 1800s, when two scientists named John Tyndall and Jean Baptiste Fourier did some lab experiments that proved CO_2 (and some other molecules) had the ability to absorb IR radiation. While the idea didn't attract much notice at the time, it intrigued Svante Arrhenius, a Swedish chemist, many years later. In the late 1800s, Arrhenius was attempting to solve the riddle of how the planet had cooled enough to produce the historical ice ages. He ended up reasoning that decreased concentrations of heat-trapping gases might have precipitated the cooling trends. Through a labor-intensive series of calculations that dazzles modern climate-modeling scientists (who have the benefit of computers to accomplish the same task), Arrhenius managed to correlate atmospheric CO_2 concentrations with predicted temperature changes. He went on—purely out of academic curiosity—to calculate the effect increasing CO_2 concentrations would have on global temperature. In this way, he essentially invented the theory of global warming by showing that higher concentrations of greenhouse gases increase the extent to which heat is trapped. This was the first time that a link was firmly established between the concentration of greenhouse gases and the strength of the greenhouse effect. The reality of anthropogenic climate change didn't particularly concern him, however—it was unimaginable at that time that we would ever produce enough carbon emissions to significantly alter atmospheric CO_2 concentration.* Since Arrhenius's day, we've gone on to develop technologies that have proved him right in several respects: historical temperatures and corresponding atmospheric CO_2 concentrations have been determined, revealing that his calculations accurately predicted the relationship between the two. We have also, however, proven him wrong in a very disturbing way—we've increased the atmospheric CO_2 concentration more profoundly thanhe anticipated was humanly possible.

13.5 MEASURING ATMOSPHERIC CARBON

While historians in Arrhenius's time knew that ice ages had been a part of the geological record, there was no specific data available on global temperature patterns and atmospheric CO_2. The collection of historical temperature and CO_2 data came with the development of technology to remove and analyze deep **ice cores**, the first of which was drilled at the Vostok Soviet research station in Antarctica.

As snow falls through the winter, air pockets composed of trapped atmospheric gases form within the snowpack. At the Earth's poles, where summer temperatures never climb high enough to allow a thaw, each year's snow is

CONCEPT CHECK:

How is our perturbation to the carbon cycle similar to our perturbation to the nitrogen cycle?

Answer: In both cases, we're accelerating the movement of an element through one part of a cycle, but not the other part (increasing reactive nitrogen, increasing atmospheric carbon).

*NOTE

It is important to note the difference between CO_2 emissions and CO_2 concentration. Emissions are what we add to the atmosphere each year, while concentration is the amount of CO_2 in the atmosphere. If we stopped emitting CO_2 tomorrow (an impossible task in any case!), the atmospheric CO_2 concentration would still be much higher than it was prior to the Industrial Revolution—and would remain so for quite some time.

Image 13.9. Vostok Station (top left), and the original Vostok research team (top right). A deep ice core is composed of visible layers made up of snow that has long since been compressed into ice (bottom). Historical atmospheric composition and temperature record data can be extracted from the cores.

simply covered over by the next, eventually compressing snowfall from earlier years into ice. Air pockets in the snow remain trapped in the ice matrix as millions of tiny bubbles. The ice at the polar regions represents thousands and millions of years worth of snow, with the earliest lying farthest beneath the surface, meaning that a drilled core provides a historical record of snow from hundreds of thousands of years ago up to the present, stacked in neat layers. Because the bubbles in the ice are composed of the same gases (in the same concentrations) that were present in the air when the snow fell, they can be used to piece together a record of atmospheric CO_2 concentrations. A deep Vostok core drilled to a layer more than 400,000 years old provides interesting information about historical CO_2 (Image 13.10). According to the Vostok core, atmospheric carbon has fluctuated on an approximately 100,000-year cycle.

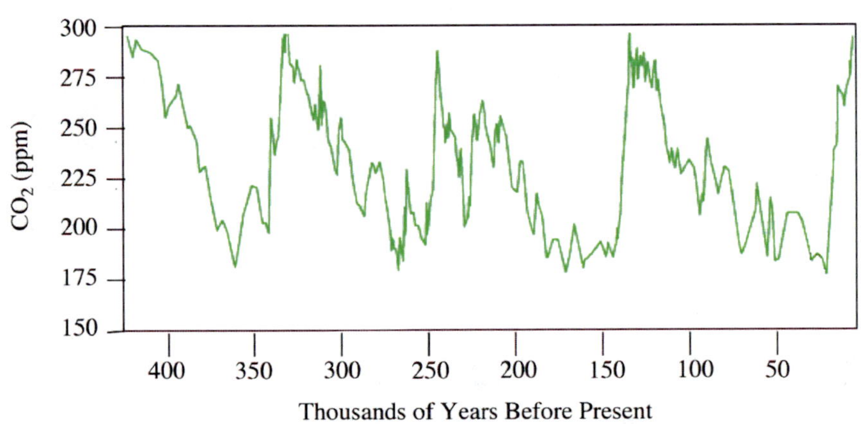

Image 13.10. The historical atmospheric CO_2 record.

There is further information to be obtained from the layers in an ice core. The atomic composition of H_2O in the ice itself allows for determination of global temperature patterns. Approximately 0.015 percent of hydrogen atoms are actually the naturally occurring, stable isotope deuterium (${}^{2}_{1}H$, sometimes just written ${}^{2}H$). Because deuterium is heavier than hydrogen, water molecules containing deuterium (about 1 in every 6700 water molecules) are slightly heavier than normal. This makes them incrementally slower to evaporate from the surface of the ice, an effect that is amplified in warmer weather. As a result, the percentage of ^2H in an ice layer correlates with the average seasonal temperature when the ice fell; higher deuterium percentages indicate warmer weather. The same Vostok core that provided a historical CO_2 record was used to produce a 400,000-year temperature record (Image 13.11).

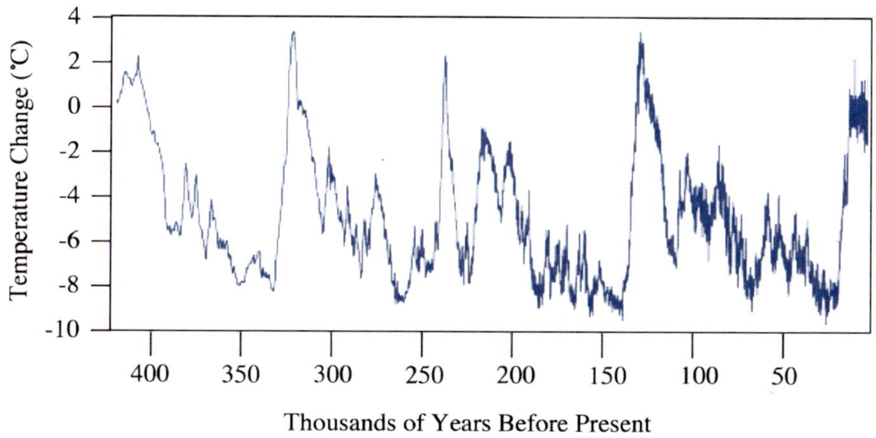

Image 13.11. The historical temperature record from Vostok ice core data. Temperatures are recorded in degrees above or below average 1990s Antarctic temperatures (the decade during which the deep core was analyzed).

A superficial glance at the temperature data reveals that it, like atmospheric carbon, fluctuates on an approximately 100,000-year time scale. Particularly interesting is that when the two graphs are superimposed upon one another, it becomes clear that the fluctuations actually occur in tandem (Image 13.12).

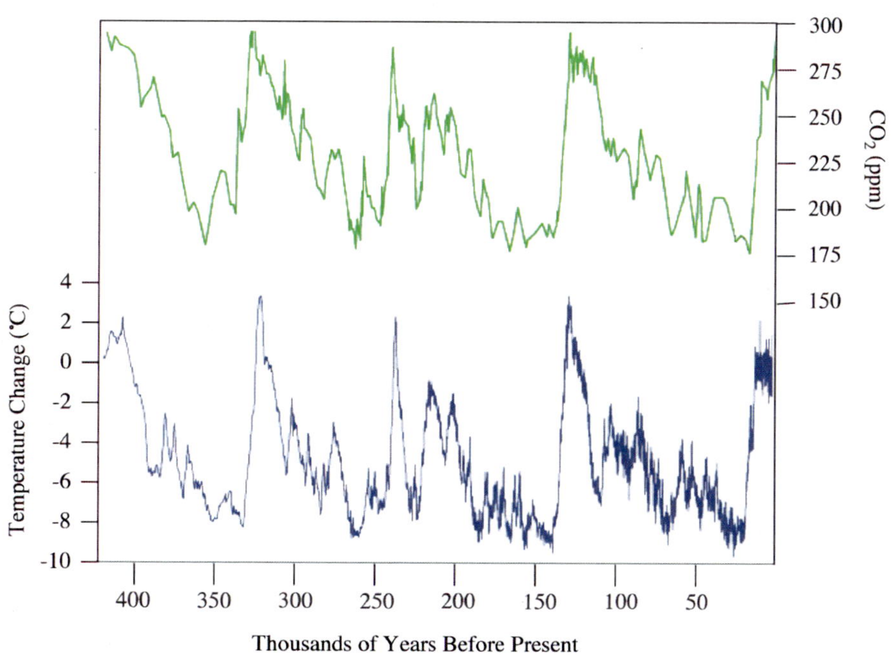

Image 13.12. Combined CO_2 and temperature records from the Vostok data.

A shrewd skeptic will note that while the CO_2 concentration and temperature data is very neatly matched, it is not perfectly synchronized: in some cases increasing CO_2 appears to slightly precede increasing temperature, while in other cases, the opposite seems to be true. As it turns out, it is logistically

CONCEPT CHECK:

Why does deuterium concentration allow determination of temperature data?

Answer: Deuterium-containing water is most concentrated in ice from snow that fell during warmest periods.

more difficult to date air bubbles than snow. While ^2H concentrations can be extracted neatly and precisely from a layer of known age (once snow falls, it neither shifts to a lower layer nor rises to a more superficial one), gas exchange between layers can occur for anywhere from 5000 to 7000 years after a layer begins to become buried by new snow. This is because the compacting process that seals bubbles permanently takes quite some time; during that time, gas exchange between layers is commonplace. For this reason, it is possible to approximate the date of air bubbles to within several thousand years, but nearly impossible to pinpoint their exact age. A skeptic further notes that the Vostok data, while interesting, is **correlational**—it is not possible to determine from this data whether increasing atmospheric CO_2 caused global temperatures to increase, or vice versa. It is even possible that a third, as-yet-unidentified factor affected both atmospheric carbon and global temperature. In any event, drawing no further conclusions as yet, the strong correlation between the two variables is interesting.

Useful as they are in yielding information about our geological past, ice cores cannot be used to generate data for more recent history. Thankfully, however, there are several hundred years' worth of global temperature data available, most of which was recorded for purposes with no connection whatsoever to the study of global warming, but which can still be put to good use in this capacity. Further, starting in the early 1900s, a few curious individuals began measuring atmospheric CO_2 from time to time. While a mild increase in carbon concentration was first noticed in 1938 (up 6 percent from the value recorded in 1900), no real attention was paid to the finding. Charles Keeling, a scientist and outdoor enthusiast, was the first researcher to track atmospheric CO_2 with any real dedication. From 1958 until his death (and thereafter taken over by other members of his research team), he monitored CO_2 concentrations from the top of Mauna Loa in Hawaii. The remote placement (and elevation) made the mountain a perfect spot for accurate measurements, as it was unaffected by local pollution and weather patterns. The record of his findings is called the **Keeling Curve** (Image 13.13).

Image 13.13. The Keeling Curve is a record of atmospheric CO_2 from 1958 to the present. The fluctuations are seasonal: photosynthesis in the spring results in a minor and temporary reduction in atmospheric CO_2. Despite these, however, a clear upward trend is noted.

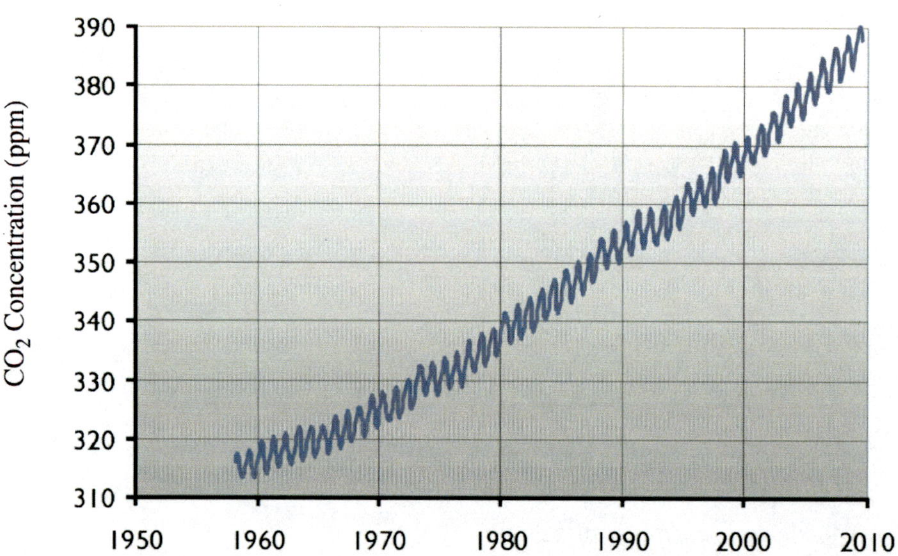

One important point worth noting is that while the Vostok data shows that atmospheric CO_2 fluctuates regularly, there is no evidence of it having been higher than approximately 280 ppm at any point in the last 400,000 years—a period that encompasses many ice ages and interglacial periods. In 1958, when Keeling began his Mauna Loa measurements, CO_2 was already over 310 ppm and was steadily climbing. Today, the atmospheric $CO_2$is near 390 ppm, and concentrations continue to increase by around 2 ppm annually (though it has been noted recently that the rate of change is itself increasing). Superimposed upon the Vostok data, the Keeling Curve appears as a disconcerting, nearly straight line (shown in red in Image 13.14).

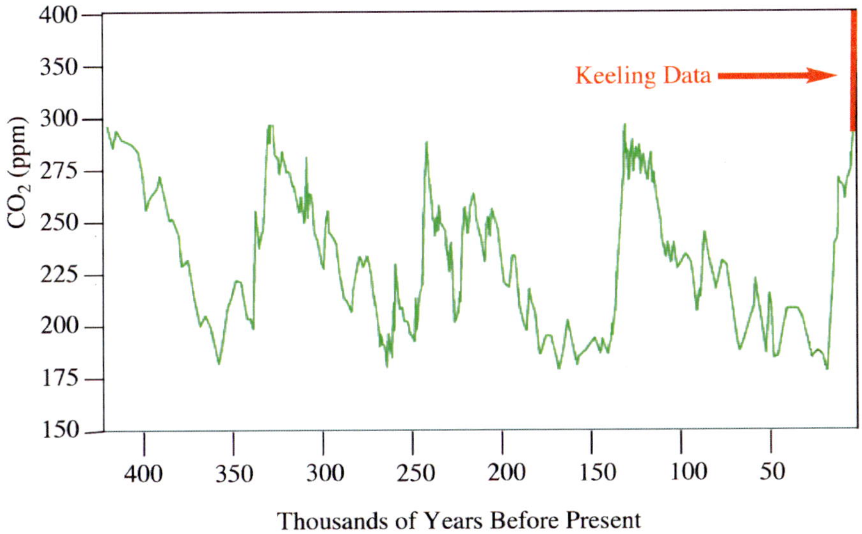

Image 13.14. What appears on the Keeling curve to be a gradual increase in CO_2 concentration is more dramatic alongside the Vostok data.

It's clear from gathered data that increasing carbon emissions (recorded by individual countries and compiled into global emissions records) have resulted in increasing atmospheric CO_2 concentrations in the last hundred or so years. Without an indication of climate change over the same period, however, there's no reason to examine the issue any further. Such indications fall into two different categories: **empirical evidence** of climate change, based upon observation and measurement, and **climate modeling**, which is used to predict future climate patterns.

CONCEPT CHECK:

How does our current atmospheric carbon dioxide concentration relate to the historical values?

Answer: It's much higher than ever recorded in the past.

13.6 EVIDENCE OF CLIMATE CHANGE

There is a significant body of empirical evidence to suggest a recent (and accelerating) warming trend. Global mean temperatures have been rising steadily for the past 150 years, with the rate of change increasing through incrementally smaller sampling ranges, meaning that the fastest warming has taken place most recently (Image 13.15).

Certain weather pattern changes are associated with warmer average global temperatures—namely increasing frequency and severity of weather events, coupled with shifting distribution of precipitation. Many of these predicted changes have been observed over the last hundred years. The National Oceanographic and Atmospheric Administration (NOAA) has collected data indicating that parts of the world (including portions of eastern North and South America, northern Europe, and north-central Asia) are significantly wetter than they were a century ago, while southern Africa, the Mediterranean, and southern Asia are much drier. More high-latitude precipitation falls as rain rather than snow, and

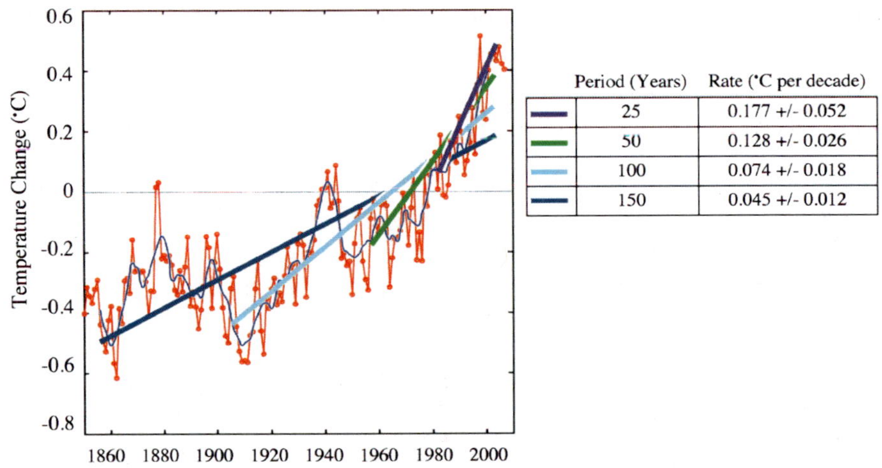

Image 13.15. This chart shows 150 years of global mean temperatures (annual averages in red, five-year means in blue). Regression lines showing average rates of change are superimposed.

heavy or severe precipitation events are far more common than they were in the early 1900s. An analysis by the National Climatic Data Center found a 14 percent increase in heavy rain events (defined as more than 2 inches in a day) and a 20 percent increase in very heavy rain events (more than 4 inches in a day) during the twentieth century.

Extreme weather events have become not only more common but more severe, particularly in the last two decades. NOAA uses a measurement called the U.S. Climate Extremes Index to track severe weather and changing patterns. The index quantifies extreme weather event severity and frequency by year, including conditions such as abnormally high and/or low temperatures, abnormally high and/or low precipitation, and high-intensity rainstorms. Higher index values indicate more extreme weather patterns—the average value through the past century has been 20. Values in the high 30s and 40s have become increasingly frequent since 1980, with particularly high values in 1998 (index value 44) and 2005 (index value 41). Hurricanes, a very specific and destructive extreme weather event, are known to develop over warm ocean surfaces. Recent studies show that sea-surface temperatures have been increasing through the twentieth century, with the warmest temperatures noted most recently. Further, the number and severity of extreme hurricanes (class 4 and 5) has increased significantly since 1970—so much so that hurricane scientists have considered introducing a sixth class for storms.

For this data to be a meaningful indicator of climate change seems to require that the last several decades of increased extreme weather be unique in recorded history. In fact, however, they are not. Hurricane intensity was also high around the 1940s. This period was preceded (and followed) by periods of reduced hurricane severity and frequency. Sea-surface temperatures closely parallel hurricane

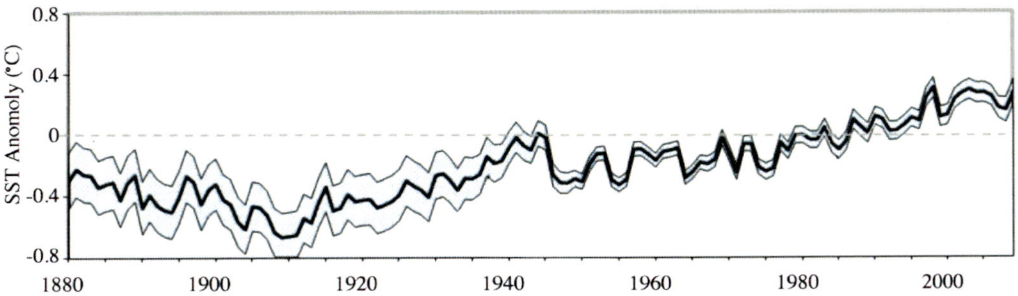

Image 13.16. Sea-surface temperature (SST) data from NOAA, 1880–2009 (dark line). The warming trend is much larger than the uncertainty (lighter lines bracketing data) in recent years.

activity for all periods, with warmer temperatures during more active periods (Image 13.16).

However, climactic events capable of producing each warming and cooling trend prior to the most recent are matters of record. Six volcanic eruptions between 1875 (Iceland) and 1912 (Alaska), including the massive eruption of Krakatoa in 1883, spewed tremendous amounts of ash and other aerosols into the lower atmosphere. These aerosols are known to have cooling effects, as they block incident solar radiation from reaching the Earth's surface. In 1963, a volcanic eruption in Indonesia produced another massive cloud of dust and aerosols—this is likely responsible for declines in global (and sea-surface) temperature noted through the early 1980s. In the periods between and following these **forcings**—environmental or anthropogenic factors that affect climate—a steady warming trend was observed.

Additional global observations include that most mountain glaciers are diminishing in size. Sea ice in the Arctic is reduced year-round, and the latest estimates suggest that it could disappear entirely as early as the summer of 2030. The shocking 2002 collapse of the Larsen B ice shelf in Antarctica—a shelf that is estimated to have formed at the end of the last ice age and disintegrated in just weeks—resulted in the loss of a mass of ice larger than the state of Delaware. Melting sea ice has resulted in an observable and incremental rise in sea level over the past century. A number of studies have noted that insect populations are beginning to alter their reproductive cycles,* with implications for several insect-borne diseases, including malaria. With warmer temperatures extending farther from the equator and an extended breeding season afforded by longer summers, mosquitoes and other disease-vector insects appear to be flourishing. With the mosquitoes' increased prevalence, malaria rates are increasing and malaria-prone areas expanding.

Of course, even if the data indicate a steady land- and ocean-warming trend over the last hundred years, the warming could be the result of increased energy input. One popular hypothesis to explain warming trends suggests that increased solar radiation (either due to a change in Earth's proximity to the Sun or increased solar activity) is responsible for the temperature change. If this is true, then the amount of energy radiated by the Earth back into

***NOTE**

Insects, with exponentially shorter life spans than other animals, respond to changes in the environment before larger organisms. Animals that feed upon insects generally follow close behind, as already noticed in several bird studies that indicate changes in migration patterns.

Image 13.17. The troposphere has experienced a warming trend during the same 30-year period (1979–2009) that the stratosphere has cooled. Surface warming (not included on image) is even more pronounced than lower tropospheric warming. Temperature change is recorded as a change in °C/Decade.

Lower Troposphere

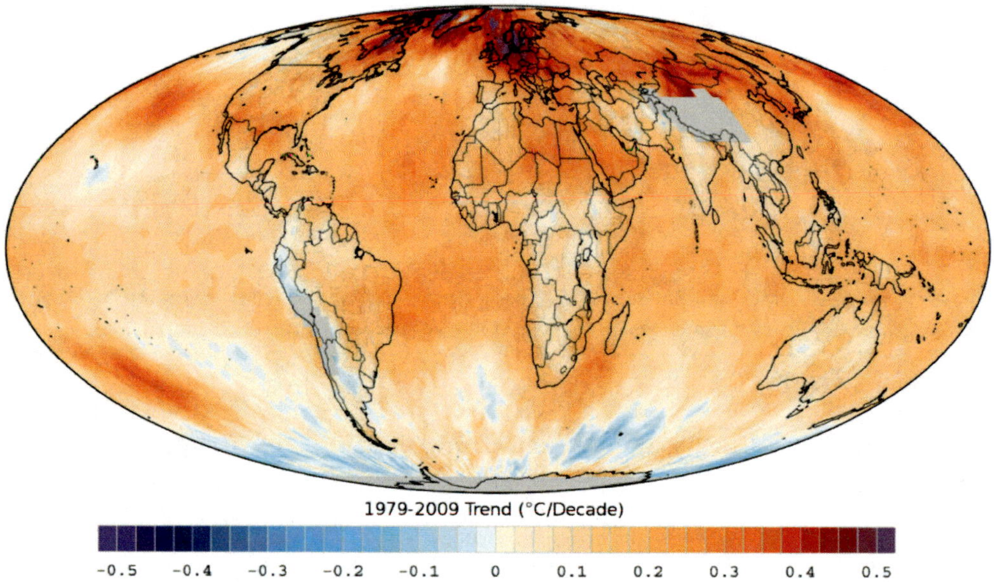

Lower Stratosphere

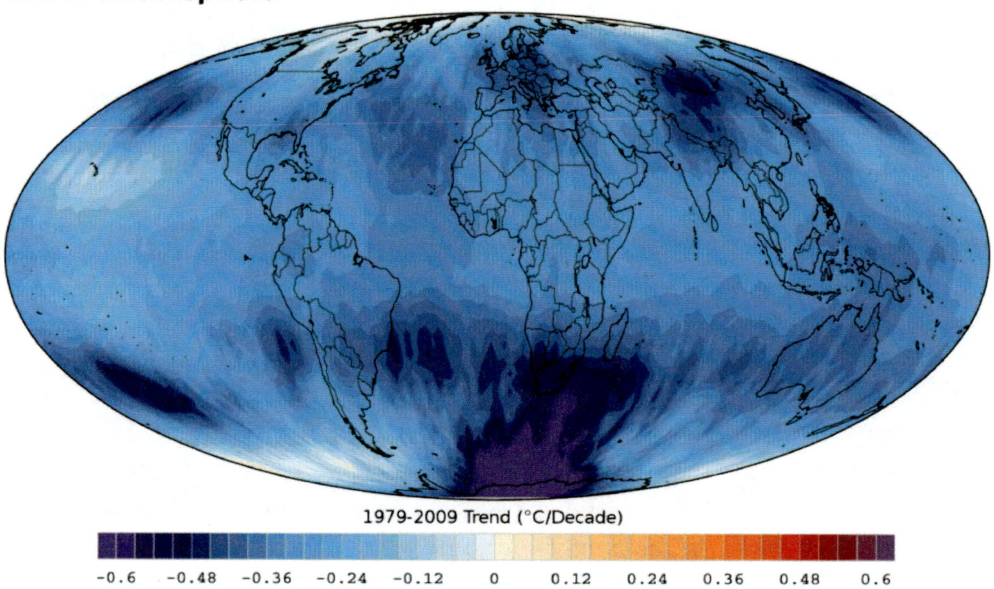

Concept Check:

Why is the information in Image 13.17 considered a "smoking gun" for the enhanced greenhouse effect?

Answer: Because if the Earth were still radiating 81 percent of heat into space, warming trends on the surface would be reflected in the stratosphere.

space should also increase—since a normal greenhouse effect maintains 81 percent of Earth's radiated heat and releases the remaining 19 percent into space, increased incident radiation would both increase the amount maintained and the amount released. Several studies, however, have shown that this is not so; in fact, the Earth is releasing less heat with each successive year. Further, while global temperatures are increasing, the average temperature of the stratosphere has been falling over the same period (Image 13.17). This cannot be explained by increased solar activity: more radiation incident upon the planet would increase the temperature of all atmospheric levels. Instead, a falling stratospheric temperature can be explained only by an increase in heat trapping at lower atmospheric levels. In other words, it can only be explained by an **enhanced greenhouse effect**—the trapping of more than 81 percent of Earth's radiated heat. Earth's decreasing radiation into space has been called the "smoking gun" of climate change: it doesn't prove anthropogenic causes for global warming, but it certainly proves that global warming is due to an enhanced greenhouse effect.

13.7 CLIMATE MODELING

With its roots in the calculations undertaken by Arrhenius in the late nineteenth century, climate modeling has the potential to be a powerful tool. Unfortunately, it's also incredibly complex, as climate may be affected by any number of factors. Volcanoes, solar activity, Earth's place in its orbit, the percentage of the Earth's surface covered by ice, composition of the atmosphere, and ocean currents are just a few of the myriad factors that have the potential to affect climate, and that must be considered by modelers. Further, forcings are often dynamic, and their effects can compound in a positive-feedback loop (see below). In the end, it's simply not possible to include every possible forcing and its predicted effect in any given climate model. This, coupled with the fact that many climate models do not include the feedback effects of forcings, means that variability in predicted climate outcomes is significant. Some models predict major climate change over the next decades and century, while others suggest that the effect will be more minor. Regardless, climate modelers agree that both increased global temperature and sea level rise will characterize the coming century. The **Intergovernmental Panel on Climate Change (IPCC)**, a United Nations panel with the mission to assess climate change research as well as its social and economic ramifications, released a report in 2007 that included projections for the twenty-first century. Conservative ("low scenario") models included in the assessment suggest a 1.8°C (3.2°F) increase in global temperature, with a range of 1.1 to 2.9°C (2.0 to 5.2°F). High scenario models suggest a 4.0°C (7.2°F) increase, with a range of 2.4 to 6.4°C (4.3 to 11.5°F). Low scenarios predict 7 to 15 inches of sea level rise during the same period, while high scenarios predict 10 to 23 inches. Many climate scientists feel that the IPCC models (even the high scenario model) may underestimate coming changes, as they do not include feedback effects, and the historical record shows that climate tends to respond to forcings in a way that amplifies, rather than mitigates, them.

These amplification, or feedback, effects can be quite significant, and are already recognized in the climate change literature. Gases are less soluble in warm liquids than in cool ones (the reason a soda goes flat sooner if it's warm). As a result, increasing global temperature—and the concomitant increasing ocean temperature—would decrease CO_2 solubility and result in release of carbon dioxide from the ocean. This would increase atmospheric concentration of CO_2, further enhancing the greenhouse effect. Soils behave similarly; they take up less CO_2 (and release more) when warm. Net loss of gas in the soil has been recorded in England and elsewhere. Tundra is yet another source of global warming positive feedback effects. It is essentially frozen mulch—organic photosynthetic material that does not decompose and release its carbon because of the exceedingly low temperature at which it is found. As tundra melts in response to increased temperature, the stored carbon is released into the atmosphere. Because of the vast quantity of carbon dioxide currently locked in tundra (around 3.6 trillion tons), defrosting of the land represents a significant potential for increased CO_2 release. Frozen peat bogs, too, are a liability on a warming planet. The sub-Arctic (Siberia) is a massive shelf of frozen peat, which contains methane ice (called *clathrate*) that was generated when the bogs were liquid, and that has since become frozen beneath the surface of the ice. The methane (estimated at 70 billion tons) is released as bogs melt, a process that is already occurring. Like CO_2, methane is a powerful greenhouse gas—in fact, it has a **global warming potential (GWP)** twelve times that of carbon dioxide. GWP is a standard measurement established by the IPCC that compares a molecule's ability to store and re-radiate IR to that of CO_2 (Table 13.3).

Compound	GWP
CO_2	1
CH_4	21
N_2O	310
CH_2FCF_3	1300

Table 13.3. GWP for a few of the many greenhouse gases. Note that N_2O is in equilibrium with NO_2 and other NOx in the atmosphere, but is the nitrogen oxide for which GWP is most commonly reported. CH_2FCF_3 is an HFC (see Chapter 10).

It's also worth noting that even without massive releases of greenhouse gases from melting tundra and peat, loss of ice alone amplifies warming. Because ice reflects sunlight to a greater extent than does water or land, a loss of ice leads to more heat energy being absorbed by the Earth, further increasing global temperatures.

13.8 CAUSALITY

At this point, we've examined large bodies of data establishing carbon dioxide and other molecules as greenhouse gases. They are important to keeping our planet warm and livable, but in high concentration, they maintain too much of Earth's heat. We've also reviewed the myriad studies, observations, and models that indicate our planet is in the midst of a clear warming trend, with significant and distressing ramifications. A skeptical analysis of these data, however, still demands proof of a causal relationship between the gases and the warming. After all, given the positive feedback discussed above, it appears possible that warming could cause a release of greenhouse gases rather than the other way around. In a sense, this argument is circular—we've established that certain gases are responsible for the greenhouse effect, so debating which comes first (the increase in gas or the increase in temperature) is tantamount to arguing about chickens and eggs. Still, there are many factors that affect climate, and we must examine each one in turn in an attempt to tease out the cause(s) of global warming.

One forcing commonly suggested as a non-anthropogenic cause of global warming is solar activity. However, while an increase in solar flares was recorded through the 1950s and 1960s, since the 1980s solar activity has been on the decline while temperatures have continued to climb more steeply than before (Image 13.18). Additionally, a portion of the mid-twentieth-century solar activity peak does coincide with a mid-century increase in temperature, but the temperature change actually preceded the sunspot activity by more than a decade. If sunspots do affect global temperature (and they are certainly considered to be a valid forcing in their own right), they aren't solely responsible for climate change—particularly not lately.

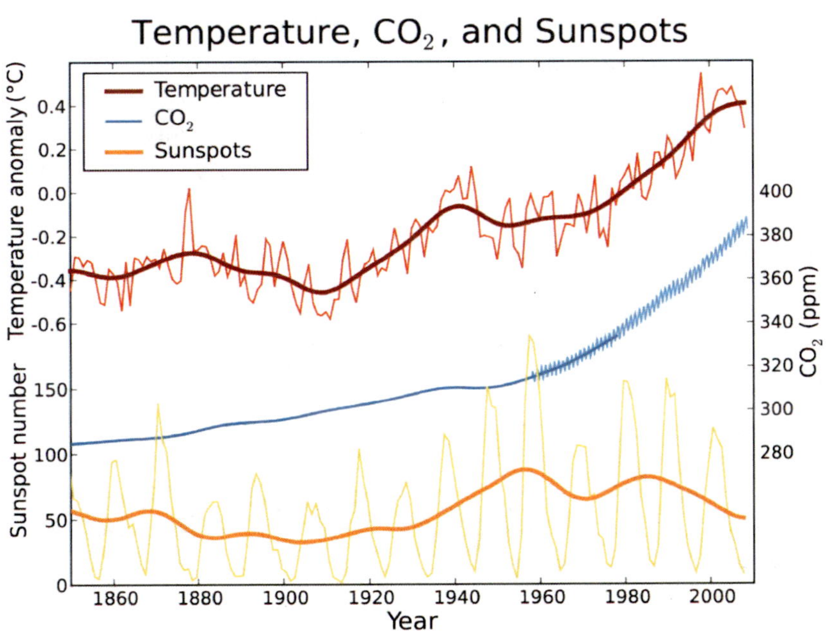

Image 13.18. Solar activity over the last 150 years (data from Solar Influences Data Center).

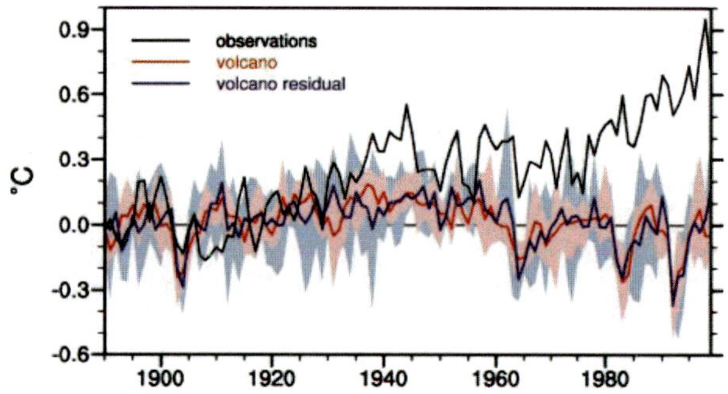

Image 13.19. Model including volcanic forcings as compared to observed data. (Image © 2004 American Meteorological Society)

A weakness of climate modeling in the traditional sense is that it's a "wait and see" science—once a model has been created, there's no way to know whether it is accurate until many years, or decades, in the future. However, it's possible to test a model (and to determine which factors are important in producing an accurate model) by simulating a past-to-present modeling scenario. That is, by programming known information about the last century's climate forcings—solar activity, volcanic activity, atmospheric composition, and the like—and allowing the computer to produce a temperature trend prediction for the period in question. A model produced in this way has the advantage of predicting temperature effects for years that have already passed, meaning that the data can be compared to known values and checked for accuracy. For instance, a model using only volcano-related forcings (no other effects on climate were considered) predicted temperature fluctuations quite out of harmony with actual observations (Image 13.19).

Similar models were produced for solar activity, ozone concentrations, aerosols, and other natural forcings, each of which predicted that temperature would fluctuate throughout the twentieth century, but not increase. Some forcings, taken alone, actually predicted a temperature decrease as the century progressed. One interesting model, however, showed a close parallel between predicted and observed temperature. The forcings used to produce the model were the collective greenhouse gases (Image 13.20).

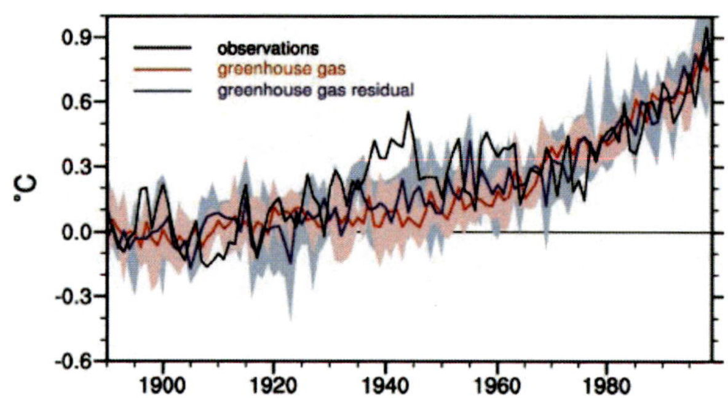

Image 13.20. Model including greenhouse gas forcings as compared to observed data. (Image © 2004 American Meteorological Society)

The same study also produced two models that combined forcings. One of these took all significant natural factors into account, while the second included all significant natural factors and anthropogenic factors. The data speak for themselves (Image 13.21).

Clearly, the best prediction of temperature change through the last century is produced through inclusion of anthropogenic factors. With forcings other than greenhouse gases failing to come close to a reasonable projection of temperature change—and greenhouse gases alone producing so remarkable a fit to observed temperature—the causal relationship between gas and temperature is established. Further, natural forcings taken alone—even including natural sources of greenhouse gases—are insufficient to model climate accurately. Only the inclusion of

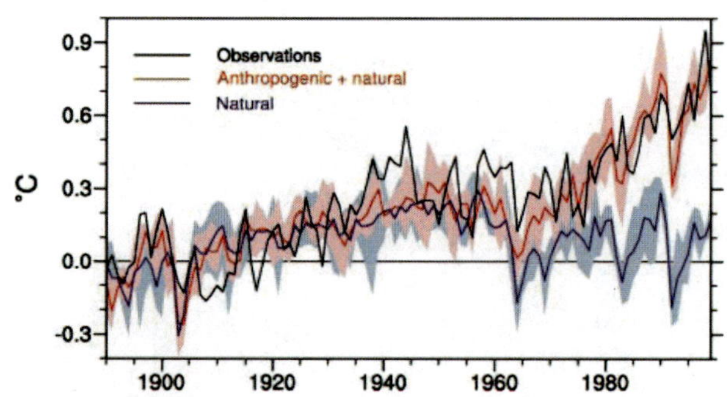

Image 13.21. Model including all significant natural forcings (purple) and model including all significant natural and anthro-pogenic forcings (red) compared to observed data. (Image © 2004 American Meteorological Society)

human-generated emissions allows for accurate modeling. If Earth's decreased heat radiation into space is the smoking gun for the enhanced greenhouse effect, then the climate model in Image 13.21 is one more smoking gun—and it has our fingerprints on it.

> **CONCEPT CHECK:**
>
> Why do retrospective models prove that we are responsible for climate change?
>
> Answer: Because accurate models of past temperatures can only be predicted when models include anthropogenic factors.

13.9 WHY FOCUS ON CO_2?

Given that there are many greenhouse gases (including water!), why is it that most discussions of global warming center on carbon dioxide? In reality, this is nothing more than a simplification that takes place in the public sector. Climate scientists and global organizations recognize that CO_2 is not the only greenhouse gas, nor is it the only one that's increasing in atmospheric concentration. Methane (emitted anthropogenically as the result of fossil fuel production, burning of biomass, and large-scale livestock operations), N_2O, HCFCs and HFCs are among the many greenhouse gases. Water falls into a somewhat different category: while it is a greenhouse gas, its concentration in the atmosphere varies greatly with temperature, so a GWP for water is generally neither calculated nor independently meaningful, and its concentration is not tracked as a measure of global warming. The fact is that all the major greenhouse gases are increasing in concentration through human activity. That CO_2 happens to be the best known and best recognized in the global population makes it the one most commonly associated with global warming in the media. Global warming reduction measures, however, take into account the effects of all greenhouse gases, and discussions of emissions reduction measures include all compounds of anthropogenic origin with global warming potential.

13.10 MAKING SENSE OF RHETORIC

Those who deny the legitimacy of global warming science sometimes point to differences in model forecasts, suggesting that this represents dissention in the scientific world (and illegitimacy of the science). In reality, while it's possible to use models such as those discussed above to tease out key forcings, retrospective models are far easier to build—and build accurately—than prospective models. The reason is twofold: First, it's impossible to know what the future holds with regard to our continued global emissions. High scenarios built from continued (and increasing) emissions will necessarily be more alarming than low scenarios, which anticipate a global response to the climate crisis. Second, the positive feedback effects upon climate are well established, but incompletely understood. It is difficult to know exactly how much polar ice melt will accelerate energy absorption. It's known that melting peat bogs will release methane, but tough to determine how much of that methane will be released unaltered, and how much will react with oxygen to form CO_2—the two gases have different GWPs, and therefore different magnitudes of impact. Prospective climate modelers must work at the intersection of science and art, combining theory with instinct and known values with anticipated human responses to produce their best scientific hypothesis for what the future holds. In reality, what is sometimes seen outside the scientific world as a disparity of outcomes is seen by scientists as a consensus of trajectories. In 2001 Donald Kennedy, the editor of *Science* (a peer-reviewed scientific journal) noted, "Consensus as strong as the one that has developed around [global warming] is rare in science."

Further, *Science* published a review article in 2004 that collated and analyzed almost 1000 climate studies, drawing the conclusion, "Politicians, economists, journalists, and others may have the impression of confusion, disagreement, or discord among climate scientists, but that impression is incorrect."

Some draw comfort from model disparity, hoping that the climatologists are wrong, and global warming will have no effect whatsoever on the planet. It seems, though, that if anything, model disparity is discomforting rather than reassuring, because it means the scientific world agrees on one thing—the Earth's climate is changing, and in a very immoderate way. How hot it will get, how wild the weather will become, and how high the sea will rise can't be known for sure. As a global community, we have several legitimate options to evaluate through risk-benefit analysis. Hoping, however, that all the models and signs are wrong, and that the evidence is being misinterpreted, is not a justifiable scientific response. Unlike risk-benefit analysis, which acknowledges and accepts some risk to achieve a desired outcome (unlike the precautionary principle, which accepts no risk), blind hope considers only the benefits of inaction, without acknowledging or evaluating the risks. In a way, our climate crisis is similar to a situation in which a driver has started to lose control of their vehicle at high speed, and has to decide quickly how to respond. Some would turn the wheel to steer into the skid, while others might decide to hit the brakes. Some would worry more about trajectory while others might be more concerned with speed. But how many would simply take their feet off the pedals, their hands off the wheel, and just hope for the best?

13.11 THE KYOTO PROTOCOL

In 1997, the Kyoto Protocol was signed by many countries around the world in an effort to reduce total greenhouse emissions. Developed countries were required to cut emissions first—the protocol required a 5 percent cut from 1990 emissions levels by 2012. While there were no restrictions put on developing countries for the first phase of the protocol, these countries nevertheless stood to benefit from reduced carbon emission: reducing emissions would afford them **carbon credits**, which they could sell to developed nations that were over-emitting. The reasons for the disparity in treatment of developed versus developing nations was outlined by the United Nations. First, the developed nations were most responsible for carbon dioxide and other greenhouse gases currently (as of 1997) in the atmosphere—they had benefited economically from the same industrialization and development processes that had resulted in emissions. As such, they were expected to respond first to the crisis. Additionally, they were expected to use some of the wealth accrued at the expense of atmospheric CO_2 concentrations to assist developing countries in establishing carbon-reducing technology. Second, the emissions from developing countries were (as of 1997) still low. Even now, while China's total emissions have eclipsed those from any other single country (6018 million metric tons of CO_2 in 2006, with the next largest emitter being the United States at 5903 million metric tons), their per capita emissions are much lower than those of developed nations (Table 13.4).

Country	Total Emissions (in millions of metric tons)	Per Capita Emissions (in tons/capita)
1. China	6018	4.58
2. United States	5903	19.78
3. Russia	1704	12.00
4. India	1293	1.16
5. Japan	1247	9.78
6. Germany	858	10.40
7. Canada	614	18.81
8. United Kingdom	586	9.66
9. South Korea	515	10.53
10. Iran	471	7.25
16. Australia	417	20.58

Table 13.4. The top ten carbon-emitting countries by total emissions. Note that while Australia's total emissions place it sixteenth in the world, its per capita emissions are highest.

Compared to the Montreal Protocol, the Kyoto Protocol hasn't enjoyed the same degree of success and global popularity. The United States, for instance, has categorically refused to sign the protocol on the grounds that it doesn't do enough to curb emissions. In particular, the United States has objected to the extent to which China and India's emissions have been allowed to grow. It's important to note that while developing nations are now becoming large emitters, it will take decades of such emissions before they have contributed as much carbon to the atmosphere as the United States and other developed nations already have. The developed nation signatories to the Kyoto Protocol suggest that leading by

Image 13.22. Signatories to the Kyoto Protocol as of 2009. Grey countries are undecided. The United States has announced it has no intention of signing the protocol.

CONCEPT CHECK:

Why did the UN feel that developing nations shouldn't have to cut emissions as early or dramatically as developed nations?

Answer: Because they weren't producing significant emissions until recently, and developed nations are mostly responsible for global warming.

example and assisting where necessary may be the best route to encouraging developing nations to reduce emissions. Reluctance on the part of the United States to sign the protocol may be a sign of distaste for its economic ramifications. Quite simply, to an even greater extent than the Montreal Protocol, the Kyoto Protocol threatens national economic security, and that has been difficult for many developed nations to accept.

Since 1997 several countries have met emissions targets: Russia, Germany, the UK, France, and Poland all decreased emissions, while the United States, China, India, and many others increased emissions drastically. Overall, global carbon emissions have increased by more than 30 percent since the protocol was initially signed.

Set to expire in 2012, the Kyoto Protocol will need to be replaced by another global treaty. There are difficult decisions in our future as a community of nations, as we once again tackle the conundrum of how to mitigate environmental damage without impacting developing nations beyond their economic capacities.

13.12 GLOBAL AND NATIONAL STRATEGIES

The failure of the Kyoto Protocol to achieve significant global cooperation leaves us in a troubling place. Clearly, developing a unified global strategy that addresses global warming and greenhouse emissions is a matter of political will as much as it is a matter of developing and implementing new technologies. There are many interesting discussions of global politics and climate change (see Sources and Further Reading), but as this is a chemistry text, we'll focus on technologies with chemical significance. Strategies to reduce carbon dioxide concentration in the atmosphere fall into two groups: carbon sequestration and fossil fuel alternatives.

Carbon sequestration is the capture of CO_2 released during combustion processes, with subsequent condensation and injection into deep geological repositories. While the technology in no way addresses the fossil fuel shortage, it attempts to curb emissions and slow the pace of global warming. Since petroleum reserves are in far shorter supply than coal, most carbon sequestration technologies being developed are for implementation in coal-fired power plants. Unfortunately, carbon emissions from traditional smokestacks are very diluted by the air, and it's nearly impossible to capture and sequester CO_2. As a mechanism for addressing this, a new technology in development seeks to combust coal in the presence of both oxygen and water, generating H_2 (the combustion of coal provides energy to produce hydrogen from water) and CO as the major products. The CO would then be converted to CO_2, which would be sequestered, while the H_2 would be utilized for energy through a mechanism addressed in a later section of the chapter. The problem with this is that it's very energetically expensive—25 percent of energy extracted from the coal must be used simply to power the plant, representing a large decrease in efficiency. Another 20 percent of the energy generated from burning a given quantity of coal is required to liquefy the CO_2, and a variable (but also significant) quantity of the energy is necessary to transport and pipe the liquid underground. The volume of liquid CO_2 is also staggering—an estimated 50 km^3 would need to be stored on a daily basis to completely eliminate global carbon emissions. Of course, the Earth's crust is a finite space—it's not difficult to imagine that at some point, its capacity for accepting CO_2 would begin to diminish. Finally, the storage containers, like nuclear waste storage containers, run the risk of developing leaks. If liquid CO_2 began to leak from its container, it would quickly turn to gas (high pressure is required to liquefy carbon dioxide at livable temperatures), and could flood the atmosphere with years or decades worth of carbon emissions in a matter of hours to days. While sequestration as a limited or auxiliary strategy for reducing emissions may have merits, it nevertheless fails to address an important issue—continued reliance on fossil fuels—while creating several new problems.

Image 13.23. Switchgrass is a tall, rapid-growth plant with biofuel potential.

The fossil fuel alternatives encompass a wide variety of technologies and materials, some of which are **alternative energy** strategies that generate energy through means other than combustion. Solar and nuclear power have been discussed elsewhere in this text (Chapters 11 and 12, respectively), and while wind and hydroelectric power have intriguing potential, they are not strictly chemistry-related, and so will not be discussed in this text. **Alternative fuels**, unlike alternative energy strategies, produce energy through combustion. The most viable alternative fuels are nonpolluting and **carbon-neutral**, meaning that they do not result in the net release of carbon dioxide into the atmosphere. One alternative fuel class with the potential to reduce both fossil fuel dependence and net carbon emissions is **biofuel**, which is combustible material from renewable sources. Biofuels, like fossil fuels, release solar energy that was originally captured and stored through the process of photosynthesis. The major difference is that biofuels rely upon organisms capturing the Sun's energy now, rather than organisms that captured energy millions of years ago. While either form of fuel yields its stored energy through a CO_2-releasing combustion reaction, when fossil fuels are burned, the carbon released was taken up from the atmosphere at the same time solar energy was stored—eons ago. As a result, adding that carbon back into the atmosphere represents an increase in current atmospheric carbon dioxide concentration. Biofuels, on the other hand, can be cultivated in fields (or in water, in the case of algae). They engage in photosynthesis, storing the Sun's energy in the form of cellulose and sugar and removing carbon dioxide from the atmosphere. When the organisms are harvested and burned (either as whole biomass or in a processed form), the carbon is re-released, meaning that CO_2 emission is in proportion to CO_2 uptake and the carbon cycle is balanced. Additionally, some biofuels have low environmental impacts compared to fossil fuels: they don't require mining or drilling, and their combustion is associated with much lower tropospheric pollutant emissions than traditional fuels. Owing to these considerations, some organisms are far more appropriate to biofuel production than others—tall trees, for instance, are nowhere near as renewable nor as carbon-neutral as rapid-growth grasses, because of their long lifetimes. Switchgrass (a rapid-growth tall grass), corn, soy, sugarcane, and some species of algae have been identified as the most plausible cultivated biofuels.

We must note, however, that simply because a biofuel is carbon-neutral and environmentally low-impact in theory doesn't make it carbon neutral in practice. For instance, the cellulose and sugar in corn can be used to produce ethanol, which is incorporated into the fuel mixture burned in many vehicles worldwide (including the United States) and can be used alone as a biofuel. Corn in the United States, however, is farmed using chemical fertilizers, which have an environmental impact all their own (see Chapter 10), and which are produced through the combustion of fossil fuels. Additionally, there is an energy cost associated with refining ethanol from corn. In fact, many researchers who have undertaken calculations to determine the energy efficiency and biofuel potential of cultivated corn note that there is less environmental impact and fewer carbon emissions if petroleum is burned directly, as opposed to being used to fuel corn growth and biofuel production. In the end, it takes more calories of energy to produce ethanol from corn than are contained in the ethanol itself, because of the energy transformed to entropy during each step of the process. Recent estimates suggest that 131,000 BTU of energy are required to produce a gallon of corn ethanol containing 77,000 BTU of energy, meaning that 70 percent more energy is required to produce corn ethanol than is yielded.

Another biofuel concern has to do with the amount of land necessary for cultivation. For instance, planting switchgrass on otherwise "useless" (nonagricultural, non-forested) land can produce sustainable, carbon-neutral biofuel. However, destroying (generally by burning) thousands upon thousands of acres of rainforest in Brazil to plant sugarcane for biofuel preempts any carbon neutrality or low impact. Since forest and rainforest are carbon sinks containing hundreds of years' worth of sequestered carbon, removing the forests not only releases all that CO_2 into the atmosphere, but also removes plants that would have otherwise continued to draw upon and store atmospheric CO_2. This is to say nothing of the loss of biodiversity and impact to soil that are effects of deforestation.

Biofuels have the potential to be important, low-impact energy sources, but care must be taken when evaluating their environmental effects. In general, if they are cultivated, they must be grown upon land with ample water (to reduce unnecessary water dependency in areas that may lack sufficient precipitation). The land should not have been deforested to provide for the fuel cultivation, nor should the organisms require chemical fertilization. Processing of raw material to cellulose or ethanol should be minimal and as energetically conservative as possible in order to minimize usable energy loss and unnecessary energy input.

Another energy option worth discussing is hydrogen, which combines with oxygen in a combustion reaction that produces water:

$$2\,H_2 + O_2 \rightarrow 2\,H_2O$$

Hydrogen fuel cells (voltaic cells) running this highly spontaneous reaction can be used to provide energy without any carbon emissions, making them attractive alternative fuels. The problem, however, is that unlike the carbon compounds synthesized by biofuel organisms, there are no renewable, natural sources of hydrogen gas—at least not in any great quantity. Instead, H_2 must be produced through the electrolysis of water. This reaction is endothermic and requires an

CONCEPT CHECK:

Why is carbon dioxide from burning grass "better" than carbon dioxide from burning petroleum?

Answer: The grass took up the carbon recently (removed it from the atmosphere), so we're not producing a net increase in atmospheric carbon dioxide by re-releasing it.

energy source. Currently, most hydrogen produced in the United States and elsewhere in the world is fueled by petroleum, nullifying the carbon-neutrality and low impact of the hydrogen fuel. Proponents of hydrogen fuel suggest that it will be more environmentally friendly when energy to produce the reaction is provided by alternative sources, such as solar or wind. Opponents, however, argue that the problem with that logic is two-fold: first, it is predicated upon further development of solar, wind, or other technologies; second, once those technologies are developed, it would be far more energy-efficient (because of the second law of thermodynamics) to use them directly as sources of electricity. The reason for this becomes clear through an examination of one proposed application of hydrogen fuel cells—as energy sources for transportation. A regular internal combustion engine burns a hydrocarbon fuel, the result of which is the production of a large amount of heat (essentially a small explosion), which is used to physically drive a piston. Hydrogen fuel cells, however, do not literally burn hydrogen—instead, hydrogen and oxygen are the anode and cathode (respectively) of a voltaic cell, and the electricity generated by the cell is used to power an electric motor. As a result, a hydrogen fuel cell is energetically inefficient compared to both traditional fuels and alternatives. Rather than burn petroleum to produce hydrogen, which is then converted into electricity that is used to power a motor (two energy transformations), it makes more sense to burn petroleum directly in a traditional engine. Rather than use alternative energy sources to generate electricity to produce hydrogen, which is then converted back into electricity that is used to power a motor (three energy transformations), it makes more sense to use the alternative energy to charge an electrical battery that is used to power a motor. Hydrogen fuel cells are sleek-sounding, and have consumer appeal, but until the technology advances to the point that hydrogen generation doesn't require many cycles of energy transformation, the energetics won't add up.

13.13 COMMUNITY AND INDIVIDUAL STRATEGIES

Politics, technologies, and global strategies aside, there are ways for communities and individuals to reduce their global impact and the emissions for which they are personally responsible. Generation of electricity consumes the largest percentage of energy used in the United States (Image 13.24). Conserving electricity, then, is an important strategy to reduce carbon emissions.

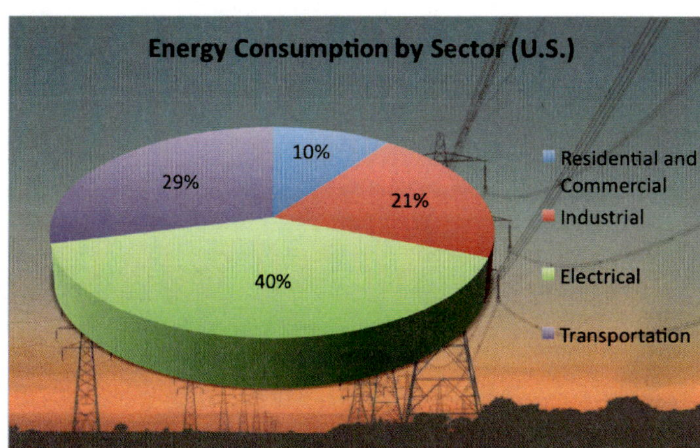

Image 13.24. Energy consumption in the United States (percentages by sector). Residential and commercial energy consumption includes non-electrical sources, such as natural gas.

Reduced consumption of electricity to heat and cool homes and buildings (even a few degrees makes a difference) significantly reduces daily electrical demand. New construction should incorporate energy-efficient and insulating building materials, while older construction can be updated in small ways that make a big difference: replacing single-pane windows with dual glass and exchanging curtains or vertical blinds for closed-cell blinds both help prevent heat loss (or gain) through windows. Most appliances (including televisions and computers) draw electricity even when switched off; unplugging these can significantly reduce unnecessary electrical consumption. Chargers for cell phones, MP3 players, and other small electrical devices draw power when plugged in, even if the device is not attached. Old refrigerators, dishwashers, washing machines, and water heaters are less energy efficient than newer models.

Transportation is another major source of energy consumption, and therefore offers many opportunities for reducing environmental impact. Using public transportation (which operates, and therefore produces emissions, whether it's used or not) is a great way to reduce individual impact, as are biking and walking. Air travel produces significant emissions under even the best of circumstances, but direct flights are less energy-intensive than flights with multiple stops. Purchasing local food not only supports the local economy, but also reduces global emissions—locally grown food need not be shipped or trucked from farms outside the state or country. The carbon cost of food transportation worldwide is tremendous. On a similar topic, organic food (which is grown without pesticides or chemical fertilizers) also reduces carbon dioxide and other greenhouse emissions, because the energy input is purely solar: no fossil fuels are burned to produce fertilizer.

Finally, because all commercially processed foods and goods require energy input, decreased reliance upon processed foods and decreased consumption of manufactured goods positively impacts the environment. In addition, it's far more energy efficient to reprocess plastic (or paper, or metal, or any other commonly recycled material) into a new form than it is to refine or produce the material from scratch. In general, the more energy transformations a particular action or material requires, the more energy it consumes, and the more carbon emissions it's responsible for producing. Bottled water, as discussed in Chapter 5, represents a tremendous energy expenditure as compared to what's available (nearly free of charge!) from the tap. A variety of sources with many more suggestions for reducing personal and community impact on the environment are listed in Sources and Further Reading.

While modern technologies have provided us with conveniences that are now hard to imagine doing without, we must remember that as with many of the topics discussed in this text, our solutions to problems have become the problems for which we now require solutions. It's easy, sometimes, to allow ourselves to become convinced that global problems require global solutions—and that we as individuals aren't able to effect change. However, a global community is nothing more than the sum of national communities, which are merely the sum of local communities, which themselves are the sum of individuals. We are all a part of the global organism, and we all affect its well-being.

13.14. THE LAST WORD—STOMPING OR TIPTOEING?

In the last section of the chapter, we discussed some strategies for reducing individual impact upon the environment. This is commonly called a **carbon footprint**. While we may want to make a difference, most of us have a tendency to feel very small and insignificant in a very big world. It's easy to start wondering whether our attempts at conservation really matter, particularly when the people around us don't appear to be making much of an effort. It's easy, too, to wonder whether the sacrifices and lifestyle changes make any sort of a dent at all in total carbon emissions. To that end, let's look at how much a conscientious individual can reduce the carbon emissions for which they are personally responsible in a year. Each of the calculations below is based upon EPA and other data, and is for a SINGLE INDIVIDUAL (averaged over whether they live alone or in a group), with carbon emissions savings given in tons of CO_2 per year:

- Efficient heating and cooling of a home = **1.5–2.5 tons**, depending upon your home state. Extremely hot or cold states have greater discrepancy between emissions for efficient versus inefficient heating and cooling.
- Installing energy-efficient lighting = **0.5 ton.**
- Using energy-efficient appliances = **1.2 tons.**
- Reducing hot water usage by turning down the thermostat on the water heater and switching the heater to vacation mode when out of town = **0.7 ton.**
- Walking, biking, or using public transportation one day a week = **1 ton or more**, depending upon your vehicle and commute.
- Reducing meat consumption from every meal to once a day = **1.8 tons.**
- Eating mostly organic food = **1.8 tons.**
- Recycling and composting whenever possible = **0.8 ton.**

None of the measures above are extreme—certainly there are greater lengths to which an individual can go in order to reduce the size of their carbon footprint. However, by focusing on just a few key points, it's possible to save the environment a total of nearly **10 tons** (that's 20000 pounds) of CO_2 annually, and that's **per person**! Furthermore, a person who engaged in all of the above would enjoy the added benefits of a lower electric bill, a little extra exercise, and a healthier diet. Frankly, those are all worthy goals in their own right, to say nothing of environmental impact!

SUMMARY OF MAIN POINTS

- Our global energy needs are increasing annually. Most are fulfilled through the burning of fossil fuels.
- Fossil fuels include coal, petroleum, and natural gas, all of which are partially decomposed organisms that captured and stored solar energy millions of years ago.
- Fossil fuels are nonrenewable and emit CO_2 when combusted.
- The greenhouse effect helps maintain the Earth's temperature: certain gases trap heat radiated by the Earth and prevent it from being lost to space.
- Infrared radiation (IR) is absorbed by molecules and induces bond bending or stretching that changes the molecule's net dipole.
- When IR is released, the molecule relaxes; in the atmosphere, this released IR (heat) is re-radiated upon the planet.
- Molecules with at least one bond dipole are greenhouse gases.
- The carbon cycle is a dynamic equilibrium whereby carbon is released to the atmosphere from various sources and taken up by various sources; human emissions perturb the carbon cycle.
- Global carbon emissions from the combustion of fossil fuels are increasing dramatically.
- Historical atmospheric and global temperature data is extracted from ice cores.
- Bubbles in ice core layers provide a record of historical atmospheric CO_2 content.
- ^2H concentration in ice core layers provide a record of historical global temperature.
- Both the global temperature and atmospheric carbon records show fluctuations through the last 400,000 years.
- Recorded data from temperature and CO_2 monitoring shows that for the last 100 years, global temperature and atmospheric CO_2 have been increasing dramatically, with the most significant increase in the last 25 years.
- Sea-surface temperatures and extreme weather events have been increasing.
- Empirical evidence—weather, insect lifecycles, bird migrations, glacial disintegrations—all suggest significant warming.
- Earth's radiation of heat into space has been decreasing over this same period. This is the "smoking gun" for an enhanced greenhouse effect.
- Due to complications associated with model-building, climate models disagree as to how much the temperature will increase over the next century.
- Many factors affect models, some of which (ice melting, tundra melting, soil CO_2 solubility) show positive feedback effects.
- CO_2 is not the only important greenhouse gas—several other gases have global warming potential.
- Sunspots, volcanoes, and other natural forcings can't explain recent climate change.
- Only models that include anthropogenic factors can accurately predict observed global temperatures over the past century.
- The Kyoto Protocol put restrictions on developed nations, and encouraged emissions capping from developing nations. It has had limited success.
- Global emission reduction strategies include carbon sequestration (which is logistically difficult and energetically expensive) and alternative energies.
- Alternative fuels may be low-impact and carbon-neutral, but their ability to reduce atmospheric carbon while providing efficient energy depends on how they are produced.
- Energy transformations result in loss of usable energy, making hydrogen fuel cells likely to be an inefficient alternative energy strategy for the time being, unless technology changes.
- Individuals and communities can help reduce carbon emissions by conserving electricity and reducing personal and commodity transportation.

- Reducing dependence upon commodities, reusing, and recycling commodities results in reduced greenhouse emissions.

Questions and Topics for Discussion

1. People who believe global warming to be a hoax are sometimes called "climate skeptics" by the media. Is this an accurate label? Why or why not?
2. What does it mean to say that the fossil fuels are nonrenewable? In light of global energy requirements, what are the ramifications?
3. *The greenhouse effect* is often used to mean the same as *climate change*. Are the concepts synonymous? Explain.
4. Why is the greenhouse effect important to our planet?
5. What is meant by a molecular vibration? Why do some vibrations absorb IR?
6. Why is our perturbation to the carbon cycle significant, even though it's relatively small compared to other flux quantities?
7. In earlier chapters, we considered the C-H bond to be essentially nonpolar. Is this completely accurate? Why or why not? Why do we make this approximation?
8. Why did Arrhenius begin examining the relationship between carbon dioxide and global temperature? Why was he unconcerned by his theory of global warming?
9. What are the limitations of ice core data? How do you think these may have fueled global warming debate?
10. Is it legitimate to say that our current temperature and CO_2 concentrations are part of a normal global fluctuation based upon ice core data? Why or why not?
11. Why can ice core data not be used to prove that CO_2 concentrations cause global climate change?
12. What challenges do climate modeling scientists face in trying to predict future temperatures?
13. Does the disagreement between climate predictions undermine arguments for global warming? Why or why not?
14. Does the empirical evidence support global warming? Why or why not?
15. Why is extreme weather considered evidence of global warming?
16. Why is stratospheric cooling evidence of the enhanced greenhouse effect?
17. What positive feedback systems amplify global warming, and how?
18. Given that there are many greenhouse gases, why do we hear so much about carbon dioxide?
19. Can sunspot activity be used to explain global warming? Why or why not?
20. Why are retrospective models powerful mechanisms for determining which forcings are contributing most to climate change?
21. What were the major goals of the Kyoto Protocol? Was it successful? Why or why not?
22. What provisions do you think a new global emissions treaty should contain?
23. Is it completely accurate to say that China is the world's largest carbon emitter? Do you think China should be required to reduce emissions?
24. What are some of the challenges associated with using carbon sequestration? Do you think the technology should be pursued? What does a risk-benefit analysis reveal?
25. How might global economics and politics be negatively affected by biofuels? Think about developing agricultural nations.
26. What are the considerations that must be taken into account in determining the environmental impact of biofuels?
27. The U.S. government heavily subsidizes corn, some of which is converted into ethanol. Is this a good sustainable energy strategy? Explain.

28. What are the considerations that must be taken into account in determining the energy efficiency of hydrogen fuel cells?
29. Can you imagine a system that would make hydrogen fuel cells a good alternative energy? What technologies would need to be developed?
30. What are ways that an individual or community can reduce the carbon emissions for which they are responsible?
31. A common environmental slogan is "reduce, reuse, recycle." How does each of these help reduce emissions?

Problems

1. What are the three fossil fuels and what are their sources?
2. Where does the energy in fossil fuels come from (originally)?
3. Which fossil fuel contains the most sulfur and nitrogen? Why is this a problem?
4. What are the possible dispositions for energy that is radiated toward Earth by the Sun? Which of these are considered part of the greenhouse effect?
5. What is the difference, numerically, between the greenhouse effect and an enhanced greenhouse effect?
6. What kind of radiation is absorbed by greenhouse gases?
7. What kinds of molecules are greenhouse gases?
8. Without drawing Lewis structures, identify the greenhouse gases among the following: H_2; CH_2Cl_2; and CHF_3.
9. Without drawing Lewis structures, identify the greenhouse gases among the following: NO_2; N_2; and CO_2.
10. Using Lewis structures, explain why H_2O is a greenhouse gas but O_2 is not.
11. Using Lewis structures, explain why CO_2 and SO_2 are both greenhouse gases, despite having different shapes.
12. Must molecules have bond dipoles to be greenhouse gases? Must they have net dipoles?
13. Why are oceans an important part of the carbon cycle?
14. What are the three major sources of U.S. energy?
15. What are the three major sources of global carbon emissions (fuels)?
16. How is temperature record data extracted from ice cores?
17. How is atmospheric CO_2 data extracted from ice cores?
18. What data is used to provide information about trends in the last century with regard to temperature and CO_2?
19. What is the Keeling Curve and why is it important to a discussion of climate change?
20. How has global temperature changed over the last 150 years?
21. How is sea-surface temperature related to hurricane frequency and severity?
22. What is the "smoking gun" of enhanced global warming?
23. What is a climate forcing?
24. Why do the IPCC predictions for climate change contain so much uncertainty?
25. What is a GWP and what does it mean quantitatively?
26. How might volcanoes have contributed to temperature patterns in the last century?
27. What do retrospective models show us about the cause of global warming?
28. What is carbon sequestration and why is it being developed mainly for coal burning?
29. What is a biofuel?
30. Why are CO_2 emissions from fossil fuels "worse" than CO_2 emissions from biofuels?
31. Based on current technology, why are hydrogen fuel cells responsible for carbon emissions?

Answers to Problems

1. Coal, petroleum, and natural gas are the fossil fuels. All three are partially decomposed organic matter: Coal is composed mostly of terrestrial organisms, where petroleum and natural gas are mostly ocean organisms.
3. Coal contains the most sulfur and nitrogen by percent mass. This makes it a more polluting combustion fuel than petroleum or natural gas.
5. The greenhouse effect is re-reflection of 81 percent of Earth's radiated energy. The enhanced greenhouse effect is re-reflection of more than 81 percent of the energy.
7. Molecules that can vibrate in a way that changes their net dipole are greenhouse gases.
9. N_2 has no bond dipole, so it's not a greenhouse gas. NO_2 and CO_2 have bond dipoles, and therefore are greenhouse gases.
11. Both CO_2 and SO_2 can vibrate in ways that change their net dipoles.

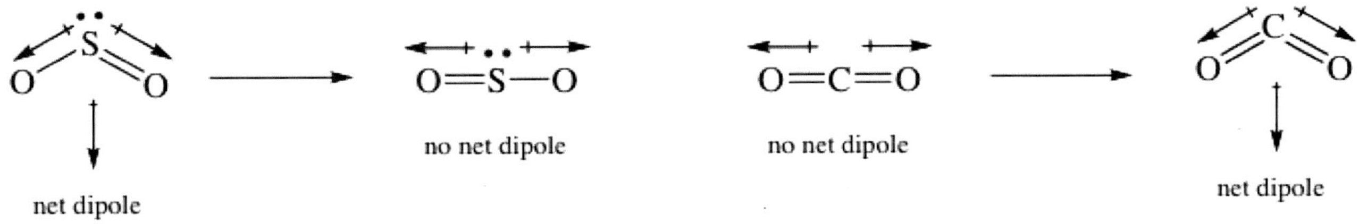

13. Oceans exchange CO_2 with the environment and sequester CO_2.
15. Petroleum, natural gas, and coal.
17. Bubbles in the ice layers contain historical atmospheric gas samples.
19. The Keeling Curve is a record of atmospheric carbon dioxide since 1958. It shows a huge increase in atmospheric CO_2 over any values in the last 400,000 years.
21. Hurricanes are more frequent and severe when sea-surface temperature is higher.
23. A climate forcing is a factor (natural or anthropogenic) that affects climate.
25. GWP is a molecule's global warming potential, as compared to carbon dioxide. A GWP of 21 means a molecule has 21x carbon dioxide's ability to cause global warming.
27. Retrospective models show that temperature change can only be explained through inclusion of anthropogenic factors.
29. A biofuel is a photosynthetic organism that is grown and combusted for energy.
31. Currently, hydrogen is produced from water using fossil fuel-derived energy, which results in carbon emissions.

Sources and Further Reading

Climate Change

http://www.ipcc.ch

www.realclimate.org

Bellamy, P. H., Loveland, P. J., Bradley, R. I., *et al.* (2005). Carbon losses from all soils across England and Wales. *Nature*, 437(7056), 245–48.

Emanuel, K. (2005). Increasing destructiveness of tropical cyclones over the past 30 years. *Nature*, 436(7051), 686–88.

Groisman, P. Y., Knight, R. W., Karl, T. R., *et al.* (2004). Contemporary changes of the hydrological cycle over the contiguous United States: trends derived from in situ observations. *Journal of Hydrometeorology*, 5(1), 64–85.

Hansen, J., Nazarenko, L., Ruedy, R., *et al.* (2005). Earth's energy imbalance: confirmation and implications. *Science*, 308(5727), 1431–35.

International Panel on Climate Change. (2007). *Climate Change 2007 Synthesis Report.* Geneva, Switzerland: Pachauri, R. K., and Reisinger, A. (eds.)

Kennedy, D. (2001). An unfortunate u-turn on carbon. *Science*, 291(5513), 2515.

Oreskes, N. (2004). Beyond the ivory tower—the scientific consensus on climate change. *Science*, 306(5702), 1686.

Paaijmans, K. P., Read, A. F., and Thomas, M. B. (2009). Understanding the link between malaria risk and climate. *Proceedings of the National Academy of Sciences in the United States of America*, 106(33), 13844–49.

Romm, J. (2007). *Hell and High Water.* New York: HarperCollins.

Slenning, B. D. (2010). Global climate change and implications for disease emergence. *Veterinary Pathology*, 47(1), SI 28–33.

Vegvari, Z., Bokony, Barta, Z., *et al.* (2010). Life history predicts advancement of avian spring migration in response to climate change. *Global Change Biology*, 16(1), 1–11.

Webster, P. J., Holland, G. J., Curry, J. A., *et al.* (2005). Changes in tropical cyclone number, duration, and intensity in a warming environment. *Science*, 309(5742), 1844–46.

Wirawan, I. M. A. (2010). Public health responses to climate change health impacts in Indonesia. *Asia-Pacific Journal of Public Health*, 22(1), 25–31.

Zimov, S. A., Schuur, E. A. G., and Chapin, F. S. (2006). Permafrost and the global carbon budget. *Science*, 312(5780), 1612–13.

Global and National Strategies

Fargione, J., Hill, J., Tilman, D., *et al.* (2008). Land clearing and the biofuel carbon debt. *Science*, 319(5867), 1235–38.

Nelson, G. C., Robertson, R. D. (2008). Green gold or green wash: environmental consequences of biofuels in the developing world. *Review of Agricultural Economics*, 30(3), 517–29.

Romm, J. (2004). *The Hype About Hydrogen.* Washington, DC: Island.

Scharlemann, J. P. W., and Laurance, W. F. (2008). Environmental science—how green are biofuels? *Science*, 319(5859), 43–44.

Individual and Community Strategies

http://www.nature.org/initiatives/climatechange/calculator/

http://www.epa.gov/climatechange/wycd/index.html

http://www.350.org/

http://www.carbonfootprint.com

Yarrow, J. (2008). *Ways to Reduce Your Carbon Footprint.* London, UK: Duncan Baird.

Glossary

INDEX—note that terms commonly appearing as abbreviations in the text are listed alphabetically by abbreviation.

Achiral—identical to its mirror image; a symmetric molecule.
Acid—a chemical that ionizes to produce H^+ in aqueous solution.
Acid rain—rain with pH < 5; produced from SOx and NOx.
Acidic—a solution of pH < 7.
Activity—a chemical's ability to be oxidized (highly active substances are easily oxidized).
Aerobic exercise (aerobic zone)—moderate-intensity efforts in which sufficient oxygen for fuel combustion is available; the heart rate zone in which such exercise occurs.
Aerosolized—to be made airborn.
Alcohol—A compound containing the –OH functional group. Also refers specifically to the compound *ethanol*.
Allotrope—alternate form of an element (i.e., O_3 is an allotrope of oxygen, where O_2 is the most common form).
Alpha particle (decay)—a helium nucleus released in nuclear decay; a decay event in which alpha particles are released.
Alternative energy—ways to generate energy through means other than combustion.
Alternative fuels—combustibles other than fossil fuels.
Amine—a nitrogen-containing functional group (these are basic).
Amino acid—a small molecule with an amine group, carboxylic acid group, and variable side chain. The structural unit of proteins.
Ammonium—a reactive nitrogen species that can be assimilated by plants.
Amphiphilic—possessing both water- and fat-soluble regions.
Amylose—a polysaccharide. Plant starch.
Anaerobic exercise (anaerobic zone)—high-intensity efforts in which sufficient oxygen is not available for fuel consumption; the heart rate zone in which such exercise occurs.
Anion—a negatively charged atom or group of atoms (polyatomic anion).
Anode—site of oxidation in an electrochemical cell.

Anthropogenic—of human origin.
Antioxidant—a chemical that ameliorates damage from free radical formation, ionizing radiation, and toxins.
AQI (Air Quality Index)—an EPA schema for quantifying air quality.
Aqueous solution—a solution in which the solvent is water.
Aquifers—underground pools of water; ground water.
Aromatic ring—a highly fragranced hydrocarbons and hydrocarbon-derivative functional group. Includes benzene.
Arthritis—degeneration of the joints; can occur from overfat.
Atherosclerosis—hardening of the arteries.
Atom—the smallest particle of an element that maintains the chemical identity of the element.
Atomic mass— the mass of an atom of an element (in amu), or the mass of a mole of an element (in grams).
Atomic mass units—the units used to measure the masses of atoms and molecules, 1 amu = 1.67×10^{-24} g.
Atomic number—the number of protons in the nucleus of an atom of an element.
Atomic symbol—the letter or letters that represent an element on the periodic table.
ATP—chemical energy currency.
Avogadro—individual who determined the number of atoms (or molecules, or anything else) in a mole.
Avogadro's number—the number of atoms/molecules/etc in a mole (6.02×10^{23})
Background radiation—unavoidable radiation from sources in the natural world.
Balanced equation—a chemical equation in which the identity and number of atoms is conserved from products to reactants.
Barometric pressure—ambient atmospheric temperature at a particular location at a particular time.
Base—a chemical that can take H^+ from water or from an acid.
Basic—a solution of pH > 7.
Battery—a spontaneous electrochemical cell (common parlance); a series of such cells (technical).
Benzene ring—a ring of six carbon atoms (and requisite hydrogens) connected with alternating single and double bonds. Also called an aromatic ring.
Beta particle (decay)—an electron released from the nucleus during nuclear decay; a decay event in which beta particles are released.
Binary acid—an acid composed of hydrogen and one other element.
Binary compound—a compound consisting of only two elements.
Bioaccumulation—the concentration of a compound in organisms

near the top of a food chain.
Biofuel—combustible material from renewable, carbon-neutral sources.
Blood glucose (blood sugar)—the concentration of glucose in the blood stream; it is carefully regulated.
BMI (body mass index)—a metric for assessing body composition; a ratio of weight to height.
Body composition—the amount of lean and fat tissues on the body. Measured to assess risk for weight-related disease.
Boiling (boiling point)—the transition from solid to liquid; the temperature at which the transition occurs.
Bond—see Chemical bond, Covalent bond, Ionic bond, Polar covalent bond.
Bond dipole—the polarization of a bond, indicated using an arrow (points toward more electronegative atom).
Bond dissociation—the breaking of a chemical bond.
Bran—the fibrous husk around a grain.
Calcification—the precipitation of calcium and magnesium salts onto fixtures and surfaces. Results from hard water.
Calorie—a measure of energy in food. If "calorie," then the amount of heat to increase 1 mL of water by 1 °C. If "Calorie," then 1 kcal.
Carbon credits—credits awarded under the Kyoto protocol to nations emitting less than target values.
Carbon dating—the use of C-14 concentration to determine the age of a sample of organic material.
Carbon fixation—reaction of carbon dioxide to produce organic molecules.
Carbon footprint—an individual's impact upon the environment; the amount of carbon emissions for which an individual is personally responsible.
Carbon monoxide—an environmental pollutant from hydrocarbon combustion, it reduces oxygen-carrying capacity of the blood.
Carbon-neutral—practices that do not result in the net release of carbon dioxide into the atmosphere.
Carbon sequestration—the capture of CO_2 released during combustion processes, with subsequent condensation and injection into deep geological repositories.
Carboxylic acid—a compound containing the –COOH arrangement of atoms.
Carcinogen—substance that causes cancer.
Catalyst—a chemical that speeds up the rate of a reaction without being consumed itself. Enzymes are biological catalysts.
Cathode—site of reduction in an electrochemical cell.
Cation—a positively charged atom or group of atoms (polyatomic

cation).

Cellular respiration—see Respiration.

Cellulose—a polysaccharide. Plant fiber.

CFCs (chlorofluorocarbons)—substituted methane and ethane molecules, also called Freons; refrigerant compounds that deplete the ozone layer.

Chapman Cycle—the cycle whereby UV light produces ozone from O_2 and vice versa. A dynamic equilibrium.

Chemical combination—a combination in which two or more atoms become chemically bonded together, acquiring a new and distinct chemical identity.

Chemical bond—made up of electrons, it holds two atoms together, forming a molecule. See Covalent bond.

Chemical equation—a symbolic representation of a chemical reaction.

Chemical reaction—the breaking and reforming of chemical bonds to yield new combinations of atoms in molecules.

Chiral—non-superimposable upon its mirror image; an asymmetric molecule.

Chlorophyll—a light-harvesting pigment involved in photosynthesis.

Chloroplast—the cellular machinery of photosynthesis.

Cholesterol—a biomolecule required for synthesis of vitamin D and maintenance of appropriate cell membrane rigidity.

***Cis* double bond**—a double bond that kinks a straight hydrocarbon chain.

Clean Air Acts—regulatory measured passed by the U.S. in the 1960's and 1970's (amended in 1990) to reduce environmental pollution.

Climate modeling—a science that uses computational and predictive techniques to predict the effects of forcings upon the climate.

Coal—fossilized remains of ancient plant life that died, partially decayed, and was then buried beneath layers of earth. A fossil fuel.

Coal scrubbing—chemical washing/filtering technique to reduce SOx emissions from coal combustion.

Coenzyme—a molecule that works with an enzyme to produce a reaction.

Cohesive—the ability or tendency of a substance to stick to itself.

Collagen—an elastic protein in the skin. Damaged by UV light.

Combustion—reaction of a compound with oxygen, yields heat.

Compound—a chemical combination of two or more elements in fixed proportions.

Computed tomography (CT)—a three-dimensional x-ray image.

Concentration—the proportion of solute to a given volume of solution.

Condensation—transition from a gas to a liquid.
Conductive—can conduct electricity.
Control rods—rods of a neutron-absorbing material designed to prevent nuclear reactions in power plants from becoming critical (growing exponentially)
Correlational—data that shows a relationship but for which causality has not been established.
Covalent bond—a shared pair (or shared pairs) of electrons that count toward the valence of both atoms involved in bonding. A physical connection between atoms.
Critical mass—the mass of fissionable fuel that is capable of producing a sustained or growing nuclear reaction.
Curie, Marie—chemist who investigated radioactivity and discovered two elements.
Current—the movement of electrons; electricity.
Dalton, John—chemist who proposed atomic theory, conservation of mass, conservation of matter.
Daughter nucleus—a product in a nuclear reaction.
DDT (dichlorodiphenyltrichloroethane)—a pesticide that is used in developing countries for malaria control. Banned for agricultural use because it is a persistant organic pollutant.
Decay—See Nuclear decay.
Deep geological repository—a proposed underground storage facility for high-level radioactive waste.
Dehydration synthesis—linking two molecules together through removal of the atomic constituents of water.
Density—the mass per unit volume of a substance.
Depleted uranium—the waste from uranium enrichment; uranium low in U-235.
Diatomic molecule—a molecule made of exactly two atoms.
Diet—properly used to mean food and drink that is taken habitually to support life and health. Improperly (but commonly) used to mean a departure from normal behavior for the purpose of weight loss.
Dietary Supplement Health and Education Act (1994)—considered herbals to be dietary supplements. 2007 amendment stipulated that herbals and supplements must be free from contamination and correctly labeled.
Dipole—a separation of charge.
Direct effect (of exercise)—the calories burned during an exercise activity.
Disaccharide—two sugar units linked together. A sugar.
Dissociation—the separation of anions and cations in aqueous solution.
DNA (deoxyribonucleic acid)—genetic material.
Doping (of silicon)—to add non-silicon elements to silicon in

order to affect electron density.
Dry-casking—method of storage in which spent nuclear fuel is kept above ground in metal containers within a concrete structure
Ductile—can be drawn into wires.
Dynamic equilibrium—a process in which no net changes to a system occur over time. A reaction system in which concentrations of chemical species remain constant, despite individual reactions taking place.
Einstein, Albert—a physicist and chemist who related mass to energy by the equation $E = mc^2$.
Electrical energy—the movement of electrons; electrical current; electricity.
Electrical potential energy—the potential for charged particles to move toward or away one another on the basis of electrostatic forces.
Electricity—the movement of electrons; electrical current; electrical energy.
Electrochemical cell—a cell in which a redox reaction takes place.
Electrode—the contact point at which oxidation or reduction occurs in an electrochemical cell.
Electrolysis—the use of electrical energy to power a non-spontaneous chemical reaction (generally one that reduces a compound to elements).
Electrolytic cell—a non-spontaneous electrochemical cell.
Electrolyte—a compound that dissociates in aqueous solution; an ionic compound, a salt.
Electronegativity—a quantitative measure of how tightly an atom holds its electrons. A measure of the extent to which an atom will pull electrons in a bond. This is a periodic trend.
Electron holes—the electron paucity in p-type as compared to n-type silicon.
Electrons—subatomic particles found in a nearly massless cloud surrounding the nucleus with mass = 0 and charge = -1. The components of covalent bonds.
Electron shells—location of electrons in an atom, lower numbered shells are closer to the nucleus.
Electrostatic forces—the attractions between oppositely charged particles and repulsions between like-charged particles.
Elements—the simplest type of matter, the substances listed on the periodic table.
Elevated post-exercise metabolic rate—the higher RMR observed for up to 24 hours following exercise.
Empirical evidence—taken from observation.
Empty calories—calories containing little nutritional benefit apart from energy.

Enantiomers—a pair of non-superimposable mirror image molecules.
Endothermic—a reaction that requires energy.
Enhanced greenhouse effect—a greenhouse effect phenomenon of greater than 81%.
Entropy—disordered energy.
Enzyme—a protein that makes a chemical reaction happen faster. A biological catalyst.
Ephedrine—a chemical compound used in diet pills; a sympathomimetic.
EMR (electromagnetic radiation)—electrical and magnetic disturbance that propagates through space; radiant energy.
Essential body fat—the amount of body fat required to maintain normal function. Around 2-4% in men, 10-12% in women.
Ester—a compound functional group made from a carboxylic acid and an alcohol. Often highly fragranced.
Ether—a functional group consisting of two hydrocarbons connected by a common oxygen. Refers specifically to the compound *diethyl ether.*
Exothermic—a reaction that releases energy.
Exposure—a measure of how much of a pollutant substance is likely to be encountered.
Fatty acid—a long-chain carboxylic acid.
Fermentation—anaerobic glucose metabolism.
First law of thermodynamics—energy of the universe is constant; energy can neither be created nor destroyed.
Food and Drug Act (1906)—required that substances transported across state lines were free from contaminants and correctly labeled.
Food, Drug and Cosmetic Act (1938)—added toxicity study requirements to the above.
Food, Drug and Cosmetic Act Amendment (1962)—increased safety requirements and added proof of efficacy to the above.
Forcings—environmental or anthropogenic factors that affect climate
Fossil fuels—fossilized terrestrial or sea organisms that contain solar energy captured through photosynthesis; coal, petroleum, and natural gas.
Free radical—highly reactive chemical with unpaired electrons.
Frequency (n)—the number of waves per unit time that pass a given point in space, measured in Hz.
Fructose—a monosaccharide commonly found in fruit and as part of the disaccharide sucrose.
Fuel assemblies—pellets of fissionable fuel assembled for use in a nuclear reactor.
Functional group—a characteristic arrangement of atoms in a

molecule that lends a predictable set of properties to the molecule.

Galvanic cell—a spontaneous electrochemical cell; a voltaic cell.

Gamma particle (decay)—a photon released from the nucleus during nuclear decay; a decay event in which gamma particles are released.

Gamma rays—very high energy EMR; ionizing radiation. A form of nuclear decay particle with no mass or charge.

Gas—a physical state of matter. Gases are compressible and take the shape of their container.

Gaseous diffusion—a separation process for uranium enrichment in which lighter isotopes pass through a filter faster than heavier ones, and are concentrated.

GI (glycemic index)—a measure of the rate at which a carbohydrate enters the bloodstream.

Glucagon—a hormone that increases blood sugar.

Glucose—a common monosaccharide.

Glycerol—a trifunctional alcohol to which fatty acids are bonded to make triglyceride.

Gravitational potential energy—the energy an object has when above the ground (the potential to fall).

Greenfreeze—a substitute refrigerant that is not ozone-depleting and not a greenhouse gas.

Greenhouse effect—the phenomenon whereby atmospheric gases trap 81% of Earth's radiant energy (heat) and re-radiate it back toward Earth.

Ground water—water found underground; aquifer water.

GWP (global warming potential)—a standard measurement established by the IPCC that compares a molecule's ability to store and re-radiate IR to that of CO_2.

Haber-Bosch process—industrial nitrogen fixation.

Half-life—the time it takes for one half of a sample of radioactive isotope to decay.

Half-reactions—individual oxidation and reduction components of a redox reaction.

Hard water—water containing dissolved calcium and magnesium; generally ground water.

HCFCs (hydrochlorofluorocarbons)—first-generation CFC substitutes; refrigerants. These are less ozone-depleting but are greenhouse gases.

HDL (high-density lipoprotein)—"good" cholesterol; a transporter that carries cholesterol from the cells to the liver for excretion.

HFCs (hydrofluorocarbons)—second-generation CFC substitutes; refrigerants. These are not ozone-depleting but are greenhouse gases.

Hemoglobin—the protein that carries oxygen from the lungs to the tissues in the blood.

High-level radioactive waste—highly-radioactive spent fuel and the waste from reprocessed spent fuel that must be stored indefinitely to prevent contamination.

Homeostasis—maintenance of a constant environment.

Hormone—a chemical used for cell-to-cell communication in the body; may be peptide or steroid.

HR_{max} (maximum heart rate)—the highest heart rate an individual can sustain.

Hydrocarbon—a compound composed of hydrogen and carbon. They combust in O_2 to produce carbon dioxide and water (idealized).

Hydrogen bond—the electrostatic attraction between molecules containing FH, OH, or NH bonds.

Hydrogen fuel cells—voltaic cells that generate electricity from the reaction of hydrogen with oxygen to produce water.

Hydrolysis—breaking a bond by adding the constituent atoms of water across the bond.

Hydrophilic—water-loving. A descriptive term for polar compounds in terms of their solubility behavior.

Hydrophobic—water-fearing. A descriptive term for non-polar compounds in terms of their solubility behavior.

Hyperglycemia—high blood sugar.

Hypoglycemia—low blood sugar.

Ice core—a sample of ice drilled for temperature, atmospheric, and other historical data. Generally taken from polar regions.

Immersion test—a metric for measuring body composition; based upon weight underwater.

Incompressible—can't be packed into a smaller volume.

Industrial Revolution—historic period during 18th and 19th century in which machine-generated power replaced human- and animal-generated power. Inception of major dependence on fossil fuel.

Inert—chemically unreactive.

Insulin—a hormone that decreases blood sugar; a hormone that causes cells to take up glucose.

Intermediate—a chemical formed in one step of a reaction series and used in a later step.

Ion—a charged atom or group of atoms (polyatomic ion).

Ion exchange—the process whereby hard water is softened by exchanging calcium and magnesium ions for sodium.

Ionic bond—the electrostatic attraction between two charged atoms or groups of atoms.

Ionizing radiation—radiation capable of breaking chemical bonds and knocking electrons out of molecules.

IPCC (International Panel on Climate Change)—United

Nations panel with the mission to assess climate change research as well as its social and economic ramifications

IR (infrared radiation)—non-ionizing EMR; heat. Trapped by greenhouse gases, absorbed by some molecular vibrations.

Irradiation (of food)—use of gamma rays to kill microbes and insects in food.

Isotope—different types of atoms of the same element. All have the same atomic number, but have varying numbers of neutrons.

Ketogenic diet—a low-carbohydrate diet that results in ketosis.

Ketone bodies—chemicals including acetone that are produced via fatty acid metabolism and can enter the bloodstream in times of starvation.

Ketosis—the systemic circulation of ketone bodies.

Kinetic energy—the energy of motion.

Lactic acid—a metabolic waste product of anaerobic metabolism.

Lactose—a disaccharide. Milk sugar.

Law of conservation of matter—atoms cannot appear, disappear, or change into atoms of other elements by any chemical means.

LDL (low-density lipoprotein)—"bad" cholesterol; a transporter that carries cholesterol from the liver to the cells.

Lead—a heavy metal pollutant. A potent neurotoxin.

Leaven (leavening)—to cause dough to rise; the agent the causes dough to rise.

Lewis dot structure—a representation of the valence electrons in an atom, or a representation of bonds and lone pairs in a molecule.

Line-angle diagram—a simplified representation of a molecule (usually used for large molecules) in which hydrogen atoms attached to carbon are implicit, and the atomic symbol for carbon is omitted.

Lipid—a hydrophobic molecule.

Liquid—a physical state of matter. Liquids are incompressible and take the shape of their container.

Lone pair—a non-bonding pair of electrons.

Macronutrient—carbohydrate, protein, and fat. Nutrients required in large amounts that provide energy.

Malleable—can be hammered into sheets.

Manhattan Project—the U.S. Army project to develop nuclear weapons.

Mass defect—the mass "lost" (converted to energy) during a nuclear reaction.

Mass number—the number of protons plus the number of neutrons in an atom's nucleus.

Matter—anything that occupies space and has mass.

Meltdown (nuclear)—a containment failure at a nuclear facility.

Melting (melting point)—the transition from liquid to gas; the

temperature at which the transition occurs.
Mercury—a heavy metal pollutant. A potent neurotoxin.
Metabolism—the sum of all chemical reactions in the body.
Metalloids—elements with physical properties in between those of metals and non-metals. Located on a diagonal line near the left side of the periodic table. Semiconductors.
Metals—elements that are shiny, ductile, conductive, and malleable. The majority of the periodic table.
Micelle—a spherical arrangement of fatty acids in water, with grease sequestered at the inside.
Micronutrient—vitamins and minerals. Important chemicals required in small amounts.
Midgley, Thomas—an inventor and mechanical engineer. Developer of tetraethyl lead and Freons.
Minerals—metal and halide ions required in various amounts for normal function.
Mitochondrion—the machinery of respiration.
Mixture—physical combinations of two or more substances, in variable proportions.
Molar mass—the mass of a mole of substance
Molarity—quantitative measure of solution concentration, moles solute per liter of solution.
Mole—a number. 1 mole = 6.02×10^{23}.
Molecular compound—a compound that does not dissociate in aqueous solution; a non-electrolyte. Composed of covalent bonds.
Molecular mass—the mass of a molecule (in amu).
Molecule—a pure substance made up of a chemical combination of two or more atoms.
Monosaccharide—a single sugar unit; the constituent unit carbohydrates.
Montreal Protocol (1987)—international treaty to phase out ODS.
Mutagen—substance that causes mutation.
Myoglobin—a chemical related to hemoglobin, it stores oxygen in the muscle (primarily type I fibers)
N-type silicon—silicon that has been doped with group 5 elements.
Natural gas—partially decomposed remains of sea organisms, found with petroleum; a fossil fuel.
Net dipole—the sum of the bond dipoles within a molecule.
Neurochemical—a compound with activity in the brain or nervous system.
Neurotoxin—a compound toxic to or damaging to brain or nerve cells.
Neurotransmitter—a chemical used for cellular communication within the nervous system.

Neutral—having no charge (for ions); neither acidic nor basic (for solutions).
Neutralize—to react an acid with a base.
Neutrons—subatomic particles found in the nucleus with mass = 1 amu and no charge.
Nitrate—a reactive nitrogen species that can be assimilated by plants.
Nitrogen dioxide—a NOx species and environmental pollutant. A respiratory irritant.
Nitrogen fixation (nitrogen-fixing bacteria)—the process of making atmospheric nitrogen reactive, accomplished by soil or root-nodule bacteria; the bacteria responsible for the process.
Nitrogen oxides—binary compounds of nitrogen and oxygen (including NOx).
Nomenclature—a system of naming.
Non-electrolyte—a molecular compound.
Non-metals—elements that are dull, brittle, and non-conductive. Located on the left-hand side of the periodic table.
Non-nutritive sweeteners—sweet compounds that provide no calories because they are indigestible.
Non-polar—a molecule with no net dipole.
Non-renewable (fuel)—a resource that is non-renewable; fossil fuels.
Normal body fat—the amount appropriate to normal function. Around 6-26% in men, 14-31% in women.
NOx—mononitrogen oxides. Important species include NO_2 and NO. Produced by the reaction of atmospheric nitrogen with oxygen at high temperature.
Nuclear chain reaction—a self-sustaining nuclear reaction in which decay particles initiate reactions in other nuclei.
Nuclear decay—a nuclear reaction in which a parent nucleus loses a small particle to become a daughter nucleus.
Nuclear equation—a symbolic representation of a nuclear reaction.
Nuclear fission—the splitting of a large nucleus into two or more pieces.
Nuclear fusion—the combination of two nuclei to form a larger nuclei, often with subsequent decay.
Nuclear reaction—a reaction in which the nuclei of atoms gain or lose subatomic particles and/or rearrange.
Nuclear reactor—an isolated reaction vessel in a nuclear power plant.
Nucleons—subatomic particles found in the nucleus (protons and neutrons).
Nucleus—the small-volume, massive center of an atom. Contains protons and neutrons.

Obesity—too much body weight relative to body size, defined in several ways (by percentage body fat or BMI). Less severe than obesity.
Octet rule—atoms are most stable when they have a full (generally 8, except in the case of row 1 elements) outer shell of electrons.
ODS (ozone-depleting substances)—substances like CFCs that deplete the ozone layer.
Omega-3 fatty acid—a fatty acid with a *cis* double bond three carbons from the omega (tail) end.
Omega-6 fatty acid—a fatty acid with a *cis* double bond six carbons from the omega (tail) end.
Organic chemistry —the study of organic compounds/the molecules of life.
Organic compound/organic molecule—a carbon based molecule.
Overweight—too much body weight relative to body size, defined in several ways (by percentage body fat or BMI). Less severe than obesity.
Oxidation—the loss of electrons.
Oxyacid—an acid containing hydrogen, oxygen, and one other non-metal.
Ozone (the molecule)—an allotrope of oxygen; a reactive molecule.
Ozone (tropospheric)—an environmental pollutant (secondary to combustion products). Formed from the reaction of NOx with sunlight. A respiratory irritant.
Ozone (stratospheric)—forms a protective layer that protects Earth from UV light.
P-type silicon—silicon that has been doped with group 3 elements.
PAHs (polycyclic aromatic hydrocarbons)—large organic molecules. Sources include combustion of petroleum and coal. Known carcinogens.
Pancreas—an organ that releases insulin and glucagon.
Parabolic trough solar thermal power plants—plants that concentrate the Sun's energy on a pipe containing molten salt to store solar energy.
Parent nucleus—a reactant in a nuclear reaction.
Partial charge (negative or positive)—the slight charge imbalance that results from the formation of a polar covalent bond.
Partial hydrogenation (partially hydrogenated oils)—a chemical process in which some double bonds are removed from fatty acids and others are changed from *cis* to *trans*. A liquid oil that has been converted to solid through the process of partial hydrogenation.
Particulate matter—includes PM_{10} and $PM_{2.5}$.

Peptide bond—the bond that links amino acids to form a protein.
Percent Daily Value—the amount of each nutrient recommended by the FDA and USDA for daily consumption.
Periodic table—an organizational scheme that groups elements by chemical properties.
Periodic trend—a physical or chemical trend that can be predicted using the periodic table. A trend that follows a pattern in groups or rows.
Perturbation—a disruption.
Petroleum—partially decomposed remains of sea organisms, found with natural gas; a fossil fuel.
pH—a measure of solution acidity or basicity. Related to concentration of hydrogen ion.
Pharmacophore—a core functional area of a drug or group of drugs that allow binding to the receptor.
Phenol—a compound containing an –OH functional group attached to a benzene ring. Refers specifically to the compound *phenol.*
Photosynthesis—storage of the Sun's energy as chemical energy (carbohydrate) through electron transfer.
Photosystem—light harvesting pigments and enzymes involved in photosynthesis.
Photovoltaic cell—a cell that transforms the Sun's energy into electrical energy.
Physical activity energy expenditure—the calories burned through daily activity.
Physical combination—a combination in which each substance maintains its individual identity and properties.
Planck's constant—a multiplier used in the wave equation relating energy to frequency. Given the symbol *h*.
Plaques—accumulations of cholesterol in the arteries.
$PM_{2.5}$—fine particles of diameter < 2.5 μm.
PM_{10}—inhalable course particles of diameter 2.5-10 μm.
Polar—a molecule with a net dipole.
Polar covalent bond—a polarized bond; the bond that exists between two atoms of unequal electronegativity.
Polar stratospheric clouds—high clouds that form over Antarctica in winter and capture chlorine; associated with rapid ozone depletion.
Polarization—the pulling of electrons within a bond; the forming of a polarized bond.
Polyatomic ion—an ion composed of two or more atoms, covalently bonded together.
Polyatomic molecule—a molecule made up of two or more atoms.
Polymer—a large molecule composed of repeating units.

Polypeptide—a large molecule composed of amino acids; a protein.
Polysaccharide—a polymer of sugar units. Starch or fiber.
Positron—a positively charged, massless particle.
Positron emission tomography—scans tracking the location of radioactive particles (in organic molecules used by the body) over time.
Potable—safe to drink, used with regard to water.
ppm (parts per million)—the number of particles of a substance of interest per million particles of sample. A measure of concentration.
Precautionary principle—a formalization of maxims such as *better safe than sorry*. The idea that it is not reasonable to take risks in cases where risk may exist.
Precipitation (precipitate)—the process whereby a salt falls out of solution and forms a solid; the solid itself.
Primary coolant—the coolant in a nuclear reactor that dissipates heat directly from fissionable fuel.
Products—substances produced in a chemical reaction.
Protein—large molecules formed as polymers of amino acids. A macronutrient; a structural and functional component of cells.
Protons—subatomic particles found in the nucleus with mass = 1 amu and charge = +1.
Pure substances—non-mixtures. Elements and compounds.
Pyruvate—the molecule produced from glucose that accepts electrons in fermentation.
Racemic mixture—a 50:50 mixture of two enantiomers.
Radiation—EMR or nuclear wave/particle rays.
Radiation sickness—a set of symptoms resulting from radiation exposure.
Radical—unpaired electron.
Radioactive decay series—a series of decay reactions that continue until a stable daughter nucleus is produced.
Radioactive particle—a small decay particle given off in a nuclear reaction.
Radioisotopes—radioactive nuclei.
Reactants—substances that react in a chemical reaction.
Receptor—a large molecule (generally on a cell membrane) that binds a substrate and responds in some physiological way.
Redox—an electron transfer reaction.
Reduction—the gaining of electrons.
Refined grains—grains which have had bran (fiber) removed.
Respiration—the process by which energy is extracted from nutrients and stored as ATP based upon electron flow.
RMR (resting metabolic rate)—the number of calories burned daily through maintenance of cellular function alone.

Salt bridge—a bridge containing spectator salt (ions) that connects the halves of an electrochemical cell to balance charge.
Salts—ionic compounds, electrolytes.
Satiety signal—something that communicates a sense of fullness and satisfaction.
Saturated fat—triglyceride containing fatty acids without double bonds.
Second law of thermodynamics—for all processes, entropy of the universe increases.
Secondary coolant—the coolant in a nuclear reactor that dissipates heat from primary coolant and uses the heat to vaporize water.
Shells—see Electron shells
Silicon p-n junction solar cell—a photovoltaic cell in which p-type and n-type silicon are juxtaposed for unidirectional electron flow.
Sink—a repository of. Generally used with regard to carbon.
Skin fold tests—a metric for assessing body composition; based upon thickness of the subcutaneous fat in several places on the body.
Sleep apnea—periodic temporary cessation of breathing during sleep. Can result from overfat.
Smog—combination of smoke and fog.
Soft water—water that does not contain high concentrations of calcium and magnesium ions. Chemically softened water or rain water.
Solid—a physical state of matter. Solids are incompressible and do not take the shape of their container.
Solute—a chemical that is dissolved to form a solution.
Solution—a mixture made up of one or more substances dissolved in another substance. A mixture. If water is the solvent, the solution is aqueous.
Solvent—a chemical that dissolves other chemicals to form a solution.
SOx—monosulfur oxides. Important species include SO_2 and SO_3. Primary anthropogenic source is coal combustion.
Spectator ion—an ionic species that is present in a reaction mixture, but not involved in the chemical reaction.
Stoichiometric equivalents—the ratios of substances involved in a chemical reaction.
Strong force—the strong attraction between nucleons over very short distances.
Substituents—atoms or groups bound to an atom or molecule of interest.
Substrate—a compound that binds to a receptor.
Sucrose—a disaccharide. Table sugar.

Sulfur dioxide—an environmental pollutant, a major source of which is coal combustion. A respiratory irritant.
Superimposable—two objects that can be placed next on top of each other so that all parts match up.
Surface water—water on Earth's surface; lakes, rivers, streams (fresh water).
Sympathomimetic—a chemical that mimics the effects of adrenaline or the sympathetic nervous system ("fight or flight").
Teratogen—a compound known to cause developmental defects in an embryo or fetus.
Tetraethyl lead—a gasoline additive invented to prevent engine knocking.
Tetrahedral—the shape of a four-sided pyramid. This shape is formed when four electron groups separate as much as possible around a central atom.
Thermic effect of food—the calories expended to digest and metabolize food.
Thermodynamics—the study of the energetics of reactions and processes.
Thought experiment—a way of finding the answer to a question that is impossible or impractical to test in real life through theoretical means.
Toxicity—a measure of the inherent harm-causing potential of a pollutant.
***Trans* double bond**—a double bond that does not kink a straight hydrocarbon chain.
Trans fat—triglyceride containing fatty acids with *trans* double bonds.
Trigonal planar—the shape of an equilateral triangle. This shape is formed when three electron groups separate as much as possible around a central atom.
Triglyceride—glycerol and three fatty acids chemically combined; storage fat.
Type 2 diabetes—insensitivity of body cells to insulin; often results from overfat.
Type I muscle fibers—fibers specializing in aerobic efforts; slow-twitch fibers.
Type II muscle fibers—fibers specializing in anaerobic efforts; fast-twitch fibers.
Unsaturated fat—triglyceride containing fatty acids with *cis* double bonds.
Uranium enrichment—the process of concentrating U-235 above what is found in nature.
Uranium tails—the waste from uranium enrichment; uranium low in U-235.
UV (ultraviolet radiation)—high energy radiation; ionizing

radiation. Filtered by the ozone layer.
UV-A—lowest-energy UV radiation.
UV-B—moderate-energy UV radiation.
UV-C—highest-energy UV radiation.
Valence—an atom's number of valence electrons.
Valence electrons—the electrons in an atom's outermost shell
van der Waals forces—the attractions between non-polar molecules. Forces are stronger between larger molecules.
Vaporization—transition from a liquid to a gas; boiling.
Vitamins—organic molecules required in small amounts for normal function.
VOCs (volatile organic compounds)—small organic molecules that vaporize easily and enter the atmosphere. Sources include solvents and building materials.
Voltaic cell—a spontaneous electrochemical cell; a Galvanic cell.
Wavelength (l)—a wave's distance crest to crest, measured in meters.
Western Paradox—the idea that Western societies are simultaneously consuming a terrible diet and putting a premium on physical perfection.
Work—the potential to move an object against a force.
Wrist, neck and waist circumference—a metric for assessing body composition based upon circumferences.
X-rays—very high energy radiation; ionizing radiation. Can pass through soft tissue, but are reflected by bone (hence their utility in diagnostic medicine).

Credits

Chapter 1

Image by Micha L. Rieser, available at http://commons.wikimedia.org/wiki/File:Maple-leaves-fall.jpg under the Creative Commons Attribution-Share Alike 3.0. Unported license. © 2008.

Image available at http://commons.wikimedia.org/wiki/File:Gartengew%C3%A4chshaus.JPG. Copyright in the public domain.

Image by Photorush, available at http://commons.wikimedia.org/wiki/File:Sequoyah_Nuclear_Power_Plant.JPG under the Creative Commons Attribution-Share Alike 3.0 Unported license. © 2008.

Image available at http://commons.wikimedia.org/wiki/File:Baking_Bread_LLCCinnamon_roll.jpg. Copyright in the public domain.

Image by Lmbuga at the Wikipedia galega project, available at http://commons.wikimedia.org/wiki/File:Ribeira_Lixo_GDFL_040825_049.jpg under the Creative Commons Attribution-Share Alike 3.0 Unported license. © 2005.

Image by Rama, available at http://commons.wikimedia.org/wiki/File:Usine_Bret_MG_1648.jpg under the Creative Commons Attribution-Share Alike 2.0 France license.

Image by Emturan, available at http://commons.wikimedia.org/wiki/File:Kyoto_Protocol_participation_map_2009.png under the Creative Commons Attribution-Share Alike 3.0 Unported license. © 2009.

Photo by Sage Ross (ragesoss.com), from Wikimedia Commons. Creative Commons Attribution-Share Alike 3.0 Unported.

Image available at http://commons.wikimedia.org/wiki/File:Pollution_de_l%27air.jpg. Copyright in the public domain.

Image available at http://commons.wikimedia.org/wiki/File:Foods.jpg. Copyright in the public domain.

Image by Scott Lefler. Permission granted by the photographer. © 2010.

Image by Chris 73, available at http://commons.wikimedia.org/wiki/File:SaltInWaterSolutionLiquid.jpg under the Creative Commons Attribution-Share Alike 3.0 Unported license. © 2004.

Image by Bios, available at http://commons.wikimedia.org/wiki/File:NaCl-zoutkristallen_op_Schott_Duran_100_ml.JPG under the Creative Commons Attribution-Share Alike 3.0 Unported license. © 2005.

Image by Kirstin Hendrickson.

Image available at http://images-of-elements.com/neon.php under the Creative Commons Attribution-Share Alike 3.0 Unported license. © 2009, Jumk.de Webprojects.

Image available at http://images-of-elements.com/mercury.php under the Creative Commons Attribution-Share Alike 3.0 Unported license. © 2009, Jumk.de Webprojects.

Image available at http://commons.wikimedia.org/wiki/File:HEUranium.jpg. Copyright in the public domain.

Image available at http://commons.wikimedia.org/wiki/File:GoldNuggetUSGOV.jpg. Copyright in the public domain.

Image by Kirstin Hendrickson.

Image by Scott Lefler. Permission granted by the photographer. © 2006.

Image by Carioca, available at http://commons.wikimedia.org/wiki/File:Cuboid_sugar.jpg under the Creative Commons Attribution-Share Alike 3.0 Unported license. © 2005.

Image by Claire H., available at http://commons.wikimedia.org/wiki/File:Greek_Marble_Statue_of_Aphrodite_Anadyomene_%28Hair-Binding%29.jpg under the Creative Commons Attribution-Share Alike 2.0 Generic license. © 2008.

Image available at http://commons.wikimedia.org/wiki/File:Glass_of_la_trappe_quadrupel.jpg. Copyright in the public domain.

Image by André Karwath (Aka), available at http://commons.wikimedia.org/wiki/File:Red_Wine_Glas.jpg under the Creative Commons Attribution-Share Alike 2.5 Generic license. © 2004.

Image available at http://commons.wikimedia.org/wiki/File:Sulfur.jpg. Copyright in the public domain.

Image by Kirstin Hendrickson.

Image by Kirstin Hendrickson.

Image available at http://commons.wikimedia.org/wiki/File:GraphiteUSGOV.jpg. Copyright in the public domain.

Image by Anton, available at http://commons.wikimedia.org/wiki/File:Graphit_gitter.svg under the Creative Commons Attribution-Share Alike 3.0 Unported license. © 2004.

Image by Gregory Phillips, available at http://commons.wikimedia.org/wiki/File:CZ_brilliant.jpg under the Creative Commons Attribution-Share Alike 3.0 Unported license.

Image by Anton, available at http://commons.wikimedia.org/wiki/File:Diamonds_glitter.png under the Creative Commons Attribution-Share Alike 3.0 Unported license. © 2004.

Image by Scott Lefler. Permission granted by the photographer. © 2010.

Image by Kirstin Hendrickson.

Image by Kirstin Hendrickson.

Image by Giovanni Dall'Orto, available at http://commons.wikimedia.org/wiki/File:Tubo_in_rame_-_Foto_di_Giovanni_Dall%27Orto.jpg. © 2005.

Image available at http://images-of-elements.com/sodium.php under the Creative Commons Attribution-Share Alike 3.0 Unported license. © 2009, Jumk.de Webprojects.

Image by Silane, available at http://commons.wikimedia.org/wiki/File:Silizium_pulver.jpg under the Creative Commons Attribution-Share Alike 3.0 Unported license. © 2004.

Image available at http://commons.wikimedia.org/wiki/File:Chlorine2.jpg. Copyright in the public domain.

Image available at http://images-of-elements.com/iodine.php under the Creative Commons Attribution-Share Alike 3.0 Unported license. © 2009, Jumk.de Webprojects.

Image by Scott Lefler. Permission granted by the photographer. © 2007.

Image available at http://commons.wikimedia.org/wiki/File:Water_molecule.svg. Copyright in the public domain.

Image by Connor Lee (WAvegetarian), available at http://commons.wikimedia.org/wiki/File:Burning_Bunsen_burner.jpg under the Creative Commons Attribution-Share Alike 3.0 Unported license. © 2003.

Image by Waugsberg, available at http://commons.wikimedia.org/wiki/File:Stacheldraht_05.jpg under the Creative Commons Attribution-Share Alike 3.0 Unported license. © 2007.

Image by Thegreenj, available at http://commons.wikimedia.org/wiki/File:Toothpasteonbrush.jpg under the Creative Commons Attribution-Share Alike 3.0 Unported license. © 2008.

Image by Kirstin Hendrickson.

Image by Kirstin Hendrickson.

Image by Scott Lefler. Permission granted by the photographer. © 2010.

Chapter 2

Image available at http://commons.wikimedia.org/wiki/File:Water_molecule.svg. Copyright in the public domain.

Image by Kirstin Hendrickson.

Image by Kirstin Hendrickson.
Image by lain, available at http://commons.wikimedia.org/wiki/File:Ear-anatomy-text-small-en.png under the Creative Commons Attribution-Share Alike 3.0 Unported license. © 2003.
Image by Kirstin Hendrickson.
Image by Kirstin Hendrickson.
Image by Kirstin Hendrickson.
Image by Kirstin Hendrickson.
Image by Kirstin Hendrickson.
Image available at http://commons.wikimedia.org/wiki/File:Avogadro_Amedeo.jpg. Copyright in the public domain.
Image by Scott Lefler. Permission granted by the photographer. © 2010.
Image by Scott Lefler. Permission granted by the photographer. © 2010.
Image by Kirstin Hendrickson.
Image by Kirstin Hendrickson.
Image by Kirstin Hendrickson.
Image by Kirstin Hendrickson.
Image by Kirstin Hendrickson.
Image available at http://commons.wikimedia.org/wiki/File:Radioactivity_symbol.png. Copyright in the public domain.
Image available at http://commons.wikimedia.org/wiki/File:Plutonium_pellet.jpg. Copyright in the public domain.
Image by Ehs2908, available at http://commons.wikimedia.org/wiki/File:Element_95_Americium_Am.jpg under the Creative Commons Attribution-Share Alike 3.0 Unported license. © 2009.
Image by Warut Roonguthai, available at http://commons.wikimedia.org/wiki/File:Vacuum_capacitor_with_uranium_glass.jpg under the Creative Commons Attribution-Share Alike 3.0 Unported license. © 2007.
Image available at http://commons.wikimedia.org/wiki/File:Red_Delicious.jpg under the Creative Commons Attribution-Share Alike 3.0 Unported license. © 1995, FirstFruits Farm.
Image available at http://commons.wikimedia.org/wiki/File:Earth_Western_Hemisphere.jpg. Copyright in the public domain.
Image by Kirstin Hendrickson.
Image by Kirstin Hendrickson.
Kenneth C. Lowe, Figure 3 from "Blood Substitutes: From Chemistry to Clinic." Journal of Materials Chemistry, Vol. 16. © The Royal Society of Chemistry, 2006. Permission to reprint granted by the publisher.

Chapter 3
Image by Kirstin Hendrickson.
Image by Kirstin Hendrickson.
Image by Kirstin Hendrickson.
Image by Kirstin Hendrickson.
Image by Kirstin Hendrickson.
Image by Kirstin Hendrickson.
Image by Kirstin Hendrickson.
Image available at http://commons.wikimedia.org/wiki/File:Lithium_paraffin.jpg. Copyright in the public domain.
Image by Superplus, available at http://commons.wikimedia.org/wiki/File:Sodium-1.jpg under the Creative Commons Attribution-Share Alike 3.0 Unported license. © 2005.
Image available at http://images-of-elements.com/potassium.php under the Creative Commons Attribution-Share Alike 3.0 Unported license. © 2009, Jumk.de Webprojects.
Image available at http://images-of-elements.com/helium.php under the Creative Commons Attribution-Share Alike 3.0 Unported license. © 2009, Jumk.de Webprojects.
Image available at http://images-of-elements.com/neon.php under the Creative Commons Attribution-Share Alike 3.0 Unported license. © 2009, Jumk.de Webprojects.
Image available at http://images-of-elements.com/argon.php under the Creative Commons Attribution-Share Alike 3.0 Unported license. © 2009, Jumk.de Webprojects.
Image available at http://images-of-elements.com/radon.php under the Creative Commons Attribution-Share Alike 3.0 Unported license. © 2009, Jumk.de Webprojects.
Image available at http://images-of-elements.com/xenon.php under the Creative Commons Attribution-Share Alike 3.0 Unported license. © 2009, Jumk.de Webprojects.
Image available at http://images-of-elements.com/krypton.php under the Creative Commons Attribution-Share Alike 3.0 Unported license. © 2009, Jumk.de Webprojects.
Image by Kirstin Hendrickson.
Image by Kirstin Hendrickson.
Image by Kirstin Hendrickson.
Image by Kirstin Hendrickson.
Image by Kirstin Hendrickson.

Chapter 4
Image by Scott Lefler. Permission granted by the photographer. © 2009.
Image by Kirstin Hendrickson.
Image by Leon Brocard, available at http://commons.wikimedia.org/wiki/File:Olive_oil_with_Balsamic_Vinegar.jpg under the Creative Commons Attribution 2.0 Generic license. © 2006.
Image by Kirstin Hendrickson.
Image by Kirstin Hendrickson.
Image by Kirstin Hendrickson.
Image available at http://commons.wikimedia.org/wiki/File:Water_drops_on_spider_web.jpg. Copyright in the public domain.
Image available at http://commons.wikimedia.org/wiki/File:Iceberg_at_Baffin_Bay.jpg. Copyright in the public domain.
Image available at http://commons.wikimedia.org/wiki/File:NCI_iced_tea.jpg. Copyright in the public domain.
Image by Kirstin Hendrickson.
Wilson Bentley, Photograph #892 from Plate XIX from: "Studies among the snow crystals during the winter of 1901-2, with additional data collected during previous winters." Monthly Weather Review, Vol. 30, No. 3. American Meteorological Society, 1902.
Wilson Bentley, Photograph #894 from Plate XIX from: "Studies among the snow crystals during the winter of 1901-2, with additional data collected during previous winters." Monthly Weather Review, Vol. 30, No. 3. American Meteorological Society, 1902.
Wilson Bentley, Photograph #895 from Plate XIX from: "Studies among the snow crystals during the winter of 1901-2, with additional data collected during previous winters." Monthly Weather Review, Vol. 30, No. 3. American Meteorological Society, 1902.
Wilson Bentley, Photograph #896 from Plate XIX from: "Studies among the snow crystals during the winter of 1901-2, with additional data collected during previous winters." Monthly Weather Review, Vol. 30, No. 3. American Meteorological Society, 1902.
Wilson Bentley, Photograph #897 from Plate XIX from: "Studies among the snow crystals during the winter of 1901-2, with additional data collected during previous winters." Monthly Weather Review, Vol. 30, No. 3. American Meteorological Society, 1902.
Wilson Bentley, Photograph #899 from Plate XIX from: "Studies among the snow crystals during the winter of 1901-2, with additional data collected during previous winters." Monthly Weather Review, Vol. 30, No. 3. American Meteorological Society, 1902.
Image by Kirstin Hendrickson.

Image by Kirstin Hendrickson.
Image available at http://commons.wikimedia.org/wiki/File:Gatorade_shower.jpg. Copyright in the public domain.
Graph by Kirstin Hendrickson. Background image by Johntex, available at http://commons.wikimedia.org/wiki/File:Gentle_waves_come_in_at_a_sandy_beach.JPG under the Creative Commons Attribution 2.5 Generic license. © 2006.
Graph by Kirstin Hendrickson. Background image by Scott Lefler. Permission for background image granted by the photographer. © 2008.
Image by Wikipedian Kbh3rd, available at http://commons.wikimedia.org/wiki/File:Ogallala_area_freshwater_ground_withdrawals_all_aquifers.png under the Creative Commons Attribution-Share Alike 3.0 Unported license. © 2009.
Image of Total Drinking Water Coverage: 2006, WHO/UNICEF Joint Monitoring Programme on Water-supply and Sanitation. © World Health Organization, 2008. Permission granted by the publisher.
Image by Scott Lefler. Permission granted by the photographer. © 2010.
Image by Kirstin Hendrickson.
Image by Scott Lefler. Permission granted by the photographer. © 2009.
Image by Nicole Koehler, available at http://commons.wikimedia.org/wiki/File:Faucet.JPG under the Creative Commons Attribution-Share Alike 3.0 Unported license. © 2009.

Chapter 5
Image by Juan Igancio Rela (JohannRela), available at http://commons.wikimedia.org/wiki/File:Pomelos_-_Grapefruits.jpg under the Creative Commons Attribution 2.0 Generic license. © 2006.
Image by Mark Sadowski, available at http://commons.wikimedia.org/wiki/File:Flickr_cc_runner_wisconsin_u.jpg under the Creative Commons Attribution-Share Alike 2.0 Generic license. © 2006.
Image available at http://commons.wikimedia.org/wiki/File:Pool-montjuic.jpg under the Creative Commons Attribution-Share Alike 3.0 Unported license. © 2008, http://www.barcel-onatips.nl/.
Image by Kirstin Hendrickson.
Image by Dan Smith, available at http://commons.wikimedia.org/wiki/File:Chocolate_chip_cookies.jpg under the Creative Commons Attribution-Share Alike 2.5 Unported license. © 2005.
Image available at http://commons.wikimedia.org/wiki/File:Ambersweet_oranges.jpg. Copyright in the public domain.
Image by Geoff Lane (Glane23), available at http://commons.wikimedia.org/wiki/File:Pringles_chips.JPG under the Creative Commons Attribution-Share Alike 3.0 Unported license. © 2008.
Image available at http://commons.wikimedia.org/wiki/File:Sirloin_steak.JPG. Copyright in the public domain.
Image available at http://commons.wikimedia.org/wiki/File:USDA_Broccolini.jpg. Copyright in the public domain.
Image by Kirstin Hendrickson.
Image available at http://commons.wikimedia.org/wiki/File:NCI_flour_tortillas.jpg. Copyright in the public domain.
Image by Fastily, available at http://commons.wikimedia.org/wiki/File:Frenchbread3000ppx.jpg under the Creative Commons Attribution-Share Alike 3.0 Unported license. © 2008.
Image by ElinorD, available at http://commons.wikimedia.org/wiki/File:Breaddough1.jpg under the Creative Commons Attribution-Share Alike 3.0 Unported license. © 2007.
Image by ElinorD, available at http://commons.wikimedia.org/wiki/File:Breaddough2.jpg under the Creative Commons Attribution-Share Alike 3.0 Unported license. © 2007.
Image by J.P.Lon, available at http://commons.wikimedia.org/wiki/File:Somerset-Cheddar.jpg under the Creative Commons Attribution-Share Alike 3.0 Unported license. © 2007.
Image by StaraBlazkova, available at http://commons.wikimedia.org/wiki/File:Emmental_015.jpg under the Creative Commons Attribution-Share Alike 3.0 Unported license. © 2007.
Image by Dominik Hundhammer, available at http://commons.wikimedia.org/wiki/File:Blue_Stilton_Quarter_Front.jpg under the Creative Commons Attribution-Share Alike 3.0 Unported license. © 2005.
Image available at http://commons.wikimedia.org/wiki/File:Skull_and_crossbones.svg. Copyright in the public domain.
Image available at http://commons.wikimedia.org/wiki/File:HAZMAT_Class_8_Corrosive.png. Copyright in the public domain.
Image by Kirstin Hendrickson.
Image available at http://commons.wikimedia.org/wiki/File:Gefahrguttransport.jpg. Copyright in the public domain.
Image by Trexer, available at http://commons.wikimedia.org/wiki/File:Weizenbier.jpg under the Creative Commons Attribution-Share Alike 3.0 Unported license. © 2005.
Image available at http://pdphoto.org/PictureDetail.php?oldpg=2479. Copyright in the public domain.

Chapter 6
Image by Michael Ertel (Wurstsalat), available at http://commons.wikimedia.org/wiki/File:Santiago30std.jpg under the Creative Commons Attribution-Share Alike 3.0 Unported license. © 2005.
Image by Richard Wheeler (Zephyris), available at http://commons.wikimedia.org/wiki/File:1GZX_Haemoglobin.png under the Creative Commons Attribution-Share Alike 3.0 Unported license. © 2007.
Image by Scott Lefler. Permission granted by the photographer. © 2010.
Image by Lamiot, available at http://commons.wikimedia.org/wiki/File:M%C3%A9taleuropNordPoussi%C3%A8rePlomb.jpg under the Creative Commons Attribution-Share Alike 3.0 Unported license. © 2002.
Image available at http://www.epa.gov/visibility/parks/bryce.html. Copyright in the public domain.
Image available at http://www.epa.gov/visibility/parks/bryce.html. Copyright in the public domain.
Image by Mila Zinkova, available at http://commons.wikimedia.org/wiki/File:Sulfur_dioxide_emissions_from_the_Halemaumau_vent_04-14-08_1.jpg under the Creative Commons Attribution-Share Alike 3.0 Unported license. © 2008.
Image by Mathiasv, available at http://commons.wikimedia.org/wiki/File:Pingyao-Kina-forurening.jpg under the Creative Commons Attribution-Share Alike 3.0 Unported license. © 2007.
Image available at http://commons.wikimedia.org/wiki/File:King_mackerel_%28_Scomberomorus_cavalla_%29.jpg. Copyright in the public domain.
Image available at http://commons.wikimedia.org/wiki/File:FEMA_-_20363_-_Photograph_by_Marvin_Nauman_taken_on_11-29-2005_in_Louisiana.jpg. Copyright in the public domain.
Image available at http://commons.wikimedia.org/wiki/File:Acid_rain_woods1.JPG. Copyright in the public domain.
National Atmospheric Deposition Program. 2009. National Atmospheric Deposition Program 2008 Annual Survey. NADP Data Report 2009-01. Illinois State Water Survey, University of Illinois at Urbana-Champaign, Champaign, IL.
Image by Nino Barbieri, available at http://commons.wikimedia.org/wiki/File:-_Acid_rain_damaged_gargoyle_-.jpg under the Creative Commons Attribution 2.5 Generic license. © 2006.
Image available at http://commons.wikimedia.org/wiki/File:1831-View-Whitechapel-Road-steam-carriage-caricature.jpg. Copyright in the public domain.
Image available at http://commons.wikimedia.org/wiki/File:Widnes_Smoke.jpg. Copyright in the public domain.
Image by N T Stobbs, available at http://commons.wikimedia.org/wiki/File:Nelson%27s_Column_during_the_Great_Smog_of_1952.jpg under the Creative Commons Attribution-Share Alike 2.0 Generic license. © 1952.
Image from the United States Environmental Protection Agency. Copyright in the public domain.
Image available at http://www.airnow.gov/index.cfm?action=aqibasics.aqi. Copyright in the public domain.
Image available at http://www.airnow.gov/index.cfm?action=aqibasics.aqi. Copyright in the public domain.
Image by Wapcaplet, available at http://commons.wikimedia.org/wiki/File:Four_stroke_cycle_intake.jpg under the Creative Commons Attribution-Share Alike 3.0 Unported license. © 2004.
Image by Wapcaplet, available at http://commons.wikimedia.org/wiki/File:Four_stroke_cycle_compression.jpg under the Creative Commons Attribution-Share Alike 3.0 Unported license. © 2004.
Image by Wapcaplet, available at http://commons.wikimedia.org/wiki/File:Four_stroke_cycle_spark.jpg under the Creative Commons Attribution-Share Alike 3.0 Unported license. © 2004.

Image by Wapcaplet, available at http://commons.wikimedia.org/wiki/File:Four_stroke_cycle_exhaust.jpg under the Creative Commons Attribution-Share Alike 3.0 Unported license. © 2004.
Image by Magnus Manske, available at http://commons.wikimedia.org/wiki/File:P1000483_Nicotiana_tabacum_%28tobacco%29_%28Solanaceae%29_Leaf.JPG under the Creative Commons Attribution-Share Alike 3.0 Unported license. © 2004.
Image available at http://commons.wikimedia.org/wiki/File:Spitkid.jpg. Copyright in the public domain.

Chapter 7

Image by Willow, available at http://commons.wikimedia.org/wiki/File:Salix_alba_004.jpg under the Creative Commons Attribution-Share Alike 2.5 Generic license. © 2007.
Bayer Aspirin Advertisement. The New York Times, February 19, 1917. Pp. 6. The New York Times Company, 1917. Copyright in the public domain.
Image by Forest & Kim Starr, available at http://commons.wikimedia.org/wiki/File:Starr_020518-0012_Cinchona_pubescens.jpg under the Creative Commons Attribution 3.0 Unported license. © 2007.
Image by Tracy Ducasse (ms. Tea), available at http://commons.wikimedia.org/wiki/File:Foxglove.jpg under the Creative Commons Attribution 2.0 Generic license. © 2005.
Image by Walter Siegmund, available at http://commons.wikimedia.org/wiki/File:PacificYew_8544.jpg under the Creative Commons Attribution 2.5 Generic license. © 2005.
Image by Kirstin Hendrickson.
Image by Kirstin Hendrickson.
Image by Kirstin Hendrickson.
Image available at http://commons.wikimedia.org/wiki/File:Elixir_Sulfanilamide.jpg. Copyright in the public domain.
Konstantinos C. Soultanis, Alexandros H. Payatakes, Vasilios T. Chouliaras, Georgios C. Mandellos, Nikolaos E. Pyrovolou, Fani M. Pliarchopoulou, & Panayotis N. Soucacos. Figure 8-A from: "Rare Causes of Scoliosis and Spine Deformity: Experience and Particular Features." Scoliosis, Vol. 2, No. 15. © BioMed Central Ltd., 2007. Image available under the Creative Commons Attribution 2.0 Generic license.
Image 7.10 is an adaptation by Kirstin Hendrickson, available under the Creative Commons Attribution-Share Alike 3.0 Unported license. Adapted material: 1. Baseball image by Tage Olsin, available at http://commons.wikimedia.org/wiki/File:Baseball_%28crop%29.jpg under the Creative Commons Attribution-Share Alike 2.0 Generic license. © 2006. 2. Image by Krzysztof P. Jasiutowicz, available at http://commons.wikimedia.org/wiki/File:Raczka4.jpg under the Creative Commons Attribution-Share Alike 3.0 Unported license. © 2005.
Image by Kirstin Hendrickson.
Image by Kirstin Hendrickson.
Image by Han-Kwang Nienhuys (Hankwang), available at http://commons.wikimedia.org/wiki/File:Amanita_phalloides.jpg under the Creative Commons Attribution 1.0 Generic license. © 2004.
Image by Michael Gäbler, available at http://commons.wikimedia.org/wiki/File:Dendrobates_azureus_%28Dendrobates_tinctorius%29_Edit.jpg under the Creative Commons Attribution 3.0 Unported license. © 2009.
Image available at http://commons.wikimedia.org/wiki/File:Castor_beans.jpg. Copyright in the public domain.

Chapter 8

Image by Scott Lefler. Permission granted by the photographer. © 2009.
Image by avlxyz (Alpha), available at http://commons.wikimedia.org/wiki/File:BBQ_Dinner.jpg under the Creative Commons Attribution-Share Alike 2.0 Generic license. © 2005.
Image by Kirstin Hendrickson.
Image by Kirstin Hendrickson.
Image by Richard Wheeler (Zephyris), available at http://commons.wikimedia.org/wiki/File:1GZX_Haemoglobin.png under the Creative Commons Attribution-Share Alike 3.0 Unported license. © 2007.
Image by Nevit Dilmen, available at http://commons.wikimedia.org/wiki/File:1K6F_Crystal_Structure_Of_The_Collagen_Triple_Helix_Model_Pro-_Pro-Gly103_04.png under the Creative Commons Attribution-Share Alike 3.0 Unported license. © 2002.
Image by Kirstin Hendrickson.
Image by Kirstin Hendrickson.
Image available at http://commons.wikimedia.org/wiki/File:Salmon.JPG. Copyright in the public domain.
Image by Badagnani, available at http://commons.wikimedia.org/wiki/File:Blackbeanstew.jpg under the Creative Commons Attribution 3.0 Unported license. © 2008.
Image by Paddy, available at http://commons.wikimedia.org/wiki/File:Boiled_eggs.jpg under the Creative Commons Attribution-Share Alike 3.0 Unported license. © 2004.
Image by Kirstin Hendrickson.
Image available at http://commons.wikimedia.org/wiki/File:Runny_hunny.jpg. Copyright in the public domain.
Image available at http://commons.wikimedia.org/wiki/File:La_Boqueria.JPG. Copyright in the public domain.
Image by Daniel J. Layton, available at http://commons.wikimedia.org/wiki/File:Homemade_potato_bread,_half.jpg under the Creative Commons Attribution-Share Alike 3.0 Unported license. © 2007.
Image by Kirstin Hendrickson.
Image by Kirstin Hendrickson.
Image by Kirstin Hendrickson.
Image by Fritzs, available at http://commons.wikimedia.org/wiki/File:Raw_cane_sugar_light.JPG under the Creative Commons Attribution-Share Alike 3.0 Unported license. © 2009.
Image by Henningklevjer, available at http://commons.wikimedia.org/wiki/File:JasmineRice.jpg under the Creative Commons Attribution-Share Alike 2.5 Generic license. © 2006.
Image by Fir0002, available at http://commons.wikimedia.org/wiki/File:Brown_rice.jpg under the Creative Commons Attribution-Share Alike 3.0 Unported license. © 2006.
Image by Alex Ex, available at http://commons.wikimedia.org/wiki/File:Italian_olive_oil_2007.jpg under the Creative Commons Attribution 2.5 Generic license. © 2007.
Image by M.Verkerk, J.J.G.Claessens, available at http://commons.wikimedia.org/wiki/File:Almonds.png under the Creative Commons Attribution-Share Alike 2.5 Generic license. © 2006.
Image available at http://pdphoto.org/PictureDetail.php?mat=pdef&pg=7525. Copyright in the public domain.
Image by Kirstin Hendrickson.
Image by Kirstin Hendrickson.
Image by Kirstin Hendrickson.
Image by Kirstin Hendrickson.
Image by Kirstin Hendrickson.
Image by Kirstin Hendrickson.

Chapter 9

Images from Centers for Disease Control and Prevention. Copyright in the public domain.
Image available at http://commons.wikimedia.org/wiki/File:USDA_Food_Pyramid.gif. Copyright in the public domain.
Image available at http://commons.wikimedia.org/wiki/File:MyPyramid1.png. Copyright in the public domain.
Image available at http://commons.wikimedia.org/wiki/File:US_Nutritional_Fact_Label_2.svg. Copyright in the public domain.
Image by Kirstin Hendrickson.
Image by Kirstin Hendrickson.
Image by Jean François Gaffard (Jeffdelonge), available at http://commons.wikimedia.org/wiki/File:Ephedra_distachya.jpg under the Creative Commons Attribution-Share Alike 3.0 Unported license.

Image by Kirstin Hendrickson.
Image by Kirstin Hendrickson.
Image by Kirstin Hendrickson.
Image by Geoff Martin, available at http://commons.wikimedia.org/wiki/File:Saccharin_warning_drpepper_gfdl.jpg under the Creative Commons Attribution-Share Alike 3.0 Unported license. © 2005.
Image by Kirstin Hendrickson.
Image by Kirstin Hendrickson.
Image by Kirstin Hendrickson.
Image by Franz Richter, available at http://commons.wikimedia.org/wiki/File:Ulrike_Maisch_6661.jpg under the Creative Commons Attribution-Share Alike 3.0 Unported license. © 2009.
Image by Paul Foot, available at http://commons.wikimedia.org/wiki/File:Dwain_Chambers_at_Olympic_Trials_2008_02.jpg under the Creative Commons Attribution-Share Alike 2.0 Generic license. © 2008.
Image available at http://commons.wikimedia.org/wiki/File:Myoglobin.png. Copyright in the public domain.
Image by Ervín Pospíšil, available at http://commons.wikimedia.org/wiki/File:Hen_with_chickens_in_native_breeding.jpg under the Creative Commons Attribution-Share Alike 3.0 Unported license. © 2006.
Image by Martin Correns, available at http://commons.wikimedia.org/wiki/File:Flying_mallard_duck_-_female.jpg under the Creative Commons Attribution-Share Alike 3.0 Unported license. © 2005.

Chapter 10
Image by Kirstin Hendrickson.
Image by Kirstin Hendrickson.
Image by WillowW, available at http://commons.wikimedia.org/wiki/File:Visible_EM_modes.png under the Creative Commons Attribution-Share Alike 3.0 Unported license. © 2006.
Image adapted by Kirstin Hendrickson, available under the Creative Commons Attribution-Share Alike 3.0 Unported license. Original image by WillowW, available at http://commons.wikimedia.org/wiki/File:Visible_EM_modes.png under the Creative Commons Attribution-Share Alike 3.0 Unported license. © 2006.
Image edited by Kirstin Hendrickson, available under the Creative Commons Attribution-Share Alike 3.0 Unported license. Original image by WillowW, available at http://commons.wikimedia.org/wiki/File:Visible_EM_modes.png under the Creative Commons Attribution-Share Alike 3.0 Unported license. © 2006.
Image by Philip Ronan, available at http://commons.wikimedia.org/wiki/File:EM_spectrum.svg under the Creative Commons Attribution-Share Alike 3.0 Unported license. © 2007.
Image by Kirstin Hendrickson.
Image by Kirstin Hendrickson.
Image by Kirstin Hendrickson.
Figure 7.1: The distribution of atmospheric ozone in partial pressure as a function of altitude, "1999 EOS Science Plan (Revised)." Pp. 311. NASA: National Aeronautics and Space Administration, 1999. Copyright in the public domain.
Image by Kirstin Hendrickson.
Image by Kirstin Hendrickson.
Image available at http://commons.wikimedia.org/wiki/File:Min_ozone.jpg. Copyright in the public domain.
Image available at http://commons.wikimedia.org/wiki/File:Largest_ever_Ozone_hole_sept2000_with_scale.jpg. Copyright in the public domain.
Ross Salawitch, Arctic Stratospheric Cloud Image from: "Arctic Ozone Loss More Sensitive to Climate Change Than Thought." http://www.jpl.nasa.gov/releases/2004/108.cfm. NASA: National Aeronautics and Space Administration, 2004. Copyright in the public domain.
Image by Kirstin Hendrickson.
Image available at "The Ozone Layer If CFCs Hadn't Been Banned." http://earthobservatory.nasa.gov/IOTD/view.php?id=38685. NASA: National Aeronautics and Space Administration, 2009. Copyright in the public domain.
Image by Kirstin Hendrickson.
Image by Kirstin Hendrickson.
Image available at http://commons.wikimedia.org/wiki/File:Nitrogen_Cycle.jpg. Copyright in the public domain.
Image by Robin Fernandes (rewbs.soal), available at http://commons.wikimedia.org/wiki/File:Humberstone.jpg under the Creative Commons Attribution-Share Alike 2.0 Generic license.
Image available at http://commons.wikimedia.org/wiki/File:DDT_WWII_soldier.jpg. Copyright in the public domain.
Image edited by Kirstin Hendrickson. Original image available at http://commons.wikimedia.org/wiki/File:Chlorophyll_ab_spectra.png. Copyright of original image in the public domain.
Image by Scott Lefler. Permission granted by the photographer. © 2007.
Image by Scott Lefler. Permission granted by the photographer. © 2007.

Chapter 11
Image by Geierunited, available at http://commons.wikimedia.org/wiki/File:Wide-lightening.jpg under the Creative Commons Attribution-Share Alike 3.0 Unported license. © 2005.
Image by Tony Boon, available at http://commons.wikimedia.org/wiki/File:Power_Lines.jpg. © 2004.
Image by MetalPhoeniX, available at http://commons.wikimedia.org/wiki/File:Four_AA_batteries.jpg under the Creative Commons Attribution 2.0 Generic license. © 2005.
Image available at http://commons.wikimedia.org/wiki/File:Copper-Architecture.jpg under the Creative Commons Attribution-Share Alike 3.0 Unported license. © 2009, Rutland Gutter Supply, Inc.
Image by S-8500, available at http://commons.wikimedia.org/wiki/File:Toyota_Toyopet_Mark_II_1900_-_ja_-e.jpg under the Creative Commons Attribution-Share Alike 3.0 Unported license. © 2007.
Egbert Andrews Millikan & Henry Gordon Gale, "Portrait of physicist Alessandro Volta." Practical Physics. Ginn and Company, 1920. Copyright in the public domain.
Image available at http://commons.wikimedia.org/wiki/File:Luigi_Galvani,_oil-painting.jpg. Copyright in the public domain.
Image available at http://en.wikipedia.org/wiki/File:Galvani-frog-legs.PNG. Copyright in the public domain.
Image by Kirstin Hendrickson.
Image by Kirstin Hendrickson.
Image by Kirstin Hendrickson.
Image available at http://commons.wikimedia.org/wiki/File:Giant_photovoltaic_array.jpg. Copyright in the public domain.
Image by Kirstin Hendrickson.
Image available at http://commons.wikimedia.org/wiki/File:Moody_Sunburst.jpg. Copyright in the public domain.
Image by Tobi 87, available at http://commons.wikimedia.org/wiki/File:Hoover_Dam-USA.jpg under the Creative Commons Attribution-Share Alike 3.0 Unported license. © 2007.
Image edited by Kirstin Hendrickson. Original image available at http://commons.wikimedia.org/wiki/File:Chlorophyll_a_b_d.svg. Copyright of original image in the public domain.
Image by Kirstin Hendrickson.
Image by Kirstin Hendrickson.
Image by Kirstin Hendrickson.
Image by Kirstin Hendrickson.
Image by Kirstin Hendrickson.
Image by Kirstin Hendrickson.

Image available at http://www.erh.noaa.gov/er/lwx/lightning/tstm-lgtg-fromside.jpg. Copyright in the public domain.
Image by Kirstin Hendrickson.

Chapter 12
Image available at http://commons.wikimedia.org/wiki/File:Trinity_%28black_and_white%29.jpg. Copyright in the public domain.
Image by Stefan Kühn, available at http://commons.wikimedia.org/wiki/File:Nuclear_Power_Plant_Cattenom.jpg under the Creative Commons Attribution-Share Alike 3.0 Unported license. © 2005.
Image available at http://commons.wikimedia.org/wiki/File:Radioactivity_symbol.png. Copyright in the public domain.
Image available at http://commons.wikimedia.org/wiki/File:JosephWright-Alchemist.jpg. Copyright in the public domain.
Image available at http://commons.wikimedia.org/wiki/File:Solvay_conference_1927.jpg. Copyright in the public domain.
Image by Kirstin Hendrickson.
Image by Kirstin Hendrickson.
Image 12.6 is an adaptation by Kirstin Hendrickson, available under the Creative Commons Attribution-Share Alike 3.0 Unported license. Adapted material: 1. Image by HowardMorland, available at http://commons.wikimedia.org/wiki/File:Little_Boy_Internal_Components_%28no_labels%29.png under the GNU Free Documentation License. © 2009. Copy of GNU Free Documentation License follows credits. 2. Image by Dake, available at http://commons.wikimedia.org/wiki/File:Fatman.svg under the Creative Commons Attribution-Share Alike 3.0 Unported license. © 2005.
Image available at http://commons.wikimedia.org/wiki/File:Trinity_explosion.jpg. Copyright in the public domain.
Image available at http://commons.wikimedia.org/wiki/File:TrinitySiteHistoricalMarkerHighwaySign.jpg. Copyright in the public domain.
Image edited by Kirstin Hendrickson. Original image available at http://commons.wikimedia.org/wiki/File:Teller-Ulam_device_3D.svg. Copyright of original image in the public domain.
Image available at http://commons.wikimedia.org/wiki/File:Nuclear_fuel_pellets.jpeg. Copyright in the public domain.
Image available at http://commons.wikimedia.org/wiki/File:Nuclear_fuel_element.jpg. Copyright in the public domain.
Simple Schematic of the Three Mile Island Unit 2 Nuclear Power Plant, "U.S. Nuclear Regulatory Commission Fact Sheet on the Accident at Three Mile Island." United States Nuclear Regulatory Commission, 2009. Copyright in the public domain.
Image by Graevemoore, available at http://upload.wikimedia.org/wikipedia/commons/thumb/4/46/LMFBR_schematics2.svg/1000px-LMFBR_schematics2.svg.png under the Creative Commons Attribution-Share Alike 3.0 Unported license. © 2008.
Image by Kirstin Hendrickson.
Image by Finlay McWalter, available at http://commons.wikimedia.org/wiki/File:Wfm_area51_map_en.png under the Creative Commons Attribution-Share Alike 3.0 Unported license.
Image by Mond, available at http://commons.wikimedia.org/wiki/File:Chernobyl_Nuclear_Power_Plant.jpg under the Creative Commons Attribution-Share Alike 3.0 Unported license. © 2007.
Image by Elena Filatova, available at www.elenafilatova.com. © 1986.
Image available at http://commons.wikimedia.org/wiki/File:Gamma_Knife_Graphic.jpg. Copyright in the public domain.
Graph by Kirstin Hendrickson, available under the Creative Commons Attribution-Share Alike 3.0 Unported license. Background image by Thomas Ihle, available at http://commons.wikimedia.org/wiki/File:Tarbosaurus_museum_Muenster.jpg under the Creative Commons Attribution-Share Alike 3.0 Unported license. © 2004.
Image by Kirstin Hendrickson.

Chapter 13
Image available at http://commons.wikimedia.org/wiki/File:World_energy_use_projections.png. Copyright in the public domain.
Image available at http://commons.wikimedia.org/wiki/File:Sources_of_US_energy_production_%28historic%29.gif. Copyright in the public domain.
Image by Kirstin Hendrickson.
Image by Kirstin Hendrickson.
Image by Kirstin Hendrickson.
Image by Kirstin Hendrickson.
Image by Kirstin Hendrickson.
Image by Mak Thorpe, available at http://commons.wikimedia.org/wiki/File:Global_Carbon_Emission_by_Type_to_Y2004.png under the Creative Commons Attribution-Share Alike 3.0 Unported license. © 2008.
Image available at http://commons.wikimedia.org/wiki/File:Russian_station_Vostok.jpg. Copyright in the public domain.
Image available at http://commons.wikimedia.org/wiki/File:GISP2_team_photo_core37.jpeg. Copyright in the public domain.
Image available at http://commons.wikimedia.org/wiki/File:GISP2D1837_crop.jpg. Copyright in the public domain.
Image 13.10 is an adaptation by Kirstin Hendrickson, available under the Creative Commons Attribution-Share Alike 3.0 Unported license. Adapted material by Maksim, available at http://commons.wikimedia.org/wiki/File:Vostok-ice-core-petit.png under the Creative Commons Attribution-Share Alike 3.0 Unported license. © 2006.
Image 13.11 is an adaptation by Kirstin Hendrickson, available under the Creative Commons Attribution-Share Alike 3.0 Unported license. Adapted material by Maksim, available at http://commons.wikimedia.org/wiki/File:Vostok-ice-core-petit.png under the Creative Commons Attribution-Share Alike 3.0 Unported license. © 2006.
Image 13.12 is an adaptation by Kirstin Hendrickson, available under the Creative Commons Attribution-Share Alike 3.0 Unported license. Adapted material by Maksim, available at http://commons.wikimedia.org/wiki/File:Vostok-ice-core-petit.png under the Creative Commons Attribution-Share Alike 3.0 Unported license. © 2006.
Image available at http://commons.wikimedia.org/wiki/File:CO2-Mauna-Loa.png. Copyright in the public domain.
Image 13.14 is an adaptation by Kirstin Hendrickson, available under the Creative Commons Attribution-Share Alike 3.0 Unported license. Adapted material: 1. Image by Maksim, available at http://commons.wikimedia.org/wiki/File:Vostok-ice-core-petit.png under the Creative Commons Attribution-Share Alike 3.0 Unported license. © 2006. 2. Image available at http://commons.wikimedia.org/wiki/File:CO2-417k.pngG. Copyright in the public domain.
Image available at http://commons.wikimedia.org/wiki/File:Instrumental_Temperature_Record.svg. Copyright in the public domain.
Image edited by Kirstin Hendrickson. Original image available at http://commons.wikimedia.org/wiki/File:Instrumental_Temperature_Record.svg. Copyright of original image in the public domain.
Image by Kirstin Hendrickson.
Image by Giorgiogp2, available at http://commons.wikimedia.org/wiki/File:RSS_troposphere_stratosphere_trend.png under the Creative Commons Attribution-Share Alike 3.0 Unported license. © 2010.
Image by Leland McInnes, available at http://commons.wikimedia.org/wiki/File:Temp-sunspot-co2.svg under the Creative Commons Attribution-Share Alike 3.0 Unported license. © 2007.
Gerald A. Meehl, Warren M. Washington, Caspar M. Ammann, Julie M. Arblaster, T. M. L. Wigley, & Claudia Tebaldi, "Combinations of Natural and Anthropogenic Forcings in Twentieth-Century Climate." Journal of Climate, Vol. 17, Issue 19. Pp. 3722. © American Meteorological Society, 2004. Permission granted by the publisher.
Gerald A. Meehl, Warren M. Washington, Caspar M. Ammann, Julie M. Arblaster, T. M. L. Wigley, & Claudia Tebaldi, "Combinations of Natural and Anthropogenic Forcings in Twentieth-Century Climate." Journal of Climate, Vol. 17, Issue 19. Pp. 3722. © American Meteorological Society, 2004. Permission granted by the publisher.
Gerald A. Meehl, Warren M. Washington, Caspar M. Ammann, Julie M. Arblaster, T. M. L. Wigley, & Claudia Tebaldi, "Combinations of Natural and Anthropogenic Forcings in Twentieth-Century Climate." Journal of Climate, Vol. 17, Issue 19. Pp. 3722. © American Meteorological Society, 2004. Permission granted by the publisher.
Image by Emturan, available at http://commons.wikimedia.org/wiki/File:Kyoto_Protocol_participation_map_2009.png under the Creative Commons Attribution-Share Alike 3.0 Unported license. © 2009.
Image available at http://commons.wikimedia.org/wiki/File:Measuring_switchgrass.usda.jpg. Copyright in the public domain.
Graph by Kirstin Hendrickson. Background image by Tony Boon, available at http://commons.wikimedia.org/wiki/File:Power_Lines.jpg. © 2004.

GNU Free Documentation License

Version 1.3, 3 November 2008

Copyright © 2000, 2001, 2002, 2007, 2008 Free Software Foundation, Inc. <http://fsf.org/>

Everyone is permitted to copy and distribute verbatim copies of this license document, but changing it is not allowed.

0. PREAMBLE

The purpose of this License is to make a manual, textbook, or other functional and useful document "free" in the sense of freedom: to assure everyone the effective freedom to copy and redistribute it, with or without modifying it, either commercially or noncommercially. Secondarily, this License preserves for the author and publisher a way to get credit for their work, while not being considered responsible for modifications made by others.

This License is a kind of "copyleft", which means that derivative works of the document must themselves be free in the same sense. It complements the GNU General Public License, which is a copyleft license designed for free software.

We have designed this License in order to use it for manuals for free software, because free software needs free documentation: a free program should come with manuals providing the same freedoms that the software does. But this License is not limited to software manuals; it can be used for any textual work, regardless of subject matter or whether it is published as a printed book. We recommend this License principally for works whose purpose is instruction or reference.

1. APPLICABILITY AND DEFINITIONS

This License applies to any manual or other work, in any medium, that contains a notice placed by the copyright holder saying it can be distributed under the terms of this License. Such a notice grants a world-wide, royalty-free license, unlimited in duration, to use that work under the conditions stated herein. The "Document", below, refers to any such manual or work. Any member of the public is a licensee, and is addressed as "you." You accept the license if you copy, modify or distribute the work in a way requiring permission under copyright law.

A "Modified Version" of the Document means any work containing the Document or a portion of it, either copied verbatim, or with modifications and/or translated into another language.

A "Secondary Section" is a named appendix or a front-matter section of the Document that deals exclusively with the relationship of the publishers or authors of the Document to the Document's overall subject (or to related matters) and contains nothing that could fall directly within that overall subject. (Thus, if the Document is in part a textbook of mathematics, a Secondary Section may not explain any mathematics.) The relationship could be a matter of historical connection with the subject or with related matters, or of legal, commercial, philosophical, ethical or political position regarding them.

The "Invariant Sections" are certain Secondary Sections whose titles are designated, as being those of Invariant Sections, in the notice that says that the Document is released under this License. If a section does not fit the above definition of Secondary then it is not allowed to be designated as Invariant. The Document may contain zero Invariant Sections. If the Document does not identify any Invariant Sections then there are none.

The "Cover Texts" are certain short passages of text that are listed, as Front-Cover Texts or Back-Cover Texts, in the notice that says that the Document is released under this License. A Front-Cover Text may be at most 5 words, and a Back-Cover Text may be at most 25 words.

A "Transparent" copy of the Document means a machine-readable copy, represented in a format whose specification is available to the general public, that is suitable for revising the document straightforwardly with generic text editors or (for images composed of pixels) generic paint programs or (for drawings) some widely available drawing editor, and that is suitable for input to text formatters or for automatic translation to a variety of formats suitable for input to text formatters. A copy made in an otherwise Transparent file format whose markup, or absence of markup, has been arranged to thwart or discourage subsequent modification by readers is not Transparent. An image format is not Transparent if used for any substantial amount of text. A copy that is not "Transparent" is called "Opaque."

Examples of suitable formats for Transparent copies include plain ASCII without markup, Texinfo input format, LaTeX input format, SGML or XML using a publicly available DTD, and standard-conforming simple HTML, PostScript or PDF designed for human modification. Examples of transparent image formats include PNG, XCF and JPG. Opaque formats include proprietary formats that can be read and edited only by proprietary word processors, SGML or XML for which the DTD and/or processing tools are not generally available, and the machine-generated HTML, PostScript or PDF produced by some word processors for output purposes only.

The "Title Page" means, for a printed book, the title page itself, plus such following pages as are needed to hold, legibly, the material this License requires to appear in the title page. For works in formats which do not have any title page as such, "Title Page" means the text near the most prominent appearance of the work's title, preceding the beginning of the body of the text.

The "publisher" means any person or entity that distributes copies of the Document to the public.

A section "Entitled XYZ" means a named subunit of the Document whose title either is precisely XYZ or contains XYZ in parentheses following text that translates XYZ in another language. (Here XYZ stands for a specific section name mentioned below, such as "Acknowledgements", "Dedications", "Endorsements", or "History.") To "Preserve the Title" of such a section when you modify the Document means that it remains a section "Entitled XYZ" according to this definition.

The Document may include Warranty Disclaimers next to the notice which states that this License applies to the Document. These Warranty Disclaimers are considered to be included by reference in this License, but only as regards disclaiming warranties: any other implication that these Warranty Disclaimers may have is void and has no effect on the meaning of this License.

2. VERBATIM COPYING

You may copy and distribute the Document in any medium, either commercially or noncommercially, provided that this License, the copyright notices, and the license notice saying this License applies to the Document are reproduced in all copies, and that you add no other conditions whatsoever to those of this License. You may not use technical measures to obstruct or control the reading or further copying of the copies you make or distribute. However, you may accept compensation in exchange for copies. If you distribute a large enough number of copies you must also follow the conditions in section 3.

You may also lend copies, under the same conditions stated above, and you may publicly display copies.

3. COPYING IN QUANTITY

If you publish printed copies (or copies in media that commonly have printed covers) of the Document, numbering more than 100, and the Document's license notice requires Cover Texts, you must enclose the copies in covers that carry, clearly and legibly, all these Cover Texts: Front-Cover Texts on the front cover, and Back-Cover Texts on the back cover. Both covers must also clearly and legibly identify you as the publisher of these copies. The front cover must present the full title with all words of the title equally prominent and visible. You may add other material on the covers in addition. Copying with changes limited to the covers, as long as they preserve the title of the Document and satisfy these conditions, can be treated as verbatim copying in other respects.

If the required texts for either cover are too voluminous to fit legibly, you should put the first ones listed (as many as fit reasonably) on the actual cover, and continue the rest onto adjacent pages.

If you publish or distribute Opaque copies of the Document numbering more than 100, you must either include a machine-readable Transparent copy along with each Opaque copy, or state in or with each Opaque copy a computer-network location from which the general network-using public has access to download using public-standard network protocols a complete Transparent copy of the Document, free of added material. If you use the latter option, you must take reasonably prudent steps, when you begin distribution of Opaque copies in quantity, to ensure that this Transparent copy will remain thus accessible at the stated location until at least one year after the last time you distribute an Opaque copy (directly or through your agents or retailers) of that edition to the public.

It is requested, but not required, that you contact the authors of the Document well before redistributing any large number of copies, to give them a chance to provide you with an updated version of the Document.

4. MODIFICATIONS

You may copy and distribute a Modified Version of the Document under the conditions of sections 2 and 3 above, provided that you release the Modified Version under precisely this License, with the Modified Version filling the role of the Document, thus licensing distribution and modification of the Modified Version to whoever possesses a copy of it. In addition, you must do these things in the Modified Version:

•A. Use in the Title Page (and on the covers, if any) a title distinct from that of the Document, and from those of previous versions (which should, if there were any, be listed in the History section of the Document). You may use the same title as a previous version if the original publisher of that version gives permission.

•B. List on the Title Page, as authors, one or more persons or entities responsible for authorship of the modifications in the Modified Version, together with at least five of the principal authors of the Document (all of its principal authors, if it has fewer than five), unless they release you from this requirement.

•C. State on the Title page the name of the publisher of the Modified Version, as the publisher.

•D. Preserve all the copyright notices of the Document.

•E. Add an appropriate copyright notice for your modifications adjacent to the other copyright notices.

•F. Include, immediately after the copyright notices, a license notice giving the public permission to use the Modified Version under the terms of this License, in the form shown in the Addendum below.

•G. Preserve in that license notice the full lists of Invariant Sections and required Cover Texts given in the Document's license notice.

•H. Include an unaltered copy of this License.

•I. Preserve the section Entitled "History", Preserve its Title, and add to it an item stating at least the title, year, new authors, and publisher of the Modified Version as given on the Title Page. If there is no section Entitled "History" in the Document, create one stating the title, year, authors, and publisher of the Document as given on its Title Page, then add an item describing the Modified Version as stated in the previous sentence.

•J. Preserve the network location, if any, given in the Document for public access to a Transparent copy of the Document, and likewise the network locations given in the Document for previous versions it was based on. These may be placed in the "History" section. You may omit a network location for a work that was published at least four years before the Document itself, or if the original publisher of the version it refers to gives permission.

•K. For any section Entitled "Acknowledgements" or "Dedications", Preserve the Title of the section, and preserve in the section all the substance and tone of each of the contributor acknowledgements and/or dedications given therein.

•L. Preserve all the Invariant Sections of the Document, unaltered in their text and in their titles. Section numbers or the equivalent are not considered part of the section titles.

•M. Delete any section Entitled "Endorsements." Such a section may not be included in the Modified Version.

•N. Do not retitle any existing section to be Entitled "Endorsements" or to conflict in title with any Invariant Section.

•O. Preserve any Warranty Disclaimers.

If the Modified Version includes new front-matter sections or appendices that qualify as Secondary Sections and contain no material copied from the Document, you may at your option designate some or all of these sections as invariant. To do this, add their titles to the list of Invariant Sections in the Modified Version's license notice. These titles must be distinct from any other section titles.

You may add a section Entitled "Endorsements", provided it contains nothing but endorsements of your Modified Version by various parties—for example, statements of peer review or that the text has been approved by an organization as the authoritative definition of a standard.

You may add a passage of up to five words as a Front-Cover Text, and a passage of up to 25 words as a Back-Cover Text, to the end of the list of Cover Texts in the Modified Version. Only one passage of Front-Cover Text and one of Back-Cover Text may be added by (or through arrangements made by) any one entity. If the Document already includes a cover text for the same cover, previously added by you or by arrangement made by the same entity you are acting on behalf of, you may not add another; but you may replace the old one, on explicit permission from the previous publisher that added the old one.

The author(s) and publisher(s) of the Document do not by this License give permission to use their names for publicity for or to assert or imply endorsement of any Modified Version.

5. COMBINING DOCUMENTS

You may combine the Document with other documents released under this License, under the terms defined in section 4 above for modified versions, provided that you include in the combination all of the Invariant Sections of all of the original documents, unmodified, and list them all as Invariant Sections of your combined work in its license notice, and that you preserve all their Warranty Disclaimers.

The combined work need only contain one copy of this License, and multiple identical Invariant Sections may be replaced with a single copy. If there are multiple Invariant Sections with the same name but different contents, make the title of each such section unique by adding at the end of it, in parentheses, the name of the original author or publisher of that section if known, or else a unique number. Make the same adjustment to the section titles in the list of Invariant Sections in the license notice of the combined work.

In the combination, you must combine any sections Entitled "History" in the various original documents, forming one section Entitled "History"; likewise combine any sections Entitled "Acknowledgements", and any sections Entitled "Dedications." You must delete all sections Entitled "Endorsements."

6. COLLECTIONS OF DOCUMENTS

You may make a collection consisting of the Document and other documents released under this License, and replace the individual copies of this License in the various documents with a single copy that is included in the collection, provided that you follow the rules of this License for verbatim copying of each of the documents in all other respects.

You may extract a single document from such a collection, and distribute it individually under this License, provided you insert a copy of this License into the extracted document, and follow this License in all other respects regarding verbatim copying of that document.

7. AGGREGATION WITH INDEPENDENT WORKS

A compilation of the Document or its derivatives with other separate and independent documents or works, in or on a volume of a storage or distribution medium, is called an "aggregate" if the copyright resulting from the compilation is not used to limit the legal rights of the compilation's users beyond what the individual works permit. When the Document is included in an aggregate, this License does not apply to the other works in the aggregate which are not themselves derivative works of the Document.

If the Cover Text requirement of section 3 is applicable to these copies of the Document, then if the Document is less than one half of the entire aggregate, the Document's Cover Texts may be placed on covers that bracket the Document within the aggregate, or the electronic equivalent of covers if the Document is in electronic form. Otherwise they must appear on printed covers that bracket the whole aggregate.

8. TRANSLATION

Translation is considered a kind of modification, so you may distribute translations of the Document under the terms of section 4. Replacing Invariant Sections with translations requires special permission from their copyright holders, but you may include translations of some or all Invariant Sections in addition to the original versions of these Invariant Sections. You may include a translation of this License, and all the license notices in the Document, and any Warranty Disclaimers, provided that you also include the original English version of this License and the original versions of those notices and disclaimers. In case of a disagreement between the translation and the original version of this License or a notice or disclaimer, the original version will prevail.

If a section in the Document is Entitled "Acknowledgements", "Dedications", or "History", the requirement (section 4) to Preserve its Title (section 1) will typically require changing the actual title.

9. TERMINATION

You may not copy, modify, sublicense, or distribute the Document except as expressly provided under this License. Any attempt otherwise to copy, modify, sublicense, or distribute it is void, and will automatically terminate your rights under this License.

However, if you cease all violation of this License, then your license from a particular copyright holder is reinstated (a) provisionally, unless and until the copyright holder explicitly and finally terminates your license, and (b) permanently, if the copyright holder fails to notify you of the violation by some reasonable means prior to 60 days after the cessation.

Moreover, your license from a particular copyright holder is reinstated permanently if the copyright holder notifies you of the violation by some reasonable means, this is the first time you have received notice of violation of this License (for any work) from that copyright holder, and you cure the violation prior to 30 days after your receipt of the notice.

Termination of your rights under this section does not terminate the licenses of parties who have received copies or rights from you under this License. If your rights have been terminated and not permanently reinstated, receipt of a copy of some or all of the same material does not give you any rights to use it.

10. FUTURE REVISIONS OF THIS LICENSE

The Free Software Foundation may publish new, revised versions of the GNU Free Documentation License from time to time. Such new versions will be similar in spirit to the present version, but may differ in detail to address new problems or concerns. See http://www.gnu.org/copyleft/.

Each version of the License is given a distinguishing version number. If the Document specifies that a particular numbered version of this License "or any later version" applies to it, you have the option of following the terms and conditions either of that specified version or of any later version that has been published (not as a draft) by the Free Software Foundation. If the Document does not specify a version number of this License, you may choose any version ever published (not as a draft) by the Free Software Foundation. If the Document specifies that a proxy can decide which future versions of this License can be used, that proxy's public statement of acceptance of a version permanently authorizes you to choose that version for the Document.

11. RELICENSING

"Massive Multiauthor Collaboration Site" (or "MMC Site") means any World Wide Web server that publishes copyrightable works and also provides prominent facilities for anybody to edit those works. A public wiki that anybody can edit is an example of such a server. A "Massive Multiauthor Collaboration" (or "MMC") contained in the site means any set of copyrightable works thus published on the MMC site.

"CC-BY-SA" means the Creative Commons Attribution-Share Alike 3.0 license published by Creative Commons Corporation, a not-for-profit corporation with a principal place of business in San Francisco, California, as well as future copyleft versions of that license published by that same organization.

"Incorporate" means to publish or republish a Document, in whole or in part, as part of another Document.

An MMC is "eligible for relicensing" if it is licensed under this License, and if all works that were first published under this License somewhere other than this MMC, and subsequently incorporated in whole or in part into the MMC, (1) had no cover texts or invariant sections, and (2) were thus incorporated prior to November 1, 2008.

The operator of an MMC Site may republish an MMC contained in the site under CC-BY-SA on the same site at any time before August 1, 2009, provided the MMC is eligible for relicensing.

Index

A

Acetylsalicylic acid. *See* Aspirin
Acid
  acid definition of 111
Acid
  Acid behavior in water 113
  Acid commonly encountered 113
  Acid types of 112
Acid rain
  acid rain anthropogenic sources of 145
  acid rain effects of 145
Acid rain
  Acid rainformation of 143
Acid Rain 143
Acids
  Acids and industrial spills 126
  Acids neutralization of 121
  Acids stomach acid 121
  Acids taste of 117
Activity 311
Adrenaline 247
Aerobic exercise 258
Aerobic metabolism 259
Air pollution. *See* individual pollutant substances
  Air pollution early regulation 147
  Air pollution EPA regulation. *See* Clean Air Act
  Air pollution London Fog 147
Air Quality Index 149
Alchemists 337
Alcohol
  Alcohol in beer and wine 127
  Alcohol production of 328
  Alcohol toxicity of 128
Allotropes 139
Alpha decay 339
Alpha particle 339
Altitude sickness 31
  Altitude sickness chemistry of 31
American diet 244
Amines
  Amines properties of 169
Amino acids 205
Amu. *See* atomic mass units; *See* atomic mass units
Amylose 211
Anaerobic metabolism 260
Anaerobic zone 260
Anions
  Anions definition of 94
Anode 316
Antacids 121
Anthocyanins 300
Antioxidants 226, 228
Aqueous solution. *See* solution
Aquifers 100
Aromatic ring. *See* benzene ring
Arthritis 238
Artificial flavor 184
Artificial sugar. *See* Sugar substitutes
Aspartame 251
Atherosclerosis 223
Atmosphere
  Atmosphere composition of 30
  Atmosphere composition of at high altitude 31
  Atmosphere regions of 29
  Atmosphere thickness of 29
Atmospheric pressure
  Atmospheric pressure elevation and 29
Atomic bomb 349
Atomic mass 32
Atomic mass units 32
Atomic number
  Atomic number definition of 42
  Atomic number representing in shorthand 42
  Atomic number uniqueness of 43
Atomic symbol
  Atomic symbol representing in shorthand 42
Atomic symbol 8
Atoms 10
ATP 324
Avogadro 33
Avogadro's number 34

B

Background radiation 360
Baking
  Baking chemistry of 123
Baking powder 123
Baking soda. *See* Carbonic acid
  Baking soda as a leavening agent 123
  Baking soda as an antacid 122
  Baking soda substitutions in baking 123
Barometric pressure. *See* atmospheric pressure; *See* atmospheric pressure
Bases 115
  Bases behavior in wataer 116
  Bases behavior in water 115
  Bases taste of 117
Batteries 316
Beer 127
Benzene ring
  Benzene ring structure of 168

Beta-carotene
Beta-caroteneand Vitamin A 227
Beta decay 341
Beta particle 341
Binary compound
Binary compounddefinition of 21
Blood glucose. *See* Blood sugar
Blood pH regulation. *See* Carbonic acid; *See* Carbonic acid
Blood sugar
Blood sugarfiber and 215
Body composition
Body compositionmetrics for 240
Body Composition 240
Body fat
Body fatexcess 238
Body Fat. *See* Fat; *See* Body composition
Body mass index (BMI) 240
Boiling
Boilingenergy and 91
Bond dipole
Bond dipolehow to draw 84
Bond dissociation 197
Bond energies 197
Bond energy
Bond energyand chemical reactions 201
Bran 215

C

Calcification 102
Calcium 228
Caloric Expenditure 256
Car batteries 317
Carbohydrate
Carbohydrateabsorption of 212
Carbohydratedigestion of 211
Carbohydratehydrolysis of 212
Carbohydrates 209
Carbon
Carbonelemental forms of 12
Carbonubiquity of 65
Carbon credits 396
Carbon Dating 364
Carbon dioxide
Carbon dioxideas a greenhouse gas 395
Carbon emissions
Carbon emissionsreducing individual 401
Carbon fixation 325
Carbon footprint
Carbon footprintreducing 402
Carbonic acid
Carbonic acidblood pH regulation 31
Carbonic acid. *See* Baking soda
Carbon monoxide
Carbon monoxidecontrol of 136
Carbon monoxidetoxicity of 136
Carbon Monoxide 135
Carotenoids 300
Catalytic converter
Catalytic converteruse of 137
Cathode 316
Cations
Cations definition of 94
Cellular respiration 325
Cellulose 211
CFCs
CFCs declining use of 289
Cheese
Cheesemanufacture of 124
Chemical bonding. *See* Covalent bond
Chemical combinations. *See* Compounds
Chemical equation
Chemical equationdefinition of 17
Chemical equationhow to balance 18
Chemical reaction
Chemical reactiondefinition of 17
Chernobyl 358
Cheyne-Stokes respirations 32
Chirality 179
Chiralityand pharmaceutical design 182
Chiralitydetermination of 179, 181
Chloride 228
Chlorophyll 300, 323
Cholesterol 222
CholesterolAHA recommendations and 223
Cholesterolatherosclerosis and 223
Cholesterolfats and 223
Cigarette smoking 153
Clean Air Act(s)
Clean Air Act(s)effectiveness of 148
Climate change
Climate changecausality and 393
Climate changeforcings and 394
Climate changemodels and 394
Climate changepositive feedback and 392
Climate changeRhetoric and 395
Climate changeScientific consensus and 396
Climate changesunspots and 393
Climate Modeling 391
Coal 375
Coal Scrubbing 147
Coenzymes 226
Color
Colorsunlight and 300
Combustion
Combustion definition of 17
Compounds 9
Computed tomography 362
Concentrating solar power 321
Concentration
Concentration definition of 81
Concentration. *See* Molarity
Conductive. *See* Metals
Conservation of matter 337
Covalent bond
Covalent bond energy and formation of 71
Covalent bond multiple bonds 62
Covalent Bond
Covalent Bond formation of 60
Curie, Marie 338
Current 308
Cyanocobalamin 226

D

Dalton, John 337
Daughter nucleus 339
Deep geological repository 357
Dehydration synthesis 206
Diatomic molecule
Diatomic moleculeelements that form 12

Diatomic molecule 10, 11
Diet
Dietfor weight loss 245. *See* Low-fat, Stimulant diet pills, Ketogenic, Low-carbohydrate
Dietwell-balanced 253, 255
Dietary Supplement Health and Education Act 186
Diethylene glycol 177
Diethyl ether
diethyl etheras an anaesthetic 171
Dipoles
Dipoles Greenhouse effect and 378
Dipoles IR and 378
Direct effect of exercise 257
Disaccharides 210
Disaccharides
Disaccharidesformation of 210
Dissociation 99
Doping 320
Double bond. *See* Covalent bond: multiple bonds
Double bonds 220
Drug Solubility 172
Ductile. *See* Metals

E

Einstein, Albert 347
Electrical energy 308
Electrical potential energy 308
Electricity 99
Electricity. *See* Electrical energy
Electrodes 315
Electrolysis 317
Electrolytes 99
Electrolytes 98. *See* Ionic compounds, Salts
Electrolytic cells 317
Electron configuration
Electron configurationplacement in shells 56
Electronegativity
Electronegativitybond polarization and 83
Electronegativitymnemonic device for 83
Electronegativitytrends in 82
Electrons
Electronsdetermining number of in neutral atom 42
Electronsmass and charge of 40
Electron shells 55
Electrostatic forces
Electrostatic forces and nuclear organization 43
Electrostatic forces definition of 41
Elements 8
Elements of life
Elements of life major and minor 53
Elements of life percent by mass 53, 54
Elements of life percent by number 54
Elevated post-exercise metabolic rate 257
Enantiomers
Enantiomers definition of 180
Energy
Energy and chemical reactions 201, 202
Energy calories 204
Energy disposition of 199, 200
Energy from food 204
Entropy 199
Enzymes 207
EPA air standards. *See* Air Quality Index
Ephedrine 247
Escherichia coli
Escherichia coliand infection 125
Essential body fat 240
Esters
Estersreaction in water 170
Esters
Estersproperties of 170
Ethanol. *See* Alcohol
Ether. *See* diethyl ether
Ethers
Ethersproperties of 171
Exercise 257

F

Fall color of leaves 299
Fat. *See* Body Fat
Fats 215. *See* Triglycerides
Fatsenergy of 215
Fatsmelting points of 218
Fatty acids 216
Fermentation
Fermentationlactic acid 327
Fermentation 326
Fermentation and food 127
Fermentation ethanolic 328
Fiber
Fiber role of 211
First law of thermodynamics 199
Fission
Fission and weapons 347
Fission atomic bombs 349
Fission energy from 351
Fixed proportions. *See* Compounds
Food and Drug Act 177
Food, Drug and Cosmetic Act 177
Food, Drug, and Cosmetic Act
Food, Drug, and Cosmetic Act 1962 Amendment to 183
Food pyramid 241
Fossil fuels 374
Fossil fuels
Fossil fuels composition of 376
Free radicals 139
Fructose 209
Functional groups 168. *See* Individual functional groups
Fusion
Fusion power generation 359

G

Galvanic Cells 313
Galvani, Luigi 313
Gamma decay 343
Gamma particle 343
Gases 91
Global Energy Dependence 374
Global energy needs 374
Global warming. *See* Climate change
Glucose 209
Glycemic index 254
Glycerol 217
Gravitational potential energy 199
Greek prefixes
Greek prefixesin naming compounds 22
Greenfreeze 293
Greenhouse effect

Greenhouse effectNormal conditions 377
Greenhouse gases 378, 392
Ground water 100
Ground wateraquifers 100
GWP 392

H

Half-life
half-life equation for 355
Half-Life 353
Half-reactions 309
Hard Water 102
Haworth projection 209
HCFCs 292
HDL 223
Herbal supplements
Herbal supplements regulation of 186
Herbal supplements toxicity of 185
HFCs 292
Homeostasis
Homeostasis blood sugar and 212
Hormones 207
Hydrocarbons
Hydrocarbons definition of 17
Hydrofluoric acid 126
Hydrogen
Hydrogen classification of 15
Hydrogen bomb 351
Hydrogen bonds
Hydrogen bondsin water 91
Hydrogen Bonds
Hydrogen Bondsformation of 88
Hydrogen fuel cells 400
Hydrolysis 208

I

Immersion test 240
Industrial Revolution 146
Infrared radiation 378. *See* IR
Interesterified fats 243
Ion exchange 102
Ionic bond
Ionic bond formation criteria 95
Ionic bond formation of 94
Ionic bond octets and 95
Ionic bond. *See* Ionic compounds
Ionic compounds. *See* Ionic bond
Ionic compoundsin aqueous solution 98
Ionic compoundsnature of 98
Ionic compoundsnomenclature of 96
Ionic compoundsNomenclature of. *See* Nomenclature
Ionic compoundspredicting charge 97
Ionic salts. *See* Ionic compounds, Electrolytes, Salts
Ions
Ions definition of 94
IPCC 391
IR 378
IR dipoles and 378
IR Greenhouse effect and 378
Iron 228
Irradiation
Irradiationhealth effects of 360
Irradiationof food 364
isotopes 340

K

Karats 13
Ketogenic diet 249
Ketone bodies 250
Ketosis 250
Kyoto Protocol 396

L

Lactic acid
lactic acid during exercise 260
Lactic acid
Lactic acid fermentation and 327
Lactobacillus acidophilus
Lactobacillus acidophilus and cheese making 124
Lactose 210
Lactose intolerance 211
Law of conservation of matter 18
LDL 223
Lead
Lead sources of 142
Lead toxicity of 143
Leavening 123
Leavening with baking powder 123
Leavening with baking soda 123
Leavening with yeast 124
Lewis dot structures. *See* Lewis structures
Lewis structures
Lewis structures of atoms 60
Lewis structures rules 63
Lewis structures symmetry and 65
Lightning 329
Line-angle diagrams
Line-angle diagrams and bond angles 165
Lipids 216
Liquid breathing medium. *See* perfluorodecalin; *See* perfluorodecalin
Liquids 91
London Fog (Great Smog) 147
Lone pairs 62
Low-Carbohydrate Diet 249

M

Macronutrients 204
Malleable. *See* Metals
Manhattan Project 349
Mass defect 346
Mass defect
Mass defectand energy 347
Mass number 32
Mass number representing in shorthand 42
Matter 7
Maximum heart rate 258
Melting
Melting energy and 91
Mercury
Mercury and seafood 142
Mercury exposure to 141
Mercury toxicity of 141
Mercury
Mercury and bioaccumulation 141
Metals 14

Methamphetamine 248
Micelles
  Micelles formation of 224
Micronutrients 225
Microwaves
  Microwaves and food 365
  Microwaves misconceptions regarding 365
Minerals 228
Mitrochondria 326
Mixtures 7
Molarity
  Molarityhow to calculate 118
Molar mass 35
Molecular compounds 99
Molecular mass 32
Molecular shapes
  Molecular shapeshow to determine geometry 67
  Molecular shapeslinear 70
  Molecular shapeslone pair effect on geometry 69
  Molecular shapestetrahedral 68
  Molecular shapestrigonal planar 70
  Molecular shapeswedged and dashed bonds 68
Molecule polarity
  Molecule polarityhow to determine 86
  Molecule polarityhydrocarbons 88
  Molecule polaritytetrahedral molecules 87
Molecules 10
  Moleculesmacroscopic representation 16
  Moleculesmicroscopic representation 16
  Moleculessymbolic representation 16
Moles 34
Monosaccharides 209
Montreal Protocol
  Montreal Protocol effectiveness of 292
  Montreal Protocol ozone and 290
Morphine
  Morphine and heroin 176
  Morphine solubility of 174
Morphine 174
Muscle types
  Muscle types and nutritional profile 263
Muscle Types 261

N

Natural flavor 185
Natural gas 375
Net dipole
  Net dipole how to determine 85
Neutralization 121
Neutrons
  Neutrons determining number of 42
  Neutrons mass and charge of 40
Nitrogen dioxide
  Nitrogen dioxideas a pollutant 138
  Nitrogen dioxidetoxicity of 138
Noble gases 59
Nomenclature
  Nomenclature definition of 21
Nomenclature
  Nomenclaturemetal and non-metal 21
  Nomenclaturenon-metal and non-metal 22
non-electrolytes 100
Non-nutritive sweeteners 251
Non-spontaneous cells. *See* Electrolytic cells
Normal body fat 240
NOx
  NOx and acid rain 144
  NOx definition of 137
  NOx formation of 137
  NOx sources of 137
N-type silicon 320
Nuclear Energy 351
Nuclear equations 339
Nuclear equations
  Nuclear equationsbalancing of 340
Nuclear fission. *See* Fission
Nuclear Fission 345
Nuclear fusion. *See* Fusion
Nuclear Fusion 350
Nuclear Medicine 362
Nuclear power
  Nuclear powerfuture of 359
Nuclear reactions 339
Nuclear reactor 352
Nuclear Waste 356
Nucleons
  Nucleons and nuclear organization 43
Nucleus
  Nucleus organization of 41
Nutritional Recommendations
  Nutritional Recommendations government standards 241
Nutrition facts labels
  Nutrition facts labels information included 242
Nutrition facts labels 242

O

Obesity. *See* Body composition
  Obesity in the U.S. 239
  Obesity social issues associated with 239
Octet rule 59
Omega-3 fats 221
Omega-6 fats 221
Organic nomenclature 169
Overweight. *See* obesity
Oxidation numbers 318
Ozone
  Ozone properties of 139
  Ozone toxicity of 139
  Ozone tropospheric 139
  Ozone tropospheric, formation of 139
Ozone hole
  Ozone hole discovery of 289

P

PAHs 139
Parabolic trough solar plants 321
Paradoxical respirations. *See* Cheyne-Stokes respirations; *See* Cheyne-Stokes respirations
Parent nucleus 339
Partial charge
  Partial charge how to indicate 85
Partially hydrogenated oils 221
Particulate matter
  Particulate matter effects of 140
Particulate Matter (PM10 and PM2.5) 140
Peptide bond 206

Percent Daily Value 242
Perfluorodecalin 46
Periodic table 8
Periodic tabledivisions of 14
Periodic trend
Periodic trenddefinition of 82
Petroleum 375
pH
pH how to calculate 120
pH of cow rumen 125
Pharmaceutical Design 168
Pharmacophores 174
Phases of matter 91
Phenols
Phenols properties of 169
Phenylketonuria 252
Phocomelia 178
Photosynthesis 322
Photovoltaic Cells 319
Physical activity energy expenditure 257
Physical combination. *See* Mixtures
Polar covalentbond
Polar covalentbond formation of 84
Polarity
Polarity solubility and 88
Polar stratospheric clouds 290
Polyatomic ions
Polyatomic ions common examples of 114
Polyatomic ions
Polyatomic ions nomenclature of 115
Polyatomic ions salts of 114
Polyatomic Ions 114
Polyatomic molecule 11
Polypeptides 206
Polysaccharides 210
Polysaccharides
Polysaccharides formation of 210
Positron 350
Positron emission tomography 363
Precautionary principle 151
Protein 205
Proteins
Proteins functions of 207
Proteins
Proteins formation of 206
Proteins in food 208
Proteins utilization of 208
Protons
Protons determining number of 42
Protons mass and charge of 40
P-type silicon 320
Pure substances. *See* Elements, Compounds
Pyruvate 327

R

Racemic mixture
Racemic mixture definition of 180
Radiation sickness 360
Radioactive Decay 339
Radioactive Decay Series 344
Radioactivity
Radioactivityatomic size and 45
Rain
Rain(acidity of acid rain). *See* Acid rain
Rainacidity of (normal) 143
Receptors 174
Redox 308
Redox 308
Redoxmnemonic for 308
Refined grains 215
Respiration 325
Resting metabolic rate 256
Risk-to-benefit ratio
Risk-to-benefit ratiopharmaceuticals and 183
RMR 256
Rutherford, Ernest 338

S

Saccharin 251
Saccharomyces cerevisiae 127
Salicin 163
Salicylic acid 163
Salt bridge 315
Salts
Salts definition of 94
Salts formation of. *See* ionic bond
Satiety signals 246
Saturated fats 218
Second law of thermodynamics 200
Semiconductors. *See* Metalloids
Serotonin 208
Silicon 320
Silicon p-n junction solar cell 320
Skepticism 373
Skin foldtest 240
Sleep apnea 238
Soap
Soapformation of 224
Soft water 102
Solar power 321
Solar poweras an alternative energy 321
Solids 91
Solubility
Solubility mnemonic device 88
Solubility of large molecules 172
Solubility of non-polar molcules 88
Solubility of polar molecules 88
Solubility ranking molecules in terms of 173
Solutes 81
Solution
Solution definition of 81
Solvents 81
SOx
SOxand acid rain 144
Spectator ions 116
Steroid Hormones 222
Stimulant diet pills
Stimulant diet pillsregulation of 248
Stimulant Diet Pills 247
Stoichiometric equivalents
Stoichiometric equivalentsdefinition of 38
Strong force 43
Subatomic Particles
Subatomic Particlestype of 40
Substrates
Substrates interaction with receptors 175
Sucralose 252
Sucrose 210
Sugars 210

Sugar substitutes
Sugar substituteshealth effects of 252
Sugar Substitutes 250
Sulfur dioxide
Sulfur dioxide source of 140
Sulfur dioxide toxcitiy of 141
Sulfur Dioxide 140
Surface water 100
Sympathetic nervous system 247
Sympathomimetic 247

T

Taste
Tastechemicals and 117
Thalidomide
thalidomideas a teratogen 178
Thalidomide 178
The Paracelsus Principle 187
Thermic effect of food 256
Thought experiment
Thought experiment definition of 35
Three Mile Island 358
Toxins
Toxinsin nature 185
Trans fat
Trans fatrecommended quantities 243
Trans fats 220
Triglycerides. *See* Fats
Triglyceridesas storage fat 217
Triglyceridescomposition of 217
Triple bond. *See* Covalent bond:multiple bonds
Tryptophan 208
Type 2 diabetes 238
Type II muscle fibers 261
Type I muscle fibers 261
Typical diet. *See* American diet

U

Unsaturated fats 219
Unsaturated fats
Unsaturated fats evolution of 220
Uranium
Uranium enrichment of 348

V

Valence electrons 57
Valence electronsgroup number and 58
Valence electronsreactivity trends and 59
Van der Waalsforces 91
Visible light 299
Vitamin A 227
Vitamin and Mineral Supplements 264
Vitamin B 226
Vitamin C 226
Vitamin D 227
Vitamin E 228
Vitamin K 228
Vitamins
Vitamins absorption of 216
Vitamins 225
Vitamins enrichment of products with 229
VOCs 138
Volta, Alessandro 313
Voltaic cells 313

W

Water
Water as a lubricant 92
Water behavior of ice 93
Water bottled 103
Water bottled, energy concerns 103
Water cohesiveness 92
Water conductivity of 100
Water density 92
Water developing nations and 101
Water distribution of 100
Water fluoride and 101
Water global distribution of 101
Water hardness and 102
Water MP and BP 92
Water polarity 86
Water potability of 100
Water purity and 104
Water purity of 100, 101
Water resistance to temperature change 92
Water snowflake structure and 93
Water softening processes 102
Water structure of solid and liquid 93
Water treatment of 101
Water United States and availability of 101
Western Paradox 237
Wine 127
work 198
Wrist, neck, and waist circumference 240

X

X-rays
X-raysin medicine 362

Y

Yeast
Yeast and alcohol 127
Yeast as a leavening agent 124
Yucca Mountain 357